Catalysis for Renewables

Edited by
Gabriele Centi and
Rutger A. van Santen

1807–2007 Knowledge for Generations

Each generation has its unique needs and aspirations. When Charles Wiley first opened his small printing shop in lower Manhattan in 1807, it was a generation of boundless potential searching for an identity. And we were there, helping to define a new American literary tradition. Over half a century later, in the midst of the Second Industrial Revolution, it was a generation focused on building the future. Once again, we were there, supplying the critical scientific, technical, and engineering knowledge that helped frame the world. Throughout the 20th Century, and into the new millennium, nations began to reach out beyond their own borders and a new international community was born. Wiley was there, expanding its operations around the world to enable a global exchange of ideas, opinions, and know-how.

For 200 years, Wiley has been an integral part of each generation's journey, enabling the flow of information and understanding necessary to meet their needs and fulfill their aspirations. Today, bold new technologies are changing the way we live and learn. Wiley will be there, providing you the must-have knowledge you need to imagine new worlds, new possibilities, and new opportunities.

Generations come and go, but you can always count on Wiley to provide you the knowledge you need, when and where you need it!

William J. Pesce
President and Chief Executive Officer

Peter Booth Wiley
Chairman of the Board

Catalysis for Renewables

From Feedstock to Energy Production

Edited by
Gabriele Centi and Rutger A. van Santen

WILEY-VCH Verlag GmbH & Co. KGaA

The Editors

Prof. Gabriele Centi
University of Messina
Department of Industrial Chemistry
and Engineering of Materials
Salita Sperone 31
98166 Messina
Italy

Prof. Rutger A. van Santen
Eindhoven University of Technology
Schuit Institute of Catalysis
Laboratory of Inorganic Chemistry
and Catalysis
P.O. Box 513
5600 MB Eindhoven
The Netherlands

■ All books published by Wiley-VCH are carefully produced. Nevertheless, authors, editors, and publisher do not warrant the information contained in these books, including this book, to be free of errors. Readers are advised to keep in mind that statements, data, illustrations, procedural details or other items may inadvertently be inaccurate.

Library of Congress Card No.: applied for

British Library Cataloguing-in-Publication Data
A catalogue record for this book is available from the British Library.

Bibliographic information published by the Deutsche Nationalbibliothek
Die Deutsche Nationalbibliothek lists this publication in the Deutsche Nationalbibliografie; detailed bibliographic data are available in the Internet at http://dnb.d-nb.de

© 2007 WILEY-VCH Verlag GmbH & Co. KGaA, Weinheim

All rights reserved (including those of translation into other languages). No part of this book may be reproduced in any form – by photoprinting, microfilm, or any other means – nor transmitted or translated into a machine language without written permission from the publishers. Registered names, trademarks, etc. used in this book, even when not specifically marked as such, are not to be considered unprotected by law.

Printed in the Federal Republic of Germany
Printed on acid-free paper

Typesetting   Asco Typesetters, Hong Kong
Printing   betz-druck GmbH, Darmstadt
Bookbinding   Litges & Dopf GmbH, Heppenheim
Wiley Bicentennial Logo   Richard J. Pacifico

ISBN 978-3-527-31788-2

Contents

Preface *XV*

List of Contributors *XIX*

1	**Renewable Catalytic Technologies – a Perspective** *1*	
	Rutger A. van Santen	
1.1	Introduction *1*	
1.2	Economic and Societal Background *1*	
1.3	Technology Options *5*	
1.4	Process Options for Biomass Conversion *13*	
1.5	Conclusions *18*	
	References *19*	
2	**Lignocellulose Conversion: An Introduction to Chemistry, Process and Economics** *21*	
	Jean-Paul Lange	
2.1	Overview *21*	
2.2	Introduction *21*	
2.2.1	Need for Renewable Energy *21*	
2.2.2	Need for Biomass Conversion *22*	
2.2.3	Biomass Composition *24*	
2.2.4	Fuels and Chemicals Composition *25*	
2.2.5	Biomass Deoxygenation *26*	
2.3	Chemistry and Processes *26*	
2.3.1	Key Reactions of Carbohydrates *27*	
2.3.2	Pyrolysis *28*	
2.3.2.1	Chemistry *28*	
2.3.2.2	Product Applications *30*	
2.3.2.3	Processes *30*	
2.3.2.4	Alternative Developments *32*	
2.3.3	Gasification *32*	
2.3.3.1	Chemistry *32*	

Catalysis for Renewables: From Feedstock to Energy Production
Edited by Gabriele Centi and Rutger A. van Santen
Copyright © 2007 WILEY-VCH Verlag GmbH & Co. KGaA, Weinheim
ISBN: 978-3-527-31788-2

2.3.3.2	Process	33
2.3.3.3	Alternative Developments: Hydrogen Production	35
2.3.4	Hydrolysis	35
2.3.4.1	Chemistry	35
2.3.4.2	Sugar Derivatives	37
2.3.4.3	Process	37
2.3.4.4	Recent Developments	38
2.3.5	Fermentation	39
2.3.5.1	Chemistry	39
2.3.5.2	Process	40
2.3.5.3	Recent Developments	41
2.4	Economics	42
2.4.1	Methodology	42
2.4.2	Fuel Production	42
2.4.2.1	Plant Costs	42
2.4.2.2	Feed Costs	43
2.4.2.3	Manufacturing Costs	44
2.4.3	Process Scale	46
2.4.4	Chemicals Production	46
2.5	Summary and Conclusions	48
	References	49

3 Process Options for the Catalytic Conversion of Renewables into Bioproducts 53
Pierre Gallezot

3.1	Overview	53
3.2	Introduction	53
3.3	The Biorefinery Concept	55
3.4	Strategies for Biomass Conversion into Bioproducts	56
3.4.1	From Biomass to Products via Degraded Molecules	56
3.4.2	From Biomass to Products via Platform Molecules	57
3.4.2.1	Identification of Main Platform Molecules	57
3.4.2.2	Selected Examples of Platform Molecule Conversion into Bioproducts	58
3.4.3	From Biomass to Products via New Synthesis Routes	65
3.4.3.1	One-pot Reaction with Cascade Catalysis	65
3.4.3.2	One-pot Conversion into a Mixture of Products	66
3.5	Concluding Remarks	70
	References	72

4 Industrial Development and Application of Biobased Oleochemicals 75
Karlheinz Hill

4.1	Overview	75
4.2	Raw Material Situation	75

4.3	Ecological Compatibility	77
4.4	Examples of Products	78
4.4.1	Oleochemicals for Polymer Applications	79
4.4.1.1	Dimer Diols Based on Dimer Acid	82
4.4.1.2	Polyols Based on Epoxides	83
4.4.2	Biodegradable Fatty Acid Esters for Lubricants	84
4.4.3	Surfactants and Emulsifiers Derived from Vegetable Oil Based Fatty Alcohols and Fatty Acids	86
4.4.3.1	Fatty Alcohol Sulfate (FAS)	88
4.4.3.2	Acylated Proteins and Amino Acids (Protein–Fatty Acid Condensates)	88
4.4.3.3	Carbohydrate-based Surfactants – Alkyl Polyglycosides	89
4.4.3.4	Alkyl Polyglycoside Carboxylate	91
4.4.3.5	Polyol Esters	92
4.4.3.6	Multifunctional Care Additives for Skin and Hair	93
4.4.4	Emollients	95
4.4.4.1	Dialkyl Carbonate	95
4.4.4.2	Guerbet Alcohols	96
4.5	Perspectives	97
	References	97
5	**Fine Chemicals from Renewables**	*101*
	Herman van Bekkum and Leendert Maat	
5.1	Introduction	101
5.2	Vanillin	103
5.3	Monoterpenes	105
5.4	Alkaloids	108
5.5	Steroids	112
5.6	Enantioselective Catalysis	113
5.7	Artimisinine	115
5.8	Tamiflu	116
5.9	Final Remarks	117
	References	117
6	**Options for Catalysis in the Thermochemical Conversion of Biomass into Fuels**	*119*
	Sascha R. A. Kersten, Wim P. M. van Swaaij, Leon Lefferts, and Kulathuiyer Seshan	
6.1	Introduction	119
6.2	Biomass as Feedstock for Fuels	120
6.3	Composition of Biomass	122
6.4	Biorefinery	125
6.5	Biomass Pretreatment	126
6.6	Thermochemical Conversion of Lignocelluloses	127

6.7	Biomass Gasification	*129*
6.7.1	Gasification of Dry Biomass	*129*
6.7.2	Catalytic Gasification of Pyrolysis Oil	*129*
6.7.3	Chemistry and Catalysis of Gasification	*130*
6.7.4	Gasification in Hot Compressed Water	*131*
6.8	Liquefaction of Biomass	*132*
6.8.1	Non-catalytic Pyrolysis	*132*
6.8.2	Catalytic Pyrolysis	*134*
6.8.3	Hydrothermal Liquefaction	*135*
6.9	Upgrading Pyrolysis Oil to Fuels	*136*
6.9.1	Decarboxylation (DCO)	*136*
6.9.2	Hydrodeoxygenation (HDO)	*137*
6.9.3	Cracking over Zeolites (FCC)	*137*
6.10	Hydrolysis	*139*
6.11	Underlying Approach for Catalyst Design	*140*
6.12	Summary	*141*
	References	*141*

7 Thermal Biomass Conversion *147*
Simone Albertazzi, Francesco Basile, Giuseppe Fornasari, Ferruccio Trifirò, and Angelo Vaccari

7.1	Introduction	*147*
7.2	Biomass Resources and Biomass Pre-treatment	*148*
7.3	Biomass Combustion	*149*
7.4	Biomass Gasification	*149*
7.5	Pyrolysis	*154*
7.6	Fuels via Thermal Biomass Conversion	*158*
7.7	Conclusions	*161*
	References	*161*

8 Thermal Biomass Conversion and NO_x Emissions in Grate Furnaces *163*
Rob J.M. Bastiaans, Hans A.J.A. van Kuijk, Bogdan A. Albrecht, Jeroen A. van Oijen and L. Philip H. de Goey

8.1	Introduction	*163*
8.2	Tunable Diode Laser Measurements of Biomass Kinetics	*164*
8.2.1	Introduction	*164*
8.2.2	Tunable Diode Laser Grid Reactor Experiments	*164*
8.2.3	Experimental Setup	*165*
8.2.4	Results	*166*
8.3	Propagation of Thermal Conversion Fronts	*169*
8.3.1	Introduction	*169*
8.3.2	Modeling Approach	*170*
8.3.3	Experiments	*173*

8.4	Gas-phase CFD Modeling of Grate Furnaces *175*	
8.4.1	Introduction *175*	
8.4.2	Description of the Model *175*	
8.4.3	Construction of Look-up Tables for Numerical Simulations and Validation *176*	
8.4.4	Application of the Combustion Model on a 2D Grate Furnace *177*	
8.5	Conclusions *180*	
	References *180*	

9 Bioethanol: Production and Pathways for Upgrading and Valorization *183*
Stephane Pariente, Nathalie Tanchoux, François Fajula, Gabriele Centi, and Siglinda Perathoner

9.1	Introduction *183*
9.2	Production, a Short Overview *188*
9.3	Uses as Biofuel *193*
9.3.1	Bioethanol as Fuel Additive *193*
9.3.1.1	Gasoline/Bioethanol Blends *193*
9.3.1.2	Diesel/Bioethanol Blends *195*
9.3.2	Bioethanol and Hydrogen *196*
9.3.3	Bioethanol for Fuel Cells *199*
9.4	Bioethanol Upgrading and Valorization *200*
9.4.1	Conversion into Fuel Components *200*
9.4.2	Conversion into Chemicals *203*
9.5	Conclusions *205*
	References *205*

10 Conversion of Glycerol into Traffic Fuels *209*
Tiia S. Viinikainen, Reetta S. Karinen, and A. Outi I. Krause

10.1	Introduction *209*
10.2	Glycerol *210*
10.2.1	Properties, Production and Use of Glycerol *210*
10.2.2	Glycerol from Biodiesel Production *211*
10.3	Etherification of Glycerol with Isobutene *212*
10.3.1	Reaction Scheme *212*
10.3.2	Etherification Catalysts *213*
10.3.3	Process Conditions *214*
10.3.4	Etherification Kinetics *216*
10.3.5	Ethers of Glycerol as Fuel Components *217*
10.4	Improvements to Biodiesel Process *218*
10.4.1	Etherification with Biodiesel Process *218*
10.4.2	Heterogeneous Biodiesel Process *219*
10.5	Reforming of Glycerol *219*
10.5.1	Aqueous Phase Reforming *220*

10.5.2	Steam Reforming	*220*
10.6	Future Aspects	*220*
	References	*221*

11 Catalytic Transformation of Glycerol *223*
Bert Sels, Els D'Hondt, and Pierre Jacobs

11.1	Introduction and Scope	*223*
11.2	Catalytic Dehydration of Glycerol and Acrolein Formation	*224*
11.3	Etherification of Glycerol via Catalytic Dehydration	*226*
11.3.1	Glycerol Oligomerization	*226*
11.3.2	Reaction of Glycerol with Alkenes	*228*
11.4	Catalytic Oxidation of Glycerol	*231*
11.4.1	Electrochemical Oxidation	*231*
11.4.2	Gas-phase Catalytic Oxidation	*233*
11.4.3	Selective Oxidation with Molecular Oxygen on Pt/Bi Catalysts	*234*
11.4.4	Selective Oxidation with Molecular Oxygen on Gold-based Catalysts	*237*
11.4.5	Selective Oxidation with Oxidants Differing from Molecular Oxygen	*240*
11.5	Catalytic Hydrogenolysis of Glycerol	*241*
11.5.1	Heterogeneous Catalytic Hydrogenolysis of Glycerol	*241*
11.5.2	Homogeneous Catalytic Hydrogenolysis of Glycerol	*248*
11.6	Glycerol Reforming and Hydrogen Production	*249*
11.7	Miscellaneous Oxidation Reactions	*250*
11.8	Conclusions	*251*
	References	*252*

12 Catalytic Processes for the Selective Epoxidation of Fatty Acids: More Environmentally Benign Routes *257*
Matteo Guidotti, Rinaldo Psaro, Maila Sgobba, and Nicoletta Ravasio

12.1	Introduction	*257*
12.2	Non-catalytic Epoxidation Systems	*259*
12.3	Homogeneous Catalytic Systems	*260*
12.4	Chemoenzymatic Epoxidation Systems	*260*
12.5	Heterogeneous Catalytic Systems	*261*
12.6	Epoxidation of FAMEs Over Titanium-based Catalysts: The Skills in Milan	*264*
12.6.1	Epoxidation of Pure C_{18} Monounsaturated FAMEs	*265*
12.6.2	Epoxidation of a Mixture of FAMEs from Vegetable Sources	*266*
12.6.2.1	HO Sunflower-, Coriander-, and Castor-oil FAME Mixtures	*266*
12.6.2.2	Soya-bean Oil FAME Mixture	*269*
12.7	Conclusions	*270*
	References	*270*

13 Integration of Biocatalysis with Chemocatalysis: Cascade Catalysis and Multi-step Conversions in Concert *273*
Tom Kieboom

13.1 Overview *273*
13.2 Introduction *274*
13.2.1 Human's Chemistry *274*
13.2.2 Nature's Chemistry *275*
13.2.3 Bio-chemo Integration *276*
13.3 Types of Cascades *277*
13.3.1 Bio-bio Cascades *278*
13.3.2 Chemo-chemo Cascades *280*
13.3.3 Bio-chemo Cascades *281*
13.4 Technologies for Cascades *289*
13.4.1 Catalytic Methods *291*
13.4.2 Reactor Design *292*
13.4.3 Compartmentalization *292*
13.4.4 Medium Engineering *293*
13.4.5 Cell Factory Design *294*
13.5 Conclusions *295*
References 296

14 Hydrogen Production and Fuel Cells as the Bridging Technologies Towards a Sustainable Energy System *299*
Frank A. de Bruijn, Bert Rietveld, and Ruud W. van den Brink

14.1 Introduction *299*
14.1.1 The Hydrogen Energy Chain *300*
14.1.2 Hydrogen Sources and Production *300*
14.1.3 Use of Hydrogen in Stationary and Mobile Applications *301*
14.2 Hydrogen Production from Natural Gas *301*
14.2.1 Conventional Hydrogen Production *302*
14.2.1.1 Hydrogen Production from Natural Gas *302*
14.2.1.2 Hydrogen Production from Other Feedstocks *304*
14.2.2 Hydrogen Production with CO_2 Capture *305*
14.2.2.1 CO_2 Capture *306*
14.3 Novel Processes for Hydrogen Production with CO_2 Capture *307*
14.3.1 Hydrogen Membrane Reactors *307*
14.3.2 Sorption-enhanced Reforming and Water-gas Shift *310*
14.4 Conclusions and Catalytic Challenges *313*
14.4.1 Electrochemical Hydrogen Production and Conversion *314*
14.4.1.1 Kinetics of the Electrochemical Hydrogen–Oxygen Processes *314*
14.4.1.2 Hydrogen Production by Water Electrolysis *315*
14.4.1.3 Proton Exchange Membrane Fuel Cells *319*
14.4.1.4 Solid Oxide Fuel Cells (SOFCs) *326*
References 333

15 Pathways to Clean and Green Hydrogen 337
Gert J. Kramer, Joep P. P. Huijsmans, and Dave M. Austgen

15.1 Introduction 337
15.2 Energy Resource Availability 339
15.3 Modes of Hydrogen Production and Distribution 340
15.4 The Cost of Hydrogen Fuel 341
15.4.1 Case Definition 342
15.4.2 Results 343
15.5 "Clean Hydrogen" and the Scope for CO_2 Reduction 345
15.5.1 Scope 345
15.5.2 Hydrogen versus Gasoline and Diesel 347
15.6 Coal and Biomass 348
15.7 Conclusions 349
References 349

16 Solar Photocatalysis for Hydrogen Production and CO_2 Conversion 351
Claudio Minero and Valter Maurino

16.1 Introduction 351
16.2 The Photocatalytic Process 354
16.2.1 Quantum Yield 357
16.2.2 Catalyst Related Losses 360
16.2.2.1 Carrier Thermalization 360
16.2.2.2 Charge Separation 361
16.2.2.3 Active Charge Separation 362
16.2.2.4 Passive Charge Separation 362
16.2.2.5 Mediated Charge Separation 366
16.2.3 Surface-related Losses 367
16.3 Photoelectrochemical Cells 371
16.4 New Materials 372
16.4.1 Crystal Structure and Activity 373
16.4.2 Visible Sensitization 375
16.5 Conclusions 377
References 379

Conclusions, Perspectives and Roadmap 387
Gabriele Centi and Rutger A. van Santen

1 Introduction 387
2 Driver for a Biomass Economy 388
3 Main Issues and Perspectives on Bioenergy and Biofuels in Relation to Catalysis 389
3.1 Biofuels 389
3.1.1 First Generation Biofuels 390

3.1.2	Second (or Next) Generation Biofuels	*392*
3.2	Biorefineries	*394*
3.3	Use of By-products Deriving from Biomass Transformation	*399*
3.4	Biomass as Feedstock for Chemical Production	*400*
3.5	Use of Solar Energy	*403*
4	Conclusions	*405*
	References	*411*

Index *413*

Preface

The increasing cost of fossil fuels and the concerns related to their environmental impact and greenhouse gas effect, as well as the need of securing energy supplies, are accelerating the transition to a bio-based economy. Various R&D tools need to be provided to realize this transition. The replacement of fossil fuel by bio-mass has been addressed in recent years worldwide. The EU, for example, has defined a target to double the share of renewable energy from 6% in 1997 to 12% by 2010 (COM 1997 599).

The use of renewable resources in manufacturing chemicals and other products, such as oils from oilseed crops, starch from cereals and potatoes and cellulose from straw and wood, is also receiving increasing attention from policy makers. Further transformation of these products leads to polymers, lubricants, solvents, surfactants and specialty chemicals for which fossil fuels have traditionally been used. However, it is necessary to extend the use of biomasses for chemical production and integrate them in the energy business to create a sustainable society. The concept of the biorefinery, initially developed in the food and paper industries, is now being applied to integrate biomass-based energy, materials and chemicals production. A biorefinery maximizes the value derived from the complex biomass feedstock by producing multiple products. Integrated production of bioproducts, especially for bulk chemicals, biofuels, biolubricants and polymers, can improve their competitiveness and eco-efficiency. Also in this case, new R&D tools should be developed to address this change with respect to oil-based energy, material and chemical economy.

Many benefits for EU industry, consumers and the environment derive from the use of renewable raw materials: (a) increased competitiveness from products having tailor-made performance compared with, or in combination with, conventional materials; (b) a more stable and secure source of supply; (c) a reduction in environmental impact; (d) new and growing markets, providing economic benefits to industry as well as; and (e) employment opportunities in processing industries and the agricultural sector.

Many improvements are still needed to make really effective use of renewable raw materials in biorefineries. Full utilization of the plants is needed instead of the current under utilization, as well as the development of processes to add value to all fractions of the plant and to valorize the by-products of other industrial systems (e.g., black liquor in the wood/paper industry, glycerol from biodiesel, whey from cheese production, etc), downstream processing strategies (low cost recovery and purification), development of closed-cycle sustainable systems, etc.

Catalysis is a core technology of the current fossil fuel based economy. Over 90% of industrial chemical processes involve catalytic steps and, also, several processes in current refineries are catalytic ones. Without continuous progress and innovation in catalysis, the current pervasive oil-based economy is not possible. Similarly, catalysis technology will also have a key role in the transition to a bio-based economy. The possibility of realizing this transition and to develop effective bio-refineries will depend on the progress made in developing new catalytic processes and concepts.

Hence, catalysis may be considered as an enabling technology for this transition and, for this reason, it is necessary to better understand the limitations and possibilities in this field. We need to define future necessary directions of R&D and the needs of fundamental and applied knowledge. In other words, there is the need to develop a roadmap for catalytic processes based on renewable feedstock. This book aims to provide an overview of the current state-of-the-art on which such a research agenda can be based.

This book originates from the workshop "Catalysis for Renewables", which was dedicated to this issue and held in Rolduc (Kerkrade, The Netherlands) on May 16–18th, 2006. The objective of the workshop was to provide strategic input to catalytic process options for the conversion of renewable feedstock into energy and chemicals. It was a brainstorming meeting aimed at defining new directions and opportunities for catalytic research in this field by integrating industrial, governmental and academic points of view. The different chapters in this book cover the various aspects reviewed during the workshop, while the concluding chapter provides a critical synthesis of the active discussion held during the workshop, with the aim of defining a Research Strategic Agenda in catalysis for renewables.

The workshop was organized by the NRSC-Catalysis (National Research School Combination Catalysis) of The Netherlands within the framework of the activities of the EU Network of Excellence (NoE) IDECAT (Integrated Design of Catalytic Nanomaterials for a Sustainable Production).

The objective of this NoE is to strengthen research in catalysis by the creation of a coherent framework of research, know-how and training between the various disciplinary catalysis communities (heterogeneous, homogeneous, and biocatalysis) with the objective of achieving a lasting integration between the main European Institutions in this area. IDECAT will create the virtual "European Research Institute on Catalysis" (ERIC) that is intended to be the main reference point for catalysis in Europe.

IDECAT focuses its research on (a) the synthesis and mastering of nano-objects, the materials of the future for catalysis, also integrating the concepts common to other nanotechnologies; (b) bridging the gap between theory and modeling, surface science, and kinetic/applied catalysis as well as between heterogeneous, homogeneous and biocatalytic approaches; and (c) developing an integrated design of catalytic nanomaterials.

Objectives of IDECAT are to:
1. create a critical mass of expertise going beyond collaboration;
2. create a strong cultural thematic identity on nano-tech based catalysts;
3. increase the cost-effectiveness of European research;
4. establish a frontier research portfolio able to promote innovation in catalysis use, especially at the SMEs level;
5. increase potential for training and education in multidisciplinary approaches to nano-tech based catalysis; and
6. spread excellence beyond the NoE to both the scientific community and to the citizen.

Next-generation catalysts should achieve zero-waste emissions and use selectively the energy in chemical reactions. They will also enable the development of new bio-mimicking catalytic transformations, new clean energy sources and chemical storage methods, utilization of new and/or renewable raw materials and reuse the waste, solving global issues (greenhouse gas emissions, water and air quality) and realizing smart devices. These challenging objectives can be reached only through a synergic interaction between the best catalytic research centers and in permanent and strong interaction with companies and public institutions. This is the scope of IDECAT.

In conclusion, this book constitutes the first step of IDECAT in developing a coherent framework of activities to create a bio-based and sustainable society through catalysis. This book is at the same time an updated overview of the state-of-the-art and a roadmap that defines new directions, opportunities and needs for R&D. Finally, we warmly thank Dr. Ad Kolen, NRSC-Catalysis (The Netherlands), whose continued support made possible both the workshop cited and this book.

Gabriele Centi
Rutger A. van Santen

List of Contributors

Simone Albertazzi
University of Bologna
Department of Chimica
Industriale e dei Materiali
Alma Mater Studiorum
Viale Risorgimento 4
40136 Bologna
Italy

Bogdan A. Albrecht
Eindhoven University of
Technology
Department of Mechanical
Engineering
P.O. Box 513
5600 MB Eindhoven
The Netherlands
Present address:
DAF Trucks N.V.
Advanced Engineering Engines
Hugo van der Goeslaan 1
P.O. Box 90065
5600 PT Eindhoven
The Netherlands

Dave M. Austgen
Shell Hydrogen LLC
700 Milam Street
Houston, TX 77002
USA

Francesco Basile
University of Bologna
Department of Chimica Industriale e
dei Materiali
Alma Mater Studiorum
Viale Risorgimento 4
40136 Bologna
Italy

Rob J.M. Bastiaans
Eindhoven University of Technology
Department of Mechanical Engineering
P.O. Box 513
5600 MB Eindhoven
The Netherlands

Herman van Bekkum
Delft University of Technology
DelftChem Tech
Self Assembling Systems
Julianalaan 136
2628 BL Delft
The Netherlands

Ruud W. van den Brink
Energy Research Centre
of the Netherlands (ECN)
Programme Unit Hydrogen
and Clean Fossil Fuels
Westerduinweg 3
1755 LE Petten
The Netherlands

Frank A. de Bruijn
Energy Research Centre
of the Netherlands (ECN)
Programme Unit Hydrogen
and Clean Fossil Fuels
Westerduinweg 3
1755 LE Petten
The Netherlands

Gabriele Centi
University of Messina
Department of Industrial
Chemistry and Engineering
of Materials
Salita Sperone 31
98166 Messina
Italy

Els D'Hondt
Katholieke Universiteit Leuven
Centrum voor Oppervlaktechemie
en Katalyse
Kasteelpark Arenberg 23
3001 Leuven (Heverlee)
Belgium

François Fajula
Institut Charles Gerhardt
UMR 5253 CNRS-ENSCM-
UM2-UM1
Equipe "Matériaux Avancés pour
la Catalyse et la Santé" (MACS)
Ecole Nationale Supérieure de
Chimie de Montpellier
8, rue de l'Ecole Normale
34296 Montpellier Cedex 5
France

Giuseppe Fornasari
University of Bologna
Department of Chimica
Industriale e dei Materiali
Alma Mater Studiorum
Viale Risorgimento 4
40136 Bologna
Italy

Pierre Gallezot
Université de Lyon
Institut de Recherche sur la Catalyse et
l'Environnement de Lyon
2, avenue Albert Einstein
69626 Villeurbanne Cedex
France

L. Philip H. de Goey
Eindhoven University of Technology
Department of Mechanical Engineering
P.O. Box 513
5600 MB Eindhoven
The Netherlands

Matteo Guidotti
CNR
Institute of Molecular Sciences and
Technologies (ISTM)
Dipartimento Chimica Inorganica,
Metallorganica e Analitica
via Venezian, 21
20133 Milano
Italy

Karlheinz Hill
Cognis GmbH
Care Chemicals Technology
Rheinpromenade 1
40789 Monheim
Germany

Joep P.P. Huijsmans
Shell Hydrogen BV
P.O. Box 162
2501 AN The Hague
The Netherlands

Pierre Jacobs
Katholieke Universiteit Leuven
Centrum voor Oppervlaktechemie
en Katalyse
Kasteelpark Arenberg 23
3001 Leuven (Heverlee)
Belgium

List of Contributors

Reetta K. Karinen
Helsinki University of Technology
Department of Chemical
Technology
P.O. Box 6100
02150 Hut
Finland

Sascha R. A. Kersten
University of Twente
Thermal-Chemical Conversion
of Biomass (TCCB)
Faculty of Science and Technology,
JMPACT
P.O. Box 217
7500 AE Enschede
The Netherlands

Tom Kieboom
Leiden University
Institute of Chemistry
p/a Stationsweg 56
2991 CM Barendrecht
The Netherlands

Gert Jan Kramer
Eindhoven University of
Technology
Faculteit Scheikundige
Technologie
SKA; STW 3.42
P.O. Box 513
5600 MB Eindhoven
The Netherlands

A. Outi I. Krause
Helsinki University of Technology
Department of Chemical
Technology
P.O. Box 6100
02150 Hut
Finland

Hans A.J.A. van Kuijk
Eindhoven University of Technology
Department of Mechnical Engineering
P.O. Box 513
5600 MB Eindhoven
The Netherlands

Jean-Paul Lange
Shell Global Solutions International
B.V.
P.O. Box 38000
1030 BN Amsterdam
The Netherlands

Leon Lefferts
University of Twente
Catalytic Processes and Materials
(CPM)
Faculty of Science and Technology,
JMPACT
P.O. Box 217
7500 AE Enschede
The Netherlands

Leendert Maat
Delft University of Technology
Biocatalysis and Organic Chemistry
Julianalaan 136
2628 BL Delft
The Netherlands

Valter Maurino
Università di Torino
Dipartimento di Chimica Analitica
Via P. Giuria 5
10125 Torino
Italy

Claudio Minero
Università di Torino
Dipartimento di Chimica Analitica
Via P. Giuria 5
10125 Torino
Italy

Jeroen A. van Oijen
Eindhoven University of
Technology
Department of Mechanical
Engineering
P.O. Box 513
5600 MB Eindhoven
The Netherlands

Stephane Pariente
Institut Charles Gerhardt
UMR 5253 CNRS-ENSCM-UM2-UM1
Equipe "Matériaux Avancés pour la Catalyse et la Santé" (MACS)
Ecole Nationale Supérieure de
Chimie de Montpellier
8, rue de l'Ecole Normale
34296 Montpellier Cedex 5
France

Siglinda Perathoner
University of Messina
Department of Industrial
Chemistry and Engineering of
Materials
Salita Sperone 31
98166 Messina
Italy

Rinaldo Psaro
CNR
Institute of Molecular Sciences
and Technologies (ISTM)
Dipartimento Chimica Inorganica,
Metallorganica e Analitica
Via Venezian 21
20133 Milano
Italy

Nicoletta Ravasio
CNR
Institute of Molecular Sciences and
Technologies (ISTM)
Dipartimento Chimica Inorganica,
Metallorganica e Analitica
Via Venezian 21
20133 Milano
Italy

Bert Rietveld
ECN (Energy Research Centre
of the Netherlands)
P.O. Box 1
1755 LE Petten
The Netherlands

Rutger A. van Santen
Eindhoven University of Technology
Schuit Institute of Catalysis
Laboratory of Inorganic Chemistry
and Catalysis
P.O. Box 513
5600 MB Eindhoven
The Netherlands

Bert Sels
Katholicke Universiteit Leuven
Centrum voor Oppervlaktechemie
en Katalyse
Kasteelpark Arenberg 23
3001 Leuven (Heverlee)
Belgium

Kulathuiyer Seshan
University of Twente
Catalytic Processes and Materials
(CPM)
Faculty of Science and Technology,
JMPACT
P.O. Box 217
7500 AE Enschede
The Netherlands

Maila Sgobba
University of Milan
Dipartimento Chimica
Inorganica, Metallorganica e
Analitica
Via Venezian 21
20133 Milano
Italy

Wim P. M. van Swaaij
University of Twente
Thermal-Chemical Conversion
of Biomass (TCCB)
Faculty of Science and Technology,
JMPACT
P.O. Box 217
7500 AE Enschede
The Netherlands

Nathalie Tanchoux
Institut Charles Gerhardt
UMR 5253 CNRS-ENSCM-
UM2-UM1
Equipe "Matériaux Avancés pour
la Catalyse et la Santé" (MACS)
Ecole Nationale Supérieure de
Chimie de Montpellier
8, rue de l'Ecole Normale
34296 Montpellier Cedex 5
France

Ferruccio Trifirò
University of Bologna
Dipartimento di Chimica
Industriale e dei Materiali
Alma Mater Studiorum
Viale Risorgimento 4
40136 Bologna
Italy

Angelo Vaccari
University of Bologna
Department of Chimica
Industriale e dei Materiali
Alma Mater Studiorum
Viale Risorgimento 4
40136 Bologna
Italy

Tiia S. Viinikainen
Helsinki University of Technology
Department of Chemical Technology
P.O. Box 6100
02150 Hut
Finland

1
Renewable Catalytic Technologies – a Perspective

Rutger A. van Santen

1.1
Introduction

The awareness that our human culture requires sustainable use of the earth's resources and its biosphere has become generally accepted in those societies that are in a relatively wealthy economic position. It also provides the intuitive justification for many that the use of renewable resources is a necessity. However, several economists (see Jaccard [1], Smil [4], Lomborg [5]) caution that the introduction or transition to renewable resources will also have economic as well as environmental costs, that sometimes reverse the preference for particular raw materials or technologies that at first sight seem to be the best.

The chapters that follow are based on lectures presented at the Idecat conference on "Catalysis for Renewables", Rolduc, The Netherlands, which had been organized to discuss different technology options and related catalytic advances or challenges. Whereas the main focus is on biomass related catalytic technologies, lectures on fossil fuel technology as well as solar and biotechnological conversion paths were also included.

This provides a perspective for comparison of technologies as well as an opportunity for technological integration.

We present a short discussion of the economic and societal backgrounds for the search of alternatives to fossil fuels or technologies that make their use environmentally sustainable.

We then compare technologies that exploit different primary energy sources and secondary energy technologies. In the final section, we provide an overview of the different process options for biomass conversion.

1.2
Economic and Societal Background

The two main societal issues that create a need to reconsider the current use of fossil fuels are the cogeneration of CO_2 in stoichiometric ratio to carbon used in the fuel and depletion of fossil oil reserves.

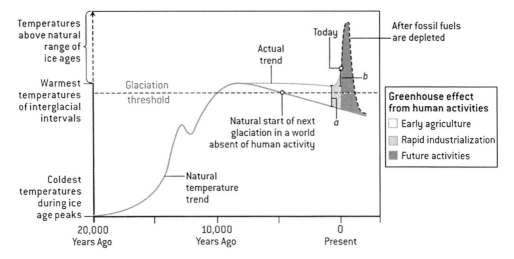

Fig. 1.1 Trends in the temperature at the earth surface as a function of ages. The greenhouse effect [2].

There is general recognition that large additional quantities of CO_2 emitted into our atmosphere contribute to the green house climate effect.

There is consensus that fossil fuel based emissions have more than exponentially increased in the last century. Whereas in 1925 the accumulated global CO_2 emissions from fossil sources were 1000 million metric tons of CO_2, in the year 2000 this was over 6000 in the same units [1]. Large emission rates that will affect the climate still continue.

Figure 1.1 gives a schematic representation of predicted trends in the earth temperature atmosphere [2].

Interestingly, this curve shows an increase in temperature when the last deep ice age period finished and a maximum in the predicted temperature, if no human activity would affect the climate. Geophysical effects, planetary motion as well as solar evolution determine that around 9000 years ago this maximum in temperature should have appeared. However, human activities shift the maximum. Initial increases in carbon dioxide due to deforestation and other agricultural activities caused a delay. The curve shown in Fig. 1.1 shows a future predicted maximum in the earth's temperature once fossil fuels are no longer available. Then CO_2 will be readsorbed from the atmosphere by various geo-biological processes. This adsorption process of CO_2 will take several hundred years. An important factor that limits the rate of this reabsorption process is the low solubility of CO_2 in water. The interesting message of Fig. 1.1 is that the greenhouse effect is of partial use to our civilization and that on very long time scales it may be actually needed.

On shorter timescales relevant to us and our grandchildren is the threat to our climate. The effects are difficult to completely predict because of the complexity of the climate system and uncertain prognoses on further increases in CO_2 emis-

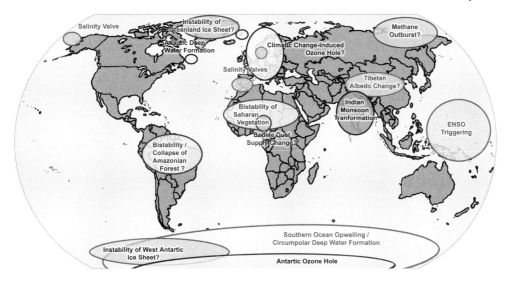

Fig. 1.2 Schellnhuber's map of global "tipping points" in climate change. (Reprinted with permission from *Nature*, 437, 1238 (2005) [3]).

sion, due to uncertainties in the economic scenarios. These are partially due to uncertainties in the reduction of energy use, uncertainties in the implementation of alternative technologies and the use of new technologies that may become developed, which are partially topic of this book. A prediction of climate changes when current rate of CO_2 emission continues is shown in Fig. 1.2.

The cost to our society to address the consequences of these changes can be very large. Therefore, technologies that will enable us to use the energy and

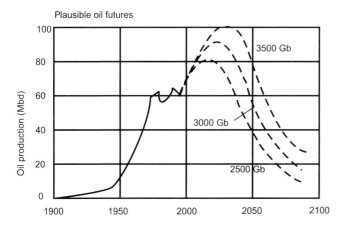

Fig. 1.3 Projection of exploitable oil resources. (V. Smil, *Energy at the Crossroads: Global Perspectives and Uncertainties* (2005). Reprinted with permission from The MIT Press, Cambridge [4]).

chemicals to maintain our current standard of living and to increase that of the many that still live under below standard conditions are very important to our society. This consideration is one of the important drivers behind the search for renewable catalytic technologies.

The other important issue is the future of fossil energy resources. As with carbon dioxide emissions, future predictions are highly uncertain. They relate to many economic factors. For instance, an increase in oil price makes oil exploitable that previously was uneconomic. Figure 1.3 shows the results of different prognoses.

The conclusion is that in the coming age there will be plenty of oil to satisfies our needs. But the oil is exhaustible and hence additional energy resources are needed. In this respect it is of interest to compare the relative amount of available primary energy resources. An estimate is given in Fig. 1.4.

One notes the large difference in amounts available, not only between available fossil resources, but also for the renewables. The amounts of coal are close to eight times those of oil, comparable to our uranium resources. The amounts of natural gas available are slightly less. If one compares these amounts with the current use of renewables as wind, solar and photosynthesis production, it is clear that fossil fuel resources will stay with us for a very long time.

The primary issue then is to either use these resources or replace them so as to reduce the emission of green house gasses. The search for processes that reduce

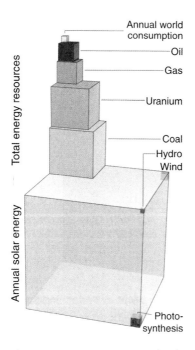

Fig. 1.4 Primary energy resource distribution [5].

energy consumption, or rather reduce wasteful use, is of course very important. This is not the topic of this book. We will, though, analyze the conversion processes from primary into secondary energy carriers.

1.3 Technology Options

Figure 1.4 indicates the relative amounts of primary energy carriers. Figure 1.5 highlights the different energy carriers that result.

We recognize the nuclear reactions of the sun as the main source of our energy and the crucial role of photosynthesis as the conversion process of solar light into fossil energy carriers and renewable biomass. We also recognize gravitational sources that result in tidal movement and the earth's thermal resources.

Interestingly, for the secondary energy carriers towards which primary energy is to be converted there is only a very limited set of options (Fig. 1.6).

Listed are only the hydrocarbons, hydrogen and electricity. These molecules and electricity are the main players we will be concerned with. Remembering our previous arguments on the primary factors that cause the drive to renewables, let us first revisit fossil energy carriers. They will only be used sustainably with respect to environment and climate when there are zero-emissions of greenhouse gasses or other detrimental gasses as NO or N_2O. Previously, we addressed the latter issues [6], here we will be primarily concerned with the greenhouse gas CO_2.

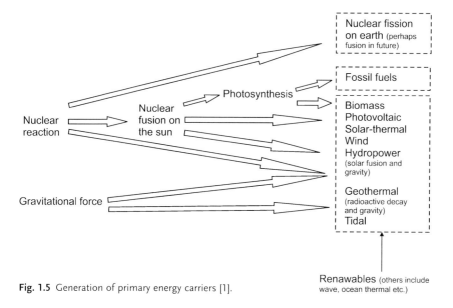

Fig. 1.5 Generation of primary energy carriers [1].

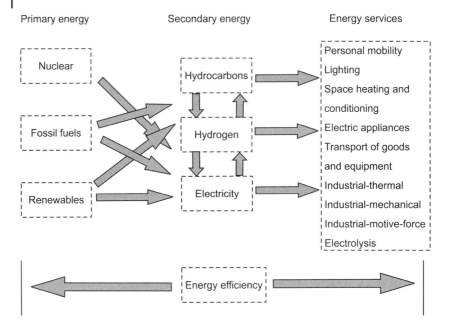

Fig. 1.6 Conversion of secondary energy carriers [1].

There are two options: To remove CO_2 using a chemical (reduction) reaction requires energy. Nuclear or solar energy are candidates that do not produce CO_2 themselves. In essence CO_2 is then reduced back to an energy carrier; renewable energy is to be used.

The other option is to capture the CO_2 and to store it. The amounts of CO_2 are huge if we continue to use the fossil fuel resource base. It is estimated to be over 6.000 gigatonnes of carbon (GtC) [1]. The most attractive opportunity for this is geological (Fig. 1.7).

It is proposed to use the depleted oil reserves or the absorption capacity of deep coal seams. Saline aquifiers are estimated to hold 3000 to 10 000 GtC. Such processes are currently explored by several oil companies and appear to be feasible.

Alternatively, one can react CO_2 with reactive rock minerals to form carbonates. There appear to be large amounts of Ca- or Mg-containing materials available, be it at the environmental cost of huge mining operations.

The additional cost to oil or gas conversion has been estimated between 10 and 25%.

Clearly, implementation of these schemes has large implications for secondary energy schemes. CO_2 sequestration will have to take place centrally at large installations and implies that hydrocarbons cannot be used as energy carriers. The alternatives are hydrogen or electricity, with important implications for catalysis. Not only production, but also efficient storage and use are important. For this reason, chapters on these topics are also included in this book.

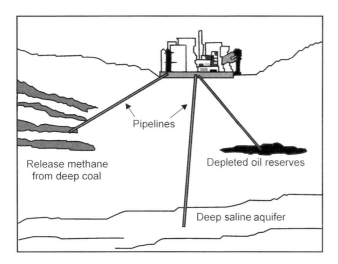

Fig. 1.7 Geological CO_2 storage options [1].

Table 1.1 Renewable energy use forecast [1].

Estimates of annual potential contribution

Renewable source (all values in EJ)	Extreme resource	Potentially economic	BAU Forecast 2000	2050	2100
Primary energy					
Hydropower	150	29–50	9	20	30
Biomass	2 900	100–300	52	120	210
Traditional biomass			(45	70	90)
Modern biomass			(7	50	120)
Wind	6 000	250–600	0.11	20	90
Solar	3 900 000	1 500–50 000	0.16	8	30
Geothermal	600 000	500–5 000	0.3	2	20
Tidal/wave/current	145	1.5	0.002	0.01	0.1
Total	4 500 000	2 300–56 000	62	170	380
Electricity generation					
Biomass	–	–	1	20	70
Solar	–	–	0.005	2	20
Geothermal	–	–	0.17	1	15

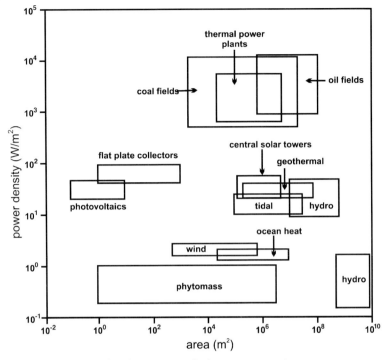

Fig. 1.8 Comparison of surface area use of solar energy versus biomass and other primary energy sources. (V. Smil, *Energy at the Crossroads: Global Perspectives and Uncertainties* (2005). Reprinted with permission from The MIT Press, Cambridge [4]).

CO_2 sequestration is not a proven technology and the long-term effects are not yet known. It can also only be an intermediate time solution.

Another option is to use nuclear energy. Whereas technologically, with the development of breeder reactors, the uranium resources can be considered non-exhaustible and reactor technology can be considered safe [4] a serious concern is the proliferation of plutonium for nuclear weapons. There is also the unproven solution for disposal of radioactive material.

There are ample predictions of future use of renewables, e.g. that shown in Table 1.1.

A comparison is made of the available resources, their principle economic exploitability and actual prognosis for their use. Again one notes the enormous potential of solar energy. Rapid growth is foreseen for biomass, solar and wind. A difference is made between traditional and modern use of biomass, because currently, especially in less developed countries, biomass is also an important energy source, but is not used in a technologically advanced way.

Wind energy can only be harvested where there are strong winds and production may vary largely in time because of fluctuations in wind intensity. This implies a need for efficient energy transportation and storage. We will not consider wind energy further in this book.

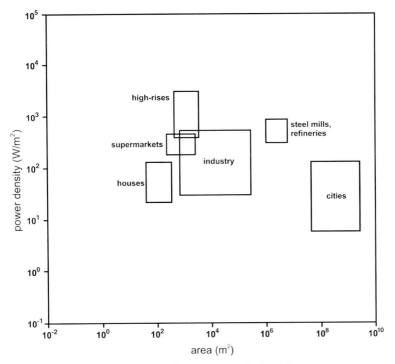

Fig. 1.9 Power density comparison of consumption of industrial, commercial and household energy. (V. Smil, *Energy at the Crossroads: Global Perspectives and Uncertainties* (2005). Reprinted with permission from The MIT Press, Cambridge [4]).

It is interesting to discuss in slightly more detail the case of solar versus biomass. The issue here is the amount of energy to be produced per unit time and unit surface area for the two cases. The important conclusion is drawn when one considers the efficiency of surface area use photovoltaics. The rate of biomass growth is too slow. Figure 1.8 illustrates these differences.

The power density of photovoltaics is 10–100× higher than that of biomass. The surface area need for the same power output is orders of magnitude less. A comparison is also made with the power densities of fossil fuel. This is important, because as one notes from Fig. 1.9 even the solar power densities do not satisfy the power need requirements of modern cities.

Two conclusions can be made at this stage of the discussion. Firstly, when surface area constraints are important, direct conversion of solar energy wins out over biomass. Estimates of an increase in biomass yields using bio-engineering approaches amount to not more than a factor of 2 [7].

Secondly, the relatively low power densities, require transport and upgrading again, with large implications for choice of energy carriers and corresponding conversion processes. Hydrocarbons may very well become energy carriers even if solar energy were to dominate primary energy sources!

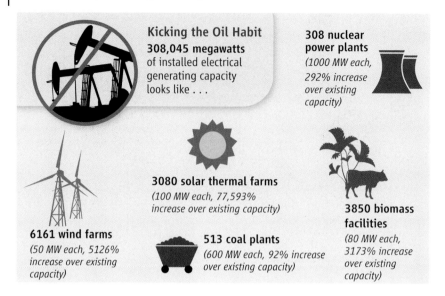

Fig. 1.10 Oil alternatives: In his State of Union speech, President George W. Bush set out a goal to find alternatives for 75% of the oil that the United States expects to be importing from the Middle East by 2025. According to the Energy Information Administration, the total then will be 5.99 million barrels of oil per day from the Persian Gulf, up from 3.7 million barrels in 2005. So the country must replace 4.49 million barrels a day, or 1.64 billion barrels a year. Here's what it would take for each power source (above) to generate an equivalent amount of electricity. (Considerably more would be needed to convert the electricity into a transportation fuel.) (E. Kintisch, J. Mervis, Science, 311, 762 (2006). Reprinted with permission from AAAS [8]).

There are several reasons why the coming age will see a dominance of biomass as renewable energy source. This in essence relates to compatibility with existing conversion technology. This is highlighted by a recent estimate of investments needed when traditional energy are substituted by alternative resources (Fig. 1.10).

Replacing oil conversion technology by solar or nuclear energy conversion units requires enormous investment. The overriding factor for the growth of biomass-based energy is the opportunity that existing facilities have to convert it with only small adaptation costs. This is reflected in cost estimates of electricity production from the different resources (Table 1.2). This cost comparison also explains the low growth of solar energy. However, the rewards may quickly fall off once large-scale implementation starts because of learning curve effects. The land use needs for both large-scale solar energy and biomass use are enormous.

Of course the kind of land needed is quite different. Whereas solar energy can be harvested on the roofs of houses or in unfertile but sunny deserts, arable land is needed for biomass conversion of sugar or alcohol (at least wood should be able to grow).

An estimate of the surface area need when photovoltaics are used is 750 by 750 square kilometers if current global electricity needs were to be covered (Fig. 1.11)

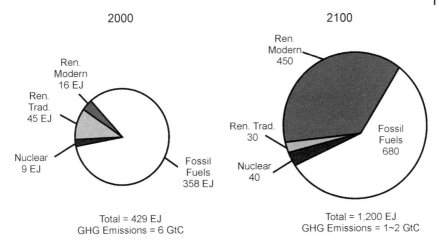

Fig. 1.11 Comparison of primary energy source in 2000 and 2100 [1].

Table 1.2 Cost comparison of electricity as a function of primary energy sources [1].[a]

	Projected electricity cost (kWh in $US 2000)						
Coal-Pc post-combustion	Coal-IGCC	NatGas-CCGT	Nuclear	Hydro	Wind	Biomass	Solar-PV
6–7.5	5.5–7	5.5–7	6–10	6–8	6–8	6–8	15–20

[a] PC = pulverized coal; IGCC = integrated gasification combined cycle; CCGT = combined cycle gas turbine; PV = photovoltaic. Assumed input prices are coal $1.5–3 GJ^{-1}, natural gas $5–7 GJ^{-1}, biomass $2–5 GJ^{-1}.

[1]. This corresponds to approximately the surface area of French and Germany combined. To replace the world coal consumption by biomass grown from wood one would need a surface area of 330×10^6 ha, which is more than the total forested land in the European Union and the United States. There is an order of magnitude difference between the two numbers in favor of solar energy [4].

The logistics requirements are quite different. Whereas biomass is optimally used when processed decentrally, the power density of solar energy requires processing at large centralized units.

Practice is quite different. The main reason for the current rapid growth of biomass conversion systems is the feasibility of using existing refinery systems and installed secondary energy systems.

Large-scale biomass production will come at a considerable cost to the society. We have already given an order of magnitude estimate for wood plantations. When biomass raw material is processed that competes with the food chain, as vegetable oils and sugars, the situation is even worse. For instance Smil [4] mentions that "if the US vehicles were to run solely on corn derived ethanol the country would have to plant corn on an area 20% larger than is currently cropland".

Such intensive use of land will make very large demands on available water supplies and soil corrosion is also an important issue. There is even a debate over the overall reduction in fossil energy use and actual contribution to CO_2 abatement for the case of ethanol production from corn [9]. Since waste arises from biomass or fossil fuel, to introduce waste products in the conversion scheme, which of course should be done, does not help in the overall scheme we discuss here. This is born out by economic predictions (Table 1.2).

For reasons mentioned above, biomass conversion will for the time being show the main growth of the renewables. However, for the longer term, photovoltaics conversion technology is the more attractive option.

Government policy and subsidies largely determine the immediate future. For instance, Dutch policy for the year 2010 is given in Table 1.3

Table 1.3 Dutch policy for 2010.

Type of measure	Relative amount (%)
Energy saving	1.3
Sustainable energy	5
Sustainable electricity	9

The European Commission wants to have a contribution of 12% energy from renewable sources to the energy budget within the EC in 2010. The relative amount of bio-fuels will increase to a level of 5.75%, this is more than twice the corresponding use of oil. The US Department of energy has set goals to replace 30% of the liquid petroleum transportation fuels with biofuels and to replace 25% of industrial organic chemicals with biomass-derived chemicals by 2025 [7].

So far little has been said on the future for raw materials to the chemical industry. Clearly, the molecular components of oil are of great value and hence the primary concern should be the limitation of fossil oil reserves in energy production. Also, the amount of oil used for the chemicals industry is orders of magnitude less than for energy use.

In addition, with the very low value of rich CO_2 streams, CO_2 could be considered an oxygen source for chemical production, whereas biomass itself could be also a source of raw materials (see below) or would be used for energy production.

1.4
Process Options for Biomass Conversion

Figure 1.12 shows a schematic representation of the proposed self-sustained integrated biomass zero-emission cycle.

The additional interesting part of Fig. 1.12 is the biorefinery, which uses biomass and waste, produces waste products CO_2 and ash, both to be recycled for the production of biofuels, heat and electricity and biomaterials. These biomaterials are highly oxygen functionalized for products such as alcohols, carboxylic acids and esters. A currently produced bioplastic is poly(lactic acid). A main cost factor is separation.

Current strategies for the production of liquid fuels from biomass are shown in Fig. 1.13

Starch and fatty acids are the main food constituents of biomass. Sugar is derived from starch by hydrolysis or directly by extraction from sugar cane or beet. Fermentation converts sugars into alcohol that can be directly used as fuel, or in principle can be used as the raw material of a biorefinery plant for further upgrading. Triglycerides, derived from oil seeds, are used to be converted into biodiesel through transesterification processes (Fig. 1.14).

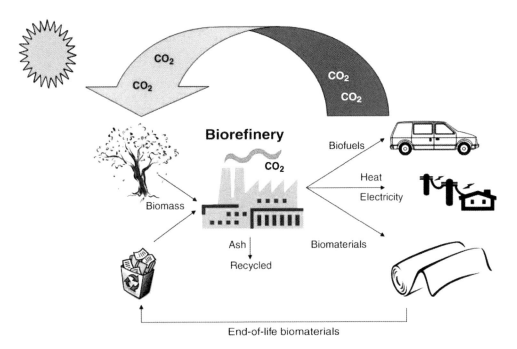

Fig. 1.12 Fully integrated agro-biofuel-biomaterial-biopower cycle for sustainable technologies (From A. J. Ragauskas et al., *Science*, 311, 484 (2006). Reprinted with permission from AAAS [7]).

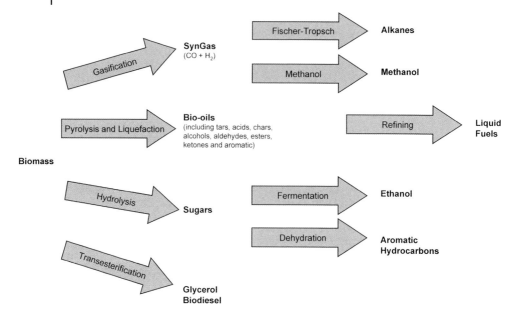

Fig. 1.13 Strategies to convert biomass. (G. W. Huber, J. A. Dumesic, *Catal. Today*, 111, 119 (2006). Reprinted with permission from Elsevier [10]).

Biodiesel Production via Catalytic Transesterification

$$\begin{array}{c} H_2C-COOR^I \\ HC-COOR^{II} \\ H_2C-COOR^{III} \end{array} + 3ROH \xrightarrow{Catalyst} \begin{array}{c} CH_2OH \\ CHOH \\ CH_2OH \end{array} + \begin{array}{c} R^I HOOR \\ R^{II} HOOR \\ R^{III} HOOR \end{array}$$

Triglyceride   Methanol   Glycerol   Biodiesel

where R^I, R^{II} and R^{III} are long-chain C_{12}–C_{32} hydrocarbons

Catalysts which are efficient at room temperature: **alkoxides, hydroxides**

Vegetable oil methyl esters (biodiesel)	Kinematic viscosity (mm²/s)	Cetane no.	Lower heating value (MJ/kg)	Cloud point (°C)	Pour point (°C)	Flash point (°C)	Density (kg/l)
Peanut	4.9	54	33.6	5	–	176	0.883
Soya bean	4.5	45	33.5	1	–7	178	0.885
Babassu	3.6	63	31.8	4	–	127	0.875
Palm	5.7	62	33.5	13	–	164	0.880
Sunflower	4.6	49	33.5	1	–	183	0.860
Standard diesel fuel	3.06	50	43.8	–	–16	76	0.855
20% biodiesel blend	3.2	51	43.2	–	–16	128	0.859

Fig. 1.14 Sources of biodiesel. (B. K. Barnwal, M. P. Sharma, *Renewable Sustainable Energy Rev.*, 9, 363 (2005). Reprinted with permission from Elsevier [11]).

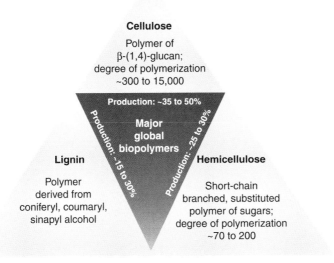

Fig. 1.15 Key global biomass resources from agricultural residues, wood, and herbaceous energy crops. (A. J. Ragauskas et al., *Science*, 311, 484 (2006). Reprinted with permission from AAAS [7]).

A few chapters of this book are devoted to the use of glycerol and alternatives to transesterification. These two processes compete with the food chain. Whereas in the short term these are viable, be it mainly through government subsidies, on the longer term these processes will have to be replaced by more amply available and non-food competing processes.

Figure 1.15 gives an overview of the main constituents of non food biomass. There are three components: Cellulose, hemicellulose and lignin. Cellulose and hemicellulose are built form sugar-type monomers, but their cost-effective isolation through enzymatic depolymerization remains a challenge.

Most lignin is now burnt for heat and power, but process options are available to depolymerize the phenolic material by thermal cracking or using base treatments [7]. In consecutive steps the products can be converted into aromatic hydrocarbon feeds.

A biorefinery scheme to produce chemicals from non-food biomass is given in Fig. 1.16.

For energy carrier and chemicals production the most important route is through glucose. For this reason, fermentation processes using genetically altered microorganisms towards conversion of hexoses of fructoses from hemicellulose are being investigated intensively.

Key products obtained from the fermentation of glucose are levulinic acid and 5-hydroxymethyl-2-furfural (Scheme 1.1).

1 Renewable Catalytic Technologies – a Perspective

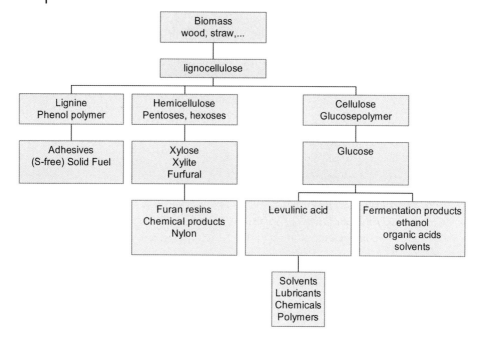

Fig. 1.16 Ligno-cellulose feedstock biorefinery.

Levulinic acid can be used in various heterogeneous and homogeneous catalytic reactions to provide monomer intermediates for polymers such as caprolactam, nylon and polyesters.

For energy production, routes towards secondary energy carriers such as H_2 and hydrocarbons are important. Sugars and alcohols can be converted with water into H_2 and CO_2 using noble metal catalysts [12]. Light hydrocarbons can also be formed. A clever biorefinery approach has also been proposed by the group of Dumesic [13]. Hydroxymethyl furfural, which is easily produced from glucose, is partially hydrogenated and dimerized through aldol condensation. The hydrocarbon that is produced by consecutive dehydration forms an oil layer on water and hence can be easily separated. Whereas the two processes in the lower part of Fig. 1.16 select a key intermediate to build specific molecules, the other two routes can be used to produce liquid energy carriers, without use of initial fermentation and separation processes.

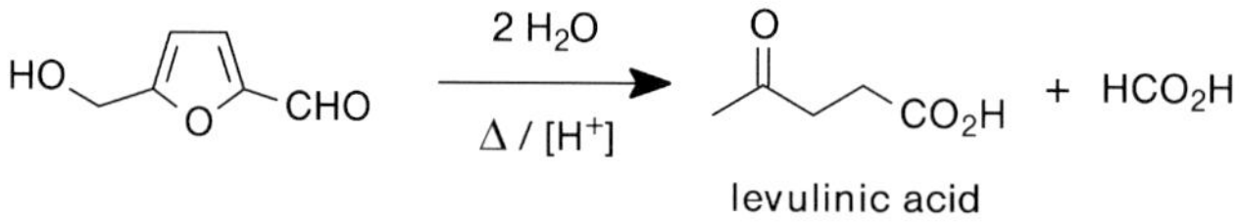

Scheme 1.1

The gasification route is most versatile and adaptable to existing processing routes. Waste or biomass is gasified towards synthesis gas. A process well known for gasification of coal or natural gas. Its main draw back is a high capital cost because of the high temperatures used and treating processes that have to take place to purify the gas and to adjust the hydrogen to carbon monoxide ratio. The product gas can be shifted with water towards hydrogen, stored and be used later in fuel cells, or be used directly in gas turbines to produce electricity. Currently the use of synthesis gas as feed for the Fischer–Tropsch process is expanding (at present for the conversion of natural gas). This produces liquid hydrocarbons. Clearly, catalytic process improvements to the many steps in this overall process are important in the context of renewables production.

Pyrolysis and liquefaction processes take an intermediate position in the sense that they maintain some larger molecular characteristics. Pyrolysis is a process in which the biomass material is quickly heated. The thermal cracking process, depolymerizes waste or dry biomass and produces a liquid of complex composition (Fig. 1.17).

This hydrocarbon mixture, rich in oxygen and aromatics, is very unstable and needs important (catalytic) treating steps before it can be used as an actual fuel. The chemistry of many of these processes is little understood and will be discussed in some of the following chapters. The advantage of this route is that it can readily be operated at a small scale. Unless the bio-oil is incinerated to produce electricity, access to a cheap hydrogen source is needed for further use.

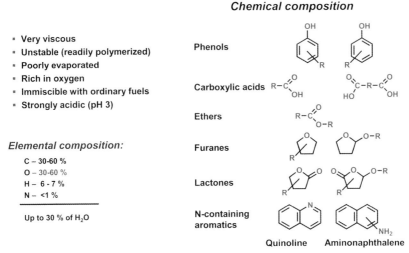

Fig. 1.17 Properties of "Bio-crude-oil".

1 Renewable Catalytic Technologies – a Perspective

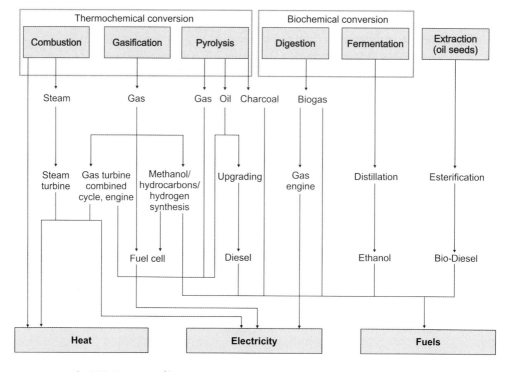

Fig. 1.18 Summary of biomass conversion routes [14].

In an alternative process, biomass is converted into oil at high pressure with water, the HTU process – again a very complex oil is produced, which, however, is less reactive because of partial hydrogenation.

Figure 1.18 summarizes the different process options for biomass conversion.

We have, so far, not considered the combustion process route. Obviously, it is a very important process route towards electricity production, with minor adaptations of present installations. The catalytic challenge relates to the purification of the emission gasses, which will have an increased NOx level.

1.5
Conclusions

The need for renewable technologies creates ample catalytic opportunities. We have seen that the techno-economic scenarios favor different technologies in the course of time. This implies on ongoing need for process innovation, for which catalysis is the chemical back bone. Clearly, progress requires a sustained integrated effort with the many other disciplines that are needed to develop the devices and the required infrastructure.

In the short to medium term, renewable fossil based energy will remain important. CO_2 sequestration and the consequential importance of hydrogen imply a large interest in hydrogen production technologies, its storage and subsequent conversion. Electrocatalysis may be expected to be of increasing importance.

There will be a parallel growth in biomass technology, stimulated to a significant extent by government subsidies. There is a significant threat to the environment if biomass resources imply expansion of agricultural activities in presently forested areas. If biomass dominates the energy sector this will be a necessity. In the longer term, waste materials, instead of food biomass, will be sustainable.

The need for catalysis in this field is very clear. Molecular conversion technologies of non-food biomass have, so far, only been implemented for fermented cellulose. Even for those products, conversion into energy carriers is not yet very efficient.

Pyrolysis and liquefaction products are in need of substantial treating processes, the chemistry of which is not yet very well understood. There is also a need and an opportunity to improve the selectivity of Fischer–Tropsch type processes.

For the longer term, there is the solar energy conversion challenge. Once solar radiation is efficiently captured it will be stored in the form of hydrogen or electricity, with major challenges again for electrocatalysis.

References

1 M. Jaccard, *Sustainable Fossil Fuels: The Unusual Suspect in the Quest for Clean and Enduring Energy*, Cambridge University Press, Cambridge (UK), **2005**.
2 W. F. Ruddiman, *Scientific American*, March **2005**.
3 *Nature*, 437, 1238 (**2005**).
4 V. Smil, *Energy at the Crossroads: Global Perspectives and Uncertainties*, The MIT Press, Cambridge (USA), **2005**.
5 B. Lomborg, *The Skeptical Environmentalist: Measuring the Real State of the World*, Cambridge University Press, Cambridge (UK), **2001**.
6 F.J.J.G. Janssen, R. A. van Santen, (eds.), *Environmental Catalysis*, Imperial College Press, London, **1999**.
7 A. J. Ragauskas et al., *Science*, 311, 484 (**2006**).
8 E. Kintisch, J. Mervis, *Science*, 311, 762 (**2006**).
9 A. E. Farrell, R. J. Plevin, B. T. Turner, A. D. Jones, M. O'Hare, D. M. Kammen, *Science*, 311, 506 (**2006**).
10 G. W. Huber, J. A. Dumesic, *Catal. Today*, 111, 119 (**2006**).
11 B. K. Barnwal, M. P. Sharma, *Renewable Sustainable Energy Rev.*, 9, 363 (**2005**).
12 R. D. Cortright, R. R. Davda, J. A. Dumesic, *Nature*, 418, 964 (**2002**).
13 G. W. Huber, J. N. Chneda, C. J. Barrett, J. A. Dumesic, *Science*, 308, 1446 (**2005**).
14 W. C. Turkenburg (ed.), Renewable energy technologies, Chap. 7, Fig. 7.1, p. 223, in: J. Goldemberg (ed.), *World Energy Assessment: Energy and the Challenge of Sustainability*, UNDP/UN-DESA/World Energy Council, New York, **2000**.

2
Lignocellulose Conversion:
An Introduction to Chemistry, Process and Economics

Jean-Paul Lange

2.1
Overview

Governments across the world are stimulating the valorization of local biomass to secure the energy supply, reduce the fossil CO_2 emissions and support the rural economy. A first generation of fuels and chemicals are presently produced from high-value sugars and oils. Meanwhile, a second generation, based on cheaper and more abundant lignocellulosic feedstock, is being developed. The present chapter addresses the various chemistries and technologies that are being explored to valorize lignocellulosic biomass. It shows the need to "deoxygenate" the biomass and review the main chemical routes for it, i.e.,

- the pyrolysis to char, bio-crude or gas;
- the gasification to syngas and its subsequent conversion into alkanes or methanol;
- the hydrolysis to sugar, furfural and levulinic acid;
- the fermentation to ethanol, biogas and biochemical.

The economics of biomass conversion needs to be considered as well, for the production costs of biofuels typically amount to $60–120 per barrel of oil equivalent. Influential factors include the cost of the biomass at the plant gate, the conversion efficiency, the scale of the process and the value of the product (e.g., fuel, electricity or chemicals).

2.2
Introduction

2.2.1
Need for Renewable Energy

Governments across the world are stimulating the utilization of renewable energies such as solar, wind, hydroelectricity and biomass. For instance, the European

Catalysis for Renewables: From Feedstock to Energy Production
Edited by Gabriele Centi and Rutger A. van Santen
Copyright © 2007 WILEY-VCH Verlag GmbH & Co. KGaA, Weinheim
ISBN: 978-3-527-31788-2

union has set an overall target of 12% share of renewable energy for 2010, with 21% share in the electricity sector and 5.75% share for biofuels [1]. Similar targets have been formulated for the USA [2]. These targets are expected to rise further in the future. The governments are driven by three major forces, which may vary in priority from country to country.

A first driver is securing the access to energy at an affordable price. The world demand for energy is expected to double from 2000 to 2050, due to a ~50% increase in world population and a comparable increase in energy consumption per capita [3–9]. Numerous energy scenarios concur in concluding that this demand cannot be satisfied by crude oil, the production of which is expected to peak well before 2050. The growing production of natural gas, coal and unconventional oil (e.g., tar sands) will compensate for the declining supply of oil but is not expected to satisfy the growth in energy demand throughout the rest of the century. Hence, nuclear and renewable energies will be required to cover about half of demand at around 2050 and about two-thirds of it by 2100.

A second driver for renewable energies is the threat of climate change that would result from the anthropogenic emission of CO_2 into the atmosphere [3, 5, 9]. The anthropogenic emission might be a fraction (~3%) of the overall CO_2 cycle; it nevertheless represents ~80% of the imbalance between the overall CO_2 emission and CO_2 fixation [3]. Numerous countries have recently signed the Kyoto protocol that was developed in 1997 to control and reduce the CO_2 emission. Efforts to reduce the consumption of fossil fuel and capture and sequester the CO_2 will help reducing CO_2 emissions in the short term. Indeed, CO_2 capture and sequestration is unavoidable for countries that want to develop their indigenous coal to secure their energy supply, such as the US, China and Australia. However, these solutions offer no real long-term answers to the overall energy question.

The third and last major driver is to develop/maintain agricultural activities and to respond to agriculture hazards. Such development could consist of promoting the production and utilization of local bio-resources for food, fuel, electricity and material production. It could also consist of the electrification of remote rural areas by means of micro-hydropower, wind or solar energy.

Interestingly, each of these major drivers also represents one of the three dimensions of sustainability, namely Profitability (affordable energy), Planet (climate change) and People (social stability). No single energy solution seems to address these three dimensions satisfactorily. It appears therefore inevitable for mankind to progress towards a complex and delicate mixture of energy sources and carriers to properly balance the three Ps of sustainable development.

2.2.2
Need for Biomass Conversion

The power sector can be supplied with various renewable sources, namely wind, solar, hydraulic power and biomass. The transportation sector, which represents 28% of the global energy consumption [7], has a limited choice, however. Biomass is the only resource in renewable liquid fuels so far. Of course, renewable electricity could be developed to drive cars or to produce H_2-fuel for powering them. Such options require very important efforts to develop the proper technologies and infrastructure to produce, distribute, store and use the fuel or energy. In contrast, the implementation of biomass-derived liquid fuels, i.e., "biofuels", is much easier. For instance, biofuels are presently being introduced in the European and US market in a smooth way by blending biofuel components into the fossil fuel pool. Other countries such as Brazil and Sweden are also exploring the use of (nearly) pure biofuels for single or dual-fuel vehicles. This approach is more costly and perturbing for the market, however, for it indeed requires large investments in the fuel distribution infrastructure as well as significant modification of the cars. It may, however, be a solution for a captive fleet such as taxis, buses and municipal vehicles.

Beyond the transportation sector, biomass is also a promising feedstock for the chemical industry. This industry accounts for 5–10% of today's oil and gas consumption. It may require an even larger fraction in the future as the demand for chemicals has outpaced that for energy in the last few decades. Recently, the chemical industry has indeed showed a significant interest in converting agricultural feedstock into chemical intermediates such lactic acid or propene-1,3-diol.

A first generation of fuels and chemicals are presently produced from sugars and vegetable oils. They are, however, competing with the food chain for their feedstock. When demand outstrips the available volumes first-generation biofuels may no longer be a sustainable option. We are already witnessing price increases for corn and vegetable oil in the US and Europe, which are due to the strong demand of biofuels. Hence, a second generation of technologies are being developed to exploit cheaper and more abundant lignocellulosic feedstock. The 3–5 Gt a^{-1} of lignocellulosic biomass left over from agricultural and forestry activities could provide 50–85 EJ a^{-1} of energy [3, 10, 11], which represents 10–20% of today's world energy demand and 5–10% of 2050's demand. The role of biomass might not be limited to this fraction, however. The total amount of biomass that is growing annually on earth and in the water has been estimated between 100 and 200 Gt a^{-1} (i.e., 1700–3400 EJ a^{-1}), which represents more than the forecasted energy demand [3, 4, 9, 10]. Biomass will likely take part in the energy mix, though its contribution might be limited to 5–15% of the total energy supply by 2050 [5, 6, 8], in part because of its limited availability.

The growth of biomass and its conversion into biofuel consumes energy and produces CO_2. This significantly impacts the efficiency of biomass as response to the first two drivers, namely (net) access to energy and CO_2 emission. Life-cycle analyses reveal a continuous improvement in the net energy balance and CO_2 emission of bio-fuels such as bio-ethanol [3, 11–13]. For instance, the amount of fossil fuel required to produce corn ethanol corresponds to as much as 60–75% of the energy content of ethanol [12, 13]. $\sim 1/3$ is consumed by agricultural practice and $\sim 2/3$ by further processing [13]. Further improvements in biomass production, collection and conversion are therefore needed to achieve real savings in fossil fuel consumption and CO_2 emission.

2.2.3
Biomass Composition

Lignocellulose is the fibrous material that forms the cell wall of a plants "architecture". It consists of three major components (Fig. 2.1): cellulose, hemicellulose and lignin [3, 14–16]. It contrasts with the "green" parts of the plants and the seeds, which are rich in proteins, starch and/or oil.

The cellulose forms bundles of fibers that provide the strength of the materials. It consists of semi-crystalline polymer chains of glucose, which are rigidly held together through intra- and intermolecular hydrogen bonds to form crystalline and amorphous segments. Glucose is a unique C_6 sugar for this purpose for it has all its hydroxyl groups in equatorial positions, which allow for a close packing of the chains and strong hydrogen bonding. Cellulose consists of some 10 000 glucoses unit and has, therefore, a molecular weight in the order of 1 MDa. The cellulose typically accounts for 40–50 wt.% of the biomass.

The hemicellulose forms the glue around and between the cellulose bundles. It consists of shorter, branched polymer chains of various C_6 and C_5 sugars. The C_6 sugars consist mainly of glucose together with mannose and galactose. The C_5 sugars include xylose and arabinose. The steric hindrance offered by the side-

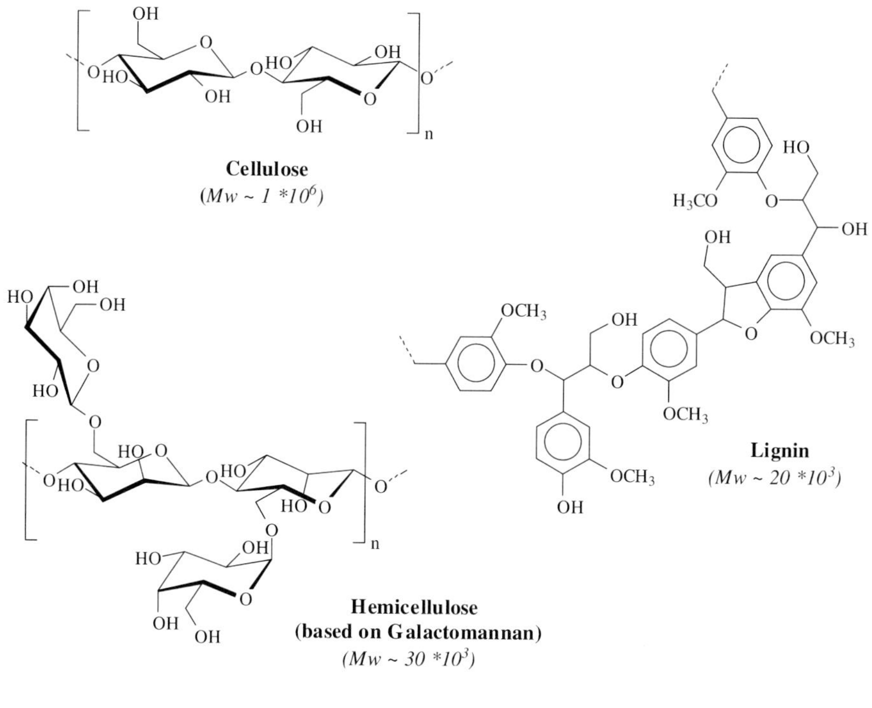

Fig. 2.1 Main components of lignocellulose. (Adapted from [13]).

chain and the axial hydroxyl groups of the various sugars does not allow the hemicellulose to form semi-crystalline domains. In contrast, it allows for numerous random contacts with other structures like the cellulose. The chain contains some 200 sugar units and has a molecular weight of some 30 kDa.

Lignin is a tri-dimensional polymer of propyl-phenol that is imbedded in and bound to the hemicellulose. It provides rigidity to the structure. Its phenol groups also provide antibacterial activity that protects the lignocellulose against microorganisms. The molecular weight of lignin seems to be up to 20 kDa, which corresponds to a network of some 120 propyl-phenol units.

Of course, the lignocellulose also contains various minor components such as proteins, terpenic oils, fatty acids/esters and inorganic materials (e.g., mainly based on N, P and K). These components will not be considered here owing to lack of space but should not be forgotten as they interfere with many processes. For sustainable biomass production the inorganic materials need to be recycled from the process to the field.

2.2.4
Fuels and Chemicals Composition

With an average O/C atomic ratio close to 1, biomass differs strikingly from gasoline and diesel, which the transportation sector is mainly relying on today. These fuels are well-defined mixtures of hydrocarbons that have evolved over several decades to provide maximum power and mileage while minimizing the impact on the environment [17]. Gasoline and diesel share a high energy density of 32–34 GJ L^{-1}. They are composed of hydrocarbons that boil over a wide temperature range, i.e., 30–200 °C for gasoline and 160–360 °C for diesel, to provide smooth combustion under various climatic conditions and avoid the formation of explosive vapor mixtures upon manipulation. Both fuels differ in the type of hydrocarbons used, however. Gasoline requires components that resist auto-ignition to allow the fuel–air mixture to ignite only with a spark. It is, therefore, preferably composed of highly branched hydrocarbons and aromatics. In contrast, diesel requires components that easily auto-ignite upon injection of the fuel in the hot compressed air. Therefore, diesel preferably consists of linear or slightly branched alkanes.

Obviously, a large variety of components, including biomass derivatives, could be added to these hydrocarbon mixtures [3, 17]. Some oxygenates can provide additional power to gasoline mixture, e.g., by improving its resistance to auto-ignition. However, they often deteriorate properties of the fuel. Oxygenates lower the energy content of the fuels and, thereby, increase its consumption. They may show a limited solubility in the hydrocarbon mixture and show incompatibilities with the materials used in fuel distribution and car fuel lines. Because of their poor solubility in hydrocarbons, light oxygenates will also lead to excessive volatility of the fuels. This is, for instance, the case for ethanol in gasoline. All these considerations show that biomass needs to be deeply deoxygenated before entering the present fuel pool.

Numerous chemical intermediates are oxygen rich. Methanol, acetic acid and ethylene glycol show a O/C atomic ratio of 1, as does biomass. Other major chemicals intermediates show a lower O/C ratio, typically between 1/3 and 2/3. This holds for instance for propene and butene glycols, ethanol, (meth)acrylic acids, adipic acid and many others. The presence of some oxygen atoms is required to confer the desired physical and chemicals properties to the product. Selective and partial deoxygenation of biomass may represent an attractive and competitive route compared with the selective and partial oxidation of hydrocarbon feedstock.

2.2.5
Biomass Deoxygenation

We have just seen that the conversion of biomass into fuel and chemicals requires deoxygenation. This can proceed via two main routes, namely the elimination of either H_2O or CO_2 (Fig. 2.2) [18]. Stoichiometric deoxygenation of glucose via H_2O and CO_2 elimination leads to charcoal and methane, respectively. These ideal reactions proceed with >95% energy efficiency: the reaction products contain >95% of the (lower) heating value of the reactant. Deoxygenation can also proceed via combined or partial H_2O and CO_2 elimination [18]. Examples are the partial deoxygenation to ethanol (+CO_2) or levulinic acid + formic acid (+H_2O) as well as the oxygen-rearrangement to acetic acid. All these reactions proceed with >90% energy efficiency. Obviously, the thermodynamics allows oxygen to be reshuffled in or removed from biomass without large energy penalty.

		Energy Efficiency:
$C_6H_{12}O_6$ → 6 H_2O + 6 [C]		93% LHV
$C_6H_{12}O_6$ → 3 CO_2 + C_3H_{12} (e.g. 3 CH_4)		95%
$C_6H_{12}O_6$ → 2 CO_2 + $C_4H_{12}O_2$ (e.g. 2 EtOH)		97%
$C_6H_{12}O_6$ → $C_6H_{12}O_6$ (e.g. 3 AcOH, 2 Lact. Ac)		93%

Fig. 2.2 Biomass deoxygenation via H_2O or CO_2 elimination (LHV: low heating value).

2.3
Chemistry and Processes

Various chemistries and processes can be applied to convert lignocellulosic materials into valuable fuels and chemicals [3, 19]. For instance, thermal reactions are exploited in the pyrolysis of biomass to charcoal, oil and/or gases and its gasification to synthesis gas and/or hydrogen. All these products can be used without further processing to provide heat, steam or power. Moreover, the synthesis gas and, in some cases, the pyrolysis oil can also be converted into high-quality liquid fuels.

Alternatively, lignocellulose can also be hydrolyzed to liberate the lignin and depolymerize the polysaccharides to sugars. The sugars can subsequently be converted into various fuel and chemical components via chemical or biological routes.

While lignocellulose is quite resistant against direct biological conversion – trees usually take years to rot away – other form of biomass readily undergo biological digestion. This is, for example, the case for green biomass or cellulosic sludge's present in municipal, industrial or consumer's wastes. Such bio-wastes can be hydrolyzed to sugars or digested to biogas, a mixture of methane and CO_2. Since we focus here on the upgrading of lignocellulosic materials, we will not consider further the conversion of these bio-wastes. The potential of high value bio-products such as vegetable oil, terpenes and starch is discussed elsewhere in this book [19–22].

2.3.1
Key Reactions of Carbohydrates

All these conversions involve complex reactions of the carbohydrate, the lignin and, eventually, between carbohydrates and lignin. In this short review, we will limit ourselves to the reactions of the main constituent, i.e., the carbohydrates.

The reactivity of carbohydrates is dominated by the reactivity of the aldehyde group and the hydroxyl on its next-neighbor (β) carbon. As illustrated by the middle row of Fig. 2.3, the aldehyde can be isomerized to the corresponding enol or be converted into its hydrate (or hemiketal) form upon reaction with water (or with an hydroxyl-group). These two reactions are responsible for the easy cyclization of sugars in five- and six-membered rings (furanose and pyranose) and their isomerization between various enantiomeric forms and between aldehyde- and ketone-type sugars (aldose and ketose).

The combination of the carbonyl group and a β-hydroxyl group allows also the sugar to (a) dissociate in two smaller aldehydes, via the retro-Aldol reaction, or (b) eliminate the OH_β, via a retro-Michael reaction.

The unsaturated retro-Michael product easily isomerizes to a dicarbonyl when the C_α also carries a hydroxyl group (X = OH in Fig. 2.3). Such components are notoriously unstable. They undergo decarbonylation to a shorter aldehyde and CO. The retro-Michael product can also be converted into a carboxylic acid via hydration of the aldehyde function (Fig. 2.3). Notably, the formation of the carboxylic acid is accompanied by the saturation of the C_β; it in fact represents an exchange reaction between the OH_β and the aldehydic H.

These retro-Aldol and -Michael reactions can, obviously, follow an isomerization of the aldose to the corresponding ketose, leading thereby to different Aldol fragments or retro-Michael products. Keto-enol exchange as well as the retro-

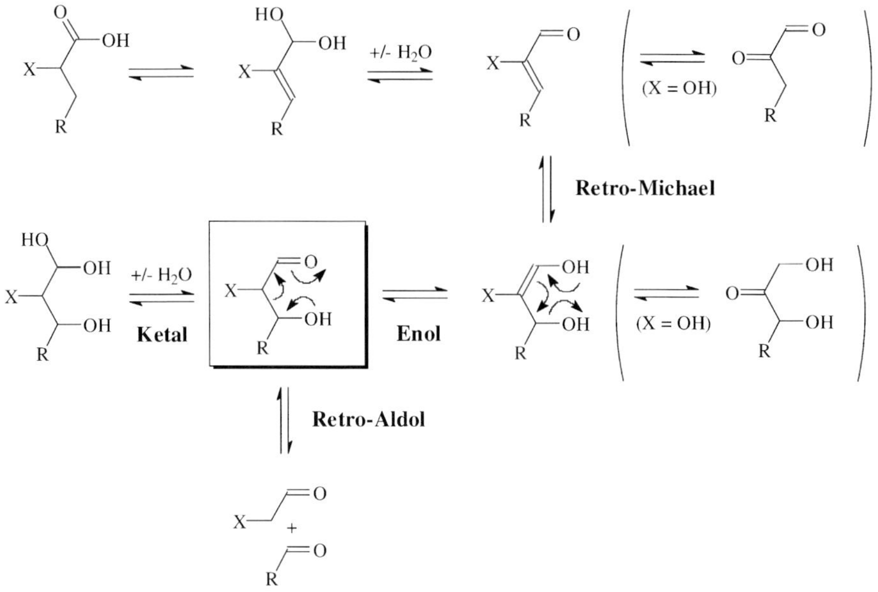

Fig. 2.3 Key reactions in carbohydrate conversion (the arrows represent the retro-Aldol and Michael reactions).

Aldol and -Michael reactions can also proceed on the reaction products of retro-Aldol and -Michael reactions. The reverse (direct) Aldol and Michael reaction can also proceed on various intermediates. Hence, these few reactions can already form a very large variety of possible products. They, indeed, account for most of the reactivity of carbohydrates discussed below, being under pyrolysis, hydrolysis or fermentation conditions.

2.3.2
Pyrolysis

2.3.2.1 Chemistry

Pyrolysis has a long history in the upgrading of biomass. The dry distillation of hardwood was applied in the early 1990s to produce organic intermediates (methanol and acetic acid), charcoal and fuel gas [3]. Today's processes can be tuned to form char, oil and/or gas, all depending on the temperature and reaction time, from ∼300 °C and hours, to 400–500 °C and seconds-minutes, to >700 °C and a fraction of a second [3, 19, 23, 24]. The process is typically carried out under inert atmosphere. We illustrate the basic chemistry of pyrolysis by focusing on the conversion of the carbohydrate components (Fig. 2.4). The reaction of the lignin will not be covered here but should obviously be considered in a real process. Interested readers could consult the literature, e.g., [25]. Pyrolysis is discussed in more details elsewhere in this book [26].

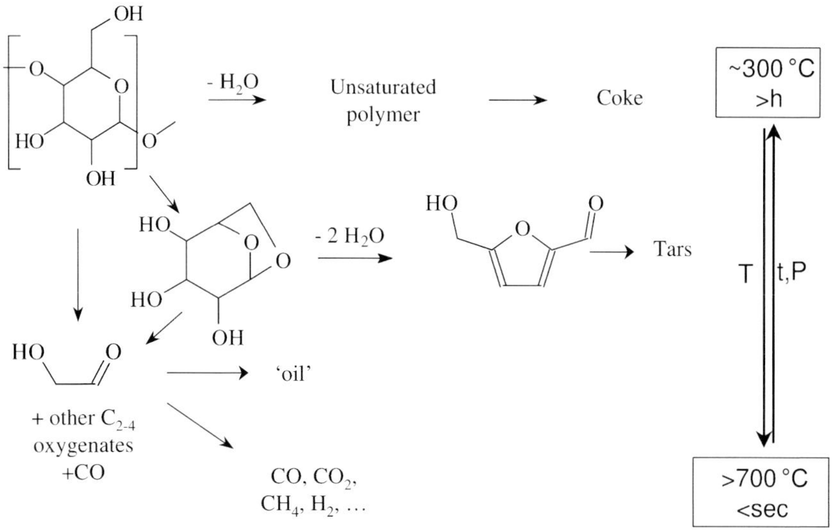

Fig. 2.4 Pyrolysis of biomass.

When heated at low temperature (<200 °C) for extended reaction times (>hours), the carbohydrate polymer partially depolymerizes to short chains of some 200 sugar units [24, 27]. In fact, the amorphous segments of the cellulose decompose preferentially while the crystalline segments remains fairly intact. The depolymerization probably proceeds through intramolecular ketal-formation and/or retro-Aldol decomposition [27]. The former reaction breaks the polymer in two fragments, one of which is terminated by an anhydro-sugar. The retro-Aldol reaction leads to two fragments that are terminated by a glycol aldehyde entity (-OCH$_2$-CHO). Upon heating at higher temperatures ($\sim$300 °C), the depolymerization is accompanied by slow dehydration to unsaturated species, most probably via retro-Michael reaction [27]. Depending on residence time and pressure, these various aldehydes and other unsaturated fragments can undergo subsequent oligomerization and elimination reactions (e.g., dehydration, decarbonylation or decarboxylation) to form unsaturated polymers and, eventually, char.

When heated at higher temperature and/or under reduced atmosphere, the depolymerization reactions mentioned above can be pushed to the point of liberating volatile species (Fig. 2.4), e.g., glycol aldehyde (HOCH$_2$-CHO, via retro-Aldol decomposition), anhydro-sugars (via intramolecular ketalization) and furan species (via retro-Michael elimination of anhydro-sugars). Unless they are efficiently removed from the medium, these volatile products will undergo oligomerization reactions that lead to the formation of tars.

At even higher temperature, the polysaccharides decompose further by extensive C–C bond breaking. This leads to the formation of C$_{2-4}$ oxygenates such as glycol aldehyde, acetic acid and hydroxyacetone (CH$_3$-CO-CH$_2$OH). The formation of these products can be rationalized by a series of reactions that include, for example, retro-Aldol and decarbonylation of ketones and aldehydes [28]. The decarbonylation reactions are obviously accompanied by the liberation of CO. These oxygenates can undergo condensation reactions to produce heavier oxygenates, oil and, eventually, tars.

If the temperature is further increased, e.g., beyond 700 °C, the C_{2-4} oxygenates further decompose to a mixed gas of moderate heating value, namely to a mixture of CO, CO_2, H_2 and CH_4. This gas resembles the town gas that was produced from coal at the beginning of the 20th century. Beyond 1000 °C, the hydrocarbons constituents of the mixed gas are further reformed, which results in the production synthesis gas, a valuable mixture of CO and H_2 with some CO_2 and water as main contaminants.

2.3.2.2 Product Applications

The char produced by pyrolysis is typically light and porous. Removal of most of the oxygen present in the biomass significantly increases its heating value, e.g., from 17 GJ t^{-1} for the lignocellulose to some 30 GJ t^{-1}. This makes the char a valuable fuel for industrial and consumer applications.

The pyrolysis oil is a very complex and multiphase mixture of low and high molecular weight components [29]. The lighter components consist of water (25%) and various organic oxygenates such as acids, hydroxy-acids/aldehydes/ketones, furans, (anhydro)sugars and phenolic lignin fragments. Pyrolysis oil is therefore highly acidic and partly water-soluble. The steam explosion that occurred upon fast pyrolysis also results in a mechanical destruction of the biomass structure and the formation of aerosols of oligomeric carbohydrates, lignin and tar fragments, which partly end up in the pyrolysis oil. Because of its multiphase structure and the reactivity of its unsaturated components, the pyrolysis oil is unstable. Deposits form slowly at ambient conditions and rapidly under mild heating. It has also a moderate heating value of ~17 GJ t^{-1} because of its high water and oxygen content. These various characteristics make pyrolysis oil a low quality fuel. It can be used to generate heating steam or power. However, it cannot be used as substitute to gasoline or diesel in the transportation sector. Further upgrading by means of hydrodeoxygenation and acid-cracking typically proceeds with moderate oil yields [19]. These applications require significant upgrading of the oil.

The gases have moderate heating value. They can be used as fuel for generating heat, steam or electricity. The synthesis gas that is processed at high temperature can also be used for further synthesis, as discussed in more details in Section 2.3.3.

2.3.2.3 Processes

Various pyrolysis processes have been reported in the literature. A popular approach, called flash pyrolysis, applies high temperature and short residence time to minimize the condensation of the volatile products. The BTG wood Pyrolysis process is a typical example.

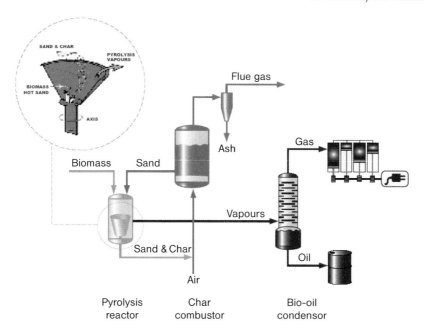

Fig. 2.5 BTG's biomass pyrolysis process [28].

The previous discussion on chemistry showed that oil components are intermediates that are susceptible to consecutive decomposition reactions, i.e., cracking and condensation. It is therefore essential to remove them from the reaction environment as soon as they are formed. In other words, the volatile components should have a very short residence time. The Pyrolysis process also requires the supply of heat to drive all these cracking and decomposition reactions. An elegant answer to these two challenges has been proposed and developed by van Swaaij et al. [30] and forms the basis of the Pyrolysis process that is commercialized by BTG, the Biomass Technology Group BV [31].

This process (Fig. 2.5) is based on a conic reactor that is fed with biomass and hot sand and operates as a cyclone to separate the volatile components from the hot sands and char as soon as they form. The hot sand is fed to the reactor to provide the required reaction heat. The pyrolysis oil is condensed out of the volatile stream and the remaining gaseous species are fired to provide the required process energy. The sand and char exit the conic reactor through the bottom and are sent to a combustion reactor, where the char is burnt to heat up the sand to the required temperature, before recycling to the pyrolysis reactor. This elegant design allows yields in pyrolysis oil of up to 70 wt.%. The process is also simple and auto-sufficient since it produces its own process energy from the char and gas by-products.

2.3.2.4 Alternative Developments

Pyrolysis can also be carried out in various liquid and/or gas mediums that function as heat carrier, dispersing medium and/or reactive medium. For instance, pyrolysis can be performed in an aqueous medium at high temperature and pressure. The hot water serves simultaneously as dispersing medium, reactant and acid catalyst. In the 1980s, Shell developed the HydroThermal Upgrading (HTU) process to convert biomass into a bio-crude of good quality [32]. The process, which is now being piloted and commercialized by Biofuel B.V., consists of a first digestion of biomass with water at 200 °C and 30 bar to produce a paste, which is subsequently converted into a bio-crude at 330 °C and 200 bar for 8 min. The resulting crude contains 10–15% oxygen and has a heating value of 30–35 GJ t^{-1}. Accordingly, the biomass is deoxygenated by CO_2 and H_2O elimination in comparable amounts, on an oxygen basis.

The aqueous-phase pyrolysis can also be assisted by catalysts and reactive gases. For instance, the PERC process [3, 33] produces a pyrolysis oil upon dissolving wood chips in a recycle pyrolysis oil, mixing the slurry with a water-Na_2CO_3 solution and treating the resulting mixture at 370 °C and 275 bar under synthesis gas atmosphere.

Several groups have investigated the potential of supercritical solvents as medium for liquefying the wood at milder temperatures but high pressures, i.e., 200–350 °C and >100 bar [19, 34–36]. In the presence of acetone, methanol or ethanol/water mixture, various types of woods have been partly dissolved or extracted. The extraction appears easiest for the lignin, followed by the hemicellulose and, eventually, the cellulose. A high yield in low-molecular weight species is favored by high temperature and moderate pressure. The product consists mainly of dissolved glucosides and anhydro-sugars, with pure cellulose as residue. It has, therefore, undergone limited deoxygenation.

These various alternatives provide a pyrolysis oil of much better quality than do flash pyrolysis processes such as the BTG one. However, they also require much higher investments that result from the use of high pressure and a corrosive reaction medium.

2.3.3 Gasification

2.3.3.1 Chemistry

As discussed above, the pyrolysis of biomass at high temperature (>1000 °C) results in the formation of synthesis gas, a valuable mixture of CO and H_2. The decomposition of carbohydrate to synthesis gas is an endothermic reaction since the heating value of product is ~125% of that of the feedstock (Reaction 1). The reaction becomes nearly thermo-neutral upon burning about 1/4 of the products. Since the thermodynamics favors the combustion of H_2 over CO, the gasification reaction resemble the theoretical Reaction (2). Indeed numerous gasification processes feed O_2 or air to drive the gasification reaction.

$$C_6(H_2O)_6 \rightarrow 6CO + 6H_2 \text{ (En. Efficiency} = 125\% \text{ LHV)} \tag{1}$$

$$C_6(H_2O)_6 + 3/2O_2 \rightarrow 6CO + 3H_2 + 3H_2O \text{ (En. Efficiency} = 95\% \text{ LHV)} \tag{2}$$

The resulting synthesis gas can subsequently be converted into methanol (Reaction 3) or "polymerized" to a mixture of hydrocarbons via the Fischer–Tropsch synthesis (Reaction 4) [37, 38]. These conversions usually require a H_2/CO molar ratio close to 2 (Reactions 3 and 4), which contrasts with the H_2/CO ratio of ~0.5 that is delivered upon biomass gasification (Reaction 2). It can therefore be suitable to adjust the H_2/CO ratio through the water-gas shift reaction (Reaction 5):

$$CO + 2H_2 \leftrightarrow CH_3OH \text{ (En. efficiency} = 79\% \text{ LHV)} \tag{3}$$

$$CO + 2H_2 \rightarrow 1/n\text{-}[CH_2]_n\text{-} + H_2O \text{ (En. efficiency} = 76\% \text{ LHV)} \tag{4}$$

$$CO + H_2O \leftrightarrow CO_2 + H_2 \text{ (En. efficiency} = 85\% \text{ LHV)} \tag{5}$$

After combining all these equations, the overall conversion of biomass into hydrocarbon or methanol adopts the stoichiometry of Reactions (6) and (7):

$$\begin{aligned} &C_6(H_2O)_6 + 3/2O_2 \rightarrow 3/n\text{-}[CH_2]_n\text{-} + 3H_2O + 3CO_2 \\ &\text{(En. efficiency} = 72\% \text{ LHV)} \end{aligned} \tag{6}$$

$$C_6(H_2O)_6 + 3/2O_2 \rightarrow 3CH_3OH + 3CO_2 \text{ (En. Efficiency} = 75\% \text{ LHV)} \tag{7}$$

These simple and nearly ideal equations reveal that the conversion of sugars into alkanes or methanol proceeds at best with carbon and energy efficiencies of 50% C or 75% LHV. Any yield loss in the gasification of conversion step results in further drop in efficiency. Reaction (6) also reveals that the oxygen is removed from the biomass, 2/3 as CO_2 and 1/3 as water.

2.3.3.2 Process

The gasification process can be fed with air or pure oxygen [3, 23]. The utilization of air obviously dilutes the synthesis gas with significant amounts of N_2 (~40%). When the synthesis gas is directly burned as fuel, the penalty of the N_2 dilution is limited to lowering the heating value of the fuel gas. Such a penalty is generally small compared with the cost advantage of using cheap air. For synthesis applications, however, the use of pure O_2 is generally cheaper as it avoids the cost of pressurizing the air to the working pressures (e.g., 40–80 bar of synthesis gas partial pressure) and/or having all down-stream equipments over-designed to accommodate the N_2 diluent. Numerous types of reactors have been proposed for gasifying biomass. The designs include fixed bed, moving beds with co- or counter-current O_2/air flow, fluidized bed or circulating beds [3]. The synthesis gas derived from biomass often contains undesirable contaminants such as tars, ash and inorganic volatile components (e.g., NH_3 and HCl) [3]. These contaminants need to be removed prior to converting the synthesis gas into valuable products. The cleaning train usually contributes significantly to the complexity and cost of the plant. The numerous aspects of biomass gasification are discussed in more length elsewhere in this volume [39].

The water-gas shift reaction, which is required to adjust the H_2/CO ratio of the synthesis gas, can be carried out in a dedicated reactor, e.g., placed between the gasification and the synthesis reactors. However, it can also be incorporated in the gasification reactor, e.g., by co-feeding water to the biomass, or in the synthesis reactor, by allowing water to the synthesis reactor and using a synthesis catalyst that is active for the water-gas shift reaction. Modern Cu-based methanol synthesis catalysts typically exhibit a high activity for the water-gas shift reaction [37, 38]. They can therefore take care of the required H_2/CO adjustment. However, they might operate at reduced rate or with reduced long-term stability when exposed to a synthesis gas with H_2/CO around 0.5. Similarly, Fe-based Fischer–Tropsch catalysts exhibit an intrinsic water-gas shift activity. They are therefore well suited for converting the CO-rich synthesis gas that is obtained upon biomass gasification. Co-based Fischer–Tropsch catalysts are more active and stable, however. Their utilization could become advantageous when combined with a dedicated water-gas shift unit. Various reactors have also been proposed to accommodate the mass and heat transfer of the methanol and the Fischer–Tropsch synthesis [37, 38]. They include fixed bed, slurry, stationary or circulating fluidized beds that are equipped with cooling coils, cold interstage feed injections or even periodic flow-reversal.

With these various considerations in mind, it is interesting to review one process design for illustration purpose, namely, the process that is under joint development by CHOREN and Shell (Fig. 2.6). This process combines the gasification technology developed by CHOREN [40] with Shell's Fischer–Tropsch technology [41]. It consists of a pyrolysis reactor, which converts the biomass (e.g., wood chips) into volatiles and char. The reactive volatiles are sent to an O_2-fed burner for initial gasification while the less reactive char is injected down-stream in the high-temperature part of the burner flame for effective gasification. The resulting

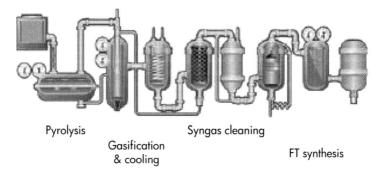

Fig. 2.6 CHOREN's Carbo-V® process [37].

synthesis gas is subsequently cooled, cleaned up, partly shifted and introduced to the Fischer–Tropsch reactor for conversion into high-quality alkanes.

The gasification segment has been demonstrated by CHOREN at 1 kt a^{-1} scale. Shell's Fischer–Tropsch technology is presently applied in Malaysia at 0.5 Mt a^{-1} for the conversion of natural gas into high-quality diesel fuel. A unit of 6 Mt a^{-1} is under construction in Qatar. In a joint effort, CHOREN, Shell and Volkswagen are building a 15 kt a^{-1} demonstration unit in Germany.

2.3.3.3 Alternative Developments: Hydrogen Production

Various gasification schemes have been conceived for the direct production of H_2 (and CO_2) instead of synthesis gas. Matsumura has reviewed the gasification of biomass with near- and super-critical water [42]. The presence of "liquid" water suppressed the formation of char but not of tars. Full gasification proceeds in the presence of metal catalysts at 350–600 °C but also in absence of any catalysts at 500–750 °C. This subject is discussed elsewhere in this book [26].

Dumesic has reviewed the "aqueous-phase reforming" of polyols to H_2 and CO_2 under milder conditions, i.e., 200–250 °C [43]. He also showed the possibility to divert the reaction towards to production of alkanes by using an acidic support for the Pt or Pd reforming catalyst.

2.3.4
Hydrolysis

2.3.4.1 Chemistry

Biomass does not necessarily require the severe conditions applied for pyrolysis or gasification to be converted. The polysaccharides are linked through ketal functions whereas the lignin contains alkyl-aryl ether links. Both types of linkages are prone to hydrolysis. Indeed, the lignin and hemicellulose readily hydrolyses at mild temperature (<150 °C) in the presence of acids or bases. The cellulose is much more difficult to hydrolyze, however, because its semi-crystalline structure hinders the catalyst to access the ether links. More severe conditions are therefore required to fully depolymerize the cellulose to its glucose constituents. Full hydrolysis is not always desired, however, as the cellulose fibers that are free from hemicellulose and lignin are valuable feedstock for the pulp and paper industry. We will not consider this industry in more detail here; the reader is referred to the literature for further reading [14]. In contrast, we focus on the complete hydrolysis of the lignocellulose to provide intermediates for fuel and chemical manufacture. For simplicity, we limit the present discussion to the hydrolysis of the polysaccharides, while neglecting the role of lignin in the first instance. The hydrolysis of lignin in acidic and basic conditions is well documented [14].

In the presence of excess water, the ketal functions of the polysaccharides can be hydrolyzed to the corresponding hemi-acetals, liberating thereby the individual sugars or, at least, their oligomers (Fig. 2.7). This hydrolysis is catalyzed by acids

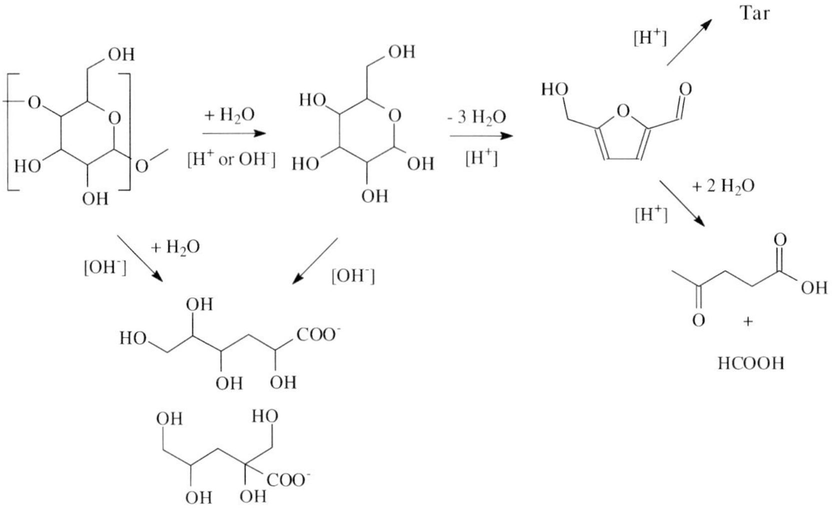

Fig. 2.7 Biomass hydrolysis in acidic and basic medium.

and bases. It is also accompanied by extensive isomerization reactions, which produce various sugars and their oligomers. At a given temperature, however, the sugars tend to undergo consecutive decomposition reactions, which vary for acidic and basic medium (Fig. 2.7). In the presence of acids, the sugars dehydrate to form furan species, most probably via retro-Michael reactions. The C_5 and C_6 sugars are converted into furfural and hydroxymethyl furfural, respectively [15, 44–47]. These furan species are also quite reactive. They undergo easy condensation or "acylation" reactions through their aldehyde function and/or unsaturated ring. These reactions lead to the undesired formation of tars. Hydroxymethyl furfural also undergoes extensive rearrangement and rehydration reactions that afford levulinic acid and formic acid [15, 44–47]. All these reactions contribute to lowering the yield in sugars during hydrolysis. These losses are more pronounced when the hydrolysis is carried out at higher temperature and higher (ligno)cellulose concentration. When present in the hydrolysis solution, lignin is also reported to react with the sugars and form adducts [48].

Older hydrolysis processes applied concentrated inorganic acids such as H_2SO_4 and mild temperature (100–120 °C) to minimize the undesired consecutive reactions [49]. The recovery and reconcentration of the acid catalyst from the product mixture turned out to be difficult and expensive. Modern processes are therefore based on diluted acid and higher temperatures (180–220 °C) [49].

In the presence of bases, monomeric as well as polymeric sugars are converted into various carboxylates salts [14]. The reaction proceeds through a retro-Michael dehydration step to form an unsaturated aldehyde, followed by rehydration of the aldehyde function and isomerization to acid (Fig. 2.3). The same reaction is also responsible for a stepwise depolymerization of polymeric sugar to carboxylate salts (peeling reaction). The formation of carboxylate salts is undesirable in pulp processes as they result in the (partial) neutralization of the base catalyst.

Beyond conventional acids and bases, (ligno)cellulose can also be hydrolyzed by enzymes [49] that also exhibit acid–basic properties. These enzymes do not catalyze the undesired decomposition of the sugars, thanks to their high specificity and, possibly also, to their very mild operating temperature. The cellulose enzymes are much bulkier than homogeneous acids or bases, however. They therefore encounter difficulties in properly reaching and hydrolyzing the hemicellulose and the lignin. Mechanical and chemical pre-treatments are usually required to open up the structure of the lignocellulose to enable the enzymes to hydrolyze its constituents.

2.3.4.2 Sugar Derivatives

In some cases, the hydrolysis reaction liberates the sugars from the biomass and converts them directly into derivatives such as furfural, hydroxymethyl furfural and/or levulinic acid. These derivatives can be further converted into various chemical intermediates. We will not discuss these further conversions as they are extensively reported in the literature, e.g., for furfural [15, 44], hydroxymethyl furfural [15, 44, 50] and levulinic acid [15, 44–47].

In other cases, the hydrolysis reaction leads to sugar mixtures, which can subsequently be converted into various derivatives through chemical or biological conversion. The fermentation of sugars will be discussed below. The chemical technologies offer a large variety of possibilities to upgrade the sugars. Sugars can be hydrogenated to C_{5-6} polyols such as xylitol, mannitol and sorbitol [15, 44], hydrogenolyzed to C_{2-3} glycols [51, 52] or further upgraded via oxidation or halogenation reactions [53]. Sugars can be used for detergent manufacture or converted into N-heterocyclic components, pyrones and aromatics [44]. The potential of sugars for producing fine chemicals is also discussed further in this book [21, 22].

2.3.4.3 Process

Various processes have been developed for hydrolyzing lignocellulose to its major constituents, i.e., to sugars and (partly) depolymerized lignin. The lignin is usually precipitated from the aqueous solution and either used as chemical feedstock or burned as process fuel. The aqueous sugar solution is then applied for fermentation to ethanol after neutralization and purification.

For illustration purposes, we consider the Biofine process for the conversion of lignocellulose into levulinic acid (Fig. 2.8) [45, 46]. The biomass is shredded to the appropriate particle size (0.5–1 cm), contacted with diluted sulfuric acid and hydrolyzed in two stages. The first stage proceeds at 220 °C and 12 s in a plug-flow reactor. The second stage is carried out at 190 °C for ~20 min in a CSTR reactor. Furfural and other volatiles are removed at the CSTR stage. Tarry materials are separated by gravitation and dried whereas levulinic acid is removed by boiling off the water. This process has been piloted at the 0.3 kt a^{-1} scale.

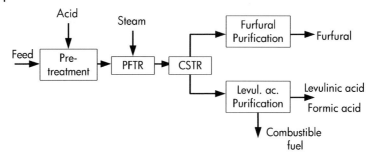

Fig. 2.8 Biofine's process to convert biomass into levulinic acid. (Adapted from [43]).

2.3.4.4 Recent Developments

Various solvents are being investigated to dissolve lignocellulosic materials. Some approaches focus on the selective depolymerization and extraction of lignin and hemicellulose as pre-treatment to produce clean cellulose fibers for subsequent fermentation or for pulping. Other approaches attempt to dissolve the whole lignocellulose with or without depolymerization. The liquefaction processes that are carried out at high temperature (>300 °C), and produce a complex oil mixture, are discussed above with the pyrolysis processes.

Acetic acid–water mixtures are also effective in delignifying lignocellulose (e.g., the Acetocell, Formacell and Acetosolv processes) and, eventually, in hydrolyzing the cellulose to sugars and furans at high severity. The addition of formic acid or HCl significantly enhances the hydrolysis and dissolution of the material, although it also leads to higher furan yields [54].

A solution of ethanol–water with diluted H_2SO_4 selectively dissolves the lignin and hemicellulose at ∼180–200 °C [55]. The resulting purified cellulose can subsequently be fermented to ethanol at high yield. Copersucar and Dedinin are developing a process for bagasse-ethanol that is based on such a pre-treatment of the bagasse [56].

Oxidants are occasionally added to the hydrolysis mixture to favor the removal of lignin. For instance, the co-feed of O_2 or air appeared beneficial to the delignification efficiency of a lime-catalyzed hydrolysis step [57]. Similar improvements are also reported for the co-feed of H_2O_2 to formic or acetic acid hydrolysis [58].

Full dissolution has been reported to proceed in ionic liquids such as butyl- or allyl-methyl-imidazolium chloride under microwave irradiation [59, 60]. The Cl-anion is claimed to be essential to favor the de-agglomeration of the cellulose by breaking its H-bonds that hold it together [61]. The cellulose can subsequently be precipitated from the ionic liquid upon addition of, for example, water, without significant depolymerization.

Detailed discussion of the classical wood pulping processes – e.g., the Sulfite and Kraft processes – is available in the literature [14]. Pre-treatments that aim to facilitate the fermentation of lignocellulosic materials are also discussed elsewhere [49, 62–64].

2.3.5
Fermentation

2.3.5.1 Chemistry

Life has developed efficient metabolisms to oxidize sugars back into H_2O and CO_2 to provide the energy required for organism growth. In the absence of oxygen, however, organisms have developed alternative metabolisms that stop at less oxidized stages and, thereby, deliver less energy. This is the case for bakers yeast (*Saccharomyces Cerevisiae*), which stops its metabolism at ethanol. Other organisms stop at other intermediates like acetic acid or convert the "waste" intermediate of congeners organisms to produce, for example, methane. These metabolism pathways can be represented like conventional organic reactions, as illustrated in Fig. 2.9 for the fermentation of glucose [3].

The conversion of glucose proceeds via its splitting into pyruvic acid and hydrogen, which is bound as NADPH. Pyruvic acid is subsequently decarboxylated to CO_2 and acetaldehyde (bound to the coenzyme-A), which is subsequently rehydrogenated to ethanol. The overall reaction delivers therefore two molecules of ethanol and two CO_2 for every glucose unit. Notice that such a simplified metabolic pathway does not display the energy fluxes, e.g., in the form of ATP/ADP interconversion.

The metabolic pathways can be diverted to other products, however. For instance, the pyruvic acid can be rehydrogenated to lactic acid. Accordingly, glucose is converted into two molecules of lactic acid, which is the building block for Cargill's polylacate polymer [65].

Alternatively, the acetaldehyde intermediates can be oxidized to acetic acid, which can be converted into CO_2 and CH_4 by metagenic organisms [3]. These

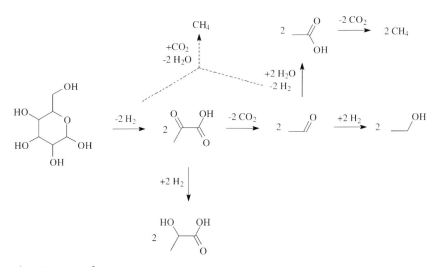

Fig. 2.9 Biomass fermentation.

metagenic organisms can also extract energy by converting the NADPH and CO_2 produced at earlier stages into CH_4. Accordingly, one glucose molecule is converted into three CO_2 and three CH_4 molecules, which is the bio-gas that emanates from dormant waters or waste landfills.

2.3.5.2 Process

We can illustrate fermentation processes using the process developed by Iogen to convert lignocellulosic materials such as wheat straw into ethanol (Fig. 2.10) [66]. The straw is chopped and milled prior to a "steam-explosion" pre-treatment to

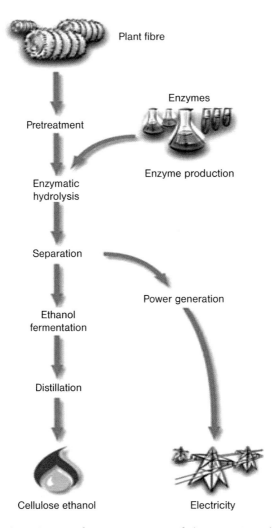

Fig. 2.10 Iogen's fermentation process of wheat straw into ethanol [63].

open-up the cellulosic structure. The resulting pulp is subsequently impregnated by an enzymatic solution obtained from the fungus *Trichoderma*, which has been specifically developed for hydrolyzing lignocellulose. Lignin is precipitated from the hydrolysis liquor and used to generate electricity to supply the plant. The hydrolysis liquor is loaded with a proprietary microorganism that ferments the glucose as well as the C_5 xylose to ethanol – the common bakers' yeast cannot ferment C_5 sugars and would, thereby, deliver lower ethanol yield. The ethanol is finally recovered and purified to specifications.

2.3.5.3 Recent Developments

Much effort is still devoted to improving the fermentation of lignocellulose to ethanol. For instance, research groups are working on improving the pre-treatment and hydrolysis processes, which are needed to liberate the sugars from the biomass without liberating components that are toxic for the fermentation microorganisms [49, 62, 63]. Alternative microorganisms are being explored and developed, which can convert the pentose into ethanol and/or are more resistant to the toxins liberated during the pre-treatment processes. Efforts are also devoted to integrate the sugar fermentation step with the upstream cellulose hydrolysis step (SSF or simultaneous saccharification and fermentation) and also with the cellulose production step (CBP or consolidated bioprocessing) [3, 62].

Microorganisms have also been developed to produce alternative products, such as lactic acid [65], propane-1,3-diol [67], 3-hydroxypropionic acid [68], butane-2,3-diol [69] and numerous other intermediates. For instance, bacteria such as the *Clostridium acetobutylicum* ferment free sugars to C_4 oxygenates such as butyric acid or butanol. They form the C_4 oxygenates by Aldol condensation of the acetaldehyde intermediates. The Weizmann process exploits this property to ferment starch feedstock anaerobically at 37 °C to produce a mixture of *n*-butanol, acetone and ethanol in a volume ratio of 70:25:5 [3].

The C_4 aldol intermediate of the Weizmann process is also key in the aerobic fermentation of sugars to poly(3-hydroxybutyric acid) or PHB ($-O[-CH(CH_3)-CH_2-COO]_n-$) [70]. This natural and biodegradable polymer is produced inside microorganisms, e.g., *Ralstonia eutropha*. A complex processing is required to extract and purify the polymer granules from the microorganism.

Other microorganisms ferment sugars to succinic acid ($HOOC-CH_2-CH_2-COOH$), a promising intermediate for numerous chemicals [71]. Technical challenges include the toxicity of the succinate for the microorganisms, the need for expensive nutrients, the undesired co-production of acetic or pyruvic acid and the cost of acidifying the succinate to succinic acid.

These new fermentation processes often require high costs for recovering the product from the fermentation broth. For instance, the production of lactic acid requires the neutralization of the product during the fermentation, to avoid acidification of the medium, and the subsequent re-acidification of the lactate [65]. Similarly, the recovery of 1-butanol implies the distillation of large amounts of water. Alternative recovery processes are therefore the subject of intensive research.

2.4
Economics

Numerous studies have discussed the economics of biomass conversion processes over the last 20–30 years. They usually concur in concluding that these processes require high oil prices to become competitive. However, their varying basis and assumptions hinder the reader in extracting a global picture of the important economic factors. We will attempt here to unravel this global picture by revisiting some these studies and subjecting them to a common set of basis and assumptions.

2.4.1
Methodology

The analyses considered here are based on various feedstock, products and process routes.

They include:

- Conversion of lignocellulose into biocrude via pyrolysis and hydrothermal treatment [30–32, 72].
- Conversion of lignocellulose into transportation fuels via pyrolysis and subsequent oil upgrading [72], via gasification and subsequent Fischer–Tropsch or methanol synthesis [3], via hydrolysis and subsequent fermentation to ethanol or subsequent conversion into ethyl levulinate [45, 46, 73].
- Conversion of lignocellulose into electricity via combustion or gasification [3, 74].
- Conversion of corn starch into ethanol via fermentation.
- Conversion of vegetable oil into diesel via transesterification of hydrogenolysis [75].
- Conversion of municipal solid waste into biogas via fermentation [76].

2.4.2
Fuel Production

2.4.2.1 Plant Costs

The cost of the biofuel plants reported in the literature appeared to follow the same general laws as those of chemical and fuels plants [77, 78]: irrespective of the technology applied, the plant cost showed a nice power-law correlation (R^2 of 0.88) with the overall energy loss of the plant over two orders of magnitude. It correlated much less (R^2 of 0.56) with capacity of the plant.

The plant cost was recalculated to a single plant size of 400 MW intake, which corresponds to an intake of $\sim$680 kt a^{-1} lignocellulose, using a scaling exponent of 0.8. This capacity is recognized as typical "large scale" wet-corn mill, though larger mills also exist [79]. When recalculated to a single scale, the capital charge (i.e., the plant cost per unit of product) appeared to decrease fairly linearly with

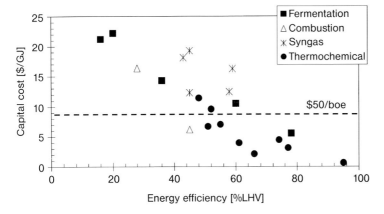

Fig. 2.11 Capital cost of biomass conversion plants (400 MW intake, 25% capital charge, $ 2005).

increasing the energy efficiency of the plant (Fig. 2.11): As in the earlier analyses [69, 70], energy losses were defined as the difference between the low heating value of "feed + energy" entering the plant and the low heating value of all saleable products exiting the plant. The energy efficiency was defined as the low heating value ratio between products and "feed + energy". Notably, the saleable products include the co-production of electricity in a few cases and the co-production of proteins in the case of corn-based ethanol. All costs were indexed to a single reference year (2005) using an average inflation of 2% per year. Notably, however the recalculated investments may not be fully comparable with one another, as the initial studies likely use different sets of assumptions. This could partly explain the scatter of the data. It might also contribute to the apparent higher cost of syngas plants compared with the other technologies. A high utilization of the feed and energy is an obvious prerequisite for an affordable plant.

2.4.2.2 Feed Costs

The technologies described above converted various feedstock. Some used lignocellulosic biomass, which is typically valued at $ 2–4 GJ^{-1} (i.e., $ 34–70 per t dry) [3, 11, 80], depending for instance on source, quality and transportation distance. Other plants were based on grains such as corn or wheat. These feedstocks would typically cost $ 4–7 GJ^{-1} (i.e., $ 65–110 t^{-1}) [11]. Finally, the vegetable oils that were upgraded in a few cases may cost $ 13–18 GJ^{-1} (i.e., $ 500–700 t^{-1}). Sugarcane and bagasse, the fibrous residue of sugar cane after sugar extraction, were not included here because of lack of reliable information. However, the prices are expected to be low, likely <$ 2 GJ^{-1}, because of the high growth rate of sugar cane and low labor cost in Brazil, the main producer of sugar cane.

Obviously, the overall feed cost depends on the price of the feed as well as its utilization efficiency. This is illustrated in Fig. 2.12. Accordingly, vegetables oils are not competitive with crude oil priced at $ 50 bbl^{-1}, whereas grains and ligno-

2 Lignocellulose Conversion:

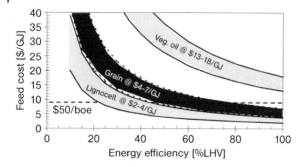

Fig. 2.12 Feed cost of biomass conversion plants.

cellulose can be competitive feedstocks, when converted with high and moderate efficiency, respectively.

2.4.2.3 Manufacturing Costs

The data discussed above can easily be used to develop a crude figure for the cost of biomass conversion processes. It suffices to consider the feed cost, for the appropriate feed and conversion efficiency (Fig. 2.12), the appropriate capital charge (Fig. 2.11) and add a contribution of ~1/3 of the capital charge for fixed cost, to account for various additional costs such as labor, maintenance, etc. The manufacturing cost is not far from the sum of these three contributions.

The economic comparison can also be visualized by plotting the processing cost, i.e., the "capital + fixed cost", against the feed of the various process alternatives (Fig. 2.13). The diagonal "eco-cost" lines represent overall manufacturing cost. Figure 2.13 displays processes that produce transportation fuels from bio feedstock (lignocellulose, starch and vegetable oil) and fossil feedstock (crude oil

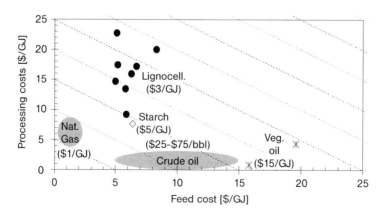

Fig. 2.13 Feed and processing cost of transportation fuels derived from lignocellulose and fossil resources.

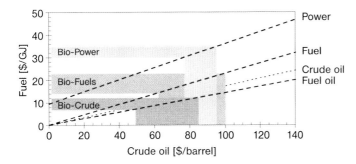

Fig. 2.14 Manufacturing cost of bio-crude, bio-fuels and bio-power from lignocellulose (400 MW or 680 kt a^{-1}, $ 2005).

and natural gas). The former are based on the data discussed above whereas the latter are adapted from the literature [37]. The figure reveals that the cost of oil refining is dominated by feed cost whereas those from natural gas conversion (e.g., MeOH or Fischer–Tropsch synthesis) are dominated by technology. Similarly, biofuels derived from vegetable oils are dictated by feed cost whereas those derived from lignocellulose or starch are dominated by technology. The diagonal lines also indicate that production of biofuels typically costs $ 15–25 GJ^{-1} whereas that of fossil fuels cost $ 5–15 GJ^{-1}. Hence, biofuels are competitive with oil refining only at high oil prices, say $ 50–75 bbl^{-1}.

Figure 2.13 did not include all the biomass conversion processes discussed above. It only considered those that produce transportation fuels. The processes that convert bio-feedstock into biocrude or electricity could not be included because their products have a different value than the transportation fuels. Such a comparison can be attempted by displaying the total manufacturing cost of bio-based products in a graph that shows typical relationships between the price of crude and that its derivatives, i.e., of fuel oil, transportation fuel and electricity. This has been done in Fig. 2.14 for the lignocellulose conversion processes.

Accordingly, biocrude may compete with fuel oil at oil price of $ 50–80 bbl^{-1}, biofuel may compete with gasoline and diesel at $ 70–110 bbl^{-1} and green-electricity is affordable at $ 80–100 bbl^{-1} oil. Although not shown here, the ratio between the processing costs (capital charge + fixed cost) and the feed cost increases from about 1:1 for biocrude, to 2:1–3:1 for biofuels and around 4:1 for green electricity. This shows that technological improvements and/or scale-up could significantly impact the competitiveness of biofuels and "green" electricity.

Such improvements could for instance be looked for in the handling and pre-treatment of solid feedstock, which require heavy and energy consuming equipment. Purification of the product can also be capital intensive. For instance, hydrolysis and fermentation technologies often result in a product that is highly diluted in water and requires expensive recovery by distillation or extraction.

Obviously, the present analysis provides only an approximated and global view of the economics of biomass conversion. It leaves room for special cases, such as those that are strongly affected by local factors such as feedstock supply, by-product value, labor cost or tax incentives. As illustration, the co-firing of lignocellulose in existing coal power plants is obviously more competitive than the biopower reported here because the plant is already written off.

2.4.3
Process Scale

The process scale of 400 MW (i.e., 680 kt a^{-1} of lignocellulose) intake selected here is in the range of chemical plants. Those indeed vary between ~100 and ~2000 kt a^{-1}. World-scale fuel manufacturing plants are much larger, however, at around ~5 Mt a^{-1} for a gas-to-liquid plants and ~20 Mt a^{-1} for an oil refinery. Although a larger scale would offer better economics, the 400 MW scale assumed here is already fairly optimistic, when considering the availability of biomass within a reasonable radius around the plant. For instance, Hettenhaus discussed the implication of supplying 1 Mt a^{-1} of biomass to a plant [80]. Such a plant would require the collection of crop residue from 1000–2000 farmers within three areas of 25 km radius. This would result in congesting road traffic of 40–80 trucks per hour to/from the three interim storage locations during the harvest period and about a third of this traffic intensity between the interim storage locations and the central plant the whole year around. (This calculation assumed that 30% of the area is cultivated; it produces on average 5 t ha^{-1} of residue besides 5 t ha^{-1} of food and is collected by trucks of 18 t that circulate 12 h day^{-1} during a 20–40 day harvest period).

Obviously, a single collection area of 25 km radius could only feed a plant of half the size assumed here. Since plant costs scale with the plant capacity raised to the power 0.65–0.8, such a two-fold smaller plant would have a 20–25% higher capital charge (per ton of product). This would result in additional processing cost and total production cost of $1–2 GJ^{-1} for biocrude plant, $2–5 GJ^{-1} for a biofuel plant and $5–6 GJ^{-1} for a power plant.

As mentioned above, wet corn-mills of >1 Mt a^{-1} capacity are already in operation [79]. Such a scale is economical for corn because it has a high density and is easily transportable. However, it is not necessarily economical for straw, corn stover and numerous crop residues, which are characterized by a low density and fibrous structures.

2.4.4
Chemicals Production

We have so far focused the economic discussion of the manufacture of biofuels. However, chemical intermediates are also promising products of biomass conversion. Four reasons can be mentioned:

1. We have mentioned above that numerous chemical intermediates contain oxygen atoms that provide valuable physical properties to the final product, being

a polymer, a detergent or a solvent. Chemical intermediates indeed exhibit a O/C atomic ratio that is between that of fuel and biomass, i.e., between 0 and 1. Obviously, bio-based chemicals could require less deoxygenation of the biomass than do biofuels.

2. Chemical intermediates are much more valuable than fuels, between 2 and 20 times more valuable on a carbon base. They should therefore allow for a larger margin.

3. Chemical intermediates are produced at a scale that is close to that of biomass conversion plants, e.g., between 100 kt a^{-1} (e.g., for acrylic acid) to 800 kt a^{-1} (e.g., for ethene diol). This has to be compared with the 20 Mt a^{-1} scale of a major oil refinery.

4. Finally, some chemical intermediates require complex and inefficient manufacturing routes when starting from crude oil. Lange indeed showed that the manufacture of several major polymers, i.e., nylon, poly(methyl methacrylate), epoxy resins and poly(urethane), makes inefficient use of natural resources [81]. They produce 2–7× more waste than product. The petrochemical industry is, therefore, devoting a large effort to develop more efficient processes, mainly based on crude oil. However, an increasing number of companies recognizes the potential of biomass for new and, possibly, more efficient routes for these materials.

The potential of combining a lower need for deoxygenation and a higher product value is illustrated in Fig. 2.15. It shows that the selective incorporation of oxygen into a hydrocarbon, as done in the petrochemical industry, is very expensive. In contrast, the bio-based alternative enjoys two advantages. Firstly, the feedstock is cheaper than crude oil, even on an energy and carbon base, as discussed above. Secondly, its selective deoxygenation has been proven to cheaper than the petrochemical route in a few cases, e.g., for ethanol and furfural. The same can be expected for other biomass derivates in the future.

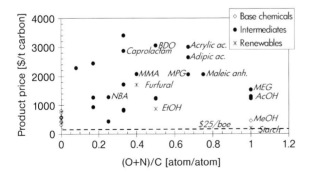

Fig. 2.15 Historical average prices of petro- and biobased-chemicals (corresponding average oil price ∼23 $ bbl^{-1}).

2.5
Summary and Conclusions

Lignocellulose biomass is a mixture of phenolic lignin and carbohydrates – cellulose and hemi-cellulose. It grows abundantly on earth and is largely available as agricultural and forestry residues. Lignocellulose can be converted via four major routes: pyrolysis, gasification, hydrolysis and fermentation.

Pyrolysis consists of heat treatment under an inert atmosphere, which, with increasing temperature, leads to dehydration to char, depolymerization to sugar and furan species and fragmentation to low molecular weight oxygenates or gas. The pyrolysis oil obtained at moderate temperature is a complex colloidal mixture of water, oxygenates (including acids) and polymeric materials. It is, therefore, a low-quality fuel. Its upgrading to high-quality fuel represents a real challenge. Alternative pyrolysis processes apply a reactive atmosphere such as synthesis gas or H_2 and/or near/super-critical solvent (e.g., water) to obtain a higher quality pyrolysis oil. Such variations obviously bring additional processing costs. Various catalysts are being explored to facilitate the pyrolysis and improve the quality of the oil.

Biomass can be gasified to synthesis gas, a valuable mixture of CO and H_2, which can be subsequently converted into high-quality fuels such as diesel-range alkanes or methanol. Processes based on gasification are complex and require large investments. Synthesis gas cleaning is an important issue. Much research effort is being devoted to improve the gasification process (e.g., favoring the gasification of tars), cleaning up the synthesis gas and improving its conversion into high-quality fuels.

Biomass can be hydrolyzed to its constituting sugars at mild temperature in aqueous solutions of acids, bases or enzymes. The sugars can undergo subsequent reactions that lead to the formation of furanic species (with acids) or carboxylate salts (with base). Organic media and ionic liquids are being explored to dissolve and/or depolymerize the lignocellulose or some constituents (e.g., lignin and/or hemicellulose) without the undesired consecutive reactions. The sugars or furanic species can be further upgraded to a large variety of fuel and chemical components by chemical and biological routes.

Finally, the sugar present in the biomass can be converted into various fuel and chemical components via fermentation processes. Efficient fermentation usually requires severe pre-treatment steps to make carbohydrate accessible for depolymerization and biological digestion. A great deal of research efforts is being devoted to such pre-treatment processes (see biomass hydrolysis) and to develop microorganisms that convert the sugar more extensively (e.g., digesting the C_5 sugars besides the easier C_6 sugars) and/or converting them into H_2, synthesis gas and various chemical intermediates (e.g., lactic acid, propane diol or succinic acid).

Biomass conversion processes are still expensive today, being competitive at crude oil prices between $50 and $100 bbl^{-1}. Lignocellulose might be a fairly cheap feedstock, cheaper than crude oil. However, its conversion requires large investments. These investments are due to the large amount of energy that is needed to run the process, e.g., for sizing/pre-treating the feedstock or for purifying the product. The large investments are also partly due to the limited scale of the biomass conversion processes, which is typically 1/10 to 1/100 of a world-scale oil refinery. The plant scale is largely dictated by the local availability of the biomass and its transportation cost over longer distances. Reduction in costs will require improved processes (e.g., more energy efficient), improved infrastructure for collecting the biomass over a large area and improved agricultural and forestry practice that provides residual biomass in significant amounts without deteriorating the soil for the long term.

References

1 European Commission: *Catal. Rev. Newsletter* 2005, 18(12), 5. [http://europa.eu.int/comm/energy/res/biomass_action_plan/index_en.htm].
2 Biomass technical Advisory Committee USA: http://www.bioproducts-bioenergy.gov/pdfs/ BioVision_03_Web.pb.
3 D. L. Klass, *Biomass for Renewable Energy, Fuels and Chemicals*, Academic Press, San Diego (1998).
4 C. Okkerse, H. van Bekkum, *Green Chem.* 1999, 107.
5 World Business Council for Sustainable Development: *Facts and Trends to 2050 – Energy and Climate Change* 2004. [http://www.wbcsd.org].
6 S. Fletcher, *Oil & Gas J.* 2005, January, 26.
7 US Department of Energy, Energy Information Administration: *International Energy Outlook 2006.* [http://www.eia.doe.gov].
8 W. E. Schollnberger, *Oil Gas Eur. Mag.* 2006, 1, 8.
9 R. van Santen, Chapter 1 of this book.
10 R. H. Whittaker, G. E. Likens, in *Primary Productivity of the Biosphere* (H. Leith, R. H. Whittaker, eds) Springer-Verlag, New York, 1975, p. 305.
11 B. E. Dale, S. Kim, in *Biorefineries–Industrial Processes and Products* (B. Kamm, P. R. Gruber, M. Kamm, eds) vol. I, Wiley-VCH, Weinheim (2006), p. 41.
12 A. E. Farrell, R. J. Plevin, B. T. Turner, A. D. Jones, M. O'Hare, D. M. Kammen, *Science* 2006, 311, 506.
13 R. Hammerschlag, *Environ. Sci. Technol.* 2006, 40, 1744.
14 E. Sjöström, in *Wood Chemistry, Fundamentals and Applications*, 2nd edn. Academic Press, San Diego (1993).
15 B. Kamm, M. Kamm, M. Schmidt, T. Hirth, M. Schulze, in *Biorefineries–Industrial Processes and Products* (B. Kamm, P. R. Gruber, M. Kamm, eds) vol. II, Wiley-VCH, Weinheim (2006), p. 97.
16 G. Brunow, in *Biorefineries–Industrial Processes and Products* (B. Kamm, P. R. Gruber, M. Kamm, eds) vol. II, Wiley-VCH, Weinheim (2006) p. 151.
17 K. Owen, T. Coley, *Automotive Fuels Handbook*, Society Automotive Engineers, Warrendale (PA), 1990.
18 L. Petrus, M. A. Noordermeer, *Green Chem.* 2006, 8, 861.
19 G. W. Huber, S. Iborra, A. Corma, *Chem. Rev.* 2006, 106, 4044.
20 K. Hill, Chapter 4 of this book.
21 H. van Bekkum, L. Maat, Chapter 5 of this book.
22 P. Gallezot, Chapter 3 of this book.
23 R. C. Brown, in *Biorefineries–Industrial Processes and Products* (B. Kamm, P. R. Gruber, M. Kamm, eds) vol. I, Wiley-VCH, Weinheim (2006) p. 227.
24 M. J. Antal, Jr., *Adv. Solar Energy* 1983, 61.
25 R. Alén, E. Kuoppala, P. Oesch, *J. Anal. Appl. Pyrolysis* 1996, 36, 137.
26 S. R. A. Kersten, W. P. M. van Swaaij, L. Lefferts, K. Seshan, Chapter 6 of this book.

27 J. A. Lomax, J. M. Commandeur, P. W. Arisz, J. J. Boon, *J. Anal. Appl. Pyrolysis* **1991**, 19, 65.
28 J. Piskorz, D. Radlein, D. S. Scott, *J. Anal. Appl. Pyrolysis* **1986**, 9, 121.
29 D. Mohan, C. U. Pittman Jr., P. H. Steele, *Energy Fuels* **2006**, 20, 848.
30 R. W. J. Westerhout, J. Waanders, J. A. M. Kuipers, W. P. M. van Swaaij, *Ind. & Eng. Chem. Res.* **1998**, 37(6), 2316.
31 BTG: http://www.btgworld.com.
32 F. Goudriaan, B. van de Beld, F. R. Boerefijn, G. M. Bos, J. E. Naber, S. van der Wal, J. A. Zeevalkink; *5th Conference on Progress in Thermochemical Biomass Conversion* (Tyrol, Austria, Sept. 17–22, **2000**), (A. V. Bridgwater, ed.) vol. 2, Blackwell Sci., Oxford, pp. 1312–1325.
33 P. L. Thigpen, W. L. Berry Jr., in *Energy from Biomass and Wastes VI* (D. L. Klass, ed.), **1982**, p. 1057.
34 P. Köll, B. Brönstrup, J. O. Metzger, *Chem. Eng. Supercrit. Fluid Cond.* **1983**, 499.
35 M. G. Poirier, A. Ahmed, J.-L. Grand-maison, S. C. F. Kaliaguine, *Ind. Eng. Chem. Res.* **1987**, 26, 1738.
36 E. Minami, S. Saka, *J. Wood Sci.* **2003**, 49, 73.
37 J.-P. Lange, *Catal. Today* **2001**, 64, 3.
38 J. A. Moulijn, M. Makkee, A. van Diepen, *Chemical Process Technology*, Wiley, New York **2001**.
39 S. Albertazzi, F. Basile, G. Fornasari, F. Trifirò, A. Vaccari Chapter 7 of this book.
40 CHOREN: http://www.choren.com.
41 S. T. Sie, M. M. G. Sneden, H. M. H. van Wechem, *Catal. Today* **1991**, 8, 371.
42 Y. Matsumura, T. Minowa, B. Potic, S. R. A. Kersten, W. Prins, W. P. M. van Swaaij, B. van de Beld, D. C. Elliott, G. G. Neuenschwander, A. Kruse, M. J. Antal Jr., *Biomass Bioenergy* **2005**, 29, 269.
43 R. R. Davda, J. W. Shabaker, G. W. Huber, R. D. Cortright, J. A. Dumesic, *Appl. Catal. B: Environ.* **2005**, 56, 171.
44 F. W. Lichtenthaler, in *Biorefineries–Industrial Processes and Products* (B. Kamm, P. R. Gruber, M. Kamm, eds) vol. II, Wiley-VCH, Weinheim (**2006**) p. 3.
45 D. J. Hayes, S. Fitzpatrick, M. H. B. Hayes, J. R. H. Ross, in *Biorefineries–Industrial Processes and Products* (B. Kamm, P. R. Gruber, M. Kamm, eds) vol. I, Wiley-VCH, Weinheim (**2006**), p. 139.
46 J. J. Bozell, L. Moens, D. C. Elliott, Y. Wang, G. G. Neuenschwander, S. W. Fitzpatrick, R. J. Bilski, J. L. Barnefeld, *Resources, Conservation Recycling* **2000**, 28, 227.
47 B. V. Timokhin, V. A. Baransky, G. D. Eliseeva, *Russ. Chem. Rev.* **1999**, 68(1), 73.
48 Q. Xiang, J. S. Kim, Y. Y. Lee, *Appl. Biochem. Biotechnol.* **2003**, 105–108, 337.
49 R. Katzen, D. J. Schell, in *Biorefineries–Industrial Processes and Products* (B. Kamm, P. R. Gruber, M. Kamm, eds) vol. I, Wiley-VCH, Weinheim (**2006**) p. 129.
50 C. Moreau, M. N. Belgacem, A. Gandini, *Top. Catal.* **2004**, 27(1–4), 11.
51 S. P. Crabtree, R. C. Lawrence, M. W. Tuck, D. V. Tyers, *Hydrcarbon Proc.* **2006** (2), 87.
52 D. C. Elliott, Final Report CRADA with Int. Polyol Chemicals Inc. and Pacific Northwest National Laboratory: *Process Optimization for Polyols Production from Glucose*, PNNL-11476, Pacific Northwest National Laboratory, Richland, Washington, **1997**.
53 D. de Wit, L. Maat, A. P. G. Kieboom, *Ind. Crops Prod.* **1993**, 2, 1.
54 R. Lehnen, B. Saake, H. H. Nimz, *Holzforschung* **2001**, 55, 199.
55 E. K. Pye, in *Biorefineries–Industrial Processes and Products* (B. Kamm, P. R. Gruber, M. Kamm, eds) vol. II, Wiley-VCH, Weinheim (**2006**) p. 165.
56 C. E. V. Rossell, D. L. Filho, A. G. P. Hilst, M. R. L. V. Leal, *Sugar Ind.* **2006**, 131(2), 105.
57 V. S. Chang, M. Nagwani, C. H. Kim, M. T. Holtzapple, *Appl. Biochem. Biotechnol.* **2001**, 94, 1.
58 K. Poppius-Levlin, R. Mustonen, T. Huovila, J. Sundquist, *Paper Timber* **1991**, 73(2), 154.
59 R. P. Swatloski, S. K. Spear, J. D. Holbrey, R. D. Rogers, *J. Am. Chem. Soc.* **2002**, 124, 4974.
60 S. Zhu, Y. Wu, Q. Chen, Z. Yu, C. Wang, S. Jin, Y. Ding, G. Wu, *Green Chem.* **2006**, 8, 325.
61 R. C. Remsing, R. P. Swatloski, R. D. Rogers, G. Moyna, *Chem Comm* **2006**, 1271.
62 L. R. Lynd, C. E. Wyman, T. U. Gerngross, *Biotechnol. Prog.* **1999**, 15, 777.

63 B. C. Saha, *J. Ind. Microbiol. Biotechnol.* **2003**, 30, 279.
64 N. Mosier, C. Wyman, B. Dale, R. Elander, Y. Y. Lee, M. Holtzapple, M. Ladisch, *Bioresource Technol.* **2005**, 96, 673.
65 P. Gruber, D. E. Henton, J. Starr, in *Biorefineries–Industrial Processes and Products* (B. Kamm, P. R. Gruber, M. Kamm, eds) vol. II, Wiley-VCH, Weinheim (**2006**) p. 381.
66 J. S. Tolan, in *Biorefineries–Industrial Processes and Products* (B. Kamm, P. R. Gruber, M. Kamm, eds) vol. I, Wiley-VCH, Weinheim (**2006**) p. 193; see also http://www.iogen.ca.
67 C. E. Nakamura, G. M. Whited, *Curr. Opin. Biotechnol.* **2003**, 14, 454.
68 R. Zvosec, 25th Symposium Biotech. For Fuels and Chemicals (Breckenbridge, Colorado, USA) **2003** [http://www.nrel.gov/biotech_symposium].
69 C. S. Gong, N. Cao, G. T. Tsao, in *Fuels and Chemicals from Biomass* (B. C. Saha, J. Woodward, eds), American Chemical Society, Washington DC, (**1997**), p. 280.
70 C. E. V. Rossell, P. E. Mantelatto, J. A. M. Agnelli, J. Nascimento, in *Biorefineries–Industrial Processes and Products* (B. Kamm, P. R. Gruber, M. Kamm, eds) vol. I, Wiley-VCH, Weinheim (**2006**) p. 209.
71 T. Werpy, J. Frye, J. Holladay, R. Bush, in *Biorefineries–Industrial Processes and Products* (B. Kamm, P. R. Gruber, M. Kamm, eds) vol. II, Wiley-VCH, Weinheim (**2006**) p. 367.
72 D. C. Elliott, D. Beckman, A. Östman, Y. Solantausta, S. B. Gevert, C. Hörnell, in *Energy from Biomass and Waste* **1990**, 13, 743.
73 C. N. Hamelinck, G. van Hooijdonk, A. P. C. Faaij, *Biomass & Bioenergy* **2005**, 28(4), 384.
74 A. C. Caputo, M. Palumbo, P. M. Pelagagge, F. Scacchia, *Biomass & Bioenergy* 28 (**2005**), 35.
75 M. Stumborg, D. Soveran, W. Craig, W. Robinson, K. Ha, in *Energy from Biomass and Waste* **1993**, 14, 721.
76 R. Isaacson, J. Pfeffer, P. Mooij, J. Geselbracht, in *Energy from Biomass and Waste* **1988**, 11, 1123.
77 J.-P. Lange, P. J. A. Tijm, *Chem. Eng. Sci.* **1996**, 51, 2379.
78 J.-P. Lange, *CatTech* **2001**, 5, 82.
79 D. L. Johnson, in *Biorefineries–Industrial Processes and Products* (B. Kamm, P. R. Gruber, M. Kamm, eds) vol. I, Wiley-VCH, Weinheim (**2006**), p. 345.
80 J. Hettenhaus, in *Biorefineries–Industrial Processes and Products* (B. Kamm, P. R. Gruber, M. Kamm, eds) vol. I, Wiley-VCH, Weinheim (**2006**) p. 317.
81 J.-P. Lange, *Green Chem.* **2002**, 4, 546.

3
Process Options for the Catalytic Conversion of Renewables into Bioproducts

Pierre Gallezot

3.1
Overview

This chapter surveys different process options to convert terpenes, plant oils, carbohydrates and lignocellulosic materials into valuable chemicals and polymers. Three different strategies of conversion processes integrated in a biorefinery scheme are proposed: "from biomass to bioproducts via degraded molecules", "from platform molecules to bioproducts", and "from biomass to bioproducts via new synthesis routes". Selected examples representative of the three options are given. Attention is focused on conversions based on one-pot reactions involving one or several catalytic steps that could be used to replace conventional synthetic routes developed for hydrocarbons.

3.2
Introduction

The use of biomass for the production of energy, chemicals and materials is one of the key issues of sustainable development. Indeed, bio-based resources are renewable and CO_2 neutral, in contrast with fossil fuels. Furthermore, owing to the rapid increase of oil price in 2005–2006 we face a new situation where the market price of crude oil (€ 0.40–0.45 kg^{-1}) is higher than that of biomass-derived molecules such as sucrose or glucose. Also, the cost of molecules derived from carbohydrates or vegetable oils is fairly stable compared with that of fossil fuels and even tends to decrease steadily with time. Because of new ecological and economic incentives, government agencies or industrial organizations worldwide are actively promoting the use of renewables for energy and chemical production. Thus, the SusChem organization in Europe has published its Implementation Action Plan 2006 [1], advocating the use of renewables as alternative feedstocks that should be processed to bioproducts within an integrated biorefinery scheme.

In addition to the incentives mentioned above, there is an additional interest to use renewables for the production of bioproducts. Thus, the molecules extracted from bio-based resources are already functionalized so that the synthesis of chemicals may require fewer steps than from alkanes, thereby decreasing the overall waste generated. Also, bio-based products may have unique properties compared with hydrocarbon-derived products, for instance biodegradability and biocompatibility. Biomass processing via clean catalytic routes involving a limited number of steps fulfils several principles of green chemistry at the same time [2]. Merely on economic grounds, products issued from biomass have a higher added value and their marketing is made easier because of their "natural" or "bio" label.

Diversified, cheaper sources of biomass rather than conventional crops are recommended for energy, chemicals and material production. Indeed, there is severe competition for the use of agricultural crops in the production of food/feed, bioproducts (chemical and polymers) and transportation biofuels (bioethanol and biodiesel). Conventional crops based on cereals and seed oils could only be a partial answer to the fuel issue because of the huge needs at stake [1]. To meet biofuel supply in a more substantial way, it is recommended to process agricultural wastes, new crops grown on marginal land, and vegetative biomass (wood, stems, leaves, etc.) consisting of fast-growing lignocellulosic plants rather than using cereals and oil seeds.

Various hurdles may hamper the development of renewables for bio-product production. The supply and composition of renewable raw materials vary with year and location of crops. This could be solved in the long term by using crops dedicated to chemical production and genetically engineered plants, giving a more constant supply and suitable composition that matches the desired products. The main issue is the high cost involved in processing renewable feedstock to chemicals. Processes employed for the synthesis of chemicals from fossil fuels improved continuously throughout over a century, resulting in a very high degree of technical and cost optimization. In contrast, processes to derive chemicals from biomass are comparatively in infancy and their cost weighs heavily on the market price of bioproducts. Accordingly, extensive R&D efforts in biotechnology, chemistry and engineering are required to reduce processing cost. As outlined in Ref. [1] a prerequisite for expending the use of biomass-derived feedstocks is the development of alternative value chains. This can be achieved by designing processing routes and catalytic systems different from those employed from hydrocarbons and adapted to the specific molecular structure of biomolecules.

The present chapter focuses on process options integrated in a biorefinery scheme that should yield bio-products at a more competitive market price and quality. Although bioconversions are essential steps to derive the platform molecules that are used subsequently for catalytic transformations, only chemocatalytic process will be examined. Selected examples of catalytic conversions illustrating different process options will be given.

An important issue, not discussed in the present chapter, is the need to assess by life cycle analysis the sustainability of processes employing biomass instead of fossil fuels. Moreover, socio-economic life cycle assessment rather than simple conventional LCA should be performed to assess the societal impact of intensive agricultural activities covering much larger land area, leaving little space available for recreation areas, increasing the water stress and impairing the biodiversity.

3.3
The Biorefinery Concept

The biorefinery concept has been developed in food and paper industries and is now going to be applied for the production of energy, chemicals and materials from renewable feedstocks [1, 3, 4]. The underlying idea is to maximize the value derived from biomass by producing energy and multiple products via well integrated processes, valorizing co-products and by-products, and optimizing the inputs (energy, water, feedstock) and outputs (energy, products, treatments of gaseous emissions and waste water). In that respect biorefineries are quite comparable to petrochemical refineries. Part of the biomass is converted into fuels via gasification (syngas and hydrogen), pyrolysis (bio-oil), fermentation (biogas) while the other part is converted by successive operations involving hydrolysis, fermentation and chemo-catalytic routes into well-identified platform molecules that can be employed as building blocks in the synthesis of chemicals and polymeric materials. A simplified scheme of biorefinery operations designed for carbohydrate or lignocellulosic materials is given in Fig. 3.1.

The biorefinery scheme was developed initially for carbohydrate-containing feedstocks. Large biorefineries are currently operating in the USA (e.g., Cargill at Blair, Nebraska) and in Europe (e.g., Roquette Frs. at Lestrem, France). The concept can be extended to produce chemicals from other renewable feedstocks. An integrated production of oleochemicals and biofuels can be achieved in biorefineries using vegetables oils as main feedstock to produce versatile platform mole-

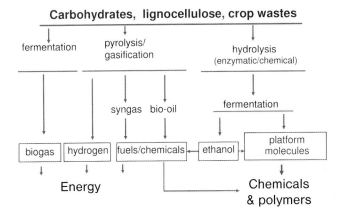

Fig. 3.1 Biorefinery scheme (simplified).

cules such as fatty acids, fatty alcohols and glycerol. Similarly, the production of fine chemicals derived from terpenes can be integrated in a biorefinery scheme.

3.4
Strategies for Biomass Conversion into Bioproducts

Within the biorefinery framework several process options can be chosen to produce bioproducts from renewable feedstocks. We have identified the following three main options that will be illustrated by selected examples.

3.4.1
From Biomass to Products via Degraded Molecules

Biomass can potentially be converted into synthesis gas by adapting the well known gasification processes based on reforming conversions. Syngas is then converted via Fischer–Tropsch synthesis into methanol or hydrocarbons which are subsequently converted into chemicals via usual synthesis routes developed for petroleum feedstock. This approach has been little developed particularly for the production chemicals. The overall sustainability of these high temperature processes has to be established. An interesting alternative to conventional combined gasification/Fischer–Tropsch processes consists of performing aqueous phase reforming of biomass yielding directly alkanes. Thus, a water solution of ethylene glycol, glycerol and sugar-derived polyols was converted into hydrogen and alkanes at 550 K under 15–50 bar pressure [5]. Different selectivities to H_2, CO_2 and alkanes were obtained, depending upon metal catalysts and support acidity. Silica-alumina supported, rhodium catalysts yielded over 50% alkanes. The reaction chain leading to liquid alkanes up to C_{15} involves the acid-catalyzed dehydration of sorbitol, yielding intermediates that are subjected to aldol condensation over solid base catalysts, and the resulting oxygenates are hydrogenated to alkanes over metal-supported catalysts [6, 7]. Although the aim of these studies was to produce transportation fuels it would have been interesting to know more about the organic products formed during the dehydration step because they could be more valuable than the hydrocarbons. In future, biomass gasification or liquid phase reforming employed for fuel production should be conducted using organic wastes or cheap and abundant vegetative biomass rather than high priced molecules such as sugar polyols obtained from cereals.

Another approach to produce chemicals via degraded molecules is the fast pyrolysis of biomass at high temperatures in the absence of oxygen. This gives gas, tar and up to 80 wt.% of a so-called bio-oil liquid phase, which is a mixture of hundreds molecules. Some of compounds produced by pyrolysis have been identified as fragments of the basic components of biomass, viz. lignin, cellulose and hemicellulose. The bio-oil composition depends upon the nature of starting materials and process conditions [8, 9]. Beside bio-oil application in energy production some valuable compounds present in bio-oils, particularly phenolic compounds issued from the degradation of lignin, can potentially be recovered,. Furfural and furfuryl alcohol are present at up to 30 and 12–30%, respectively. Phenol-formaldehyde resins have been prepared from bio-oils containing a high fraction of phenolic compounds [10, 11]. Although bio-oils are presently of interest for heat and power generation, their use as a source of chemicals may become attractive in future. The present situation is very much like the early days of coal and petrochemical industry, i.e., a great deal of research and process development is still needed.

At any rate, gasification or pyrolysis of biomass are energy demanding processes so that the production of energy or chemicals via these routes should be submitted to life cycle analysis to assess their sustainability. For chemists involved in the synthesis of bio-products employed as specialties and fine chemicals, this approach is not attractive because the highly functionalized molecules obtained from renewables are first deeply degraded, even down to C_1 and H_2 species in the case of gasification, before being subjected to conventional chemical synthesis routes in order to be functionalized again into valuable chemicals.

3.4.2
From Biomass to Products via Platform Molecules

3.4.2.1 Identification of Main Platform Molecules
As far as carbohydrates are concerned, several platform molecules are already well identified and currently employed to produce specialties and fine chemicals. Monosaccharides such as glucose, and disaccharides such as sucrose, which are easily obtained with great purity from various carbohydrate-containing crops are well-known platform molecules in sugar chemistry. Lactose, with a production of 6×10^5 t a^{-1}, is also a useful platform molecule obtained from the cheese industry. Most other platform molecules are obtained via fermentation from glucose or other carbohydrates using continuously improved processes with new genetically modified bacteria or yeasts [12]. The followings biomass-derived platform molecules are potentially useful building blocks for chemical synthesis: aspartic acid, 1,4-diacids (succinic, fumaric and malic), ethanol, glutamic acid, glucaric acid, 2,5-hydroxymethylfurfural, 2,5-furan dicarboxylic acid, 3-hydroxypropionic acid, 2-hydroxypropionic acid, itaconic acid, levulinic acid, 1,3-propanediol. The list includes the top 12 platform molecules identified by the US department of Energy [13]. In future, the challenge will be to produce fermentable sugars from cellulose and hemicelluloses, which are available in huge amounts from vegetative biomass. This involves progress in the chemical and enzymatic hydrolysis of cellulose [12].

Vegetable oils or triglycerides obtained from the seeds of various plants are the source of a wide variety of fatty acid esters and derivatives (fatty acids and alcohols) with different molecular structure (chain length, number and position of

C=C bonds) that can be used as platform molecules, as well as glycerol, which is a co-product of triglycerides transesterification.

The three main platform molecules employed in terpene chemistry are α-pinene and β-pinene, which are extracted from turpentine oil ($350\,000$ t a^{-1}) a co-product of paper pulp industry, and limonene extracted from citrus oil ($30\,000$ t a^{-1}).

3.4.2.2 Selected Examples of Platform Molecule Conversion into Bioproducts

Terpenes Terpenes [α-pinene (**1**), β-pinene (**2**), and limonene (**3**)] are employed in the synthesis of flavors and fragrances (F&F), although these compounds are often obtained by catalytic routes from hydrocarbons.

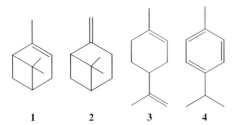

    1   2   3   4

p-Cymene (**4**), a precursor of *p*-cresol and various F&Fs, was obtained by dehydrogenation of α-pinene at 300 °C in a continuous fixed-bed flow reactor in the presence of 0.5-wt.% Pd/SiO$_2$ [14]. Under similar conditions, but starting from limonene, *p*-cymene was obtained in 97% yield and the catalytic activity was stable for 500 h on stream [15].

The liquid phase alkoxylation of limonene (**3**) with C$_1$–C$_4$ alcohols to 1-methyl-4-[α-alkoxy-isopropyl]-1-cyclohexene (**5**) was carried out both in batch and continuous fixed-bed reactor at 60 °C on various acidic catalysts (Scheme 3.1) [16]. The best yields were obtained in batch (85%) or continuous reactor (81%) using a β-type zeolite with SiO$_2$/Al$_2$O$_3$ = 25.

Scheme 3.1

Carbohydrates

Glucose and Fructose Sucrose (total production 130×10^6 t a^{-1}) and starch (40×10^6 t a^{-1} used in industry) are two major sources of glucose and fructose. A much larger supply of glucose or other fermentable sugars is expected to be obtained at lower cost in the future by advanced enzymatic processing of cellulose [12]. It is not the purpose of the present chapter to review the numerous catalytic reactions starting from glucose to derive valuable chemicals. Reviews have been given by van Bekkum and Besemer [17] and by Lichtenthaler and Peters [18]. The trend is to develop heterogeneous processes where the catalysts are recycled many times or used in continuous reactor for a long time on stream. Hydrogenation of glucose to sorbitol [19] and oxidation of glucose to gluconic acid [20] are illustrative examples of very selective catalytic conversion (yield > 99%) where the catalysts are used many times with negligible loss of activity and selectivity.

There is great interest in converting glucose available in large supply into C$_5$ and C$_4$ polyols that are little present in biomass and find many applications in food and non-food products. Thus, glucose can be converted into arabitol by an oxidative decarboxylation of glucose to arabinonic acid followed by hydrogenation to arabitol (Scheme 3.2). The main pitfall is to avoid dehydroxylation reactions leading to deoxy-products not compatible with the purity specifications required for arabitol. Aqueous solutions (20 wt.%) of arabinonic acid have been hydrogenated on Ru-catalysts in batch reactor [21].

The selectivity was enhanced by adding small amounts of anthraquinone-2-sulfonate (A2S), which decreased the formation of deoxy by-products. Thus, by adding 260 ppm of A2S with respect to arabinonic acid the selectivity to deoxy-products decreased from 4.2 to 1.6%. A2S acted as a permanent surface modifier since the catalyst was recycled with the same selectivity without further addition of A2S. The highest selectivity to arabitol was 98.9% at 98% conversion, with a reaction rate of 73 mmol h^{-1} g$_{Ru}$$^{-1}$ at 80 °C.

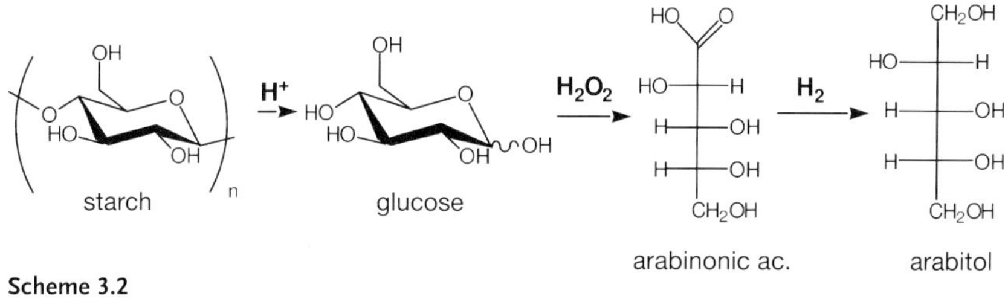

Scheme 3.2

Lactose Abbadi et al. [22] have studied the oxidation of lactose, a co-product of the milk industry, on PtBi/C catalyst at pH 7. Lactobionate was formed transiently and subsequently converted into 2-keto-lactobionate with a final yield of

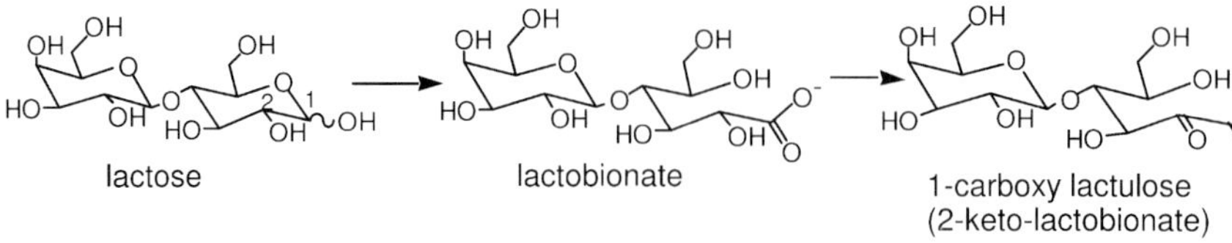

Scheme 3.3

lactose → lactobionate → 1-carboxy lactulose (2-keto-lactobionate)

ca. 80% (Scheme 3.3). Starting from lactobionate without pH control, 2-keto-lactobionate was obtained with 95% selectivity, but the oxidation reaction stopped at 50% conversion due to the poisoning of Pt-Bi/C catalysts.

Carboxylic Acids Obtained by Fermentation of Carbohydrates Lactic (2-hydroxy-propionic) acid obtained by fermentation of glucose and polysaccharides is used by NatureWorks (Cargill/Dow LLC) to prepare polylactide (PLA), a biodegradable or recyclable polymer with a potential production of 140 000 t a^{-1} (Scheme 3.4) [23]. This and other potential useful reactions from lactic acid have been reviewed by Datta and Henry [24].

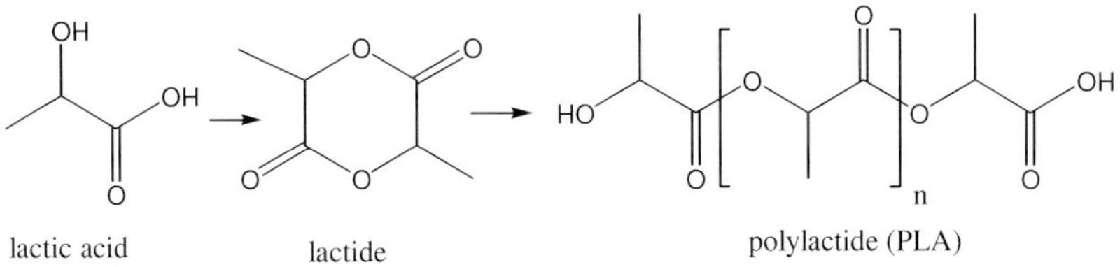

lactic acid → lactide → polylactide (PLA)

Scheme 3.4

3-Hydroxypropionic acid obtained by fermentation from glucose could also be a good candidate to produce various chemicals by catalytic routes [25].

Levulinic acid is obtained by hydrolysis of cellulose-containing biomass. R&D is actively conducted at DuPont Co. to employ levulinic acid for the synthesis of pyrrolidones (solvents and surfactants), α-methylene-γ-valerolactone [monomer for the preparation of polymers similar to poly(methyl methacrylate)], and levulinic acid esters (fuel additives) [26].

Furan Derivatives Catalytic processes used to obtain furan derivatives from carbohydrates and the catalytic routes from furan intermediates to chemicals and polymers have been reviewed by Moreau et al. [27]. Some of the main reactions are summarized in Fig. 3.2. From fructose or carbohydrates based on fructose (sucrose, inulin), the first transformation step is dehydration to 5-hydroxymethylfurfural (HMF). Fructose dehydration at 165 °C was performed in the presence of

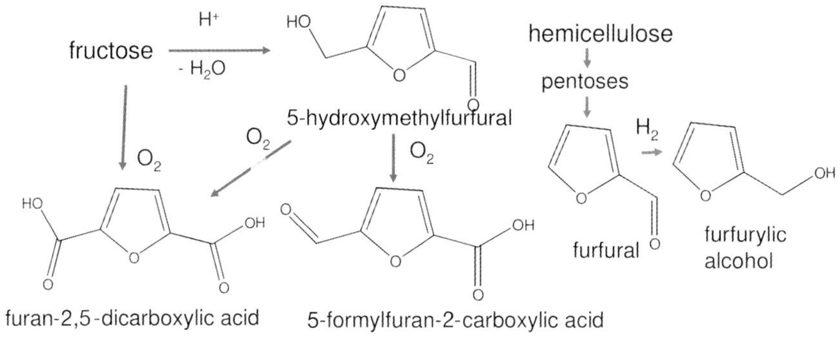

Fig. 3.2 Catalytic reactions on furan derivatives.

dealuminated mordenite (Si/Al = 11) with a selectivity of 92% at 76% fructose conversion [28]. Starting from raw inulin hydrolysates for better process economy, the selectivity to HMF was up to 97% at 54% conversion [29].

The hydrogenation of HMF in the presence of metal catalysts (Raney nickel, supported platinum metals, copper chromite) leads to quantitative amounts of 2,5-bis(hydroxymethyl)furan used in the manufacture of polyurethanes, or 2,5-bis(hydroxymethyl)tetrahydrofuran that can be used in the preparation of polyesters [30]. The oxidation of HMF is used to prepare 5-formylfuran-2-carboxylic acid, and furan-2,5-dicarboxylic acid (a potential substitute of terephthalic acid). Oxidation by air on platinum catalysts leads quantitatively to the diacid. [32]. The oxidation of HMF to dialdehyde was achieved at 90 °C with air as oxidizing in the presence of V_2O_5/TiO_2 catalysts with a selectivity up to 95% at 90% conversion [33].

Furfural is obtained industrially (200 000 t a^{-1}) by dehydration of pentoses produced from hemicelluloses. Furfurylic alcohol is obtained by selective hydrogenation of the C=O bond of furfural, avoiding the hydrogenation of the furan ring. Liquid phase hydrogenation at 80 °C in ethanol on Raney nickel modified by heteropolyacid salts resulted in a 98% yield of furfuryl alcohol [31].

Fatty Acid Esters and Fatty Alcohols Fatty acid esters are obtained by transesterification of triglycerides (vegetable oils) or by esterification of fatty acid with alcohol or polyols. Fatty alcohols are obtained by hydrogenation of esters on metal catalysts. Fatty acid esters and fatty alcohols are useful platform molecules to prepare surfactants, emulsifier, lubricants and polymers.

Surfactants Fatty acid esters of glycerol are efficient surfactants obtained either by transesterification of triglycerides with glycerol (glycerolysis) or by esterification of fatty acids with glycerol. The challenge in both cases is to obtain, selectively, glycerol monoesters that are non-ionic surfactants with a good hydrophilic/hydrophobic balance. Glycerolysis reactions have been conducted on basic oxides to replace liquid bases. Glycerolysis of rapeseed oil on MgO catalysts gave a 63%

yield of monoglyceride [34]. Glycerol monoester has been prepared by esterification of fatty acid with glycerol, using acidic solids instead of sulfuric acid [35–42].

Fatty acid esters of sugars are also very important biodegradable and biocompatible surfactants that are prepared either by transesterification of methyl ester with sugar on basic catalysts or by esterification of fatty acids with sugar on acidic catalysts. Liquid acids and bases have been replaced by enzymatic catalysis with lipase, giving a higher yield of monoester [43, 44], but solid catalysts have not been used extensively so far.

Alkylglucosides are a class of valuable commercial surfactants, particularly for cosmetics applications because of their biocompatibility. They are obtained by acetalization of carbohydrates with fatty alcohols in the presence of acid catalysts. Zeolites and MCM-41 have been used as acidic catalysts to achieve the acetalization of glucose with alcohols of different chain lengths [45, 46]. Shape selectivity effects decrease the amount of oligomers formed and the activity and selectivity can be controlled with the Si/Al ratio.

Lubricants Fatty acid esters could be suitable lubricants, but their resistance to oxidation and tribological properties need to be improved. This can be achieved by epoxidation of the fatty acids followed by alcoholysis of the epoxide (Scheme 3.5).

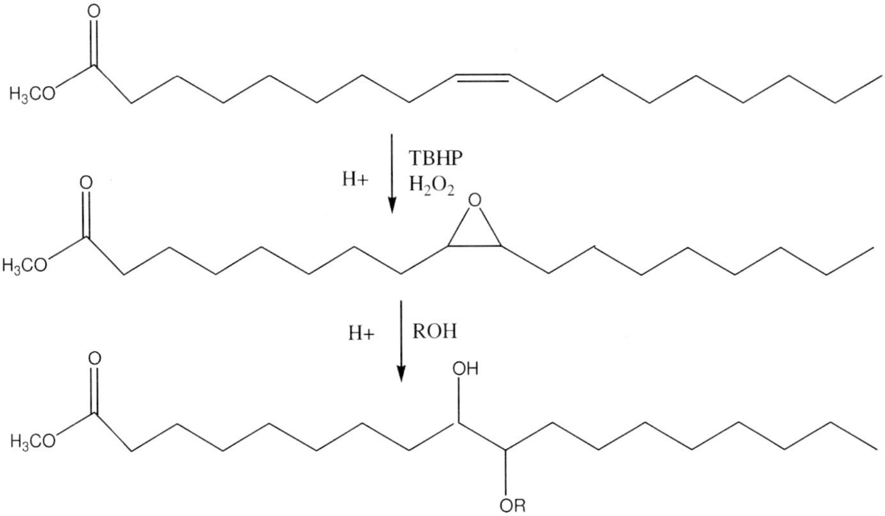

Scheme 3.5

The epoxidation of fatty acid methyl esters (FAME) is traditionally conducted in strong acidic media, e.g., with peracetic acid in sulfuric acid solutions. These reactions can be conducted by an environmentally benign route, however, in the presence of acidic solids. Thus, a mixture of FAME from sunflower oil has been epoxided with *tert*-butyl hydroperoxide (TBHP) at 363 K in the presence of Ti-MCM-41 catalysts to yield 98% conversion with 85% selectivity for mono-epoxy compounds [47]. In the same way Rios et al. [48] have used different Ti-MCM-41 materials, with pores diameters ranging from 1.9 to 4.1 nm, and amorphous Ti/SiO$_2$.catalysts with different Ti-dispersion to perform methyl oleate epoxidation with TBHP at 70 °C. Selectivities > 95% were obtained whatever the structure of the supporting material, provided titanium is well dispersed. Alcoholysis with different alcohols of epoxidized FAME has been studied on acidic resins of various structure and acid strength [49, 50]. The addition of methanol on epoxidized methyl oleate at 60 °C in the presence of Nafion entrapped in silica (SAC13) or of Amberlyst15, a sulfonated styrene-divinylbenzene copolymer, resulted in complete conversion with selectivity higher than 98%.

Polymers Catalytic reactions involving C=C bonds are widely used for the conversion of unsaturated fatty compounds to prepare useful monomers for polymer synthesis. Catalytic C–C coupling reactions of unsaturated fatty compounds have been reviewed by Biermann and Metzger [51]. Metathesis reactions involving unsaturated fatty compounds to prepare ω-unsaturated fatty acid esters have been applied by Warwel et al. [52]. Ethenolysis of methyl oleate catalyzed by ruthenium carbenes developed by Grubb yields 1-decene and methyl 9-decenoate (Scheme 3.6), which can be very useful to prepare monomers for polyolefins, polyesters, polyethers and polyamide such as Nylon 11.

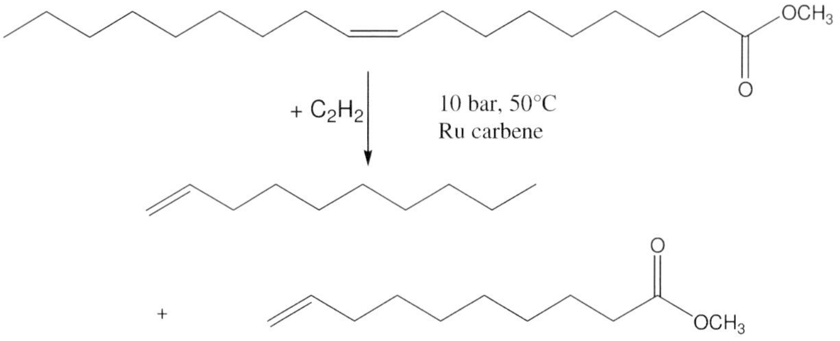

Scheme 3.6

Glycerol Glycerol is now mainly obtained as a co-product of triglyceride transesterification in the production of fatty acid esters employed as bio-diesel, Glycerol should find new outlets to optimize the economy of biodiesel production and to rebalance supply and demand [53]. As far as the production of chemicals is concerned, the main outlet of glycerol is the production of glycerol esters employed as emulsifiers, surfactants and lubricants. Fatty acid esters of glycerol are obtained either by transesterification of triglycerides with glycerol (glycerolysis) or by esterification of fatty acids with glycerol. The challenge in both cases is to obtain glycerol monoesters selectively.

The selective oxidation of glycerol leads to various valuable oxygenates (glyceric, tartronic, mesoxalic, and hydroxypyruvic acids, dihydroxyacetone). Besson and Gallezot [54] have shown that they can be obtained by oxidation with air of aqueous solutions of glycerol in the presence of carbon-supported platinum and palladium catalysts. The selectivity can be tuned by promotion of the noble metals with bismuth or by operating under controlled pH. Thus, Garcia et al. [55] have found that the oxidation of glycerol at basic pH on palladium and platinum catalysts yielded 70% glycerate. Fordham et al. [56] have found that glyceric acid oxidation on 5%-Pt-1.9%-Bi/C catalyst yielded 74% hydroxypyruvic at 80% conversion at acidic pH (3–4), but on the same catalyst under basic conditions (pH 10–11) a 83% yield of tartronate was obtained at 85% conversion. Abbadi and van Bekkum [57] have obtained 93% selectivity for hydroxypyruvic acid at 95% conversion of glyceric acid on 5%-Bi-5%-Pt/C catalyst without pH regulation. More recently, the oxidation of glycerol was conducted in the presence of gold catalysts in basic medium [58, 59]. The selectivity was shown to depend critically upon the size of gold particles [60].

Polyglycerols obtained by the dehydration of glycerol (Scheme 3.7) are employed as surfactants, lubricants, cosmetic, food additives, etc. Their esterification with fatty acids leads also to valuable emulsifiers or metal-working fluids. Zeolites have been used to take advantage of their shape selectivity effect to minimize oligomer formation, as described in two patents [61, 62]. A fair compromise between activity and selectivity has been obtained by Clacens et al. [63] using cesium-impregnated mesoporous MCM-41.

Acrolein has been obtained in 38% yield by glycerol dehydration at 360 °C, 25 MPa in the presence of zinc sulfate [64].

Glycerol can be selectively dehydroxylated to either 1,2-propanediol (1,2-PDO), a chemical that can advantageously replace ethylene glycol as anti-freezing agent, or 1,3-propanediol (1,3-PDO), which when copolymerized with terephthalic acid

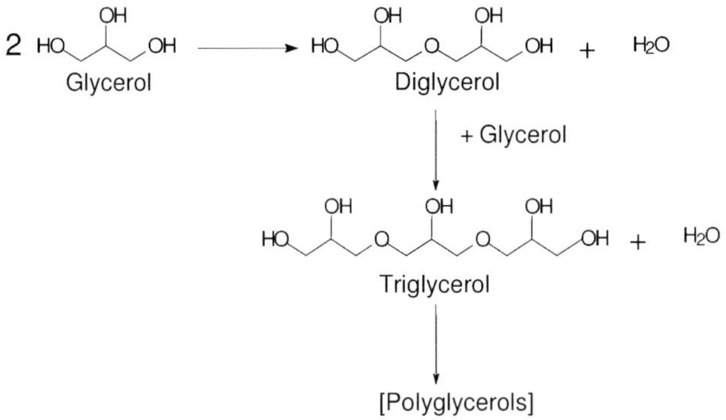

Scheme 3.7

gives polyesters with unique mechanical properties. 1,3-PDO is currently produced by catalytic routes from ethylene oxide (Shell route) or acrolein (Degussa-DuPont route). The microbial production of 1,3-PDO is under development by DuPont-Genencor to produce 1,3-PDO from glucose [65]. Chaminand et al. [66] have studied the hydrogenolysis of aqueous solutions of glycerol at 180 °C under 80 bar H_2-pressure in the presence of supported metal catalysts in an attempt to produce selectively 1,2- and 1,3-PDO. The best selectivity (100%) to 1,2-PDO was obtained by hydrogenolysis of water solution of glycerol in the presence of CuO/ZnO catalysts. To control the selectivity toward 1,3-PDO the reaction was conducted with rhodium catalysts with tungstic acid added to the reaction medium. The best selectivity to 1,3-PDO (1,3-PDO/1,2-PDO = 2) was obtained by operating in sulfolane. Hydrogenolysis of glycerol has also been conducted on Ru/C catalysts in the presence of Amberlyst resin, yielding mainly 1,2-PDO [67].

3.4.3
From Biomass to Products via New Synthesis Routes

The processing cost to convert biomass into valuable products can be greatly reduced under the following process conditions:

1. Decreasing the number of reaction steps via a one-pot reaction associating two or more catalytic steps. This can be achieved by multistep reactions carried out by cascade catalysis without intermediate product recovery, thus decreasing the operating time and reducing considerably the amount of waste produced.

2. Converting renewable feedstocks into a mixture of products that can be used as such in the synthesis or formulation of end-products. This approach is widely used in food and feed industries where there is no requirement to prepare specific molecules from bio-resources but rather mixtures of triglycerides, carbohydrates and proteins.

Approach (1) is very similar to classical synthesis routes except that the number of steps is reduced. In contrast approach (2) is quite different from chemical synthesis where the isolated products synthesized are used as such or in the formulation of end-products.

3.4.3.1 One-pot Reaction with Cascade Catalysis

There are several examples of one-pot reactions with bifunctional catalysts. Thus, using a bifunctional Ru/HY catalyst, water solutions of corn starch (25 wt.%) have been hydrolyzed on acidic sites of the Y-type zeolite, and glucose formed transiently was hydrogenated on ruthenium to a mixture of sorbitol (96%), mannitol (1%), and xylitol (2%) [68]. Similarly a one-pot process for the hydrolysis and hydrogenation of inulin to sorbitol and mannitol has been achieved with Ru/C catalysts where the carbon support was preoxidized to generate acidic sites [69]. Ribeiro and Schuchardt [70] have succeeded in converting fructose into furan-2,5-dicarboxylic acid with 99% selectivity at 72% conversion in a one-pot reaction over a bifunctional acidic and redox catalyst consisting of cobalt acetylacetonate encapsulated in sol–gel silica.

p-Cymene (4), a precursor of p-cresol and various F&Fs, can be obtained by dehydrogenation of α-pinene at 300 °C in a continuous fixed-bed flow reactor in the presence of 0.5-wt.% Pd/SiO$_2$, as mentioned in the subsection on terpenes in Section 3.4.2.2 [14]. Interestingly, p-cymene has been produced under similar reaction conditions with nearly 100% yield from a mixture of di-pentene isomers (Sylvapine DP-378), showing that even unpurified raw materials can still be converted into the desired product in one step [71].

Cascade catalysis without recovery of intermediate products may require more than two steps, involving enzymatic, homogeneous, and heterogeneous catalysis. Several examples of this approach have been given [72, 73]; one of the most representative consists of a four-step conversion of glucoside into aminodeoxysugar without intermediate product recovery.

3.4.3.2 One-pot Conversion into a Mixture of Products

Conversion of Sugars into Polyols Deoxyhexitols consisting of C$_6$ diols, triols, and tetrols are well suited to replace polyols derived from petrochemistry for applications in polyester and polyurethane manufacture. Investigations on controlled hydrogenolysis of carbohydrates have been performed mainly to produce C$_2$–C$_3$ polyols rather than higher molecular weight polyols [74, 75]. In the framework of the European program STARPOL, starch hydrolysates were converted by combined hydrolysis–hydrogenation in a reactor loaded with Ru/HY catalysts into sorbitol (Fig. 3.3). Then, in a second reactor, depending upon reaction conditions, sorbitol was converted either by dehydroxylation into C$_4$–C$_6$ products [76] or by

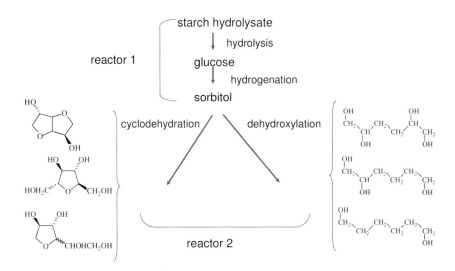

Fig. 3.3 Preparation of polyols (left-hand side: cyclic ethers; right-hand side: dehydroxyhexitols) from starch hydrolysates via a two-step process.

cyclodehydration into cyclic polyols [77]. The dehydroxylation was performed under mild conditions to favor selectivity to deoxyhexitols. Copper-based catalysts, which have a low activity for hydrogenolysis of C–C bonds, were employed to treat 20 wt.% aqueous sorbitol solutions in the temperature range 180–240 °C. Reactions carried out in the presence of 33% CuO–65% ZnO catalyst at 180 °C under H_2-pressure yielded 73% $C_4{}^+$ polyols, and, more specifically, 63% deoxyhexitols. Cyclodehydration of sorbitol and mannitol was carried out at 250 °C under 80 bar of H_2, in the presence of 0.5% Pd/C catalyst and propionic acid that can be recovered by distillation at the end of reaction (Fig. 3.3). The reaction yielded cyclic ethers with the following selectivity at 90% conversion: 37.5% of isosorbide, 37.5% of 2,5-anhydromannitol, and 25% of 1,4-anhydrosorbitol (Fig. 3.3) [77]. The mixture of polyols, obtained either by dehydroxylation or cyclodehydration, was effectively employed to synthesize alkyd resins and make decorative paints with performances comparable to commercial ones.

Oxidation of Starch and Other Polysaccharides There is a challenge to convert polysaccharide polymers such as starch or cellulose into valuable end-products via one-pot process. This is difficult because these natural polymers are insoluble and partially crystallized and heterogeneous catalysts cannot be employed with these solid substrates. Hydrophilic starch obtained by partial oxidation is used in paper and textile industries and can be potentially applied in various applications, e.g., for the preparation of paints, cosmetics, and super-absorbents. Oxidation occurs at the C_6 primary hydroxyl group or at the vicinal diols on C_2 and C_3, involving a cleavage of the C_2–C_3 bond to give carbonyl and carboxyl functions (Scheme 3.8). Several transition metal catalysts based on Fe, Cu or W salts have been proposed to activate H_2O_2 but the concentration of metal ions was quite high and they were retained by carboxyl functions in the modified starch [78].

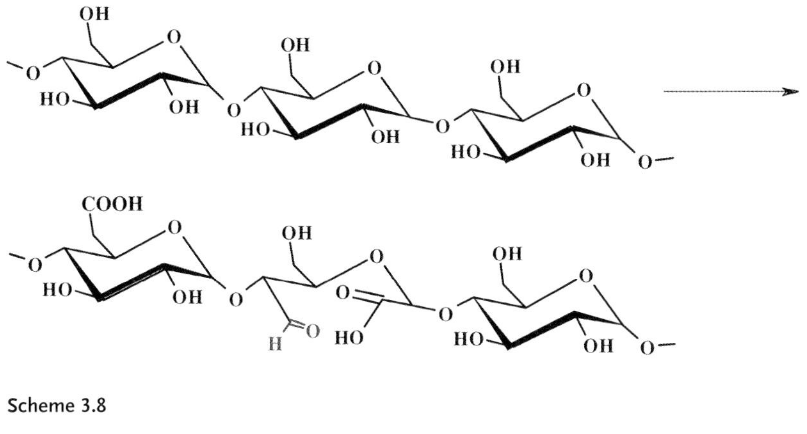

Scheme 3.8

Instead, native starch has been oxidized with H_2O_2 in the presence of soluble organometallic complexes to meet specific hydrophilic/hydrophobic properties needed for end-products to be used in paper, paint and cosmetic industries [79–

Table 3.1 Oxidation of starch in aqueous suspension with H_2O_2 in the presence of iron phthalocyanine.[a] Effect of substrate/catalyst ratio.

Anhydro-glucose unit (AGU): Fe ratio	DS_{COOH}[b]	$DS_{C=O}$[b]
25800:1	0.70	3.20
12900:1	2.00	10.40
6450:1	2.00	9.00

[a] Reaction conditions: 58 °C; pH 7; reaction time: 7 h; molar ratio H_2O_2:AGU = 1:2.1.
[b] Degree of substitution expressed per 100 anhydroglucose units (AGU).

81]. The water-soluble iron tetrasulfophthalocyanine (FePcS) complex, which is cheap and available on an industrial scale, was also a very active and selective catalyst for the oxidation reaction. Starches of different origin (potatoes, rice, wheat, corn) were oxidized by H_2O_2 following two operating modes, viz. oxidation in aqueous suspension and oxidation by incipient wetness.

The oxidation of starch in aqueous suspension with H_2O_2 in the presence of iron phthalocyanine gives both carboxylic and carbonyl groups (Table 3.1). The best yields were obtained with a molar ratio 12900/1 (0.0078 mol%), but the oxidation was still quite efficient with 0.0039 mol% of catalyst [25800 per anhydroglucose unit (AGU)/catalyst ratio]. The oxidized starch had almost the same final Fe-content as the initial potato starch. Still, the efficiency of this method in view of scaling up was limited by comparatively low activity and product isolation problems.

Native starch has been oxidized by the incipient wetness method by adding a small volume of water containing the dissolved catalysts to starch powder under continuous mixing, followed by addition of hydrogen peroxide to the impregnated solid under mixing. With a substrate/catalyst ratio of only 25800/1 the oxidation yielded 1.5 carboxyl and 5.6 carbonyl functions per 100 AGU. The process was applied with success to the oxidation of starches of different physical and chemical properties (amylose/amylopectin ratio, granule size, temperature of gelatinization) obtained from different crops (potato, wheat, rice, corn). It has been further extended to cellulose, inulin and guar gum, giving a high degree of substitution (up to $DS_{COOH} = 26.5$ and $DS_{CHO} = 11.6$ for cellulose).

This catalytic system was very flexible because by simple modification of the reaction conditions it was possible to prepare oxidized polymers with the desired level of carboxyl and carbonyl functions. No waste was formed because the process did not involve any acids, bases or buffer solutions. The incipient wetness process is very easy to scale up. Hydrophilic starch was prepared in batches of 150 L and incorporated successfully in paint formulations. Good results were also obtained with *in vitro* and *in vivo* tests for cosmetic formulation. Interestingly, this is a rather unique example of a heterogeneous catalytic process involving a soluble catalyst and a solid substrate.

Modification of Starch and Other Saccharides by Grafting Hydrophobic Chains

To prepare more hydrophobic starches for specific applications, the partial substitution of starch with acetate, hydroxypropyl, alkylsiliconate or fatty-acid ester groups has been described in the literature. A new route, however, consists of grafting octadienyl chains by butadiene telomerization (Scheme 3.9) [79, 82, 83]. The reaction was catalyzed by hydrosoluble palladium-catalytic systems prepared from palladium diacetate and trisodium tris(*m*-sulfonatophenyl)phosphine (TPPTS).

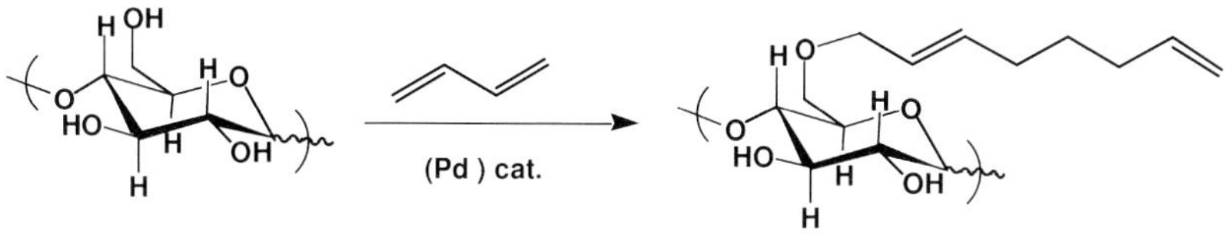

Scheme 3.9

The reaction was first conducted with success on sucrose [82]. The degree of substitution (DS) obtained was controlled by the reaction time. Thus, under standard conditions (0.05% Pd(OAc)$_2$/TPPTS, NaOH (1 M)/iPrOH (5/1), 50 °C) the DS was 0.5 and 5 after 14 and 64 h reaction time, respectively. The octadienyl chains were hydrogenated quantitatively in the presence of 0.8-wt.% [RhCl(TPPTS)$_3$] catalyst in a H$_2$O–EtOH (50/10) mixture, yielding a very good biodegradable surfactant (surface tension of 25 mN m^{-1} at 0.005% concentration in water) [84]. Telomerization reaction was also conducted with success on other soluble carbohydrates such as fructose, maltose, sorbitol and β-cyclodextrin.

The transposition of this reaction to starch [79, 83] was challenging because this substrate is insoluble in water at room temperature and gelatinizes above ca. 70 °C. The DS should be kept low enough because modified starch should not be too hydrophobic, and for obvious economic reasons the catalyst/starch ratio should be kept low. Figure 3.4 shows that the DS depends upon the amount

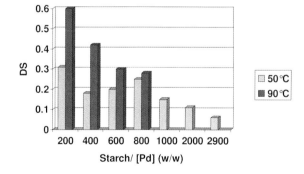

Fig. 3.4 Modification of starch by butadiene telomerization. Influence of the catalyst mass and temperature on the degree of substitution (DS).

of catalyst and temperature. Modified starch with DS = 0.06 obtained with 0.03% palladium at 50 °C meets the specification required for use as thickener for decorative paints. No palladium was detected in the modified polymer when the reaction was conducted in the presence of 0.05% palladium.

The etherified starch was further transformed by hydrogenation of the double bonds to yield the corresponding linear octyl groups using [RhCl(TPPTS)$_3$] catalyst soluble in EtOH–H$_2$O mixtures. Complete hydrogenation was obtained at 40 °C under 30 bar of H$_2$ after 12 h using 0.8-wt.% Rh-catalyst [84]. Other catalytic transformations such as double bond oxidation and olefin metathesis could possibly be used to prepare other modified starches for various applications.

3.5
Concluding Remarks

So far, bioproducts are derived mainly from vegetables oils and carbohydrates. These raw materials issue from grains harvested primarily for food and feed and their productivity is low per area of cultivated land. In contrast, the non-grain portion of biomass, i.e., agricultural wastes (cobs, stalk, stovers) and vegetative biomass (trees, leaves, etc.) are hardly used despite their much larger availability. Vegetative biomass is based on lignocellulosic materials built on the intimate mixture of cellulose, hemicellulose and lignin that are difficult to process. Hemicelluloses are more easily hydrolyzed than celluloses and yield valuable pentoses such as xylose and arabinose that are potentially an important untapped renewable source of chemicals. Cellulose polymers are built on glucoside units connected via β-1,4-glycosidic linkages instead of α-linkages in the case of starch; they are difficult to hydrolyze by chemical or enzymatic means, and the depolymerization of lignin is even worse. The development of an extended use of lignocellulosic materials for producing cost competitive bio-products must await progresses in depolymerization processes relying on improved biotechnologies. Nonetheless, chemicals such as vanillin can be obtained by catalytic oxidation of lignosulfonate [85]. In the meantime, pyrolysis of lignocellulosic materials yielding bio-oils can be used to extract chemicals.

Recent roadmaps on the use of renewable resources, such as "SusChem Implementation Action Plan 2006" [1], pledge the increasing use of biomass for the production of energy, chemicals and materials. Chemical companies worldwide have started extensive R&D efforts to develop bio-based chemicals. Although the market price of renewable feedstocks is now comparable to that of fossil fuels, their processing cost is much higher. Synthesis routes from hydrocarbons have been improved for over a century whereas biomass processing is, comparatively, in its infancy. Alternative value chains have to be developed to decrease the cost and to increase the quality of end-products, because even if they may prefer bio-based products, consumers do not want to pay more and have lower quality products. Life cycle analysis taking into account economic and societal issues due to the increased occupation of land have to be performed to validate biomass processing options and justify the use of biomass in the place of fossil feedstocks.

In view of the diversity and complexity of renewable feedstocks and of the potentially numerous bio-products at stake, integrated eco-efficient processes should be conducted in biorefineries. The biorefinery framework maximizes the value derived from biomass feedstocks by producing multiple products, valorizing by-products and co-products, balancing energy production vs. energy consumption, and optimizing inputs and outputs, including waste treatment.

Within the biorefinery scheme we have identified three process options to produce chemicals by catalytic routes:

1. The degradation of biomass by gasification or pyrolysis leading to syngas and bio-oils, respectively. This approach provides primarily fuels rather than starting materials for bio-product synthesis. However, this is a possible route to bioproducts provided life cycle analysis demonstrates its validity in terms of economy and ecology. Partial catalytic degradation of biomass such as that obtained in aqueous phase reforming brings new opportunities.

2. The catalytic conversion of platform molecules produced by bioconversion of renewables into bioproducts. This is already the basis of many industrial processes, leading to important tonnages of chemicals and polymers from carbohydrates and triglycerides and fine chemicals from terpenes. This approach needs to be extended and process efficiency should be strengthened by designing more active and selective catalysts.

3. New synthesis routes based on one-pot reactions have to be developed to reduce processing costs drastically. Pure isolated products can be obtained by cascade catalysis involving two or more steps. A much larger gain in process economy should be obtained if a mixture of products suitable for a particular application, e.g., in paper, paint, polymer and cosmetic industries, can be prepared in a one-pot process starting from raw materials such as starch, cellulose and triglycerides. This is a current approach in food industry. Examples have been given in this chapter of the direct transformation of starch into a mixture of products that can be used as such to manufacture end-products.

To meet the challenge posed by the future increasing use of renewables, a large integrated research effort in chemistry, biochemistry, and genetics as well as in chemical and biochemical engineering will be required both in industrial and academic research centers. Biocatalysis is of primary importance, particularly in the early stage of processing to obtain platform molecules that will be converted into valuable chemicals by chemo-catalytic processes using either homogeneous or heterogeneous catalysts or a combination of both. Whatever the process options chosen, catalysis in all its forms will have a major role to play. At present, comparatively few researchers from industry and Academia are working in this area of catalysis. In view of the importance of environmental and economic challenges to meet, the workforce in Academia should be strengthened to develop formation and research.

References

1 http://www.suschem.org/media.php?mId=4727.
2 P.T. Anastas, J.C. Warner, *Green Chemistry: Theory and Practice*, Oxford University Press, New York, **1998**, p. 30.
3 http://www.eere.energy.gov/biomass/
4 J.J. Bozell in *Feedstocks for the Future*, J.J. Bozell, M.K. Patel (Eds.), ACS Symposium Series 921, American Chemical Society, Washington, **2006**, p. 1.
5 R.R. Davda, J.W. Shabaker, G.W. Huber, R.D. Cortright, J.A. Dumesic, *Appl. Catal. B* **2005**, 56, 171.
6 G.W. Huber, R.D. Cortright, J.A. Dumesic, *Angew. Chem.* **2004**, 43, 1549.
7 G.W. Huber, J.N. Chheda, C.J. Barrettt, J.A. Dumesic, *Science* **2005**, 308, 1446.
8 A.V. Bridgwater, G.V.C. Peacoke, *Renew. Sustain. Energy Rev.* **2000**, 4, 1.
9 D. Mohan, C.U. Pittman Jr., P.H. Steele, *Energy & Fuels* **2006**, 20, 848.
10 H. Pakdel, C. Roy, *Bioresour. Technol.* **1996**, 58, 83.
11 C. Amen-Chen, B. Riedl, C. Roy, *Holzforschung* **2002**, 56, 281.
12 Th. Wilke, K.D. Vorlop, *Appl. Microbiol. Biotechnol.* **2004**, 66, 131.
13 T. Werpy, G. Petersen, DOE/GO-102004–1992, August 1, **2004**.
14 D.M. Roberge, D. Buhl, J.P.M. Niederer, W.F. Hölderich, *Appl. Catal. A: General* **2001**, 215, 111.
15 D. Buhl, D.M. Roberge, W.F. Hölderich, *Appl. Catal. A: General* **1999**, 188, 287.
16 K. Hensen, C. Mahaim, W.F. Hölderich, *Appl. Catal. A: General* **1997**, 149, 311.
17 H. van Bekkum, A.C. Besemer, *Chem. Sustainable Develop.* **2003**, 11, 11.
18 F.W. Lichtenthaler, S. Peters, *C.R. Chimie* **2004**, 7, 65–90.
19 N. Nicolaus, P. Gallezot, A. Perrard, *J. Catal.* **1998**, 180, 51.
20 M. Besson, F. Lahmer, P. Gallezot, P. Fuertes, G. Flèche, *J. Catal.* **1995**, 152, 116.
21 L. Fabre, P. Gallezot, A. Perrard, *J. Catal.* **2002**, 208, 247.
22 A. Abbadi, K.F. Gotlieb, J.B.M. Meirberg, H. Van Bekkum, *Appl. Catal. A: General* **1997**, 156, 105.
23 P.R. Gruber, Proc. 2nd Int. Conf. on Green and Sustainable Chemistry, Washington DC, June 20–24, **2005**, paper 149.
24 R. Datta, M. Henry, *J. Chem. Technol. Biotechnol.* **2006**, 81, 1119.
25 D.C. Cameron, NSF workshop, June 23–24, Washington 2005 http://www.cbe.iastate.edu/nsfbioren/presentations/Cameron_Cargill.pdf.
26 L. Manzer in *Feedstocks for the Future*, J.J. Bozell, M.K. Patel (Eds.), ACS Symposium Series 921, American Chemical Society, Washington, **2006**, p. 40.
27 C. Moreau, M. Belgacem, A. Gandini, *Top. Catal.* **2004**, 27, 11.
28 C. Moreau, R. Durand, S. Razigade, J. Duhamet, P. Faugeras, P. Rivalier, P. Ros, G. Avignon, *Appl. Catal. A: General* **1996**, 145, 211.
29 C. Moreau, R. Durand, C. Pourcheron, S. Razigade, *Ind. Crops Prod.* **1994**, 3, 85.
30 V. Schiavo, G. Descotes, J. Mentech, *Bull. Soc. Chim. Fr.* **1991**, 128, 704.
31 P. Vinke, H.E. van Dam, H. van Bekkum, *Stud. Surf. Sci. Catal.*, G. Centi, F. Trifiro (Eds.), Elsevier, Amsterdam, **1990**, Vol. 55, p. 147.
32 C. Moreau, R. Durand, C. Pourcheron, D. Tichit in *Heterogeneous Catalysis and Fine Chemicals IV*, H.U. Blaser, A. Baiker, R. Prins (Eds.) *Stud. Surf. Sci. Catal.*, Vol. 108, Elsevier, Amsterdam, **1997**, p. 399.
33 B.J. Liu, L.H. Lu, B.C. Wang, T.X. Cai, K. Iwatani, *Appl. Catal. A: General* **1998**, 171, 117.
34 A. Corma, S. Iborra, S. Miquel, J. Primo, *J. Catal.* **1998**, 173, 315.
35 H.E. Hoydonckx, D.E. De Vos, S.A. Chavan, P.A. Jacobs, *Top. Catal.* **2003**, 27, 83.
36 M.S. Machado, J. Perez-Pariente, E. Sastre, D. Cardoso, A.M. de Gueremu, *Appl. Catal. A: General* **2000**, 203, 321.
37 N. Sanchez, M. Martinez, J. Aracil, *Ind. Eng. Chem. Res.* **1997**, 36, 1529.
38 S. Abro, Y. Pouilloux, J. Barrault, *Heterogeneous Catalysis and Fine Chemicals IV*, H.U. Blaser, A. Baiker, R. Prins (Eds.), *Stud. Surf. Sci. Catal.*, Vol. 108, Elsevier, Amsterdam, **1997**, p. 539.
39 Y. Pouilloux, S. Abro, C. Vanhove, J. Barrault, *J. Mol. Catal. A: Chem.* **1999**, 149, 243.
40 W.D. Bossaert, D.E. De Vos, W. Van Rhijn, J. Bullen, P.J. Grobet, P.A. Jacobs, *J. Catal.* **1999**, 182, 156.

41 I. Diaz, C. Marquez-Alvarez, F. Mohino, J. Perez-Pariente, E. Sastre, *J. Catal.* **2000**, 193, 295.
42 I. Diaz, F. Mohino, J. Perez-Pariente, E. Sastre, *Appl. Catal. A: General* **2003**, 242, 161.
43 A.T.J.W. de Goede, M. van Oosterom, M.P.J. van Deurzen, R.A. Sheldon, H. van Bekkum, F. van Rantwijk, *Biocatalysis* **1994**, 9, 145.
44 R.T. Otto, H. Scheib, U.T. Bornscheuer, J. Pleiss, C. Syldatk, R.D. Schmid, *J. Mol. Catal. B: Enzym.* **2000**, 8, 201.
45 M.A. Camblor, A. Corma, S. Iborra, S. Miquel, J. Primo, S. Valencia, *J. Catal.* **1997**, 172, 76.
46 M.J. Climent, A. Corma, S. Iborra, J. Miquel, S. Primo, F. Rey, *J. Catal.* **1999**, 183, 76.
47 M. Guidotti, N. Ravasio, R. Psaro, E. Gianotti, L. Marchese, S. Coluccia, *Green Chem.* **2003**, 5, 421.
48 L.A. Rios, P. Weckes, H. Schuster, W.F. Hölderich, *J. Catal.* **2005**, 232, 19.
49 L.A. Rios, P.P. Weckes, H. Schuster, W.F. Hölderich, *Appl. Catal. A: General* **2005**, 284, 155.
50 W.F. Hölderich, L.A. Rios, P.P. Werke, H. Schuster, *J. Synthetic Lubric.* **2004**, 20, 289.
51 U. Biermann, J. Metzger, *Top. Catal.* **2004**, 27, 119.
52 S. Warwel, P. Bavaj, M. Rüschgen Klaas, B. Wolff in *Perspektiven Nachwachsender Rohstoffe in der Chemie* VCH, Weinheim, **1996**, p. 119.
53 S. Claude, *Fett/Lipid* **1999**, 101, 101.
54 M. Besson, P. Gallezot, *Catal. Today* **2000**, 57, 127.
55 R. Garcia, M. Besson, P. Gallezot, *Appl. Catal. A: General* **1995**, 127, 165.
56 P. Fordham, M. Besson, P. Gallezot, *Appl. Catal. A: General* **1995**, 133, L179.
57 A. Abbadi, H. van Bekkum, *Appl. Catal. A: General* **1996**, 148, 113.
58 S. Carrettin, P. McMorn, P. Johnston, K. Griffin, G.J. Hutchings, *Chem. Comm.* **2002**, 696.
59 S. Carrettin, P. McMorn, P. Johnston, K. Griffin, C. Kiely, G.A. Attard, G.J. Hutchings, *Top. Catal.* **2003**, 27, 131.
60 S. Demirel-Gülen, M. Lucas, P. Claus, *Catal. Today* **2005**, 102, 166.
61 US Pat 5349094, assigned to Henkel, **1992**.
62 US Pat 5635588, assigned to Unichema, **1997**.
63 J.M. Clacens, Y. Pouilloux, J. Barrault, *Appl. Catal. A: General* **2002**, 227, 181.
64 L. Ott, M. Bicker, H. Vogel, *Green Chem.* **2006**, 8, 214.
65 G.M. White, B. Bulthius, D.E. Trimbur, A.A. Gatenby, Patent WO 99 10356, **1999**.
66 J. Chaminand, L. Djakovitch, P. Gallezot, P. Marion, C. Pinel, C. Rosier, *Green Chem.* **2004**, 6(8), 359.
67 T. Miyazawa, Y. Kusunoki, K. Knimori, K. Tomishige, *J. Catal.* **2006**, 240, 213.
68 P.A. Jacobs, H. Hinnekens, EP 0 329 923 for Synfina-Oleofina.
69 A.W. Heinen, J.A. Peter, H. van Bekkum, *Carbohydr. Res.* **2001**, 330, 381.
70 M.L. Ribeiro, U. Schuchardt, *Catal. Comm.* **2003**, 4, 83–86.
71 W.F. Hölderich, *Catal. Today* **2000**, 62, 115.
72 A. Bruggink, R. Schoevaart, T. Kieboom, *Org. Proc. Res. Dev.*, **2003**, 622.
73 R. Schoevaart, T. Kieboom, *Top. Catal.* **2004**, 27, 3.
74 C. Montassier, J.C. Ménézo, L.C. Hoang, C. Renaud, J. Barbier, *J. Mol. Catal.* **1991**, 70, 99.
75 C. Montassier, J.M. Dumas, P. Granger, J. Barbier, *Appl. Catal. A: General* **1995**, 121, 231.
76 B. Blanc, A. Bourrel, P. Gallezot, T. Haas, P. Taylor, *Green Chem.* **2000**, 89.
77 T. Haas, O. Burkhardt, M. Morawietz, A. Vanheertum, A. Bourrel, DE 19749202, assigned to Degussa, **1999**.
78 P. Parovuori, A. Hamunen, P. Forssell, K. Autio, K. Poutanen, *Starch/Stärke* **1995**, 47, 19.
79 A. Sorokin, S. Kachkarova-Sorokina, C. Donzé, C. Pinel, P. Gallezot, *Top. Catal.* **2004**, 27, 67.
80 A. Sorokin, S. Sorokina, P. Gallezot, *Chem. Comm.* **2004**, 2844.
81 A. Sorokin, S. Kachkarova-Sorokina, P. Gallezot, WO Patent 2004/007560 A1, **2004**.
82 V. Desvergnes-Breuil, C. Pinel, P. Gallezot, *Green Chem.* **2001**, 3, 175.
83 C. Donzé, C. Pinel, P. Gallezot, P. Taylor, *Adv. Synth. Catal.* **2002**, 344, 906.
84 C. Pinel, C. Donzé, P. Gallezot, *Catal. Comm.* **2003**, 4, 465.
85 H.R. Bjorsvik, *Org. Proc. Res. & Dev.* **1999**, 3, 330.

4
Industrial Development and Application of Biobased Oleochemicals

Karlheinz Hill

4.1
Overview

In concepts for new products the performance, product safety, and product economy criteria are equally important. They are taken into account already when the raw material base for a new industrial product development is defined. Here, renewable resources have often been shown to have advantages compared with fossil feedstock. Over the years it has been demonstrated that the use of vegetable fats and oils in oleochemistry allows the development of competitive, powerful products that are both consumer- and environmentally-friendly. Products from recent developments fit with this requirement profile.

In polymer applications derivatives of oils and fats, such as epoxides, polyols and dimerizations products based on unsaturated fatty acids, are used as plastic additives or components for composites or polymers like polyamides and polyurethanes. In the lubricant sector oleochemically-based fatty acid esters have proved to be powerful alternatives to conventional mineral oil products. For home and personal care applications a wide range of products, such as surfactants, emulsifiers, emollients and waxes, based on vegetable oil derivatives has provided extraordinary performance benefits to the end-customer. Selected products, such as the anionic surfactant fatty alcohol sulfate have been investigated thoroughly with regard to their environmental impact compared with petrochemical based products by life-cycle analysis. Other product examples include carbohydrate-based surfactants as well as oleochemical based emulsifiers, waxes and emollients.

4.2
Raw Material Situation [1–3]

The sources of oils and fats are various vegetable and animal raw materials (e.g., tallow, lard), with the vegetable raw materials soybean, palm, rapeseed and sun-

4 Industrial Development and Application of Biobased Oleochemicals

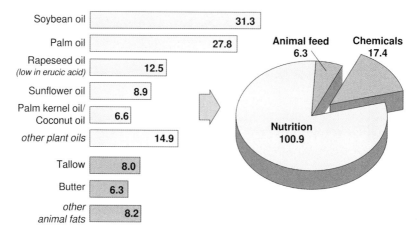

Fig. 4.1 World production of oils and fats in 2003 (in million tonnes) and main uses. (Source: Oil World).

flower oil being the most important ones regarding the amounts involved (Fig. 4.1). Of the approximately 125 million tonnes of fats and oils produced worldwide in 2003 by far the largest share was used in human foodstuffs. For oleochemistry, 17.4 million tonnes were available (incl. soap and biodiesel production). The composition of the fatty acids contained in the oil (fatty acid spectrum) determines the further use of the oils. Special attention must be given to coconut oil and palm kernel oil (lauric oils) because of their high share of fatty acids with a short or medium chain length (mainly 12 and 14 carbon atoms: C_{12}, C_{14}). For example, these are particularly suitable for further processing to surfactants for washing and cleansing agents as well as cosmetics. Palm, soybean, rapeseed and sunflower oil as well as animal fats such as tallow contain mainly long-chain fatty acids (C_{18}, saturated and unsaturated) and are used as raw materials for polymer applications, surfactants, and lubricants. Figure 4.2 compares the composition of a typical lauric oil (coconut oil) with that of sunflower oil.

Fats and oils are available in relatively large quantities. In recent years the amounts produced have continuously increased by ca. 3% per year. Price development – similar to crude oil – has not been constant but has sometimes undergone drastic changes. Parallel to the price increases in crude oil in the 1970s caused by a shortage of supply (oil crisis), price jumps in the fats and oils sector have also occurred; this has been particularly noticeable with coconut oil, which in those days was the most important source of short-chain fatty acids. In this market segment palm kernel oil, a second source of short-chain fatty acids, has had a stabilizing effect as regards raw material availability and price. In the medium and long terms it must be assumed that fats and oils will be offered at competitive prices [3, 4]. However, the increasing demand for biofuel and bioenergy will have to be considered more closely in the future.

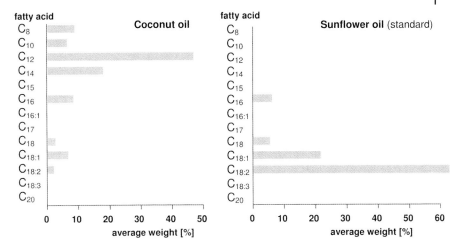

Fig. 4.2 Composition of coconut and sunflower oil.

4.3
Ecological Compatibility

Based on results from cycle analyses (Section 4.4.3) and ecological and toxicological studies for selected cases one can assume that products based on renewable resources usually are more ecologically compatible than petrochemical-based substances – an important criterion in the development of a new product, just as price and performance are [5]. However, this general assumption has to be proven for each new product. Therefore, ecological compatibility plays a decisive role in all research and development projects. Basically, it covers two different aspects: remaining in the environment and the effects on the environment (Table 4.1). Various criteria are used to evaluate these two aspects. By exposure analysis the expected environmental concentration of a particular substance (in wastewater, in exhaust gases or in a sewage treatment plant) is estimated, taking into consideration the amount of the substance produced and its biodegradation behavior. The effect on the environment, e.g., toxicity to organisms such as fish, algae or microorganisms, is determined by a series of standardized testing methods. The two results are then compared. If the expected environmental concentration of a particular substance is less than the amount at which negative effects can no longer be determined then the product is ecologically compatible [6]. Apart from the ecological investigations, toxicological tests and microbiological and dermatological investigations are also carried out. In the framework of a successful marketing strategy all parameters that are relevant to product safety for environment and consumer are evaluated at each product development stage (selection of raw materials, development of test formulations and application tests,

Table 4.1 Evaluation of environmental compatibility of chemicals.

	Method	Result
Basic information		
Environmental fate	Biodegradation tests	
Environmental effect	Ecotoxicity tests	
Criteria		
Environmental concentration	Exposure analysis, biodegradation	PEC: Predicted environmental concentration
Ecotoxicological concentration	Analysis of effects	PNEC: Predicted no effect concentration
Evaluation	Comparison PEC vs. PNEC	PEC < PNEC: environmentally compatible

process development, development of packaging, testing consumer satisfaction in test markets).

4.4
Examples of Products

Oils and fats are triglycerides that typically consist of glycerine and saturated and unsaturated fatty acids. There are a few exceptions to this rule, such as castor oil, a glycerol triester of 12-hydroxyoleic acid (ricinoleic acid). Chemically, triglycerides offer two reactive sites, the double bond in the unsaturated fatty acid chain and the acid group of the fatty acid chain. With regard to product development based on triglycerides, most derivatization reactions are carried out at the carboxylic group (>90%) whereas oleochemical reactions involving the alkyl chain or double bond represent less than 10% (Fig. 4.3).

For most of the further uses oils and fats must be split into the so-called oleochemical base materials: fatty acid methyl esters, fatty acids, glycerol and, as hy-

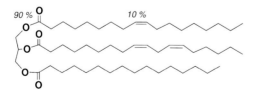

Fig. 4.3 Reactive sites in triglycerides.

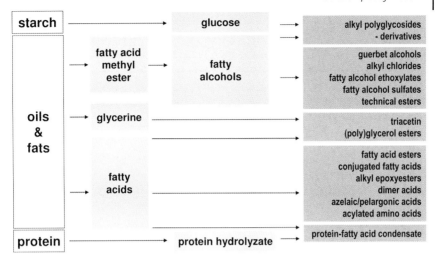

Fig. 4.4 Industrial processing of natural oils and fats and selected product derivatives.

drogenation products of the fatty acid methyl esters, fatty alcohols (Fig. 4.4) [1]. In the following, innovative products, which are derived from glycerides, fatty acids or fatty alcohols, are presented: oleochemicals for polymers, esters for lubricants, and surfactants, and emulsifiers and emollients for home and personal care applications.

4.4.1
Oleochemicals for Polymer Applications

The use of oleochemicals in polymers has a long tradition. One can differentiate between the use as polymer materials, such as linseed oil and soybean oil as drying oils, polymer stabilizers and additives, such as epoxidized soybean oil as plasticizer, and building blocks for polymers, such as dicarboxylic acids for polyesters or polyamides (Table 4.2) [7]. Considering the total market for polymers of ca. 150 million tonnes in 1997 the share of oleochemical based products is relatively small – or, in other terms, the potential for these products is very high. Without doubt there is still a trend in the use of naturally derived materials for polymer applications, especially in niche markets. As an example, the demand for linseed oil for the production of linoleum has increased from 10 000 tonnes in 1975 to 50 000 tonnes in 1998 (coming from 120 000 tonnes in 1960!) [8a]. Epoxidized soybean oil (ESO) as a plastic additive has a relatively stable market of ca. 100 000 tonnes year^{-1} [8b].

Several years ago research was undertaken to use oleochemicals to build up matrices for natural fiber reinforced plastics [9]. The use of natural fibers, such as flax, hemp, sisal, and yucca is of increasing interest for various applications, among them the automotive and public transportation industries, where the com-

Table 4.2 Oleochemicals for polymers – selected examples [7].

	Product/use	Source
Polymer materials		
Polymerized soybean oil, castor oil	Drying oils	Soybean oil, castor oil
Polymerized linseed oil	Linoleum	Linseed oil
Polymer additives		
Epoxides	Stabilizers, plasticizers	Soybean oil
Soaps (Ba/Cd, Ca/Zn)	Stabilizers	Stearic acid
Fatty acid esters, – amides, waxes	Lubricants	Rapeseed oil
Building blocks for polymers		
Dicarboxylic acids	Polyamides, polyesters,	Tall oil, soybean oil,
Ether-/ester polyols	alkyd resins, polyurethanes	castor oil, sun-flower oil, linseed oil, oleic acid

posites could be used in door pockets, covers, instrument panels, sound insulation [9a]. Other applications could be in the manufacturing of furniture. In this field Cognis was coordinating a research project, funded by the Federal Ministry of Food, Agriculture and Forestry (BML) and the National Agency for Renewable Resources (FNR) – project partners being the German Aerospace Center (DLR) and Wilkhahn. The objective was the development of a matrix system with a high content of renewable raw materials (70–75%) and comparable or better performance than purely petrochemical based matrices. Various oleochemical based monomers, such as epoxidized oils, maleinated oils, polyols and amidated fats were investigated and tested, including the manufacture of prototype products based on epoxidized oil (TRIBEST®) [9b]. In the meantime, Alstom has used this technology in the manufacture of urban transportation systems (Hamburger Hochbahn) [9c].

Oleochemical based dicarboxylic acids – azelaic, sebacic, and dimer acid (Figs. 4.5 and 4.6) – amount to ca. 100 000 tonnes year^{-1} as components for polymers. This is about 0.5% of the total dicarboxylic acid market for this application, where phthalic and terephthalic acids represent 87%. The chemical nature of these oleochemical derived dicarboxylic acids can alter or modify condensation polymers, and, used as a co-monomer, will remain a special niche market area. Some of these special properties are elasticity, flexibility, high impact strength, hydrolytic

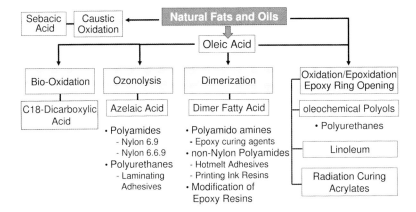

Fig. 4.5 Building blocks for polymers based on natural oils.

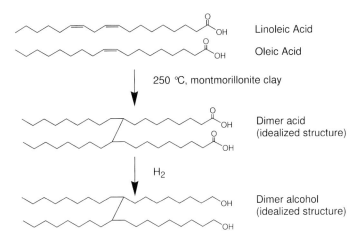

Fig. 4.6 Dimerization of unsaturated fatty acids.

stability, hydrophobicity, lower glass transition temperatures, and flexibility [10]. The crucial reactions in the development of building blocks for polymers based on oils and fats are caustic oxidation, ozonolysis, dimerization, (aut)oxidation, epoxidation, and epoxy ring opening. All of these reactions, except biooxidation [10b], are carried out at the double bond of an unsaturated fatty acid or glyceride. Figure 4.5 summarizes the end products and areas of applications derived from these reactions. In the following, recent developments in the field of diols and polyols for polyurethanes are presented in more detail [11].

4.4.1.1 Dimer Diols Based on Dimer Acid [11, 12]

Dimerization of vegetable oleic acid or tall oil fatty acid (TOFA) yields dimer acids, originally introduced in the 1950s by General Mills Chemicals and Emery (both now Cognis Corp.). The reaction is very complex, resulting in a mixture of aliphatic branched and cyclic C_{36}-diacids (Dimer acid) as the main product besides trimer acids and higher condensed polymer acids on one hand, and a mixture of isostearic acid and unreacted oleic and stearic acid on the other hand. Hydrogenation of dimer acid methyl ester or dimerization of oleyl alcohol leads to dimer alcohols (dimer diols) (Fig. 4.6). Oligomers based on dimer diol are industrially manufactured by acid-catalyzed dehydration of dimer diol. The reaction can be easily monitored by determination of the amount of water produced. Oligomers in the molecular weight range 1000–2000 Da are commercially available by this route (Table 4.3). Another method used to produce oligomers is the transesterification of dimer diol with dimethyl carbonate. The resulting dimer diol polycarbonate has an average molecular weight of 2000 Da. Both types of oligomers, ethers and carbonates, show improved chemical stability compared with dimer diol polyesters.

Owing to their improved stability towards hydrolysis and oxidation, dimer diol polyethers (and dimer diol polycarbonates) are used as soft segments in the preparation of thermoplastic polyurethanes. Polyurethanes prepared from such oleochemical building blocks are very hydrophobic and show the expected stability.

Table 4.3 Specifications of high molecular weight aliphatic diols [11].

	Dimer diol[a]	Dimer diol[a]	12-Hydroxystearyl alcohol stearyl alcohol	Decane-1,10-diol
Outer appearance	Yellow liquid	Colorless liquid	White flakes	White flakes
Hydroxyl value	180–200	180–210	345–360	625–645
Viscosity (25 °C, mPa s)	3500–4300	1800–2800	Solid	Solid
Melting point (°C)	–	–	61–65	68–73
Composition				
Monomer (%)	13	2	–	–
Dimer (%)	68	>96	–	>98
Trimer (%)	19	2	–	–
Trademark	Sovermol 650NS	Sovermol 908	Sovermol 912	Sovermol 110

[a] Molecular weight = 1000–2000 by oligomerization of dimer diol.

The products were almost unaffected when stored either in 60% sulfuric acid or 20% sodium hydroxide solution at 60 °C for 7 weeks. There is no significant change of weight of the testing sticks. For comparison, ester-based polyurethanes as a standard were destroyed completely under these testing conditions after one week (sulfuric acid) and two weeks (sodium hydroxide). Soft segments based on dimer diol ethers are used to prepare saponification resistant TPU-sealings, which allow the contact with aggressive aqueous media at elevated temperature. A typical field of application is in nutrition technology [13].

4.4.1.2 Polyols Based on Epoxides [13, 14]

Low molecular weight liquid epoxy polyol esters or ethers that can be employed as polyols for polyurethane systems are obtained by reaction of epoxidized oils with low molecular weight mono- or polyfunctional alcohols or acids. Depending on the reaction conditions, either polyols with high OH-functionality (complete reaction) or epoxy polyol esters with remaining epoxy groups (partial conversion) are obtained (Fig. 4.7). Although hydroxyl-functional triglyceride oils have found applications in casting resins and adhesives the carboxyl groups of the triglyceride backbone are not fully resistant against hydrolytic attack, particularly by alkali. To overcome this specific behavior fatty acids are used as starting oleochemicals in the preparation of (epoxy) polyols. In principal three categories of products are obtained via this route: fatty acid derivatives with reactive OH groups in the ester position only, e.g., glycerol monostearate, fatty acid derivatives with reactive OH groups in the ester position and in the hydrocarbon chain, e.g., glycerol monoricinoleate, and fatty acid derivatives with reactive OH-groups in the hydrocarbon chain only. These new polyols are of low molecular weight and relatively low viscosity. They offer outstanding hydrolytic stability against both alkali and acids and very high chemical resistance towards corrosive solvents, such as super fuel. They also offer significantly improved mechanical properties compared with hydroxyl functional oils after reaction with aliphatic isocyanates [11].

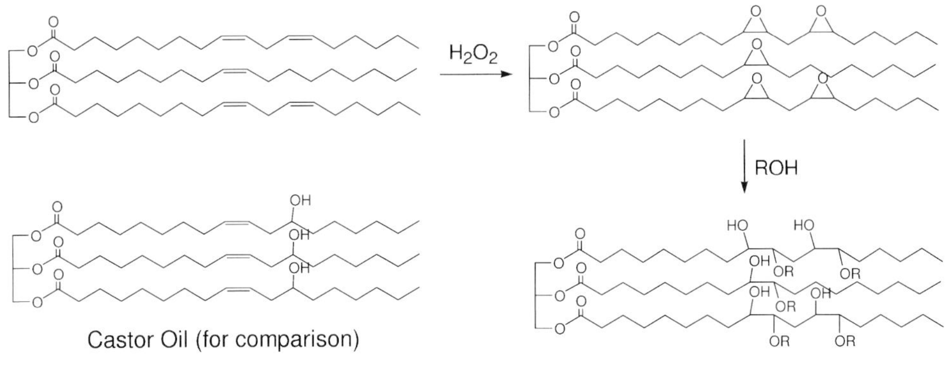

Fig. 4.7 Oleochemical polyols for polyurethanes (brand name: Sovermol®).

Oleochemical polyols have an average molecular weight of 250–2500 Da. Owing to their relatively low viscosity and their compatibility with methylene di(phenylisocyanate) (MDI) they are particularly suitable for solvent-free, two pack, full solids polyurethane systems, to be applied as thin decorative or protective coating by brush, roller or spraying. They can also be applied in thick coatings, bearing even high filler loads. In industrial flooring applications, self-leveling polyurethane or epoxy/polyurethane multilayer systems offer good chemical and mechanical properties and benefits such as minimal shrinkage, high mechanical strength and durability, and favorable cost of installation. They are broadly used for wear and crack resistant floorings on parking decks, for concrete protection in assembly areas, as well as in large kitchens, slaughterhouses and groceries, due to the ease of cleaning. Oleochemical polyols can also be used to bind porous filler materials, like perlite, and rubber particles for applications of composites in construction, soil protection, sport tracks, and playing fields [8, 11].

4.4.2
Biodegradable Fatty Acid Esters for Lubricants [15]

Apart from being used as "bio-diesel", fatty acid esters, which are obtained from fatty acids and alcohols, are becoming increasingly interesting as biodegradable replacements for mineral oils. In some application areas such as chain-saw oil, gearbox oils, hydraulic oils and lubricants for crude oil production these oleochemical products have already proved themselves.

Esters for lubricant applications are divided into five groups: monocarboxylic acid esters (monoesters), dicarboxylic acid esters (diesters), glycerol esters, polyol esters, and complex esters.

Monoesters are obtained by reacting carboxylic acids with alkyl chain lengths from C_8 to C_{22} and branched or linear alcohols. The typical diesters are obtained, for example, from adipic, sebacic, azelaic, or dimer fatty acids by reaction with butanol, ethyl hexanol, isodecanol, isotridecanol or Guerbet alcohol (Section 4.4.4). Using renewable resources as a base, sebacic acid is obtained from castor oil by oxidation with lead(II) oxide as catalyst. Azelaic acid and dimer fatty acids are obtained from oleic acid by technical processes; the first by ozone cleavage with pelargonic acid being formed as a coupling product, and the latter by thermal dimerization (see also previous section). Although the diesters already possess an excellent lubricating effect their thermal stability is surpassed by the polyol esters. These products are based on polyols with a quaternary carbon atom – for this reason glycerol esters form a separate class of products (Fig. 4.8). Complex esters are formed by esterification of polyols with mixtures of mono, di, and tricarboxylic acids and are oligomer mixtures, which from a technical application viewpoint are characterized by their high shear stability [15a].

The decisive fact is that the specially designed fatty acid esters that are used as replacements for mineral oil products not only have ecologically compatible properties but also a comparable or even better performance than that of conventional products. That this is possible can be demonstrated very clearly by an example

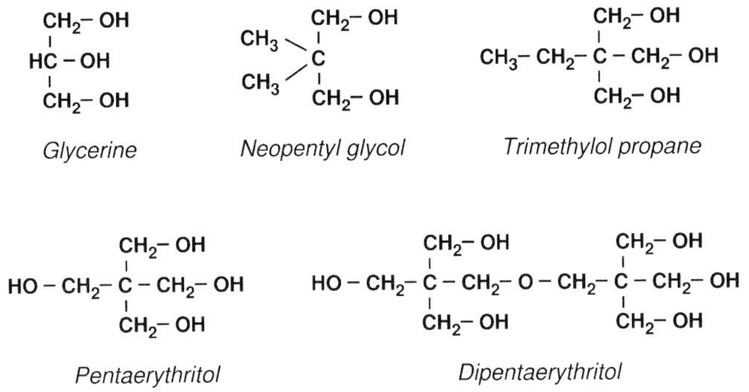

Fig. 4.8 Polyols used for the manufacturing of complex esters.

from the crude oil production sector. In coastal drillings (e.g., in the North Sea) the demands placed on the lubricants (drilling fluids) are particularly high. The drilling fluid is pumped to the surface together with the drill cuttings and after coarse separation disposed of directly into the sea. Apart from the good lubricating effect the biodegradability assumes a particular importance in this application. A specially developed fatty acid ester (Petrofree®) fulfils not only the requirements regarding biodegradability, but also has a better lubricating effect than products based on mineral oils [16].

Current developments refer to the use of specially designed fatty acid esters in a wide range of applications as biodegradable lubricants. Meanwhile, environmentally friendly alternatives are available for almost all mineral oil-based products. In Europe, the long-term potential is estimated to be 10–20% of the total market (500 000–1 000 000 tonnes year^{-1}, Table 4.4) [15b]. In 1997, 40 000 tonnes

Table 4.4 European potential market of biodegradable lubricants (1000 tonnes year^{-1}). According to Ref. [15b].

Application	Total	Biodegradable lubricants
Automotive oils	2305	250
Hydraulic oils	750	200
Turbine oils	200	20
Compressor oils	65	25
Industrial gear oils	200	10
Metal working oils	500	10
Demolding oils	110	110
Chain-saw oils	60	60
Process oils	600	200
Lubricating greases	100	100

of biodegradable lubricants were sold in Germany alone (4.5% of the total market) [15c]. An increase of this share is the aim of various measures taken by government and authorities. The success of these efforts will, finally, also depend on the statutory regulations that should govern the use of environmentally-friendly products [17].

4.4.3
Surfactants and Emulsifiers Derived from Vegetable Oil Based Fatty Alcohols and Fatty Acids

The basic way in which surfactants act is determined by their structure. With their hydrophilic head and hydrophobic tail, surfactant molecules interpose themselves between water and water-insoluble substances. By enriching themselves at the boundaries that water forms with air or oil they lower its surface tension; as ingredients in soaps and washing agents they make contact with soiled material in this way. When dissolved in water at higher concentrations these molecules group themselves together to form spherical structures (micelles); their inwards-pointing hydrophobic groups surround soil particles and keep these in solution. Surfactants are generally classified as being anionic, cationic, nonionic or amphoteric surfactants, depending on the type and charge of the hydrophilic groups (Fig. 4.9) [18a].

Surfactants are used in a wide range of fields. By far the most important application is the washing and cleansing sector as well as textile treatment and cosmetics; these use over 50% of the total amount of surfactants. Surfactants are also used in the food sector, in crop protection, in mining, and in the production of paints, coatings, inks, and adhesives. The basic manufacturing routes to important surfactants are laid out in Fig. 4.9. It is true that the most important surfactant from the amount produced apart from soap is still the petrochemical-based

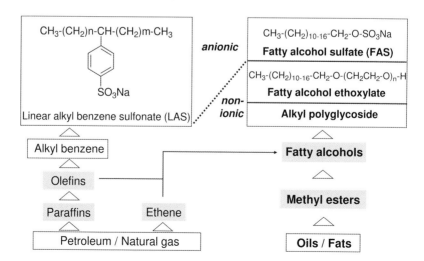

Fig. 4.9 Production of surfactants and examples of products.

Table 4.5 Global surfactant consumption in 2000[a]

Surfactant class	1000 tonnes
Soap	8800
Anionic surfactants	
Alkyl benzene sulfonates	3400
Fatty alcohol ether sulfates	1000
Fatty alcohol sulfate	500
Nonionic surfactants	
Alcohol ethoxylates	800
Alkylphenol ethoxylates	700
Cationic surfactants	810
Amphoteric surfactants	180
Others (incl. carbohydrate-based surfactants)	2900

[a] *Soap, Perf.&Cosm.*, November 2000, p. 51; Colin A. Houston & Associates, Inc.

alkyl benzene sulfonate; however, in recent years a continuous trend towards surfactants based on renewable resources has become apparent. The total worldwide market amounts to ca. 19.2 million tonnes (2000, incl. soap). The amounts involved, broken down into the individual surfactant classes, are summarized in Table 4.5, whereas Fig. 4.10 shows the regional distribution [18b].

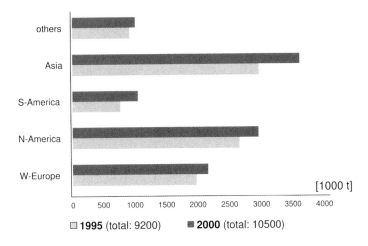

Fig. 4.10 Worldwide surfactant market.

4.4.3.1 Fatty Alcohol Sulfate (FAS) [5a, b]

Fatty alcohol sulfate has long been known and has been used as surfactant in various products, e.g., in detergents. It is produced directly from fatty alcohol by reaction with sulfur trioxide (SO_3) gas (1–8% v/v in air or nitrogen) in a falling-film reactor. The crude product is then neutralized with aqueous sodium hydroxide, using a buffer if necessary (Fig. 4.11). Certain product forms, such as aqueous solutions or granulates, are typically produced according to the requirements from the market. Fatty alcohol sulfates are readily biodegradable (no metabolites in the degradation) under both aerobic and anaerobic conditions.

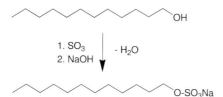

Fig. 4.11 Synthesis of fatty alcohol sulfate (FAS).

Since FAS can be produced either from vegetable oil based or petrochemical-based fatty alcohol (Fig. 4.9), both types have been evaluated in a life-cycle analysis with a positive overall result for the natural based product. With vegetable-based fatty alcohol sulfate, the analysis starts with the harvesting of the oil fruits (palm kernels or coconuts) and their processing to isolate the desired plant oil. Subsequent transesterification and hydrogenation of the methyl ester intermediates lead to the fatty alcohols, which are finally sulfated to produce the desired product. Based on this analysis the environmental impact of vegetable oil based fatty alcohol sulfate compared with the petrochemical based product is as follows:

- 70% less use of fossil resources,
- 50% less emissions to the atmosphere,
- 15% less waste,
- 50% more emission to water (low toxic waste water from small, decentralized oil plants).

4.4.3.2 Acylated Proteins and Amino Acids (Protein–Fatty Acid Condensates) [19]

In the development of the protein–fatty acid condensates it was possible to combine the renewable resources fatty acids (from vegetable oil) and protein, which can be obtained from both animal waste (leather) as well as from many plants, to construct a surfactant structure with a hydrophobic (fatty acid) and a hydrophilic (protein) part (Fig. 4.12). This was carried out by reacting protein hydrolysate with fatty acid chloride under Schotten–Baumann conditions using water as solvent. Products are obtained that have an excellent skin compatibility and, additionally, a good cleaning effect (particularly on the skin) and, in combination with other surfactants, lead to an increase in performance. For instance, even small additions of the acylated protein hydrolysate improve the skin compatibility. An

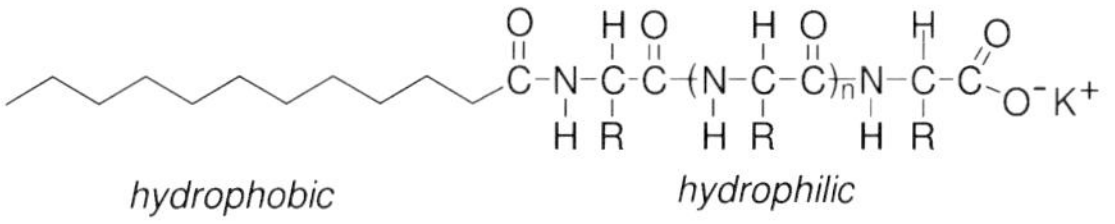

Fig. 4.12 Structure of protein hydrolyzate fatty acid condensates.

explanation for this protective effect could lie in the amphoteric behavior of the product. There is an interaction between the protein–fatty acid condensate and skin collagen. This could lead to the formation of a protective layer, which reduces the excessive attack of surfactants on the upper layers of the skin, their strong degreasing effect and the direct interaction of anionic surfactants with the skin. Comparable reaction conditions were applied to develop acylated amino acids – in particular acyl glutamates have found broad uses in recent years (Fig. 4.13).

In the personal care market, fatty acid derivatives of proteins and amino acids (glutamic acid) are mainly used in mild shower and bath products, mild shampoos, surfactant-based face cleansers, cold-wave preparations and fixatives, baby wash formulations, as well as special emulsifiers for leave-on products.

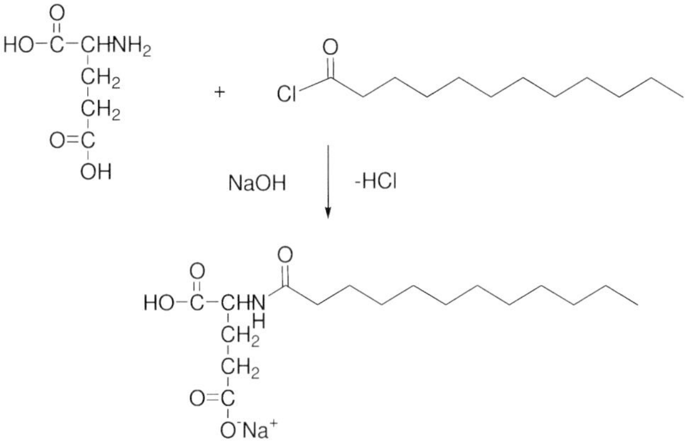

Fig. 4.13 Synthesis of acyl glutamate.

4.4.3.3 Carbohydrate-based Surfactants – Alkyl Polyglycosides [20]

The development of surfactants based on carbohydrates and oils is the result of a product concept based on the exclusive use of renewable resources. In industry, saccharose (sucrose), glucose and sorbitol, which are available in large amounts and at attractive prices, are used as the preferred carbohydrate raw materials.

The selective functionalization of saccharose and sorbitol with fatty acids for the construction of a perfect amphiphilic structure cannot be realized in simple technical processes because of the polyfunctionality of the molecule. This is why the products offered on the market contain different amounts of mono-, di- and tri-esters and are, therefore, only suitable for particular applications, e.g., as emulsifiers for foodstuffs and cosmetics or, in the case of the sorbitan esters, also in technical branches such as explosives and in emulsion polymerization.

The ideal raw material for selective derivatization is glucose. Reaction with fatty alcohol produces alkyl glucosides; N-methylglucamides are prepared by reductive amination with methylamine and subsequent acylation. Both products have proved to be highly effective surfactants in washing and cleansing agents. The alkyl glucosides have also additionally established themselves in the cosmetic products sector, as auxiliaries in crop protection formulations and as surfactants in industrial cleansing agents, and today can already be said to be the most important sugar surfactants based on the yearly production amounts.

Alkyl polyglycosides have long been known but only now, following several years' research, has it been possible to develop reaction conditions that allow manufacture on a commercial scale. The structure on which these compounds are based corresponds exactly to the surfactant model described above. The hydrophobic (or lipophilic) hydrocarbon chain is formed by a fatty alcohol (dodecanol/tetradecanol) obtained from palm kernel oil or coconut oil. The hydrophilic part of the molecule is based on glucose (dextrose) obtained from starch (Fig. 4.14).

The chemical challenge to process technology was to find reaction conditions that allowed fatty alcohol to react directly with glucose on a commercial scale and at an acceptable cost. To realize as environmentally-friendly a method as possible the use of solvents was rejected. Method development was successfully completed and Cognis was the first company to offer alkyl polyglycosides on an industrial scale at the required quality. Currently, Cognis has a capacity of ca. 50 000 tonnes year^{-1} available for the manufacture of this class of compounds (other manufacturers include BASF, Akzo Nobel, Kao, SEPPIC, and LG). By combining vegetable oil and sugar as raw materials it has for the first time become possible

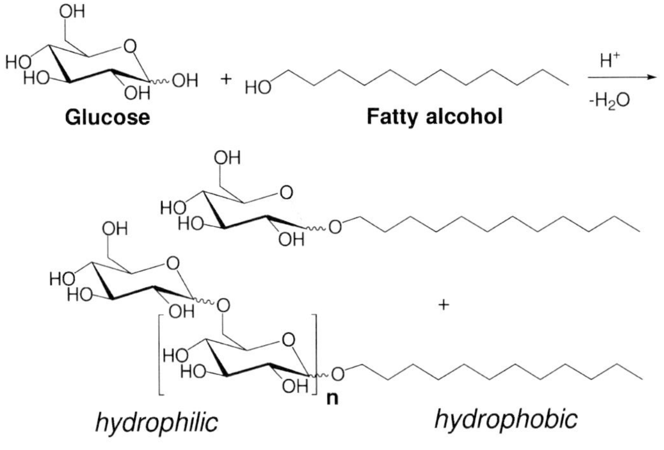

Fig. 4.14 Synthesis of alkyl polyglycosides.

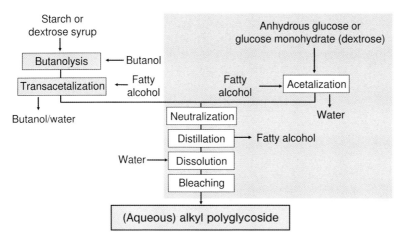

Fig. 4.15 Manufacturing processes for alkyl polyglycosides.

to offer commercially important amounts of nonionic surfactants that are completely based on renewable resources (Fig. 4.15).

Unique properties had previously been determined for alkyl polyglycosides, particularly in combination with other surfactants. For example, the use of alkyl polyglycosides in light-duty detergent or shampoo formulation means that the total amount of surfactants can be reduced without sacrificing any performance. In other combinations a particularly stable and fine foam can be produced that protects sensitive textiles during the washing process. Toxicological and ecological laboratory investigations have also produced favorable results. Alkyl polyglycosides have a good compatibility with the eyes, skin and mucus membranes and even reduce the irritant effects of surfactant combinations. On top of this they are completely biodegradable, both aerobically and anaerobically. The relatively favorable classification (for surfactants) into class I under the German water hazard classification (WGK I) results from this.

4.4.3.4 Alkyl Polyglycoside Carboxylate [21]

Alkyl polyglucoside carboxylate (INCI-name Sodium Lauryl Glucose Carboxylate (and) Lauryl Glucoside, Plantapon® LGC SORB) is a new anionic surfactant with excellent performance for personal care cleansing applications. In shampoo and shower bath formulations the anionic surfactant shows a good foaming behavior. In body wash applications it improves sensorial effects. These properties make Plantapon® LGC SORB suitable for several cosmetic applications, e.g., mild facial wash gel, mild baby shampoo, mild body wash for sensitive skin, wet wipes, and special sulfate-free shampoo applications.

A new industrial process based on the reaction of sodium monochloroacetate with aqueous alkyl polyglycoside (without additional solvents) enables the manufacturing of this product in an economically and ecologically favorable way (Fig. 4.16).

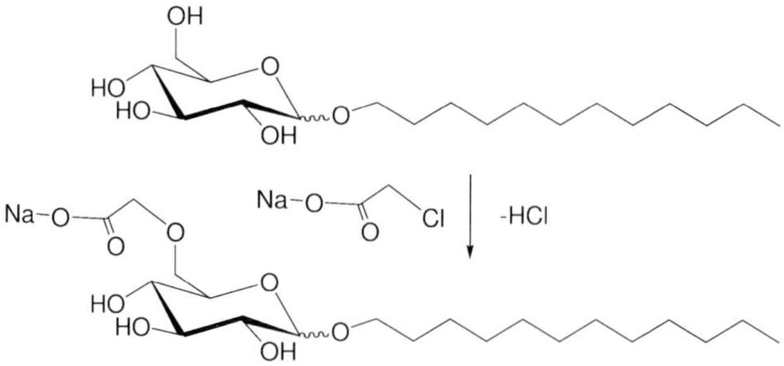

Fig. 4.16 Synthesis of alkyl polyglycoside carboxylate.

4.4.3.5 Polyol Esters [22]

Polyglycerol Esters Emulsifiers based on glycerol or polyglycerol are a class of products well known in the market and used particularly in products for personal care and the food area. Further development will focus on the design and optimization of specific emulsifier formulations. The combination of different types of emulsifiers can lead to new uses for mono- and diglycerides. Polyglycerol esters are obtained by esterification of polyglycerol, which is produced by the oligomerization of glycerol under basic conditions with fatty acids. The properties of the various products can be adjusted by the type of polyglycerol used, on one hand, and by the chain lengths and chain type of the fatty acid used on the other hand. Selected types of products with different molecular weight are shown in Table 4.6. For example, Dehymuls® PGPH is recommended to be used in body lotions. It

Table 4.6 Properties of selected polyglycerol esters as emulsifiers for personal care and food technology.

INCI name	Trade name	Properties
Polyglyceryl-3-diisostearate	Lameform® TGI	W/O – emulsifier; MW = 725; pale yellow liquid
Polyglyceryl-2-dipolyhydroxystearate	Dehymuls® PGPH	W/O – emulsifier for lotions and creams; MW $\Rightarrow$ 3000; yellow, cloudy, viscous
PEG-4-polyglyceryl-2-stearate	Lamecreme® DGE 18	O/W–emulsifier
Pentaerythritol distearate	Cutina® PES	Consistency wax with sensorial benefits
Polyglycerol stearate (E 475)	Polymuls® 4G	Food – emulsifier
Polyglycerol polyricinoleate (E 476)	Polymuls® PGPR	Food – emulsifier

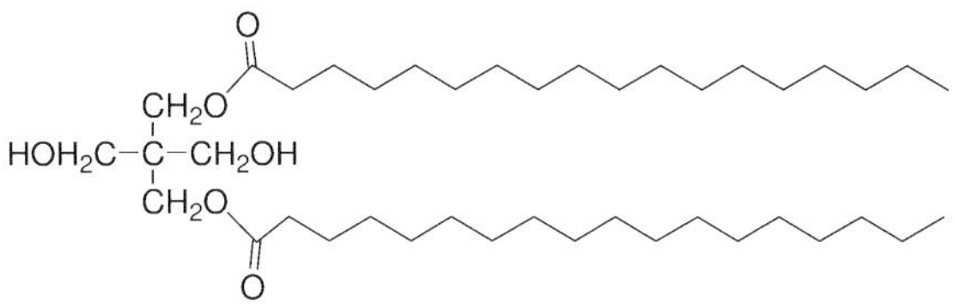

[Chemical structure of pentaerythritol distearate]

Fig. 4.17 Chemical structure of pentaerythritol distearate (main component in Cutina® PES).

leaves the skin with a smooth, non greasy, and well cared feeling, spreads easily, and absorbs quickly.

Pentaerythritol Ester   As with glycerol esters, the esters are produced by esterification of pentaerythritol with the desired fatty acids. For example, under defined reaction conditions and use of stearic acid in defined concentration, pentaerythritol distearate has been recently developed as an off-white wax with very weak odor (Cutina® PES). This type of product is offered as co-emulsifier and consistency factor for cosmetic products with high sensorial elegance and can be applied in various formulations (Fig. 4.17).

Emulsifier Compound Based on Polyglycerol Ester and Alkyl Polyglycoside   The requirements for modern emulsifiers not only include outstanding performance but also compatibility with modern emulsification techniques and balanced sensory feeling. One product that fulfils these requirements is a compound based on glycerine, alkyl polyglycoside and polyglyceryl-2-dipoly-hydroxystearate (Eumulgin® VL 75). In combination with selected emollients it allows the preparation of O/W emulsions with high quality and stability (small droplet sizes). In addition, due to the liquid appearance of the product, and, as a consequence, the possibility for cold processing, the manufacturing time and costs for the preparation of emulsions is significantly reduced (Fig. 4.18).

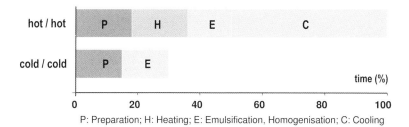

Fig. 4.18 Capacity increase by cold emulsification processes using Eumulgin® VL 75 as emulsifier.

4.4.3.6 Multifunctional Care Additives for Skin and Hair [23]

The intelligent combination of alkyl polyglycoside and glyceryl oleate resulted in a new product that combines emulsifying and cleansing properties with outstand-

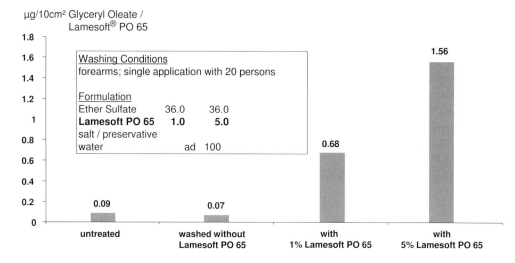

Fig. 4.19 Evaluation of lipid layer enhancing effect.

ing care effects, such as enhancing of the skin lipid layer. This effect is proven in a standardized test by washing the fore arm, rinsing, drying and extracting the lipid layer on the skin with ethanol pads. The lipid content is measured by quantitative analysis of glycerol oleate in the extract (Fig. 4.19). Table 4.7 summarizes technical data of the product (Lamesoft® PO 65).

For applications in hair care products it was found that a compound based on dioctyl ether from coconut- or palm-kernel oil based octanol (Cetiol® LDO) enables the formulation to be silicon oil free. Cetiol® LDO is a highly efficient hair care additive and is particularly suitable for the use in hair cleansing preparations to improve the tactile hair feeling and hair gloss. Cetiol® LDO, in combination with wax esters and cationic polymers, also offers benefits in silicon-free body cleansing preparations with regard to improved sensorial skin properties.

Table 4.7 Technical data of Lamesoft® PO 65.

Application	Multifunctional care additive for clear and pearlescent cleansing preparations
Composition	Glyceryl oleate (lipid) Coco glucoside (surfactant)
Dry residue	65–70%
Lipid	ca. 30%
pH (5%)	3.0–3.5
Viscosity (Brookfield, 23 °C)	max. 12000 mPa s

Table 4.8 Selected emollients.

Structure	Emollient	INCI name[a]
Ester	Cetiol® A	Hexyl laurate
	Cetiol® LC	Coco-caprylate
	Eutanol® G 16 S	Hexyldecyl stearate
	Cetiol® V	Decyl oleate
	Cetiol® J 600	Oleyl erucate
	Myritol® 318	Caprylic/capric triglyceride
Guerbet alcohols	Eutanol® G 16	Hexyl decanol
	Eutanol® G	Octyl dodecanol
Hydrocarbons	Cetiol® S	Diethylhexyl-Cyclohexane
Ethers	Cetiol® OE	Dicaprylyl ether
Carbonates	Cetiol® CC	Dicaprylyl carbonate

[a] "Capryl" is a trade name for "octyl".

4.4.4
Emollients [24]

The physicochemical nature of the oil phase components in a cosmetic emulsion, the emollients, determines the skin-care effects, such as smoothing, spreading, sensorial appearance. Test methods have been developed to characterize and classify the numerous emollients available on the market, such as silicones, paraffins, and oleochemical-based products. The latter include glycerides, esters, alcohols, ethers, and carbonates with tailor-made structures, depending on the performance needed (Table 4.8). However, especially with regard to additional effects, there is still a demand for new products with unique performance properties.

4.4.4.1 Dialkyl Carbonate

One example of a new class of compound in this area is dioctyl carbonate (**1**, Table 4.9). The product is synthesized by the trans-esterification reaction of octanol and dimethyl carbonate in the presence of alkali catalyst (Fig. 4.20). Dioctyl carbonate (Cetiol® CC) is a dry emollient with excellent dermatological compatibility and a comprehensive and convincing performance profile for various appli-

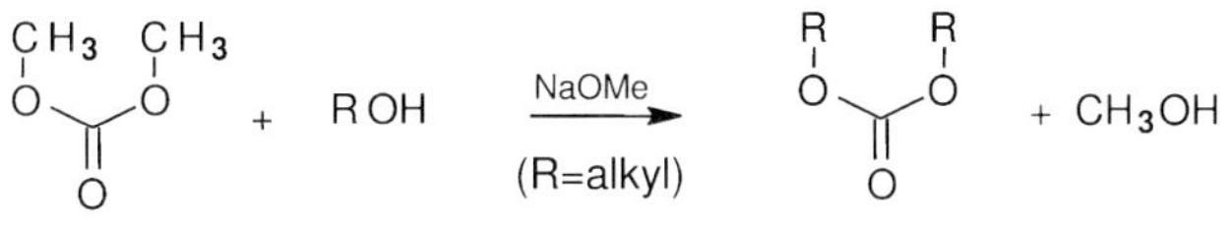

Fig. 4.20 Synthesis of dialkyl carbonates.

4 Industrial Development and Application of Biobased Oleochemicals

Table 4.9 Main properties of dioctyl carbonate.

Chemical structure	
	(structure 1)

	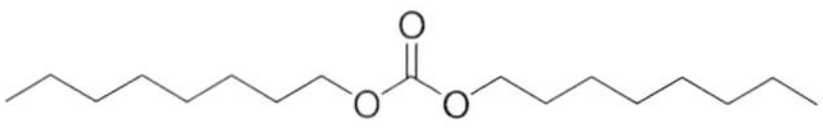
INCI	Dicaprylyl carbonate
Spreading value	1600 mm² in 10 min
Sensory feeling	Dry, volatile silicon oil like
Skin compatibility	Excellent
Stability	Hydrolysis stable
Origin	Vegetable oil (fatty alcohol)
Biodegradable	Yes

cations in the personal care segment. Described are the good emulsifiability, the outstanding behavior in deo/AP formulations, the solubilizing and dispersing ability for sun care and last but not least the special sensorial feeling, which is provided to the final formulation (Table 4.9).

4.4.4.2 Guerbet Alcohols [25]

In 1899 R. C. Guerbet discovered the self-condensation reaction of alcohols, which, via the aldehyde as an intermediate, lead to branched structures (2-alkyl alcohols) (Fig. 4.21) – the Guerbet alcohols. Starting with fatty alcohols from vegetable sources, such as octanol and decanol, the corresponding C_{16} and C_{20} alcohols are produced (2-hexyldecanol and 2-octyldecanol, respectively). The reaction is carried out under alkali catalysis and high temperatures (>200 °C). Over the years, both products have proven to be efficient emollients, but are also used for other applications, such as plasticizers or components for lubricants (Fig. 4.21).

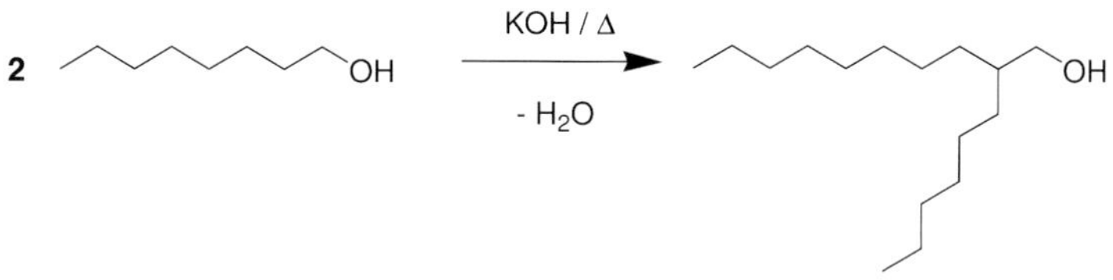

Fig. 4.21 Synthesis of the Guerbet alcohol 2-hexyldecanol (Eutanol® G 16).

4.5
Perspectives

The successful development of environmentally compatible and powerful products in the sense of a sustainable development has been demonstrated by various examples of recent product innovations from oleochemistry. It can be assumed that, in future, further possibilities for using renewable resources will be intensely investigated. Here the combination of different types of vegetable raw materials to form new products and intelligent product concepts to meet market and consumer needs will be a challenge for research and development. However, the future availability of natural fats and oils as the base raw materials for oleochemistry will depend on the scope of implementation of the Bioenergy and Biofuel Strategy as part of the Kyoto protocol, too. The subsidized use of vegetable oils for bioenergy and biofuel production is completely contradictory to their established use in nutrition and to future developments of oleochemical-based products with high value added for industrial use. Therefore, edible oils and fats should not be part of this biomass regulation and subsidies for biofuel and bioenergy in general should be more flexible [26].

Acknowledgments

APG, Cetiol, Cutina, Dehymuls, Emulgade, Eumulgin, Eutanol, Lamecreme, Lameform, Lamesoft, Myritol, Polymuls, Sovermol, and Tribest are registered trademarks of Cognis; Petrofree is a registered trademark of Baroid Drilling Inc.

References

1 (a) H. Baumann, M. Bühler, H. Fochem, F. Hirsinger, H. Zoebelein, J. Falbe, *Angew. Chem. Int. Ed. Engl.* **1988**, 27, 41; (b) A. Behler, M. Biermann, N. Huebner, L. Zander, A. Westfechtel, 95th AOCS-Meeting, Cincinnati, USA, **2004**.

2 H. Eierdanz (Ed.), *Perspektiven nachwachsender Rohstoffe in der Chemie*, VCH, Weinheim, **1996**.

3 (a) B. Brackmann, C. Hager, The statistical world of raw materials, fatty alcohols and surfactants, Proceedings 6th World Surfactant Congress, Cesio 2004, Berlin, June 20–23, **2004**; (b) P. Renaud, Natural-based fatty alcohols: Completely in line with the future?, Proceedings 6th World Surfactant Congress, Cesio 2004, Berlin, June 20–23, **2004**; (c) G. Kreienfeld, G. Stoll, Surfactants in consumer products and raw material situation – a brief survey. In: K. Hill, W. von Rybinski, G. Stoll (Eds.), *Alkyl Polyglycosides – Technology, Properties and Applications*, VCH, Weinheim, **1997**, p. 225.

4 (a) G. Röbbelen, Pflanzliche öle als rohstoffbasis – Potential und veränderungen in der verfügbarkeit. In: *Tagungsband 3. Symposium Nachwachsende Rohstoffe – Perspektiven für die Chemie, Schriftenreihe des Bundesministeriums für Ernährung, Landwitschaft und Forsten*, Landwirtschaftsverlag, Münster, **1994**, p. 115; (b) K. Hill, *Agro-Food-Industry Hi-Tech*, **1998**, 9(5), 9.

5 (a) F. Hirsinger, F. Bunzel, Ökobilanz von Fettalkoholsulfat – Petrochemische versus oleochemische Rohstoffe. In: H. Eierdanz (Ed.), *Perspektiven Nachwachsender Rohstoffe in der Chemie*, VCH, Weinheim, **1996**, p. 228; (b) M. Stalmans, et al., *Tenside Surf. Det.*, **1995**, 32, 84; (c) M. K. Patel, A. Theiß, E. Worrell, *Resources, Conservation and Recycling*, **1999**, 25, 61.

6 J. Steber, *Textilveredlung* **1991**, 26, 348.

7 R. Höfer, Anwendungstechnische aspekte der verwendung natürlicher öle und ihrer derivate in der polymer-synthese und-verarbeitung, In: H. Eierdanz (Ed.), *Perspektiven Nachwachsender Rohstoffe in der Chemie*, VCH, Weinheim, **1996**, pp. 91–106.

8 (a) B. Schulte, B. Schneider, Linoleum: Traditionelle und moderne problemlösung für den fußboden auf basis nachwachsender rohstoffe, In: H. Eierdanz (Ed.), *Perspektiven Nachwachsender Rohstoffe in der Chemie*, VCH, Weinheim, **1996**, pp. 338–344; (b) M. W. Formo, Industrial use of soybean oil, in *Proceedings of the 21st World Congress of the International Society of Fat Research (ISF)*, The Hague, P. J. Barnes & Associates, Bridgewater, 1995, pp. 519–527.

9 (a) D. Schäfer, Einsatz und potential naturfaserverstärkter kunststoffe in der automobilindustrie, in *Gülzower Fachgespräche: Nachwachsende Rohstoffe – Von der Forschung zum Markt*, Workshop 25./26.05.1998, Fachagentur Nachwachsende Rohstoffe e.V., Gülzow, **1998**, pp. 27–47; (b) B. Dahlke, H. Larbig, H. D. Scherzer, R. Poltrock, *J. Cell. Plastics*, **1998**, 34, 361; (c) J. Neubauer, Der Schritt in die industrialisierung – Nachwachsende rohstoffe im schienenfahrzeugbau, Conference presentation at Konstruktionswerkstoffe aus nachwachsenden Rohstoffen, Sept. **2002**, Braunschweig.

10 (a) R. Fayter, Technical reactions for production of oleochemical monomers, In: H. Eierdanz (Ed.), *Perspektiven Nachwachsender Rohstoffe in der Chemie*, VCH, Weinheim, **1996**, pp. 107–117; (b) D. L. Craft, R. C. Wilson, D. Eirich, Y. Zhang, Candida CYP52A2A promoter and uses in increasing gene expression in yeast for dicarboxylic acid production, PCT Int. Appl. W02002008412, **2002** (Cognis).

11 R. Höfer, J. Bigorra, *Green Chem.*, **2004**, 6, 418.

12 A. Heidbreder, R. Höfer, R. Grützmacher, A. Westfechtel, C. W. Blewett, *Fett/Lipid*, **1999**, 101, 418.

13 J. Möschel, Thermoplastische polyurethane sowie ihre verwendung, DE-PS 19512310, **1995** (Parker-Prädifa).

14 (a) B. Gruber, R. Höfer, H. Kluth, A. Meffert, *Fat Sci. Technol.*, **1987**, 89, 147; (b) P. Daute, R. Gruetzmacher, R. Höfer, A. Westfechtel, *Fat Sci. Technol.*, **1993**, 95, 91.

15 (a) F. Bongardt, A. Willing, *J. Synth. Lubricat.*, **2003**, 20–1(April), 53; (b) J. Legrand, K. Dürr, *Agro-Food-Industry Hi-Tech*, **1998**, 9(5), 16; (c) Th. Mang, *Fett/Lipid*, **1998**, 100, 524.

16 C.-P. Herold (Cognis Deutschland GmbH & Co.KG), personal communication.

17 Bundesministerium für Ernährung, Landwirtschaft und Forsten, *Bericht über den Einsatz Biologisch Schnell Abbaubarer Schmierstoffe und Hydraulikflüssigkeiten und Maßnahmen der Bundesregierung*, Bonn, Germany, **1996**.

18 (a) J. Falbe (Ed.), *Surfactants in Consumer Products: Theory, Technology, Applications*, Springer, Heidelberg, **1987**; (b) W. Dolkemeyer, Surfactants on the Eve of the Third Millennium, 5th World Surfactant Congress, Cesio 2000, Florence May 29–June 02, 2000.

19 A. Sander, E. Eilers, A. Heilemann, E. von Kries, *Fett/Lipid*, **1997**, 99, 115.

20 (a) M. Biermann, K. Schmid, P. Schulz, *Starch/Stärke*, **1993**, 45, 281; (b) J. Knaut, G. Kreienfeld, *Chim. Oggi*, **1993**, 41; (c) K. Hill, W. von Rybinski, G. Stoll (Eds.), *Alkyl Polyglycosides – Technology, Properties and Applications*, VCH, Weinheim, **1997**; (d) W. von Rybinski, K. Hill, *Angew. Chem. Int. Ed.*, **1998**, 37, 1328.

21 A. Behler, W. Hensen, W. Seipel, Alkyl Polyglycoside Carboxylate – A New Anionic Surfactant, Proceedings 6th World Surfactant Congress, Cesio 2004, Berlin, June 20–23, 2004.

22 C. Mitchell, A. Ansmann, S. Bruening, U. Issberner, S. Nefkens, Formulation

design by texture modifications, in *Conference Proceedings 51ˢᵗ SEPAWA Kongress*, Würzburg, SEPAWA-Vereinigung der Seifen-, Parfüm- und Waschmittelfachleute e.V., Ludwigshafen, **2004**, p. 283.

23 (a) T. Morris, M. Hansberry, W. Seipel, C. Nieendick, *Cosmetics & Toiletries*, **2004**, 119(5), 79; (b) W. Seipel, N. Boyxen, Moderne formulierungskonzepte mit care-effekten, in Conference Proceedings 51ˢᵗ SEPAWA Kongress, Würzburg, SEPAWA-Vereinigung der Seifen-, Parfüm- und Waschmittelfachleute e.V., Ludwigshafen, **2004**, p. 70.

24 (a) H. Tesmann, Nachwachsende rohstoffe in der kosmetik, in Ref. [2], pp. 31–39; (b) R. Kawa, A. Ansmann, B. Jackwerth, M. Leonard, *Parfüm. Kosmet.*, **1999**, 80, 17; (c) Th. Förster, U. Issberner, H. Hensen, *J. Surfact. Detergents*, **2000**, 3, 345; (d) B. Jackwerth, Cetiol® CC – The new benchmark for dry emollients, In-Cosmetics **2000**, Barcelona, April 2000.

25 (a) R. C. Guerbet, C. R. Hebd, *Seances Acad. Sci.*, **1899**, 128, 5118; (b) K. S. Markley, *Fatty Acids* (Vol. 2), Interscience Publishers, New York, **1961**, p. 1353.

26 H. Sauthoff, *Bioenergy and Biofuels – Opportunity or Threat for Oils and Fats Trading?*, Handout FOSFA "Contact Day", London, September 8, **2005**.

5
Fine Chemicals from Renewables

Herman van Bekkum and Leendert Maat

5.1
Introduction

For a long time, natural products have guided chemists to most interesting and complex structures [1] and inspired them to numerous new chemical reactions. Apart from their immense value as nutrients and building materials, they play a major role in man's daily life, particularly as medicines, fibers, flavors, fragrances, coloring materials and, increasingly, as building blocks for other useful chemicals. The term renewable is used when the natural sources are cultivated and regularly harvested.

In the search for unique properties, chemists have isolated pertinent compounds, for which they have revealed the structures and in many cases developed syntheses. For many, in general, simpler compounds, industrially feasible preparations have also been worked out. However, for most of the more complex structures, nature proves to be much more efficient and cheaper, especially in the field of chiral compounds. Therefore, nature is still the supplier of many natural products, which are useful as such or as starting materials for other chemicals and auxiliaries in new chemical reactions.

Not long ago the term "natural products" was used to cover only products obtained from trees, plants, and animals. However, nowadays the term also covers products made by microorganisms (bacteria, yeasts, fungi) from natural carbon sources such as glucose (denoted as fermentation processes). The qualification "natural" is also allowed for products obtained by enzymatic conversion of natural products and for chemicals and materials produced by genetically modified plant cells. Recently, the names white, red and green biotechnology have been introduced to indicate industrial, healthcare, and plant biotechnology, respectively.

Classical examples of industrial biotechnology include the manufacture of ethanol, lactic acid, citric acid, and glutamic acid. The share of renewables in the feedstock of the chemical industry is expected to increase substantially in the years to come [2–4]. A newcomer here is propane-1,3-diol (DuPont/Tate & Lyle), with the start-up of industrial fermentation foreseen within one year.

Catalysis for Renewables: From Feedstock to Energy Production
Edited by Gabriele Centi and Rutger A. van Santen
Copyright © 2007 WILEY-VCH Verlag GmbH & Co. KGaA, Weinheim
ISBN: 978-3-527-31788-2

In particular, in the field of fine chemicals, natural products – in the broadest sense – will serve as functional molecules and are often appropriate starting structures for other target systems.

A fine example of such a semi-synthesis is the preparation of the anti-cancer drug paclitaxel (Taxol ®), a relatively scarce compound from *Taxus brevifolia*. Here, the natural and better accessible 10-deacetylbaccatin III, isolated from the leaves of *Taxus baccata*, provides the complicated ring system of paclitaxel, including all substituents with the right stereostructure (Scheme 5.1). In just four reaction steps [5] paclitaxel is obtained from 10-deacetylbaccatin III.

10-Deacetylbaccatin III   Paclitaxel

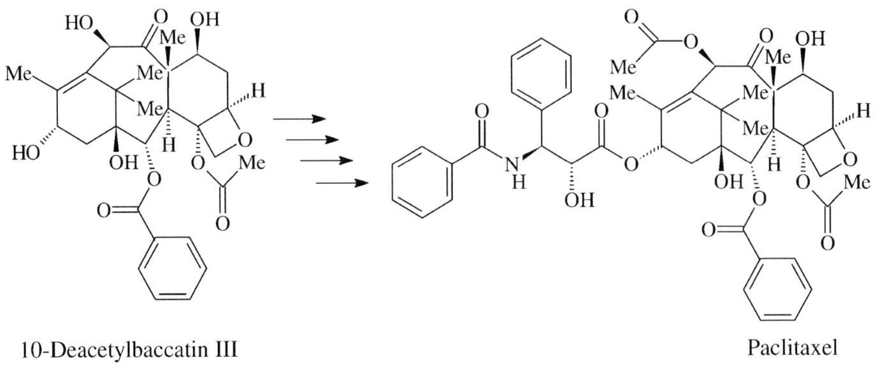

Scheme 5.1 Synthesis of paclitaxel from 10-deacetylbaccatin III.

Sometimes, natural supplies and synthetic production lines are in competition. An example is β-carotene (**1**), traditionally used as food colorant, but nowadays increasingly applied as a health ingredient: anti-oxidant and vitamin A precursor [6].

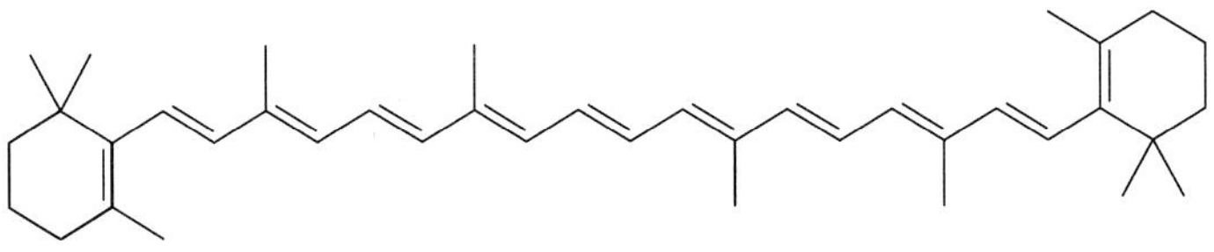

β-Carotene (**1**)

Four industrial β-carotene approaches are in operation:

1. Processing crude palm oil (contains 500–700 ppm carotenoids) to β-carotene concentrates [7].
2. Isolation from the alga *Dunabiella salina*.
3. Fermentation using the fungus *Blakeslea trispora* [8].
4. Organic synthesis.

In the organic synthesis (Roche, BASF) two C_{20} molecules, retinal and retinulidene-triphenylphosphorane, are coupled according to the Wittig reaction to give β-carotene in high yield [9]. A disadvantage is that triphenylphosphine oxide, formed in a stoichiometric amount, cannot be recycled economically to triphenylphosphine.

As to the all-*trans* β-carotene content, the synthetic and fungal preparations are by far the purest [8]. The products from algae and palm oil contain substantial amounts of *cis*-β-carotene (the central double bond has a cis configuration) and of α-carotene (ring double bond shifted one position). The DSM fermentation process seems to be in a good position here.

Sometimes natural fine chemicals are by-products in bulk products refining. Examples are: (a) lecithin and steroids in vegetable oil refining; (b) betaine, pectin and raffinose in sugar manufacture; (c) quinic acid in quinine extraction of the bark of *Cinchona* trees; (d) chitin and the red pigment asthaxanthin in lobster and shrimp processing; and (e) lanolin, lanosterol and cholesterol in sheep wool purification.

Many more preparations of fine chemicals from renewables have been worked out, and even more are being developed. The following sections present several examples that catch the eye. Examples that are now an integral part of modern chemistry, with a future in which the use of chemicals that are harmful to the environment will be avoided and where novel processes will minimize the amounts of waste products.

5.2
Vanillin

An important flavoring compound is vanillin (**2**) with an estimated total production of about 14 000 t a^{-1}.

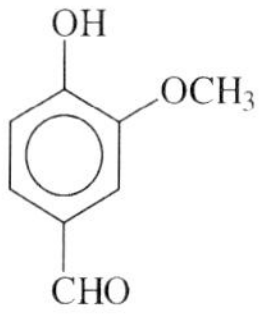

Vanillin (**2**)

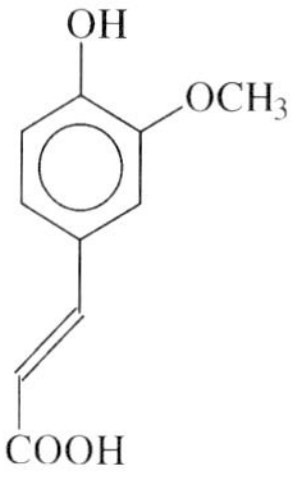

Ferulic acid (**3**)

Natural vanilla beans are expensive and the vanillin content is just 1.5–2 wt.%. Harvests of the beans fluctuate, as do the prices. The world production of vanilla beans in 2005 [10] was 1275 t (of which 700 t alone was by Madagascar/Malagasy). In 2004 the production was 1975 t. Halfway through 2004 the price was around $60 kg^{-1}$; however, at the end of 2003 the price of vanilla beans had peaked at $500 kg^{-1}$. Upon extraction of the cured beans a viscous liquid is obtained with a rich odor and flavor profile, and with many applications.

The Norwegian company Borregaard makes vanillin by chemo-oxidation of cheap natural lignin. In view of the complexity of the lignin structure this is a remarkable achievement. This semi-synthetic vanillin is, however, not allowed to carry the qualification of "natural". Borregaard is the only global producer of lignin-vanillin or "wood-sourced vanillin", with a volume of about 2000 t a^{-1}. They obtain 4 kg vanillin from one ton of wood. This vanillin is more expensive than synthetic vanillin, but the flavorist needs lower doses of the "extra round flavor".

Recently, a two-step enzymatic conversion of ferulic acid (3) leading to natural vanillin was disclosed, with a price indication of about $700 kg^{-1}$.

Several syntheses exist for vanillin. A process recently developed by Rhodia seems to be superior [11]. The process (Scheme 5.2) involves four catalytic steps starting from phenol: aromatic ring hydroxylation, O-methylation, hydroxymethylation, and oxidation. The process combines elegance and precision in organic synthesis.

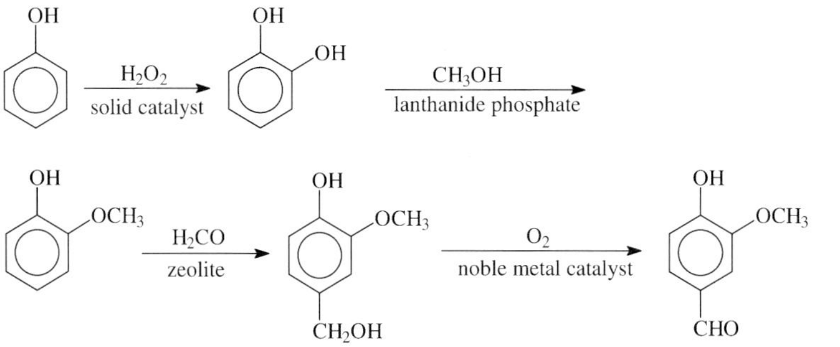

Scheme 5.2 Rhodia vanillin process.

Catechol, the first oxidation product, can also be prepared starting from glucose [12], which classifies the new vanillin process as green in two ways.

The price of synthetic vanillin has dropped to about $10 kg^{-1}$ due to increased production in China, which is now the number one supplier. In view of the very large price difference, analytical methods (isotope analysis) have been developed to distinguish between natural and synthetic vanillin.

Vanillin is the starting point for several chemicals. Thus capsaicin, the pungent principle of red pepper (*Capsicum annuum*), used in pepper-spray and pain killing ointments, is made in two steps from vanillin.

5.3
Monoterpenes

From their carbon skeletons terpenes can be seen as isoprene (C_5) oligomers. Terpenes are classified according to their number of isoprene units as: monoterpenes (C_{10}), sesquiterpenes (C_{15}), diterpenes (C_{20}), etc. β-Carotene (**1**) is a tetraterpene.

Many oxygenated monoterpenes (alcohols, carbonyl compounds, esters) serve as fragrances. Here inexpensive natural starting compounds are α-pinene (**4**), β-pinene (**5**) and limonene (**6**), with production volumes of about 18 000, 12 000, and 30 000 t a^{-1}, respectively.

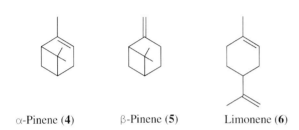

α-Pinene (**4**)   β-Pinene (**5**)   Limonene (**6**)

The two pinenes are obtained from Crude Sulfate Turpentine (CST), which is a side product of the sulfate cellulose process from pine trees. Limonene is present in orange and lemon peels [which provide different enantiomers(!)], and is a cheap by-product of the citrus industry.

Catalytic conversions in the monoterpene field have been reviewed recently [13–15]. There is an ongoing transition from conventional homogeneous catalysts (mineral acids, zinc halides) to solid Brønsted and Lewis acid catalysts. Thus, limonene can be alkoxylated with lower alcohols using zeolite H-Beta as the catalyst [16] at room temperature already, with high selectivity and conversion (Scheme 5.3). The alkoxy compounds are applied as fragrances with, dependent on the length of R, characteristic odors.

Scheme 5.3 Alkoxylation of limonene.

Heteropoly acids such as $H_3PW_{12}O_{40}$ (PW) are good catalysts for the hydration of limonene and other monoterpenes. PWs can be used as homogeneous catalysts in solution or supported on, for example, silica or MCM-41 materials. In aqueous acetic acid limonene gives, in the presence of PW, mainly α-terpineol (**7**) and α-terpinyl acetate (**8**) [17].

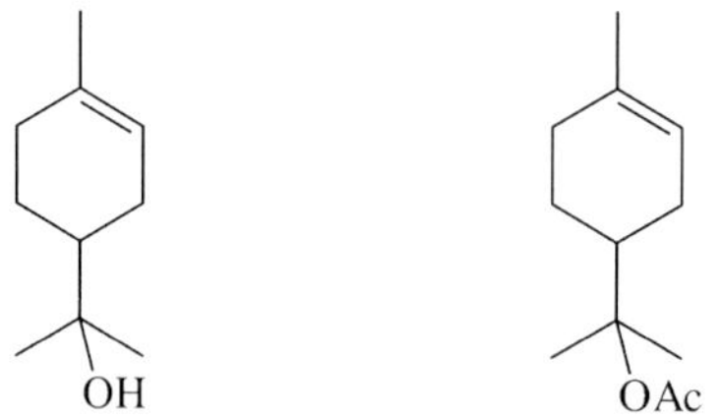

α-Terpineol (**7**)   α-Terpinyl acetate (**8**)

A quite different limonene conversion is the dehydrogenation into *p*-cymene, thus giving a "green" aromatic. Pd-catalysts give yields of >95% [18]. *p*-Cymene can be oxidized to the hydroperoxide, which can be rearranged to *p*-cresol, a valuable chemical.

The chemical networks around α- and β-pinene are versatile. The four-membered ring is strained; upon H^+-catalysis an alkyl shift takes place and the bornyl cation is formed, which can enter various reactions. In contrast, the four-membered ring remains intact upon epoxidation or hydrogenation.

Upon thermal treatment the four-membered ring transforms into a diene. Thus, β-pinene is industrially converted into the noncyclic triene myrcene. This type of ring opening also plays a role in a semi-synthetic route to linalool starting from α-pinene (Scheme 5.4).

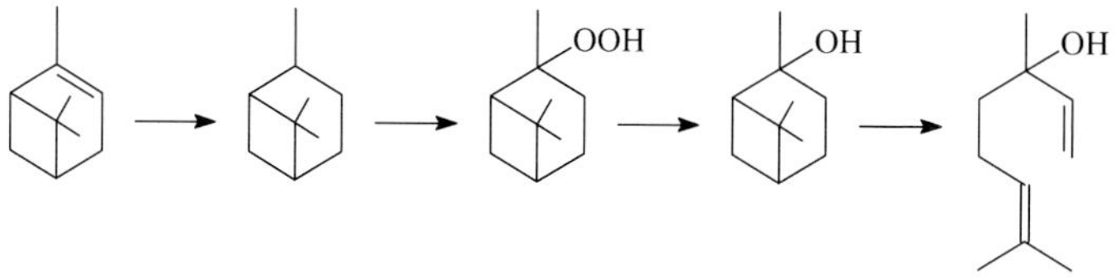

α-Pinene (**4**)   Linalool

Scheme 5.4 Linalool synthesis from α-pinene.

Linalool is an important fragrance and fragrance precursor (esters) with an annual production of over 15 000 t. Some 50% is made by semi-synthesis from α- and β-pinene, the other part is made synthetically starting from isobutene via 6-methylhept-5-en-2-one (**9**). Addition of an acetylene fragment followed by partial hydrogenation (Pd) leads to linalool.

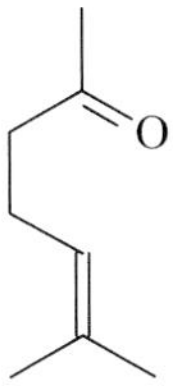

6-Methylhept-5-en-2-one (**9**)

Another large volume monoterpene is (−)-menthol, a compound that belongs to a family of eight stereoisomers. Only (−)-menthol (1R,2S,3R-configuration) possesses the characteristic peppermint odor and exerts the unique cooling sensation on the skin. Most (−)-menthol is obtained by freezing peppermint or cornmint oil, followed by recrystallization. Besides this natural menthol some 20% or 3000 t a^{-1} is made by (semi)-synthesis.

The synthetic process starts with the isopropylation of *m*-cresol to yield thymol. After catalytic hydrogenation a mixture of stereoisomers is obtained from which (±)-menthol is isolated. The process requires much separation and recycling work. In contrast, the semi-synthetic process of Takasago (Scheme 5.5) leads essentially to stereopure (−)-menthol [19].

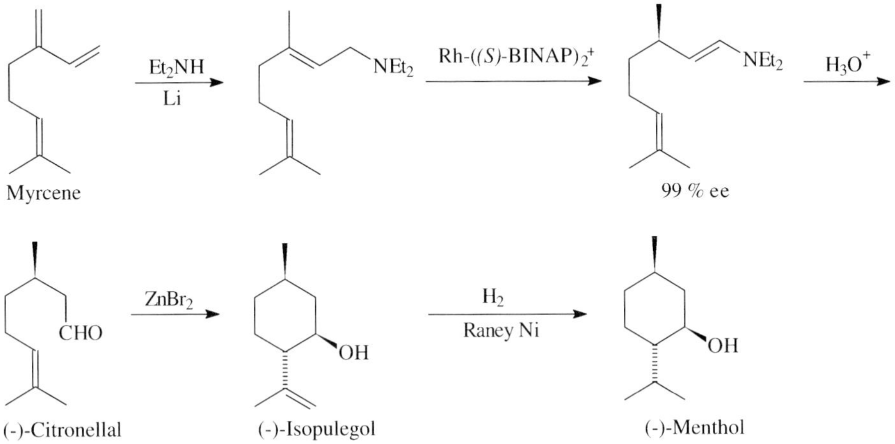

Scheme 5.5 Takasago (−)-menthol process.

Key steps are the enantioselective isomerization of diethylgeranylamine and the diastereoselective ring closure of (+)-citronellal to (−)-isopulegol. Mechanistically, the isomerization is well understood. As to the cyclization, notably, the most stable (all-equatorial) diastereoisomer is formed.

In the Takasago process ZnBr$_2$ is used as the Lewis acid catalyst for the ring closure. Recently, zeolite Sn-Beta has been reported [20] as a heterogeneous and recyclable catalyst. Moreover, the two final steps – ring closure and hydrogenation – can be combined by applying H-Beta-supported iridium as the catalyst [21]. Even a three-step cascade: citral–citronellal–isopulegol–menthol has been reported [22]. Also geraniol can be converted into menthol in a "one pot" synthesis (Cu/Al_2O_3 catalyst) [23]. The first step is then isomerization of the allylic alcohol fragment, leading to citronellal. Notably, geraniol and citronellal are the main components of citronella oil.

Another opportunity to combine two reaction steps towards a "one pot" synthesis is the epoxidation of α-pinene and the isomerization of the epoxide to campholenic aldehyde (Scheme 5.6). Zeolite Ti-Beta seems adequate to deal with both steps as a catalyst [24]. Campholenic aldehyde is the starting material for several sandalwood fragrances.

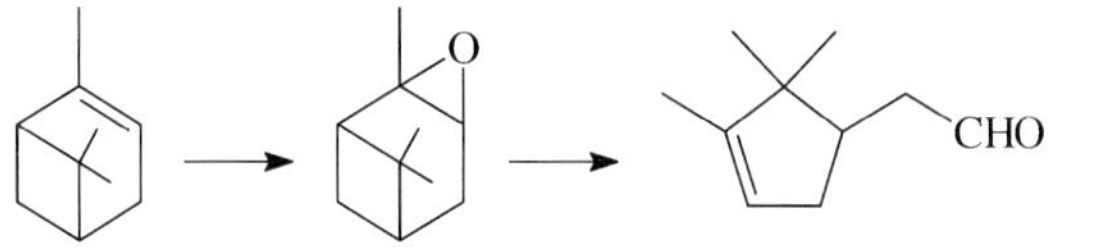

α-Pinene (4)   Campholenic aldehyde

Scheme 5.6 Conversion of α-pinene into campholenic aldehyde.

Finally, we mention here recent progress made in the Bayer–Villiger oxidation. Zeolite Sn-Beta (1.6 wt.% Sn) was found to be an excellent catalyst [25]. Thus, the monoterpene dihydrocarvone gives – with Sn-Beta and H_2O_2 – exclusively the lactone (Scheme 5.7), whereas m-chloroperbenzoic acid and Ti-Beta/H_2O_2 give the epoxide as the main product.

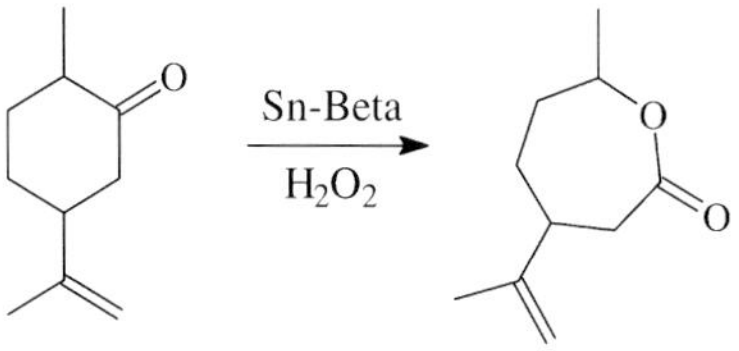

Scheme 5.7 Bayer–Villiger oxidation of dihydrocarvone.

5.4
Alkaloids

Alkaloids, nitrogen-containing compounds generally found as secondary metabolites in plants, are also classical examples of renewables. In contrast to terpenes, they show a great variety in molecular structure, and the different classes of alkaloids are usually based on their basic ring systems. Many pharmaceutically active

Table 5.1 Production in tonnes of morphine and derivatives in 2004.

Alkaloids

Morphine 355 (301 was converted into codeine)
Codeine 299 (about 30 natural codeine, rest synthesized from morphine)
Thebaine 79

Derivatives (see Scheme 5.9)

Hydrocodone 32	Dihydromorphine 2.2
Hydromorphone 2.5	Ethylmorphine 0.9
Dihydrocodeine 31.4	Oxycodone 32

alkaloids are directly applied as medicines, e.g., morphine, codeine, ephedrine, atropine, physostigmine, vinblastine, vincristine, quinine, colchicine, and galanthamine. However, bio-activity can often be improved by conversion into more effective derivatives with fewer unwanted side-effects.

Morphine (**10**) and codeine (**11**), constituents of opium, are the most interesting alkaloids found in nature. Morphine is also the oldest alkaloid isolated, in 1805, by the German pharmacist Sertürner from opium, the sun dried latex of *Papaver somniferum*. The structure of morphine with its so-called morphinan skeleton, once called the acrobat under the alkaloids, was finally elucidated in 1952 by the first total synthesis performed by Gates and Tschudi. Many syntheses would follow [26], but all morphine used today, whether legal or illicit, originates in the natural source *P. somniferum* or its extract, opium. The latex may contain up to 20% morphine. Most legal morphine is converted into the anticough medicine codeine (Table 5.1) by treatment with trimethylanilinium methoxide, whereas almost all illicit morphine is acetylated to the diacetate heroin.

R = H, Morphine (**10**)
R = Me, Codeine (**11**)

According to the 2004 statistics on narcotic drugs by the International Narcotic Control Board the main producers of legal morphine are Australia (123 t) Turkey (56.6 t) and France (56.1 t). Producers of medicinal codeine are UK (71 t), USA (64 t), France (41 t) and Australia (31 t).

To date there is no reported synthesis that would show promise for large-scale manufacturing, except perhaps two similar synthetic routes worked out by

Beyerman/Bosman/Maat in Delft and by Rice et al. at NIH, USA. The possibility of a fully synthetic supply of morphine is nevertheless important, as the availability of the natural compound depends on the political stability of a few regions in the world.

The Delft synthesis makes use of an acid-catalyzed ring closure – in fact an intramolecular aromatic alkylation – of a 1-(3,5-dihydroxy-4-methoxybenzyl) isoquinoline derivative that is prepared starting from (natural) gallic acid. One of the hydroxyl groups is removed via a Pd/C hydrogenation of the benzyl ether. Other catalytic steps play an important role; some steps were improved recently [27]. The crucial step in the Rice synthesis makes use of a 1-(2-bromo-5-hydroxy-4-methoxybenzyl)isoquinoline derivative that is also cyclized in an acid-catalyzed ring closure to the morphinan skeleton, followed by catalytic removal of the bromo substituent (Scheme 5.8).

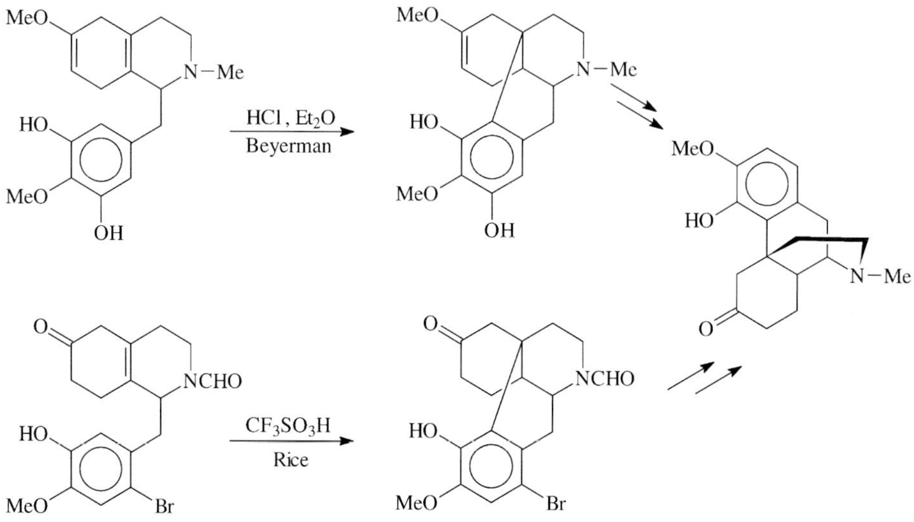

Scheme 5.8 Acid-catalyzed ring closures in the routes to morphine, according to Beyerman and Rice.

There are two types of pharmaceutically important derivatives: (a) Compounds with a hydroxyl substituent at position 14, such as in oxycodone and the antagonists naloxone and naltrexone, and (b) Diels–Alder adducts such as etorphine and buprenorphine, where the latter compounds are all derived from another opium alkaloid, (−)-thebaine (**12**) (Scheme 5.10). Because thebaine is a rather scarce alkaloid, several syntheses have been investigated. Quite recently, Australian scientists have been able to modify *P. somniferum* in such a way that thebaine is now a main alkaloid, so that it is becoming better available [28].

The opioid oxycodone, 14β-hydroxydihydrocodeinone (Scheme 5.9), is finding increasing application in clinical medicine as both an analgesic and an antitussive, as well as an intermediate synthon in the preparation of naloxone and nal-

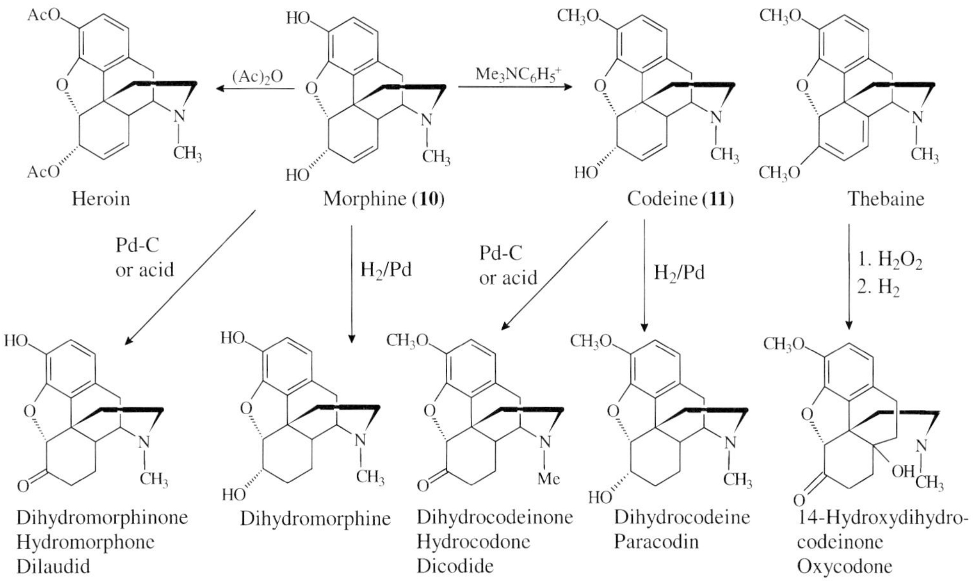

Scheme 5.9 Synthesis of different morphine derivatives.

troxone. The classical introduction of the 14-hydroxyl substituent starts from thebaine, using either hydrogen peroxide or peracids in organic acids. Apart from the low abundance of natural thebaine, the harsh reaction conditions and the need for extensive purifications make this route less attractive.

A useful introduction of the 14-hydroxyl substituent by co-catalyzed oxidation of certain morphine derivatives with air (oxygen) has been patented by Linders, a former student of the Delft alkaloid group, and Vrijhof (Organon/Diosynth) [29].

For the synthesis of Diels–Alder adducts a morphinan-6,8-diene system, as present in thebaine, is indispensable. Older publications started from thebaine and methyl vinyl ketone (but-3-en-2-one), yielding, after a Grignard reaction with propylmagnesium bromide, etorphine (**13**), a 6,14-endoethenomorphinan that is over 1000× as active as morphine and is used in veterinary medicine (Scheme 5.10).

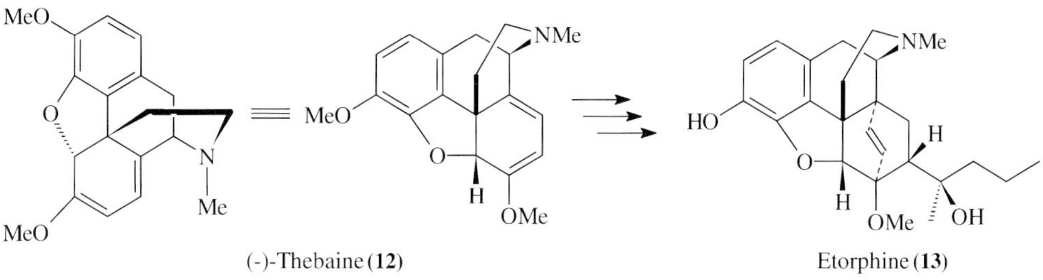

Scheme 5.10 Preparation of etorphine from (−)-thebaine.

In Delft, several new morphinan-6,8-dienes have been studied, yielding Diels–Alder adducts with interesting pharmacological properties [30]. In cases where the Diels–Alder reaction suffered from polymerization the use of microwave heating proved to be advantageous.

Another interesting alkaloid is tubocurarine chloride (14), with a bisbenzylisoquinoline structure. It is the active principle of tubocurare, an arrow poison used by Indians in South America and medicinally used as a muscle relaxant. However, the source, the leaves of the tropical rainwood liane *Chondodenron tomentosum*, is not easily accessible and the compound exhibits unwanted side-effects. Investigations showed that the basic structure can be replaced by an appropriate steroid skeleton with two nitrogen substituents at the right distance (see next section).

Tubocurarine chloride (14)

5.5
Steroids

Steroids, compounds with a cyclopenta[a]phenanthrene skeleton (15), include a wide range of natural products such as sterols (e.g., cholesterol), sex hormones, adrenocorticoid hormones, cardiac glycosides and vitamin D [31]. Sterols are steroids having a hydroxyl group at position 3 of the basic skeleton. Steroids can be found both in plants and in animals.

Steroid skeleton (15) Diosgenin (16) Dehydroepiandrosterone (17)

For a long time several steroid hormones have been synthesized from diosgenin (16) isolated from *Dioscoria* species growing in Mexico. Nowadays, China has overtaken the market by cultivating the plants and converting the isolated diosgenin into dehydroepiandrosterone (**17**). This is also a starting material for better muscle relaxants that have replaced tubocurarine chloride. The most important product now is rocuronium bromide (**18**, Esmeron®), with a production of 0.5–1 t a^{-1}. A synthesis of rocuronium bromide proceeds from **17** to a diene with double bonds at positions 2–3 and 16–17, then di-epoxidation takes place and, consecutively, pyrrolidine and morpholine are added. Finally, the system is mono-acetylated and N-alkylated with allyl bromide.

Rocuronium bromide (**18**)

Recently, Organon workers have discovered a derivative of γ-cyclodextrin (Sugammadex) that rapidly removes – by inclusion – rocuronium bromide (**18**) from the receptor sites, thus accelerating surgical after-care [32].

Plant sterols such as sitosterol and camposterol, as by-products from vegetable oils at prices of about € 15 kg^{-1}, are also important starting materials for the production of steroid hormones. A new application is the cholesterol lowering property of these sterols esterified with fatty acids (with a production of about 10 000 t a^{-1}). They can be found in the margarine "Becel pro-active" of Unilever. A Finnish equivalent is Benecol, which contains stanols such as sitostanol and campostanol, sterols having the 5,6-double bound hydrogenated, also esterified with fatty acids [33].

5.6
Enantioselective Catalysis

An obvious way to target chiral compounds is to start with a compound in which the chiral center is already present. Here natural products and derivatives offer a rich pool of generally inexpensive starting materials. Examples include L-hydroxy and amino acids. Sometimes, just one out of many chiral centers is predestined to remain, as in the synthesis of vitamin C from D-glucose, or in the preparation of (S)-3-hydroxy-γ-butyrolactone from lactose.

The other approach is to apply enantioselective catalysts. First of all, nature's ingenious catalysts, the enzymes, should be mentioned. For a review of this field, including economical considerations, see Chapter 7 of Ref. [19]. In several synthetic areas, e.g., esterification/hydrolysis/transesterification, enzymes now play an important role. Notably, in recent years the performance of some enzymes could be improved [34] by so-called "directed evolution", i.e., controlled substitution of particular amino acids of the enzymes by other amino acids.

Several examples exist of the application of chiral natural N-compounds in base-catalyzed reactions. Thus, L-proline and cinchona alkaloids have been applied [35] in enantioselective aldol condensations and Michael addition. Techniques are available to heterogenize natural N-bases, such as ephedrine, by covalent binding to mesoporous ordered silica materials [36].

Natural compounds are also applied as chiral ligands in enantioselective homogeneous metallo-catalysts. A classical example is the Sharpless epoxidation of primary allylic alcohols with *tert*-butyl hydroperoxide [37]. Here the diethyl ester of natural (R,R)-(+)-tartaric acid (a by-product of wine manufacture) is used as bidentate ligand of the Ti(IV) center. The enantiomeric excess is >90%. The addition of zeolite KA or NaA is essential [38], bringing about adsorption of traces of water and – by cation exchange – some ionization of the hydroperoxide.

Modification of catalytic metal surfaces with chiral natural compounds has led – for particular combinations – to results varying from good to excellent in enantioselective hydrogenations. The field has been reviewed [39].

The best studied systems are the Raney Ni/tartaric acid/NaBr combination, for the hydrogenation of β-functionalized ketones, and the Pt- and Pd-on-support/cinchona alkaloid systems for the enantioselective hydrogenation of α-functionalized ketones.

The Ni–tartaric acid system was discovered by Izumi et al. [40] and gives, under optimized conditions, e.e.s >90% in the hydrogenation of β-keto esters. Even with simple ketones good results are obtained, e.g., 2-octanone: 80% e.e. [41]. As yet the mechanism of the enantioselectivity is not well understood.

Much work [42] has been devoted to cinchona alkaloid modified Pd and Pt catalysts in the enantioselective hydrogenation of α-keto esters such as ethyl pyruvate (Scheme 5.11). Optimal formulation and conditions include: supported Pt, the inexpensive (−)-cinchonidine, acetic acid as solvent, 25 °C and 10–70 bar H_2. Presently, the highest e.e. is 97.6% [to (R)-ethyl lactate].

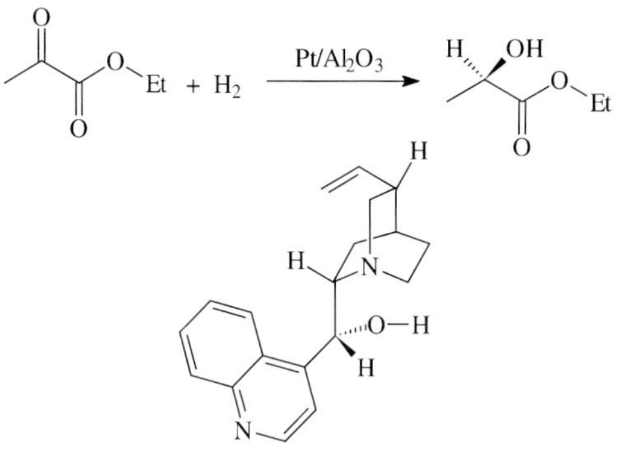

Scheme 5.11 Enantioselective hydrogenation of ethyl pyruvate.

The required amount of the cinchonidine modifier is amazingly small, indicating strong adsorption (with the quinoline ring system) onto the Pt-surface. The other part of the modifier molecule then forms an umbrella under which the substrate can be accommodated on the Pt-surface. Interaction of the protonated modifier nitrogen with the carbonyl group of the substrate is another accepted mechanistic detail. Force-field calculations indicate that the surface complex leading from pyruvate to (R)-lactate is energetically favored compared with the pro-(S)-complex.

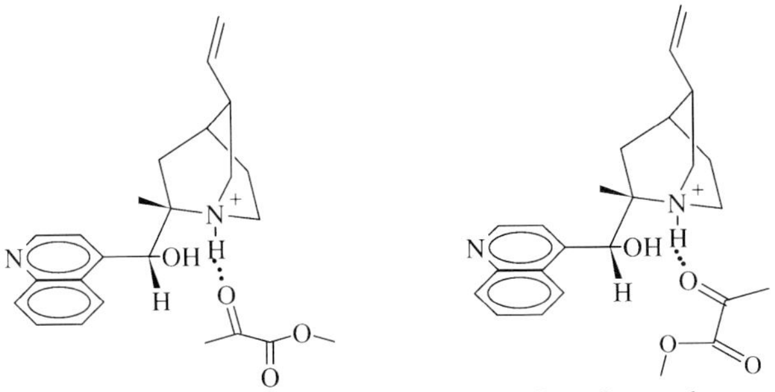

Pro-(R)-complex

Pro-(S)-complex

5.7
Artimisinine

Malaria, a tropical disease caused by protozoan parasites of the genus *Plasmodium*, has been a major concern for centuries and has now extended to a great deal of the world's population, killing every year 1–2 million people. Different medicines are in use to cure or to prevent malaria. The classical natural medicine quinine was soon replenished with synthetic compounds such as primaquine, chloroquine and mefloquine. However, a major problem is still an increasing resistance towards these compounds.

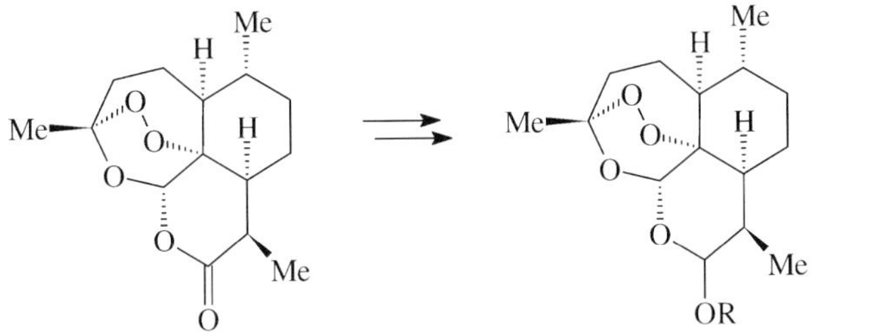

Artimisinine (**19**)

Dihydroartimisinine, R = H
Artesunate, R = CO(CH$_2$)$_2$COONa
Artemether, R = Me
Arteether, R = Et

Scheme 5.12 Artimisinine and derivatives.

A new medicine that seems to be promising in this field is artimisinine (**19**) isolated from *Artimisia annua*. Extracts of this plant have been known for centuries in Chinese folk medicine under the name of Qinghaosu or Qinghao. For solubility reasons and hydrolysis stability, artimisinine is converted via the dihydro compound into water-soluble artesunate or oil soluble artemether and arteether (Scheme 5.12) [43].

Erratic availability and high costs of the natural compound make further investigations for cheaper natural or synthetic endoperoxide-based antimalarials necessary. Meanwhile, a Belgian company, Dafra Pharma, Turnhout, is bringing the natural product and its derivatives onto the market [44].

5.8
Tamiflu

The neuramidase inhibitor oseltamivir phosphate was discovered by Gilead Sciences and developed by Roche Pharmaceuticals under the name of Tamiflu® (Scheme 5.13) to be used as an orally active antiviral compound for prevention and treatment of influenza infections. Because of the recent emergence of the avian flu, the demand for Tamiflu has gained momentum. Two industrially feasible syntheses are known, starting from (−)-shikimic acid and (−)-quinic acid, respectively (Scheme 5.13) [45].

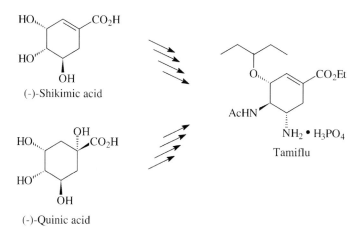

Scheme 5.13 Technical syntheses of Tamiflu.

Shikimic acid is isolated from Chinese star anis, *Illicium verum*, and quinic acid is extracted from the bark of *Cinchona* trees. Although both compounds can be found in many other plants, their isolation and purification are cumbersome. A fermentation production process of shikimic acid from other renewables such as glucose seems to be successful [46].

However, the two syntheses of Tamiflu, involving two potentially hazardous azide-containing intermediates, need further improvements. A recently published short and enantioselective pathway starting from acrylic acid and 1,3-butadiene looks promising [47].

5.9
Final Remarks

The examples given above demonstrate that natural products play an important role in various areas of fine chemicals. By genetic modification, yields of desired compounds might be further enhanced. Moreover, improved separation techniques will contribute to optimal crop use.

Sometimes, isolation from natural sources, fermentation technology, and stepwise organic synthesis are in competition for the market.

Catalytic methods, chemo- as well as bio-catalysis, are of vital importance in the conversion of natural products into derivatives (semi-synthesis). In chemo-catalysis conventional catalysts, such as mineral acids, are being replaced by recyclable solid catalysts. Further progress is also expected in cascade processes in which synthesis steps are combined to "one pot" methods.

Acknowledgements

We are indepted to Dr. Hans H. Bosman for data on the production of opiates and Dr. Joannes T.M. Linders for some useful suggestions.

References

1 *Römpp Encyclopedia, Natural Products*, Eds. W. Steglich, B. Fugmann, S. Lang-Fugmann, Thieme, Stuttgart, **2000**.
2 C. Okkerse, H. van Bekkum, *Green Chem.*, **1999**, 1, 107.
3 *Roadmap Biomass Technol. US*, December 2002.
4 T. Werpy, G. Peterson et al., *Top Value Added Chemicals from Biomass*, NREL Report, **2004**.
5 A. Kleemann, J. Engel, B. Kutscher, D. Reichert, *Pharmaceutical Substances*, Thieme, Stuttgart, **1999**, p. 1432.
6 F. Mirasol, *Chem. Market Rep.*, 20 March **2006**, p. 32.
7 C.K. Ooi, Y.M. Choo, S.C. Yap, Y. Basiron, A.S.H. Ong, *J. Am. Oil Chem. Soc.*, **1994**, 71, 423.
8 DSM Food Specialties Booklet, Delft, *Natural Beta-carotene from* B. trispora.
9 P. Laszlo, *Organic Reactions, Simplicity and Logic*, Wiley, Chichester, **1995**, p. 535.
10 R. Brownell, *Perfumer Flavorist*, March/April, **2006**, p. 24.
11 R.A. Sheldon, H. van Bekkum, *Fine Chemicals through Heterogeneous Catalysis*, Wiley-VCH, Weinheim, **2001**, p. 8.
12 K.M. Draths, J.W. Frost, in *Green Chemistry*, P.T. Anastas, T.C. Williamson, Eds., Oxford University Press, Oxford, **1998**, p. 150.
13 K.A.D. Swift, *Top. Catal.*, **2004**, 27, 143.
14 N. Ravasio, F. Zaccheria, M. Guidotti, R. Psaro, *Top. Catal.*, **2004**, 27, 157.
15 J.L.F. Monteiro, C.O. Veloso, *Top. Catal.*, **2004**, 27, 169.

16 W.F. Hölderich, M.C. Laufer, in *Zeolites for Cleaner Technologies*, M. Guisnet, J.-P. Gilson, Eds., Imperial College Press, London, **2000**, p. 301.
17 P.A. Robles-Dutenhefner, K.A. da Silva, M.R.H. Siddiqui, I.V. Kozhevnikov, E.V. Gusevskaya, *J. Mol. Catal. A*, **2001**, 175, 33.
18 W.F. Hölderich, D.M. Roberge, in *Fine Chemicals through Heterogeneous Catalysis*, Eds. R.A. Sheldon, H. van Bekkum, Wiley-VCH, Weinheim, **2001**, p. 427.
19 R.A. Sheldon, *Chirotechnology*, Marcel Dekker, Inc., New York, **1993**.
20 A. Corma, M. Renz, *Chem. Commun.* **2004**, 550.
21 F. Iosif, S. Coman, V. Parvulescu, P. Grange, S. Delsarte, D. de Vos, P. Jacobs, *Chem. Commun.* **2004**, 1292.
22 A.F. Trasarti, A.J. Marchi, C.R. Apesteguía, *J. Catal.* **2004**, 224, 484.
23 F. Zaccheria, N. Ravasio, A. Fusi, M. Rodondi, R. Psaro, *Adv. Synth. Catal.*, **2005**, 347, 1267.
24 P.J. Kunkeler, J.-C. van der Waal, J. Bremmer, B.J. Zuurdeeg, R.S. Downing, H. van Bekkum, *Catal. Lett.*, **1998**, 53, 135.
25 A. Corma, L.T. Nemeth, T. Laszlo, M. Renz, S. Valencia, *Nature*, **2001**, 412, 423.
26 J. Zezula, T. Hudlicky, *Synlett*, **2005**, 388.
27 G.J. Meuzelaar, M.C.A. van Vliet, L. Maat, R.A. Sheldon, *Eur. J. Org. Chem.*, **1999**, 2315, and references cited therein.
28 J. Bradbury, *Drug Discovery Today*, **2005**, 10, 5.
29 J.T.M. Linders, P. Vrijhof, Process of C-14 oxidation of morphine derivatives, PCT Int. Appl., **2003**, WO 2003/018588 A2.
30 L. Maat, R.H. Woudenberg, G.J. Meuzelaar, J.T.M. Linders, *Bioorg. Med. Chem.*, **1999**, 7, 529.
31 F.J. Zeelen, *Medicinal Chemistry of Steroids*, Elsevier, Amsterdam, **1990**.
32 J. Bom et al., *Angew. Chem.*, **2002**, 114, 274.
33 G.R. Thompson, S.M. Grundy, *Am. J. Cardiol.*, **2005**, 96, 3D.
34 M.T. Reetz, *Angew. Chem. Int. Ed.*, **2001**, 40, 284.
35 (a) H. Wynberg, *Top. Stereochem.*, **1986**, 16, 87; (b) M. Marigo, K.A. Jørgensen, *Chem. Commun.*, **2006**, 2001.
36 M.-J. Jin, M.S. Sarkar, V.B. Takale, S.-E. Park, *Bull. Korean Chem. Soc.*, **2005**, 26, 1671.
37 T. Katsuki, K.B. Sharpless, *J. Am. Chem. Soc.*, **1980**, 102, 5974.
38 Y. Gao, J.M. Klunder, J.M. Hanson, H. Masamune, S.Y. Ko, K.B. Sharpless, *J. Am. Chem. Soc.*, **1987**, 109, 5765.
39 T. Mallot, A. Baiker in *Fine Chemicals through Heterogeneous Catalysis*, R.A. Sheldon, H. van Bekkum, Eds., Wiley-VCH, Weinheim, **2001**, p. 449.
40 T. Harada, Y. Imachi, A. Tai, Y. Izumi, *Stud. Surf. Sci. Catal.*, **1982**, 11, 377.
41 T. Osawa, T. Harada, A. Tai, *Catal. Today*, **1997**, 37, 465.
42 J.T. Wehrli, A. Baiker, D.M. Monti, H.-U. Blaser, *J. Mol. Catal.*, **1990**, 61, 207.
43 A. Robert, F. Benoit-Vical, O. Dechy-Cabaret, B. Meunier, *Pure Appl. Chem.*, **2001**, 73, 1173.
44 H. Platteeuw, Dafra Pharma, Belgium, personal communication, 2005.
45 U. Jahn, *Nachrichte Chem.* **2006**, 54, 524.
46 M. Krämer, J. Bongaerts, R. Bovenberg, S. Kremer, U. Müller, S. Orf, M. Wubbolts, L. Raeven, *Metabol. Eng.*, **2003**, 5, 277.
47 Y.-Y. Yeung, S. Hong, E.J. Corey, *J. Am. Chem. Soc.*, **2006**, 128, 6310.

6
Options for Catalysis in the Thermochemical Conversion of Biomass into Fuels

Sascha R. A. Kersten, Wim P. M. van Swaaij, Leon Lefferts, and Kulathuiyer Seshan

6.1
Introduction

Progress towards a sustainable energy supply is without doubt one of the biggest challenges that mankind has ever faced. Energy scenarios [1–3] project that the world's annual energy consumption will increase steeply from the current 500 Exa (10^{18}) joules (EJ) per annum to 1000–1500 EJ by 2050. At least for the coming 50 years, sustainable energy sources alone will not be able to fulfill the world's energy demand. Fossil fuels will continue to dominate, and CO_2 emission abatement will become more and more important. It is forecasted that in 2050 ca. 400 EJ of energy per annum has to originate from sustainable sources to fulfill the needs of that generation. To achieve this, non-fossil energy systems such as solar (thermal, photovoltaic), indirect solar (biomass, water, wind, thermal gradients) and nuclear (fusion, fission, and geothermal) will be developed, optimized and implemented. At this point in time, it cannot be said unequivocally which "source-technology-product" combination(s) will dominate under the prevalent economical, social and environmental systems of 2050. It is recognized, however, by governmental bodies [4–6] and large industries [7] that biomass is a relevant sustainable candidate for the replacement of fossil sources, especially when it comes to the production of fuels for (non)stationary applications [e.g., gasoline, diesel, kerosene, oxygenates, heavy fuel oil, (S)NG, H_2]. These fuels are essential for our present society and it is of paramount importance for the world's economy and stability that a fuel supply is guaranteed in the future.

The present chapter discusses the options for application of catalytic technology in the thermochemical conversion of lignocellulosic biomass into fuels. It is envisaged that catalysis will play an important role in the production of biofuels, just as catalysis plays a major role in the conversion from fossil feeds into fuels currently. The development of catalytic cracking, isomerization and hydro-treating technologies have been key expertise for shaping the mineral-oil-refining industries during the 20th century into its present position, exploiting all fractions of mineral oil for producing fuels and chemicals. Thus, application of the knowledge and technology currently available is essential for a smooth and economical transition. Adaptation of biomass as a feedstock in (a) the current processes for chemicals and fuels or (b) new conversion processes is by no means automatic and poses new challenges in catalyst design. Importantly, the use of lignocellulosic waste material (not digestible by humans) or energy crops as feedstock prevents competition with the food chain. Next to thermochemical routes, biochemical conversions also show promise for the production of bio-fuels. Investigations into these routes are dominated by expertise in the areas of biochemistry, biotechnology, microbiology etc., and are outside the scope of this chapter. Most likely, both thermochemical and biochemical conversions processes will contribute in an integrated manner to the concept of the bio-refinery to produce fuels and chemicals.

Recently, Corma and coworkers [8] published an exhaustive descriptive review on the synthesis routes of transportation fuels from biomass, and Ragauskas et al. [9] sketched their vision on the path forward for bio-fuels and biomaterials. Furthermore, a review by Lange et al. [10] appears in this book.

The present chapter discusses aspects, known by the authors, of (a) biomass as feedstock, (b) the concept of bio-refinery, (c) thermochemical routes from lignocellulosic biomass to fuels, and (d) the contribution of catalytic technology. The main focus will be on the catalytic conversion of fast pyrolysis oil into fuels with regard to problems encountered currently and the challenges for future research and development.

6.2
Biomass as Feedstock for Fuels

Biomass, via its photosynthesis, has provided energy for life for the longest period of its existence. Industrial processes that take-in biomass can be integrated with the natural photosynthesis/respiration cycle of vegetation. If used in this manner, biomass is a renewable energy source and, by its utilization, overall much less CO_2 is added to the atmosphere compared with the fossil fuel counterpart processes. When combined with CO_2 sequestration, biomass based processes can actually lower the CO_2 concentration in the atmosphere [11].

Ethically, only biomass that is not competing with the food chain should be used for the production of fuels, chemicals, power or heat. This competition can be avoided by first using the abundant residues from forests (e.g., leaves, timber residues) and agriculture (e.g., stems, straws, husks, bunches) and subsequently energy crops (e.g., algae, specially engineered short rotational crops) if the residues are not sufficient or are too expensive to collect or to process. Agricultural and forestry wastes are estimated to be the energy equivalent to half of the current world's oil production [12]. The potential of special energy crops is estimated to be in the range 50–250 EJ per annum [13]. Under strict conditions, such as closure of the mineral balance at the biomass production side, water balance control, and not making use of any fossil fuels and fossil-based utilities (fertilizers),

Table 6.1 Comparison of fossil and biomass derived fuel equivalents.

Fossil	Biomass alternative (100% bio-based replacement and/or blending component for fossil fuel)
Natural gas	CH_4
LPG ($C_3 + C_4$)	DME (dimethyl ether)
Gasoline (C_6–C_9)	Butanol Ethanol Methanol MTHF (methyltetrahydrofuran) MTBE (methyl tertiary-butyl ether) Deoxygenated and refined primary bioliquids
Diesel (C_{14}–C_{16})	Fatty acid esters (methyl = FAME, ethyl = FAEE) Levulinic acid esters (methyl, ethyl) DME Ethanol Fischer–Tropsch diesel (from bio-based synthesis gas) Deoxygenated and refined primary bioliquids
Kerosene (C_{10}–C_{14})	Fischer–Tropsch wax (from bio-based synthesis gas)
Heavy fuel oil	Primary bioliquids (pyrolysis oil, hydrothermal liquefaction oil)
Coal	Solid biomass, charcoal

the use of biomass is completely sustainable. Technical and non-technical facilitators and barriers for large-scale industrial biomass use are discussed in detail by Van Swaaij et al. [11].

Biomass is built-up mainly out of the elements C, H and O, just like many of our current fuels and chemicals, although the C-H-O ratio differs significantly, as will be discussed later. The proof of principle for the conversion of biomass into single components or mixtures that can be blended with fossil fuels or that can replace fossil fuels has been delivered already. Table 6.1 lists fossil fuels that are in use today and their possible biomass derived fuel equivalents.

The list contains components that are considered as 100% bio-based replacements and/or blending components for fossil fuels. At the time of writing it is not clear if each listed component can completely replace the corresponding fossil fuel or if it is only suitable for blending.

Blends of petroleum diesel and gasoline with, respectively, bio-diesel (fatty acid esters produced by trans-esterification of vegetable oil) and bio-ethanol (from fermentation of sugar cane/corn) are already approved by the automotive industry. However, the present industrial processes for bio-ethanol and bio-diesel interfere with the food chain by consuming sugars and vegetable oils, often have a too low fuel yield per acre, and require a too high (fossil) energy input [14]. An interesting development in that respect is the cellulose-ethanol demonstration plant from

IOGEN that produces bio-ethanol from straw by a combination of thermal, chemical and biochemical techniques [15, 23; Vol. I, Chapter 7). Synthesis gas (CO & H_2) [16] and methane-rich gas [17, 18] have been produced at demonstration scale from lignocellulosic biomass via entrained flow gasification and methanation in hot compressed water, respectively. Laboratory-scale research on thermochemical conversions has shown the possibility of converting liquid lignocellulosic biomass [19, 20] into gasoline and diesel precursors. These technologies are in an embryonic stage of development and require extensive research to identify the best routes and to develop the required catalysts and reactors.

6.3
Composition of Biomass

Biomass contains more oxygen than fossil sources (30–50 wt.%, see Fig. 6.1). In Fig. 6.1, (biomass and fossil) sources and fuels (fuel blends) are positioned in the O/C–H/C plane.

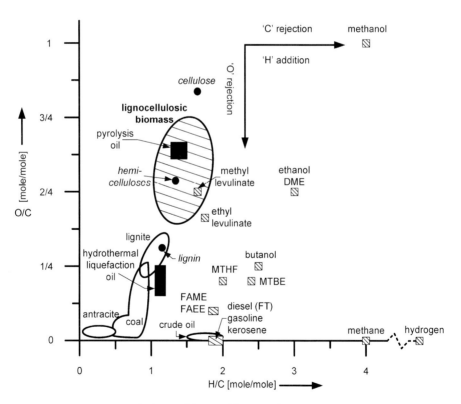

Fig. 6.1 Composition (dry basis) of fossil and biomass feedstocks and fuels derived from them. Arrows indicate current/possible upgrading routes.

Liquid hydrocarbon fuels (gasoline, diesel, kerosene, heavy fuel oil) have a typical composition of [$CH_{1.8-2}$] and they are made from the hydrogen-deficient fossil crude [$CH_{1-1.5}$] via two routes: (a) hydrogen addition (hydro-processing) and (b) carbon rejection (coking, FCC). Production of liquid hydrocarbon fuels from biomass requires, in addition to hydrogen addition or carbon rejection, oxygen removal (Fig. 6.1). Oxygenated fuels from lignocellulosic biomass such as DME, MTBE, esters, and alcohols need less oxygen exclusion. Hydrogen manufacturing from biomass requires, obviously, complete carbon and oxygen rejection. Bio-based methane can be produced by combined hydrogen addition and oxygen removal.

Along with carbon, hydrogen and oxygen, lignocellulosic biomass also contains hetero elements such as alkali and other metals. The amounts of these ashes vary over a broad range, from 30–50 wt.% in chicken litter to 1–3 wt.% in wood. Moisture is always present in lignocellulosic biomass and can be up to 80 wt.% in some cases. Detailed information on the composition of biomasses can be found in data bases, e.g., Phyllis [21] from the Dutch Energy Research Foundation (ECN). Table 6.2 lists the compositions of some typical biomasses.

The organic fraction of lignocellulosic biomass is built-up of cellulose, lignin, hemi-cellulose (Fig. 6.2), and extractives. Cellulose is a long-chain homo-polymer of β [$(C_6H_{10}O_5)_n$, $n > 3000$]. Lignin is a macromolecule with a molecular mass of higher than 1000 gram mol^{-1} and is built-up randomly from ordered aromatic substructures. Hemicellulose is a co-polymer of any of the following monomers: xylan, glucuronoxylan, arabinoxylan, glucomannan, and xyloglucan ($n > 500$). The cell walls of plants, crops, and trees are made of insoluble cellulose fibers meshed into a matrix of hemicelluloses. Lignin fills the spaces in the cell wall between cellulose and hemicelluloses. It confers mechanical strength to the cell wall

Table 6.2 Analysis (typical values in wt.%, dry) of selected biomasses.

Component	Biomass				
	Poplar (hardwood)	Pine (softwood)	Corn stover	Rice husk	Cotton seed hulls
C	50.8	52.9	47	36	33
H	6.4	6.3	5.5	5	6
O	41.8	40.7	41	40	59
N	0.3	0.1	1.5	2	0.1
Ash	0.7	1.0	5	17	1.9
Cellulose	45	46	35	33	53
Hemicelluloses	30	20	37	29	18
Lignin	22	32	18	20	25
Extractives	2.1	2	5	1	2.1

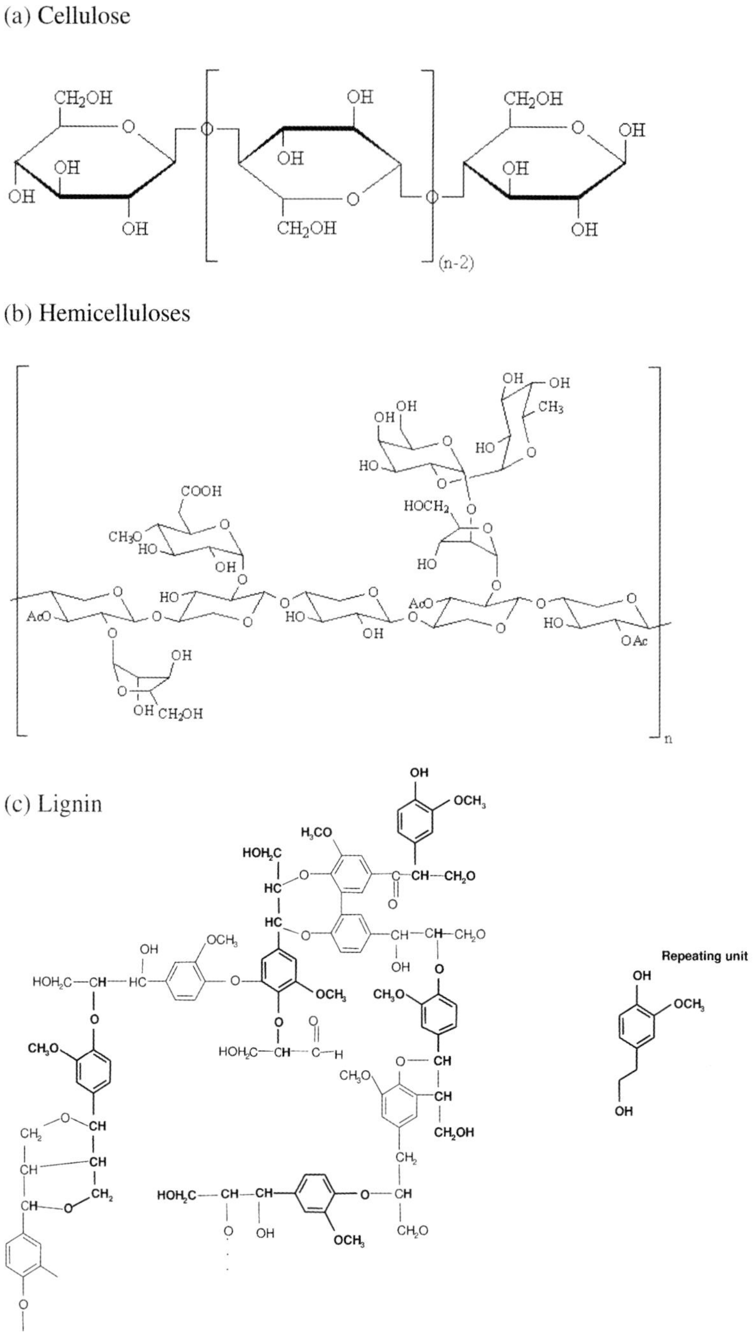

(a) Cellulose

(b) Hemicelluloses

(c) Lignin

Fig. 6.2 Structural formulas of cellulose (a), example of hemicelluloses (b), and an example of lignin (c), showing a repeating unit as possible model compound.

and, therefore, the entire biomass. Extractives are non-structural components and are deposited in cell cavities or infiltrated into the cell wall structure and occur as complex mixtures of tannins, flavonoids, stilbenes, resin- and fatty-acids, waxes, sterols, and simple sugars.

6.4 Biorefinery

There are already some industrial biomass-based processes operational (e.g., bio-diesel, bio-ethanol, bio-based heat and power). However, in general these processes cannot compete economically with their fossil counterparts, because yields of target products are not maximized and by-products often have a very low or even a negative value. An integrated concept aiming at full utilization of biomass, in which different fractions of biomass are converted in large-scale plants (economy of scale) or in standardized small-scale units (economy of numbers) in an economically optimal product state is termed a "Biorefinery". A biorefinery might, for example, produce one or several "low-volume, high-value" chemical products and a "low-value, high-volume" liquid transportation fuels. The high-value products enhance profitability while the high-volume fuel helps to meet energy needs. To start-up the biorefinery concept it is essential to integrate and to partner-up with existing industries and markets. This lowers the required capital investments and offers guaranteed markets for the products. Integration can be at the level of the products only, e.g., by producing biomass-based blending components for fossil transport fuels, or by co-processing biomass in existing refineries and chemical plants. In a later stage, 100% biomass based products can gradually replace the functionality of existing fossil products or can be linked with new applications. By producing multiple products from different fractions of biomass, a biorefinery takes advantage of the differences in the constituents of biomass and of the specific characteristics of intermediate energy carriers (e.g., pyrolysis oil and charcoal). A bio-refinery may include thermal, chemical and biological conversion processes and its development requires input from various disciplines, viz. process technology, (bio-)chemistry, bio-technology, catalysis, (micro-)biology and separations. As mentioned earlier, we will discuss the challenges of the thermochemical conversion and catalytic processing of lignocellulosic biomass.

Several biorefinery schemes have been proposed during the last 5 years [9, 22, 23]. They differ in the choice of feedstock, the proposed technologies and the targeted products. The following concepts have been considered: the Lignocellulosic Feedstock Biorefinery, the Whole Crop Biorefinery, the Green Biorefinery, and the Two Platform Biorefinery [23]. At present, we do not have detailed flow schemes envisaged for the different types of proposed biorefineries, because many of the involved technologies are still in the early stages of development and the optimal product state has not yet been identified. Therefore, the description of a biorefinery must remain abstract. Figure 6.3 is our conceptual outline of a bio-refinery that takes in raw biomass from the fields, extracts food and feed from it, and uses the remaining lignocelluloses to produce fuels, chemicals, heat and power.

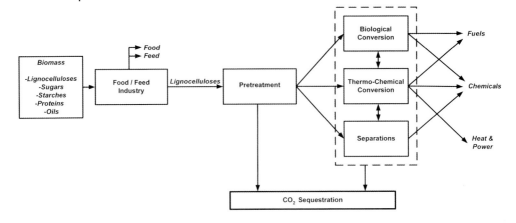

Fig. 6.3 Conceptual outline of a biorefinery.

6.5
Biomass Pretreatment

Lignocellulosic biomass needs pretreatment before it can be directed to the first process step in a series towards fuels. Drying to less (<25 wt.%) water content is required for processes such as gasification, and combustion. To simplify feeding, raw solid lignocelluloses must be homogenized by mixing, grinding, palletizing, liquefaction, or slurry preparation. Grinding becomes much cheaper when the biomass is first torrefied. Torrefaction is a roasting process that breaks down the fibrous structure of biomass, making the material more brittle. By liquefaction the volumetric energy density of biomass is increased by about a factor 5 and the produced liquid (20 MJ L^{-1}) can be more easily processed (incl. contact with catalyst) than solid biomass (4 MJ L^{-1}). Liquefaction ($T > 400\ °C$) can be done at ambient pressure [24] or at elevated pressure [25, 26]. For processes at ambient pressures (pyrolysis) the feed is dry while for high pressure processes a "biomass in water slurry" is fed (hydrothermal liquefaction). Interesting results have been achieved by dissolving (liquefying) lignocelluloses in a mixture of 90 wt.% ethylene glycol and 10 wt.% ethylene carbonate at ambient pressure and 150 °C [27]. In Section 6.9 the liquefaction processes are discussed in more detail.

Fractionation is considered as starting point for the production of chemicals and fuels from lignocelluloses via the production of sugars. A good review on the fractionation techniques is given by Huber et al. [8]. Here, only two interesting developments are mentioned: (a) Organosolv pulping is a fractionation method where organic solvents (e.g., ethanol) are used to facilitate lignin extraction [28]; (b) Zhang and Lynd [29] have developed a new fractionation method based on acid treatment, solvent extraction, and Organosolv operating at very mild temperatures of ca. 50 °C.

6.6
Thermochemical Conversion of Lignocelluloses

Basically, there are three thermochemical routes (Fig. 6.4) for the conversion of lignocelluloses into fuels:

1. **g**asification followed by catalytic upgrading of the produced gases;
2. liquefaction of biomass followed by refining;
3. **e**xtraction of sugars and sugar derivates followed by their catalytic conversion.

Figure 6.4 also shows some integration options (dotted arrows) to achieve full utilization of the feedstock, e.g., (a) aqueous phase by-product streams of hydrolysis can be used for the production of H_2-rich gas or methane via conversion in hot compressed water and (b) lignin, separated before hydrolysis, can be gasified (entrained flow) to synthesis gas. Options for the integration of the thermochemical lignocelluloses biorefinery with other biorefinery concepts are legion. The lignocellulosic waste of other biorefinery concepts (e.g., stems from the sugar cane based bio-ethanol process) can be used as feedstock for all primary conversions in Fig. 6.4.

In Section 6.7 onwards the processes listed in Fig. 6.4 are discussed while focusing on the production and upgrading of pyrolysis oil. The route involving oxygen-blown high-temperature gasification followed by catalytic upgrading of the produced synthesis gas is not discussed here. It includes technologies that are commercially available for coal, natural gas, and heavy oil feeds. The authors expect that only minor modifications are required to adapt these technologies for biomass feedstock materials. In this way bio-based top fuels can be produced in the near future. However, there is always the question of the wisdom of first breaking up the lignocelluloses polymers to the smallest possible molecules (CO and H_2) from which then a synthesis is started again up to C_{20} components.

In thermochemical conversion of biomass, temperature is a key parameter. At lower temperatures (<300 °C) only catalytic processes (e.g., acid-catalyzed hydrolysis) are possible. Conversion into various oxygenates, such as acids (e.g., levulinic acid), heterocyclic hydrocarbons (furans), alcohols (phenols) via sugars with promising yields has been shown to be possible. Lignin is not or hardly decomposed in this regime. In the low temperature processes most use is made of the composition of biomass by keeping much of the functionality of the sugar building blocks intact. However, pretreatment is required to make the fibers accessible (e.g., by steam explosion), because native lignocelluloses is inert for hydrolysis at these temperatures. In addition, the reactions are slow and require often homogeneous catalysts. In the mid-temperature range ($300 < T < 700$ °C) complete conversion of lignocellulosic biomass is possible. Temperature, contact time and use of catalysts, and water dilution level determine the type of products and their aggregation state. Liquefaction processes (pyrolysis and hydrothermal liquefaction) yield a multicomponent liquid product containing oxygenates, some permanent gases, and a solid that consists of the remainders of the fiber structure of the feedstock. Applying catalysis in aqueous environment can give

128 | 6 Options for Catalysis in the Thermochemical Conversion

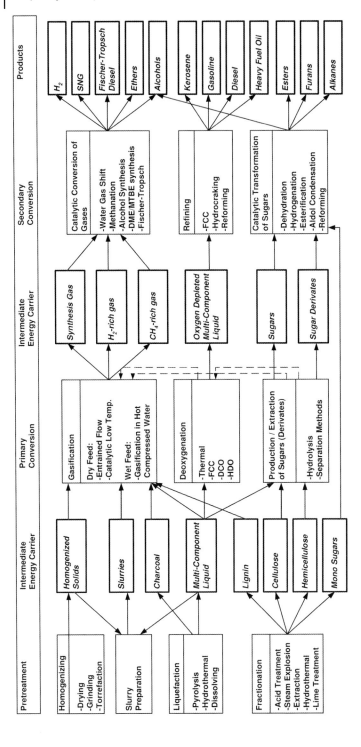

Fig. 6.4 Routes for the production of fuels from lignocellulosic biomass.

only gaseous products (H_2 and CH_4). At high temperature ($T > 700\,°C$) gasification is also possible for dry biomass. Gasification is uncontrolled and in the absence of a catalyst, up to $1300\,°C$, methane is always produced. The product gas reaches equilibrium above $1300\,°C$.

6.7 Biomass Gasification

6.7.1 Gasification of Dry Biomass

Gasification was the first technology to be considered for industrial biomass conversion. Complete and good reviews on biomass gasification are available [30–34]. Non-catalytic biomass gasifiers ($T < 950\,°C$) typically produce a fuel gas (CO, CO_2, CH_4, C_{2+}, H_2, H_2O, tars, N_2) that requires extensive upgrading before it can be used in sophisticated down-stream technology applications such as turbines and catalytic conversions. Entrained flow non-catalytic biomass gasification ($T > 1300\,°C$) has been demonstrated for both liquid [16] and solid feeds. These processes are currently being used on a large-scale for fossil-derived synthesis gas and can be put into service for bio-based synthesis gas with no or minor modifications (especially for co-feeding applications). Synthesis gas can be converted into hydrogen (water-gas shift, over supported Cu-catalysts and Fe catalysts in two stages), alcohols (Cu-Zn catalysts), DME (CuO), MTBE (zeolites), and synthetic diesel (Fischer–Tropsch, over Co or Fe catalysts) by catalytic processes. Producing CH_4 from synthesis gas is thermodynamically not favorable, because a low temperature exothermic process (methanation) needs to be coupled with a high temperature endothermic process (gasification). SNG (synthetic natural gas) from methanation of bio-based fuel gas is an interesting route to make domestic and industrial heating more sustainable. Intermediate gasification temperatures ($950 < T < 1300\,°C$) are extremely unfavorable because the ash becomes partly molten, a situation that is almost impossible to handle in a reactor. One escape from this forbidden temperature region is using ultralow ash containing bio-liquids as feedstock [35].

6.7.2 Catalytic Gasification of Pyrolysis Oil

Czernick and coworkers [36, 37] have shown that the fraction of pyrolysis oil that dissolves in water (cellulose and hemicelluloses derivates) can be gasified with a nickel catalyst to hydrogen-rich gas at around $800\,°C$. The steam over carbon ratio used [10–20] was, however, unrealistically high. Recently, Van Rossum et al. [38] introduced a new catalytic reactor concept using a commercial nickel-based pre-reforming and reforming catalyst. In a continuous bench scale unit of $0.5\,kg\,h^{-1}$ pyrolysis oil (whole oil) intake, they produced hydrogen-rich gas ($H_2 = 63\,vol.\%$,

CO = 25 vol.%, CO_2 = 12 vol.%) not containing any hydrocarbons and a low tar (200 mg Nm^{-3}) content at 800 °C and S/C (steam over carbon ratio) = 1.5. Problems associated with pyrolysis oil gasification are similar to those of biomass gasification. Gasification of the tar fraction and conversion of methane formed are important challenges. Both require highly active and stable steam/autothermal reforming catalysts.

6.7.3
Chemistry and Catalysis of Gasification

Ideal stoichiometric reaction equations for, respectively, synthesis gas, hydrogen-rich gas and methane-rich gas production by gasification are given by reactions (1) to (3):

$$C_6H_8O_4 + 2H_2O \rightarrow 6CO + 6H_2 \tag{1}$$

$$C_6H_8O_4 + 8H_2O \rightarrow 6CO_2 + 12H_2 \tag{2}$$

$$C_6H_8O_4 + 2H_2O \rightarrow 3CO_2 + 3CH_4 \tag{3}$$

The application of catalysts would lower the operation temperature as compared with the entrained flow process. Lower operation temperatures would not only decrease capital cost, but would also increase the thermal efficiency. Such processes require catalysis. Ross et al. [39] elaborate the criteria for an effective biomass gasification catalyst to be (a) effective to gasify/remove tars, (b) capable of reforming methane, (c) resistant to deactivation by coke/oligomer deposition and sintering, (d) easy to regenerate, (e) robust (mechanically strong) and, most importantly, (f) cheap. Dedicated efforts to develop catalysts for biomass gasification are in their infancy, and the strategy till now has been to use catalysts (a) off the shelf, commercial, not so cheap, methane steam reforming catalysts, (b) cheaper materials, dolomite-based clays, alkali salts (Na, K, chlorides). For solid feed stocks some success has been achieved with natural (dolomite & olivine) and nickel catalysts, although operational problems remain and the tars and hydrocarbons can be removed only partly [31, 32, 39, 40]. It is possible to remove tars and hydrocarbons from fuel gas with down-stream processes, i.e., catalytic wet (steam) and dry (CO_2) reforming [41]. Supported Ni catalysts form the best option but catalyst stability still remains a major set back due to coking. Further, with conventional steam reforming catalysts (Ni, Pt based) hydrogen or syngas selectivity is not an issue; however, the catalyst activity is the limiting factor and require higher temperatures of operation [42]. These catalysts also require high steam/carbon ratios [10–20] to operate even for short times on stream. High steam usage makes the process energy intensive and inefficient. For comparison, the steam/C ratio for commercial steam reforming of methane is <3 [43]. If syngas or hydrogen is the targeted product, consecutive conversion of methane formed requires high temperatures, despite the presence of a catalyst. Impurities contain-

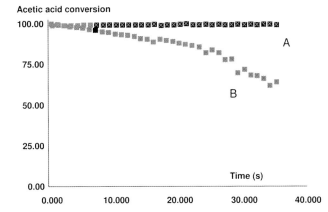

Fig. 6.5 Influence of the presence of traces of oxygen on the stability of the catalyst for steam reforming of acetic acid over Pt/ZrO$_2$ catalyst (300 °C, SV 14000 h^{-1}). (A) 500 ppm oxygen in the feed and (B) no oxygen present.

ing N, S, Cl and alkalis in the feed stock can also affect both catalyst performance and product quality.

The design of stable catalysts for the efficient gasification of biomass should take into account the ability of the catalysts to depolymerize deposits because suppressing oligomerization is nearly impossible as it occurs almost on any surface. One possibility is to remove the coke/deposits that deactivate the catalysts, via gasification with steam. For this, high activity reforming catalysts need to be developed, the idea being that the coke-forming precursors are also gasified and catalyst stability is improved. The rate-limiting step in steam reforming is normally the activation of water [44]. Thus, new catalyst (metal/support) combinations to maximize the availability of activated water on the catalyst, and to help gasification of coke oligomers as well, are essential. Another option is to carry out gasification in the presence of steam and/or oxygen just as with typical autothermal reforming. The role of oxygen in this case is to help to combust coke/oligomer and keep the catalytic sites clean. However, the catalyst should selectively combust coke and not CO or H$_2$. This is certainly both a catalyst and reactor design issue. Preliminary experiments show that at low concentrations of oxygen selective combustion of coke can be achieved (Fig. 6.5) [45].

6.7.4
Gasification in Hot Compressed Water

The high moisture content of the wet biomass streams (e.g., slurries of grass or algae, aqueous by-products of biochemical biomass conversions) makes conventional thermochemical technologies inefficient due to the high-energy requirement for water removal. Reforming (gasification) in hot compressed water

($P = 150–250$ bar, $T = 250–600\ °C$) is considered as a promising technique to convert such wet streams into a gas that is rich in either hydrogen or methane (see Section 6.7.3 for reaction equations), depending on the operating conditions and applied catalysts [17, 18, 46–48]. Feed stocks are either homogeneous liquids or slurries.

In hot compressed water ($P > 200$ bar), the heat effect associated with water evaporation is marginal compared with that at ambient conditions. Therefore, by practicing counter-current heat exchange between the feed stream and the reactor effluent, high thermal efficiencies can be reached despite the low dry matter content of the feedstock [48]. The process is still in the research phase, although some pilot plants [46] are already operational to start the process development. At laboratory- and demo-scale, methane-rich gas has been produced by using Raney nickel [18] and Ru/TiO_2 [17, 49] and hydrogen-rich gas has been produced with activated charcoal [47] and with Raney Ni-Sn [50]. The potential of heterogeneous catalysis is clearly demonstrated for reforming of biomass in hot compressed water. Hot compressed water, especially when supercritical, is acidic and a good solvent for most organic chemicals. The latter characteristic is especially useful to dissolve coke/coke precursors and keep the catalyst surface clean and extend life time. However, this also often leads to leaching of the catalytic active phase. Other challenges that are ahead are instable performance, ignorance of deactivation phenomena and chemical instability of support materials [50]. In addition, fundamental knowledge of the catalytic mechanism and reliable structure–performance relationships are missing.

6.8
Liquefaction of Biomass

6.8.1
Non-catalytic Pyrolysis

Fast pyrolysis is a high temperature (ca. 500 °C) process in which biomass is rapidly converted into vapors, gases and charcoal, in the absence of oxygen. After cooling and condensation, a dark brown organic liquid is formed from the vapors. Short contact times (<2 s) maximizes liquid yield. Such a fast pyrolysis process produces a liquid intermediate energy carrier (often called bio-oil or fast pyrolysis oil) at a scale that matches the local logistics of biomass transportation and storage (1–5 tons h^{-1}). Fast pyrolysis oil can be stored and transported over long distances using existing, or slightly modified, fossil oil infrastructure. The latter feature allows the decoupling of the locations where biomass is available and where the bio-based products are needed and produced at large scale. Owing to the relatively low process temperature ($\sim 500\ °C$), minerals and metals remain mostly in the charcoal and can thus be recovered at the biomass production site and returned to the soil. As a result, pyrolysis oil contains significantly less mineral and metal components than the solid biomass feedstock it is produced from.

Fig. 6.6 Fast-pyrolysis oil poured out of a flask.

The literature on biomass fast pyrolysis is quite extensive and excellent research and technology reviews are available [51–55]. For an optimal fast pyrolysis process in terms of organic liquid yield the temperature is around 500 °C; the biomass particle size should be small (≤5 mm).

Kinetic parameters of fast pyrolysis were derived while assuming a single process for the decomposition of wood, including three parallel first-order decay reactions for the formation of the product classes. This is the so-called "Shafizadeh" scheme [56]. The three lumped product classes are: permanent gas, liquids (bio-oil, tar), and char; a classification that has become standard over the years. The produced vapors are subject to further degradation to gases, water and refractory tars. Charcoal, which is also being formed, catalyzes this reaction and therefore needs to be removed quickly [57].

Fast pyrolysis oil is an acidic viscous dark brown liquid (Fig. 6.6) containing oxygenated hydrocarbons, water and small carbonaceous particles including some minerals.

The organic phase includes aliphatic as well as aromatic acids, alcohols, esters, ethers, sugars and extractives (Table 6.3). Approximately 70 wt.% of the oil has been identified. The molecular weight of the individual components ranges from 18 up to 2000 gram mol^{-1}. In some cases, the oil is an emulsion at microscopic level [58].

Fast pyrolysis oil has almost the same elemental composition as the biomass itself; hence it can be seen as a kind of liquid wood. It can be transported, can be pressurized and processed more easily than solid biomass. One of the major difficulties in the catalytic conversion of solid biomass is achieving efficient contact between the heterogeneous catalyst (which is most of the times a solid) and the biomass itself. In this context, bio-oil provides more options for easier catalytic conversion. However, pyrolysis is a very complex and the oil is a difficult to handle chemical mixture. Complete vaporization, for instance, is not possible because part of the components start to decompose and polymerize upon heating

Table 6.3 Composition of typical bio-oil.

Component	wt.%
Acetic acid	1–32
Formic acid	1–20
Hydroxyacetaldehyde	1–13
Furfural alcohol (2-hydroxymethylfuran)	1–5
Acetol (1-hydroxy-2-propanone)	2–8
Syringol (2,6-dimethoxyphenol)	1–5
Phenol	0–4
Methanol	1–3
Tars (polycyclic aromatics)	2–7

Table 6.4 Physical characteristics and elemental composition of bio- and fossil-oils [53].

Characteristic	Pyrolysis oil	Heavy fuel oil
Water content (wt.%)	15–35	0.1
C (wt.%, dry)	50–64	85
H (wt.%, dry)	5.2–7	11.1
O (wt.%, dry)	35–40	1.0
N (wt.%, dry)	0.05–0.4	0.3
S (wt.%, dry)	0.05–0.3	2.3
Heating value (MJ kg^{-1})	16.5–19	40
Viscosity (cp at 50 °C)	40–150	180
pH	2.4	–

before having a chance to evaporate. Table 6.4 lists typical properties of a bio-oil and petroleum crude oil.

The key problems in application of bio-oil as a fuel are related to the oxygen content. First, the heating value decreases due to the presence of oxygenates. Second, organic acids and phenols cause the oil to be corrosive. Third, too many oxygenates would prevent miscibility with hydrocarbons and, fourth, reactive oxygenates tend to oligomerize, causing chemical instability.

6.8.2
Catalytic Pyrolysis

Biomass pyrolysis in the presence of a catalyst, *in situ*, is considered as one of the options to overcome the problem characteristics of bio-oil mentioned above. Selective removal, by decarboxylation, of carboxylic acids (formic, acetic acids) will decrease the acidity of bio-oil. Selective deoxygenation of organic fractions (aldehydes, unsaturates, etc.) that undergo easy condensation/oligomerization reactions can help in the stabilization of the oil.

Several researchers have shown that alkali present in the feedstock influences the yields and compositions of the pyrolysis products [56, 59]. An interesting result was reported by Brown and coworkers [60] who found that addition of $(NH_4)_2SO_4$ as catalyst to the pyrolysis of dematerialized (alkali free) corn stover resulted in a pyrolysis oil that contained 23 wt.% levoglucosan (normally 1–3 wt.% levoglucosan is present in pyrolysis oil). Levoglucosan is a component from which various fuel blends and chemicals can be produced.

Deoxygenation reactions are catalyzed by acids and the most studied are solid acids such as zeolites and clays. Atutxa et al. [61] used a conical spouted bed reactor containing HZSM-5 and Lapas et al. [62] used ZSM-5 and USY zeolites in a circulating fluid bed to study catalytic pyrolysis (400–500 °C). They both observed excessive coke formation on the catalyst, and, compared with non-catalytic pyrolysis, a substantial increase in gaseous products (mainly CO_2 and CO) and water and a corresponding decrease in the organic liquid and char yield. The obtained liquid product was less corrosive and more stable than pyrolysis oil.

Factors that have to be taken into account while designing catalysts are (a) the bulky nature of organic molecules (large molecules up to 2000 g mol^{-1}) that escape from the biomass matrix, (b) the need to control the extent of pyrolysis/cracking and (c) selective scission of bonds, i.e., C–C > C–O > C–H to help maximize oxygen removal as CO_2. Thus, texture (pore size, geometry etc.) and acidity (strength, concentration of acid sites) are the two important parameters for design. Large-pore zeolites (Fuajasites) whose acidity can be manipulated by easy ion exchange with alkalis (Na, K) and weaker acids such as amorphous silica–aluminas are possible candidates for *in situ* catalytic pyrolysis [63].

Most importantly, biomass pyrolysis will be carried out at remote locations, and in distributed manner. Thus, the catalysts should be cheap and simple to use. Acidic clays, silica aluminas and H-FAU type zeolites are relatively cheap and robust materials, can be mixed easily with heat carriers, and used for pyrolysis. Efficient contact between the solids (catalyst and biomass) to maximize catalytic action is one of the challenges that need to be overcome.

6.8.3
Hydrothermal Liquefaction

Liquefaction of wet biomass streams is done by hydrothermal processes at elevated pressures. The feed stocks for these high pressure liquefaction processes are slurries of biomass particles and water. Feeding these slurries into the high pressure equipment, at reasonable costs, is an important hurdle in the process development. For example, the reported operating conditions for high pressure liquefaction are in the range 280–360 °C and 90–250 bar [25, 26]. Under these conditions, biomass is converted, in a complex sequence of chemical reactions, into various compounds. Upon cooling, the reactor effluent consists of three different phases: a water phase, a hydrophobic phase and a gas phase. By extraction the hydrophobic reaction product can further be separated into a solvent (e.g., acetone) soluble and a solvent insoluble part. The hydrophobic phase is considered the main reaction product and has considerably lower oxygen content than the feedstock (typically 10–20 vs. 45 wt.% of the feed). It is reported that the solvent-soluble hydrophobic product can be upgraded into diesel/gasoline range fuels by means of catalytic hydrocracking more easily than pyrolysis oil [64].

6.9
Upgrading Pyrolysis Oil to Fuels

Reviews on the status of pyrolysis oil upgrading until 1996 are those by Bridgwater [65, 66]. The goal of the early research on deoxygenation of pyrolysis oil was to produce fractions that could be blended directly to the gasoline and diesel pool. It soon became clear that, although diesel and gasoline range products were produced, the paraffins, olefins, naphthenes, aromatics ratio was not of the required specification, as a result of which further refining was required [67]. Nowadays the goal is to produce a liquid fuel precursor that can be refined. In the near future, co-feeding of these products at strategic points (e.g., FCC, hydrocracking, reforming) in a mineral oil refinery seems the most feasible option.

Deoxygenation can be done by decarboxylation, cracking, hydrodeoxygenation. The conceptual stoichiometric equations of these processes are:

$$\text{Decarboxylation (DCO): } C_6H_8O_4 \rightarrow C_4H_8 + 2CO_2 \quad (4)$$

$$\text{Hydrodeoxygenation (HDO): } C_6H_8O_4 + 4H_2 \rightarrow C_6H_8 + 4H_2O \quad (5)$$

$$\text{Cracking (CRA): } C_6H_8O_4 \rightarrow C_{4.5}H_6 + H_2O + 1.5CO_2 \quad (6)$$

Only full decarboxylation (Reaction 4) produces a paraffin-like product ($H/C = 2$), all the other methods produce more aromatic fuel precursors and need full hydrogenation (Reaction 7) if paraffins are aimed for.

$$\text{Hydrogenation (HYG): } C_6H_8O_4 + 7H_2 \rightarrow C_6H_{14} + 4H_2O \quad (7)$$

6.9.1
Decarboxylation (DCO)

Deoxygenation by full decarboxylation is the best route to make fuel precursors from bio-oil, because paraffin is produced and expensive hydrogen is not required. Decarboxylation of bio-oil has been tried over zeolites, yielding an aromatic product with a too low yield and excessive coke formation (Section 6.9.3). Selective decarboxylation of organic acids makes the bio-oil less acidic and corrosive. If acids can be removed selectively as CO_2, it would also improve the energy content of the resultant bio-oil. Removal of oxygen as CO_2, as against water, retains hydrogen content and hence the higher energy content of the bio-oil. Using an H-Y zeolite results in complete removal of formic acid, *in situ*, during pyrolysis [68]. Preliminary studies show that it is also possible to achieve selective deoxygenation as CO_2 by choosing proper catalysts. The actual oxygen removal by only decarboxylation of the acids is, however, not sufficient. New catalytic processes for deeper decarboxylation of pyrolysis oil are required to make the production of liquid fuel precursors from pyrolysis oil economically feasible. Some work on decarboxylation of model compounds has been reported [69–71].

6.9.2
Hydrodeoxygenation (HDO)

At the start of the bio-liquid upgrading research in the late 1970s, hydroprocessing was considered an obvious choice because of the existing knowledge on hetero atom (S, N, O) removal from petroleum products. Hydrotreating (standard commercial process based on Ni-Mo or Co-Mo based catalysts are available) can completely de-oxygenate pyrolysis oils to yield gasoline and diesel range hydrocarbons.

However, the cost of hydrogen required for this makes the route currently unattractive. One ton of biomass would require stoichiometrically 62 kg of hydrogen [65] and the products found until now still need to be refined before they can be added to the diesel or gasoline pool. Otherwise, commercial catalyst and process experience is available for developing a process and promising results have been obtained with pyrolysis oil as feedstock [64–67, 72, 73]. In our opinion, complete HDO of pyrolysis oil is a dead end, especially with the hydrogen shortage in refineries and the demand for hydrogen in fuel cell applications in the future. HDO might still be interesting as the last step in a series of deoxygenation processes (e.g., DCO followed by HDO) for production of a bio-liquid that can be refined (co-fed in a petroleum refinery or as such). A challenge for catalysis is the design of a catalyst that combines DCO and HDO actions with a minimum of hydrogen consumption. This would imply minimizing hydrogenolysis and formation of gaseous alkanes. Extensive commercial knowledge and experience is available at the moment (hydrotreating is one of the largest commercial catalytic processes currently) in the design of suitable catalysts, typical examples are bimetallic Pt-Pd supported on zeolites.

6.9.3
Cracking over Zeolites (FCC)

Since the early 1980s, zeolites have been considered for the upgrading of biomass-derived fluids into aromatic fuels [74, 75]. Researchers of the Univerité Laval (Québec, Canada) performed pioneering work on model compounds (phenols and furans) [76, 77]. A recent study investigating the transformation of alcohols, phenols, aldehydes, ketones, and acids on a HZSM-5 zeolite has been reported by Gayubo et al. [78–80]. The general conclusion of the work on model compounds is that the individual components in biomass derived liquids show great differences with respect to reactivity and coke formation, which can be severe. For upgrading of fast pyrolysis oils by zeolites two concepts have been applied at laboratory scale, viz. (a) downstream cracking of the pyrolysis vapors [81] and (b) cracking of the liquid pyrolysis product [82, 83].

Up to now, zeolite cracking of pyrolysis oil has been studied only by passing it over fixed beds [74, 82, 83]. Temperatures in the range 340–500 °C were used. All researchers found large amounts of carbonaceous deposits (10–30 wt.% of the feed) on top of the fixed bed. It was reasoned that these deposits were formed out of the heaviest compounds (lignin derivates) of the feed [1], which cannot be evaporated. Using a reactor concept with a mobile catalyst phase and an advanced atomization system may reduce the amount of these deposits considerably. In addition to the carbonaceous deposits on top of the fixed bed, coke formation on the catalyst was also observed (5–15 wt.% of the feed) and large amounts of water and gases were produced (mainly CO, CO_2, ethylene, propylene and butane). About 15–20 wt.% of the feed was converted into organic liquid products that consisted of 70–90 wt.% aromatic hydrocarbons, 0–5 wt.% aliphatic hydrocarbons, and 5–30 wt.% oxygenates. Hence, the organic liquid product needs further refining to produce conventional transportation fuels. Owing to catalyst deactivation the fraction of oxygenates increased at longer run (space) times [81, 84]. It was proven for HZSM-5 that deactivation by coking is reversible, but that dealumination by water causes irreversible deterioration of the acidity and hence activity. Except for less formation of carbonaceous deposits, work on pyrolysis vapors [81] afforded the same insights as the work on pyrolysis oil.

In conclusion, the development of upgrading technology for biomass derived liquids using zeolites is still in an embryonic stage. The main challenges for catalyst development are to avoid (a) deep cracking and formation of gas, which reduces liquid yields; (b) deep deoxygenation yields an aromatic product; and (c) severe catalyst deactivation due to oligomer/coke formation. Previous work has shown that the use of commercially available zeolites leads to a low liquid product yield. Deep cracking and the formation of gas may be controlled by manipulating the strength and concentration of acid sites in zeolites. Modification of acidity of H-Y zeolites with Na shows indeed that liquid yields can be affected [85]. The formation of aromatics cannot be avoided, as it arises from the low hydrogen content of biomass/bio-oil. Thus, suitability of this product as an additive to gasoline will depend on legislation. From this point of view, a selective deoxygenation, which leaves behind part of the oxygen in the bio-oil, may be a more attractive option. The resulting mixture should be made suitable for blending with hydrocarbon fuels by further processing. Formation of coke and catalyst deactivation is not a major problem. An FCC type operation, where continuous regeneration of the catalyst generates the energy required to run the endothermic cracking process, can be easily adopted as long as not too much coke is formed (otherwise enhanced gasification of coke is required).

6.10
Hydrolysis

Native cellulose is resistant to catalytic hydrolysis because it is protected by a matrix of lignin and hemicelluloses. Consequently, pretreatment is required to make the cellulose accessible and to extract the lignin (Section 6.5). Hydrolysis is the depolymerization of cellulose and hemicelluloses into mono sugars via the reaction with water using acid catalysis at 40–250 °C ([23], Vol. I, Chapter 6 and [86]). C_6 (Glucose, mannose etc.) and C_5 (xylose, arabinose etc.) sugars are the primary products, which are subject to further degradation. The rate of the sugar (monomer, oligomer, and polymer) degradation depends on the temperature, acid concentration and type of acid. For cellulose the idealized stoichiometric reaction transformation is given by:

$$\text{Cellulose} \rightarrow \text{Glucose} \rightarrow \text{5-HMF} + H_2O \rightarrow \text{Levulinic acid} + \text{Formic acid}$$

The development of economically viable hydrolysis processes for lignocelluloses has just started. Production of mono C5 and C6 sugars from lignocelluloses paves the way for development of sophisticated chemical process for the manufacture of, for example, *n*-alkanes (see below). An example of a hydrolysis process that integrates the primary conversion to mono sugars with the further conversion of these mono sugars into target components is the Biofine process. In this process, methyl-tetrahydrofuran is produced from levulinic acid made from acid hydrolysis of lignocelluloses ([23], Vol. I, Chapter 7). Catalysis may help in increasing the reaction rate and optimizing the yields of target products. The development of heterogeneous catalysts would be beneficial, because the homogeneous acid used in the present process is very corrosive, requiring expensive acid recovery units. Similar developments in catalysis of hydrocarbon alkylation point to large-pore zeolites (H-FAU, H-BEA) as possible starting catalysts [87].

Recently a very comprehensive report on the pathways from sugars to chemicals and fuels has been issued by the US department of energy [88]. The report identifies the twelve most promising building blocks (Table 6.5) that can be produced from sugars via biological and chemical conversions. These building blocks can be subsequently converted into several chemicals and fuels.

Another interesting example of a sugar route is the conversion of (a) sorbitol into hexane by acid-catalyzed dehydration to tetrahydrofuran followed by (b) aldol condensation over a solid base catalyst and (c) hydro-conversion over Pd, Pt on SiO_2-Al_2O_3 acid supports to give diesel range hydrocarbons, as recently reported by Dumesic et al. [89]. The development of the routes from sugars to alkenes is very promising and the proof of principle has been delivered. However, much more research is required to develop feasible processes. Conversion of cellulose into butanol, an additive to gasoline, is commercially targeted. Efforts in the USA aim to develop efficient cellulose conversion technology by 2012. Developments in these areas will help in the futuristic process of direct conversion of lignocellu-

Table 6.5 Top twelve building blocks from sugars according to the US department of energy [88].

1,4 succinic, fumaric and malic acids	itaconic acid
2,5 furan dicarboxylic acid	levulinic acid
3 hydroxy propionic acid	3-hydroxybutyrolactone
aspartic acid	glycerol
glucaric acid	sorbitol
glutamic acid	xylitol/arabinitol

lose via cellulose into fuels and chemicals. The reader is referred to reports on this topic for e.g. [90].

6.11
Underlying Approach for Catalyst Design

Currently, there is a debate going in the scientific community regarding the choice of methodology in catalyst development. In reality, research activities at the moment fall into three categories: (a) catalytic biomass related conversions using off the shelf commercial catalysts and the complex feed; (b) use of catalysts that are suitable for similar, though not identical, conversions in fossil oil upgrading; and (c) the use of model organic compounds with well-defined catalysts. Obviously, short term results are best obtained with approach (a) and it should be stressed that this type of research has been done over the last decade; the results have not been sufficient so far, as discussed earlier. The response to this situation can be twofold. Either rapid testing of many catalysts can be envisaged (high-throughput testing) but it is obvious that the type of experiments are even more difficult to mimic in high-throughput mode than for "standard" heterogeneous catalysis. Or, catalysts are to be improved based on a thorough knowledge of the fundamental processes on the catalyst. This approach should be inspired according to methodology (b) and requires the acceptance of methodology (c). Based on the history of the development of catalytic technology for refining of mineral oil, it is fair to state that a mix of these approaches will be indispensable to move forward rapidly, by involving the expertise available in both companies and academia.

Well-defined reaction and catalytic systems enable the development of fundamental knowledge that helps in with long term and more exhaustive problem solving possibilities. Thus the debate is not whether studies on model compounds are useful or not, but the choice of model compounds themselves. These should represent the nature and characteristic of the biomass fraction that one is studying. Not surprisingly, efforts to identify representative model compounds are gaining attention. 4-Hydroxyphenylpropane derivatives such as coniferyl, coumaryl or sinapyl alcohols can be representative for lignin, as the latter is made by the dehydrogenative polymerization of these compounds (see Fig. 6.2c). Similar studies are now appearing [91]. Model systems that represent the full complexity of the cellular structure of lignocellulosic biomass are not available. They are, however, necessary for catalysis research on pretreatment and primary conversions as it is already known that superposition of cellulose, hemicelluloses and lignin behavior does not mimic lignocelluloses [92].

6.12
Summary

A smooth transition, from the current fuel/energy scenario to a future dominated by the demand for sustainability, is essential to guarantee World's future economy and stability. Proper and timely development of technology to achieve this is therefore of paramount importance. Catalyst and reactor technology to convert fossil oil into the fuels needed by society today are mature. However, the transition from fossil to lignocellulosic biomass based feedstocks brings in new challenges both for catalyst and reactor engineering developments. To meet these demands, it is essential to adapt the knowledge available for making fossil fuels to lignocelluloses based fuels. Additionally, development of efficient processes will also demand new concepts for catalysts and reactor technologies. The focus of catalysis development should be on pretreatment (e.g., fractionation) and primary conversion processes (e.g., deoxygenation of pyrolysis oil) of lignocelluloses. These processes require cheap and robust catalysts that can cope with the fouling conditions caused by the complex feedstock materials. Secondary conversions processes take in much simpler feeds (e.g., sugars) and can, accordingly, make use of dedicated complex catalysts. Most importantly, teaming up of catalysis, reactor engineering and process engineering at an early stage is needed for the development of biomass-based processes for the generation of fuels.

References

1 International Energy Agency (IEA), *Energy to 2050 Scenarios for a Sustainable Future*, **2003**, ISBN: 92–64–01904–9.
2 European Commission, *Energy and Transport Trends to 2050–Update 2005*, **2006**, ISBN: 92–79–02305–5.
3 Global Business Environment Shell International 2001, *Exploring The Future: Energy Needs, Choices and Possibilities–Scenarios to 2050*, **2001**.
4 European Commission, Directive 2003/30/EC of the European Parliament and the Council of 8 May 2003 on the promotion of the use of biofuels of other renewable fuels for transport, *Official J. Eur. Union*, L 123, **2003**, 42–46.
5 European Commission, *An EU Strategy for Biofuels*, 2006, COM(2006) 34 final.
6 Van Geel, P.L.B.A., *Beleidsbrief Biobrandstoffen*, 15 March **2006**, KvI2006223088.
7 Petrus, L., Van Wechem, H., Van koolwaterstof naar koolhydraat, *Shell Venster*, March/April **2006**, 18–22.

8 Huber, G.W., Iborra, D., Corma, A., Synthesis of transportation fuels from biomass, *Chem., Catal., Engin., Chem. Rev.*, **2006**, 106, 4044.

9 Ragauskas, A.J., Williams, C.K., Davison, B.H., Britovsek, G., Cairney, J., Eckert, C.A., Frederick Jr., W.J., Hallet, J.P., Leak, D.J., Liotta, C.L., Mielenz, J.R., Murphy, R., Templer, R., Tschaplinski, T., The path forward for biofuels and biomaterials, *Science*, **2006**, 311, 484.

10 Lange, J.P., Chapter 2 of this book.

11 Van Swaaij, W.P.M., Prins, W., Kersten, S.R.A., Strategies for the future of biomass for energy, industry and climate protection. In: W.P.M. van Swaaij, T. Fjällstrom, P. Helm, A. Grassi (Eds.), *Second World Biomass Conference; Biomass for Energy, Industry and Climate Protection*, **2004**, ISBN: 88-89407-04-2, published by ETA-Florence and WIP-Munich, Italy.

12 Groeneveld, M.J., (Shell), personal communication.

13 Berndes, G., Hoogwijk, M., Van den Broek, R., The contribution of biomass in the future global energy supply: a review of 17 studies, *Biomass Bioenergy*, **2003**, 25, 1.

14 Katan, M., Rabbinge, R., Van Swaaij, W.P.M., Toekomst voor biodiesel is illusie, *Het Financiële Dagblad*, July 7, 2006.

15 www.choren.com.

16 Henrich, E., Weirich, F., Pressurized entrained flow gasifiers for biomass, *Environ. Eng. Sci.*, **2004**, 21, 53.

17 Elliott, D.C., Hart, T.R., Neuenschwander, G.G., Chemical processing in high pressure aqueous environments 8. Improved catalysts for hydrothermal gasification, *Ind. Eng. Chem. Res.*, **2006**, 45, 3776.

18 Vogel, F., Waldner, M., Catalytic hydrothermal gasification of woody biomass at high concentrations, In: A.V. Bridgwater, D.G.B. Boocock (Eds), **2006**, *Science in Thermal and Chemical Biomass Conversion*, 1001, ISBN: 1-872691-97-8, CPL Press. Tall Gables, The Sydings, Speen, Newbury, Berks RG14 1RZ, UK.

19 Vitelo, S., Seggiana, M., Frediani, P., Ambrosini, G., Poloti, L., Catalytic upgrading of pyrolytic oils over different zeolites, *Fuel*, **1999**, 78, 1147.

20 Samolada, M.C., Baldauf, W., Vasalos, I.A., Production of bio-gasoline by upgrading biomass flash pyrolysis liquids via hydrogen processing and catalytic cracking, *Fuel*, **1998**, 77, 1667.

21 www.ecn.nl/phyllis.

22 Kamm, B., Kamm, M., Principles of biorefineries, *Appl. Microbiol., Biotechnol.*, **2004**, 64, 137–145.

23 Kamm, B., Gruber, P.R., Kamm, M., *Biorefineries–Industrial Processes and Products*, **2006**, ISBN: 3-527-31027-4, Wiley-VCH, Weinheim.

24 Bridgwater, A.V., Meier, D., Radlein, D., An overview of fast pyrolysis of biomass, *Org. Chem.*, **1999**, 30, 1479.

25 Goudriaan, F., Peferoen, D.G.R., Liquid fuels from biomass via a hydrothermal process, *Chem. Eng. Sci.*, **1990**, 45, 2729.

26 Beckman, D., Elliott, D.C., Comparisons of the yields and properties of the oil products from direct thermochemical biomass liquefaction processes, *Can. J. Chem. Eng.*, **1985**, 63, 99.

27 Yu, F., Liu, Y., Pan, X., Lin, X., Liu, C., Chen, P., Ruan, R., Liquefaction of corn stover and preparation of polyester for liquefied polyol, *Appl. Biochem. Biotechnol.*, **2006**, 129–132, 574.

28 Pan, X., et al. Bioconversion of hybrid poplar to ethanol and co-products using an organosolv fractionation process: optimization of process yields, *Biotechnol. Bioeng.*, **2006**, 94, 851.

29 Zhang, Y.-H.P., Lynd, L.R., Novel lignocellulose fractionation featuring modest reaction conditions and reagent recycling. The 231st ACS National Meeting, Atlanta, Georgia, March 30, 2006.

30 Beenackers, A.A.C.M., Van Swaaij, W.P.M., Gasification of biomass, a state of the art review, In: A.V. Bridgwater (Ed.), **1984**, *Proceedings 1st European Workshop on Thermochemical Processing of Biomass*, Butterworths, London, p. 91.

31 Maniatis, K., Progress in biomass gasification, an overview, In: A.V. Bridgwater, *Progress in Thermochemical Biomass Conversion*, **2001**, ISBN: 0-632-05533-2, Blackwell Science Ltd., Oxford, p. 1.

32 Stassen, H.E.M., Prins, W., Van Swaaij, W.P.M., Thermal conversion of biomass into secondary products; the case of

gasification and pyrolysis, In: W. Palz, J. Spitzer, K. Maniatis, K. Kwant, P. Helm, A. Grassi (Eds.), **2002**, *Twelfth European Conference on Biomass for Energy, Industry and Climate Protection*, ISBN: 3-936338-10-8, ETA-Florence (Florence) and WIP-Munich (Munich), p. 38.

33. Higman, C., Van der Burgt, M., *Gasification*, **2003**, ISBN: 0-7506-7707-4, Elesevier Science, Amsterdam.

34. Kersten, S.R.A., van der Drift, B., Prins, W., van Swaaij, W.P.M., Experimental fact-finding in CFB biomass gasification for ECN's 500 kW$_{th}$ pilot-plant, *Ind. Eng. Chem. Res.*, **2003**, 42, 6755.

35. Wang, X., Cents, A.H.G., Kersten, S.R.A., Prins, W., Van Swaaij, W.P.M., A filter-assisted fluid bed reactor for integrated production and cleaning of bio-oil. In: W.P.M. van Swaaij, T. Fjällstrom, P. Helm, A. Grassi (Eds.), **2004**, *Second World Biomass Conference. Biomass for Energy, Industry and Climate Protection*, ISBN: 88-89407-04-2, ETA-Florence (Florence) and WIP-Munich (Munich), p. 709.

36. Czernik, C., French, R., Feik, C., Chornet, E., Hydrogen by catalytic steam reforming of liquid byproducts from biomass thermoconversion processes, *Ind. Eng. Chem. Res.*, **2002**, 41, 4209.

37. Garcia, L., French, R., Czernik, S., Chornet, E., Catalytic steam reforming of bio-oils for the production of hydrogen: effects of catalyst composition, *Appl. Catal. A.*, **2000**, 201, 225.

38. Van Rossum, G., Kersten, S.R.A., Van Swaaij, W.P.M., Catalytic and non-catalytic gasification of pyrolysis oil, submitted to *Ind. Eng. Chem. Res.*, available online: DOI: 10.1021/ie061337y.

39. Sutton, D., Keller, B., Ross, J.R.H., Review of literature on catalysts for biomass gasification, *Fuel Proc. Technol.*, **2001**, 73, 155.

40. Gil, J., Caballero, M.A., Martín, J.A., Aznar, M.P., Correla, J., Biomass gasification with air in a fluidized bed: Effect of the in-bed use of Dolomite under different operating conditions, *Ind. Eng. Chem. Res.*, **1999**, 38, 4226.

41. Aznar, M.P., Caballero, M.A., Gil, J., Martín, J.A., Correla, J., Commercial steam reforming catalysts to improve biomass gasification with steam-oxygen mixtures. 2. catalytic tar removal, *Ind. Eng. Chem. Res.*, **1998**, 37, 2668.

42. Takanabe, K., Aika, K.-i., Seshan, K., Lefferts, L., Sustainable hydrogen from biooil — Steam reforming of acetic acid as a model oxygenate, *J. Catal.*, **2004**, 227, 101.

43. Rostrup Nielson, J.R., Catalytic steam reforming, In: J.R. Anderson, M. Boudart (Eds.), **1984**, *Adv. Cat. Sci. & Technol.*, Springer Verlag, Berlin, Vol. 4.

44. Takanabe, K., Aika, K., Seshan, K., Lefferts, L., Sustainable hydrogen from biooil — Steam reforming of acetic acid as a model oxygenate, *J. Catal.*, **2004**, 227, 101-108.

45. Takanabe, K., Aika, K.-i., Inazu, K., Baba, T., Seshan, K., Lefferts, L., Steam reforming of acetic acid as a biomass derived oxygenate: Bifunctional pathway for hydrogen formation over Pt/ZrO$_2$ catalysts, *J. Catal.*, **2006**, 243, 263.

46. Matsumura, Y., Minowa, T., Potic, B., Kersten, S.R.A., Prins, W., Van Swaaij, W.P.M., Van de Beld, L., Elliott, D.C., Neuenschwander, G.G., Antal Jr, M.J., Biomass gasification in near and supercritical water: Status and prospects, *Biomass Bioenergy*, **2005**, 29, 269.

47. Antal Jr, M.J., Allen, S.G., Schulman, D., Xu, X., Divilio, R.J., Biomass gasification in supercritical water, *Ind. Eng. Chem. Res.*, **2000**, 39, 4040.

48. Kersten, S.R.A., Prins, W., Van Swaaij, W.P.M., Reactor design considerations for biomass gasification in hot compressed water, In: W.P.M. van Swaaij, T. Fjällstrom, P. Helm, & A. Grassi (Eds.), **2004**, *Second World Biomass Conference. Biomass for Energy, Industry and Climate Protection*, ISBN: 88-89407-04-2, ETA-Florence (Florence) and WIP-Munich (Munich), p. 777.

49. Kersten, S.R.A., Potic, B., Prins, W., Van Swaaij, W.P.M., Gasificaton of model compounds and wood in hot compressed water. *Ind. Eng. Chem. Res.*, **2006**, 45, 4169.

50. Cortright, R.D., Davda, R.R., Dumesic, J.A., Hydrogen from catalytic reforming of biomass-derived hydrocarbons in liquid water, *Nature*, **2002**, 418, 964.

51 Bridgwater, A.V., Peacocke, G.V.C., Fast pyrolysis processes for biomass, *Renewable Sustainable Energy Rev.*, **2000**, 5, 1.

52 Bridgwater, A.V., Meier, D., Radlein, D., An overview of fast pyrolysis of biomass, *Org. Geochem.*, **1999**, 30, 1479.

53 Czernik, S., Bridgwater, A.V., Overview of application of biomass fast pyrolysis oil, *Energy Fuels*, **2004**, 18, 590.

54 Scott, D.S., Majerski, P., Piskorz, J., Radlein, D., A second look at fast pyrolysis of biomass – the RTI process, *J. Anal. Appl. Pyrolysis*, **1999**. 51, 23.

55 Kersten, S.R.A., Wang, X., Prins, W., Van Swaaij, W.P.M., Biomass pyrolysis in a fluidized bed reactor. Part 1: literature review and model simulations. *Ind. Eng. Chem. Res.*, **2005**, 44, 8773.

56 Shafizadeh, F., Chin, P.S., Thermal deterioration of wood, *ACS Symp. Ser.*, **1977**, 43, 57.

57 Boroson, M.L., Howard, J.B., Longwell, J.P., Peters, W.A., Heterogeneous cracking of wood pyrolysis tars over fresh wood char surfaces, *Energy Fuels*, **1989**, 3, 735.

58 Garcìa-Pérez, M., Chaala, A., Pakdel, H., Kretschmer, D., Highes, P., Roy, C., The complex structure of bio-oils, In: W.P.M. van Swaaij, T. Fjällstrom, P. Helm, A. Grassi (Eds.), **2004**, *Second World Biomass Conference. Biomass for Energy, Industry and Climate Protection*, ISBN: 88-89407-04-2, ETA-Florence (Florence) and WIP-Munich (Munich), p. 725.

59 Agblevor, F.A., Besler, S., Inorganic compounds in biomass feedstocks. 1. Effects on the quality of fast pyrolysis oils, *Energy Fuels*, **1996**, 10, 293.

60 Brown, R.C., Radlein, D., Piskorz, J., Pretreatment processes to increase pyrolytic yield of Levoglucosan from herbaceous biomass, chemicals and materials from renewable resources, *ACS Symp. Ser.*, **2001**, 784, 123.

61 Atutxa, A., Aguado, R., Gayubo, A.G., Olozar, M., Bilbao, J., Kinetic description of the catalytic pyrolysis of biomass in a conical spouted bed, *Energy Fuels*, **2005**, 19, 765.

62 Lapas, A.A., Samolada, M.C., Iatridis, D.K., Voutetakis, S.S., Vasalos, I.A., Biomass pyrolysis in a circulating fluid bed reactor for the production of fuels and chemicals, *Fuel*, **2002**, 81, 2087.

63 Deshmukh, S., Bramer, E., Leferts, L., Seshan, K., Catalytic flash pyrolysis, *Ind. Engg. Chem. Res.*, in preparation.

64 Elliot, D.C., Baker, E.G., Hydrotreating biomass liquids to produce hydrocarbon fuels, In: Klass, D.L. (ed.), **1987**, *Energy from Biomass and Waste*, publ. IGT, Chicago, p. 765.

65 Bridgwater, A.V., Production of high grade fuels and chemicals from catalytic pyrolysis of biomass, *Catal. Today*, **1996**, 29, 285.

66 Bridgwater, A.V., Catalysis in thermal biomass conversion, *Appl. Catal. A.*, **1994**, 106, 5.

67 Baldauf, W., Balfanz, U., Rupp, M., Upgrading of flash pyrolysis oils and utilization in refineries, *Biomass Bioenergy*, **1994**, 7, 237.

68 Seshan, K., unpublished results.

69 Vonghia, E.V., Boocock, D.G.B., Konar, S.K., Leung, A., Pathways for the deoxygenation of triglycerides to aliphatic hydrocarbons over activated alumina, *Energy Fuels*, **1995**, 9, 1090.

70 Kubičkoví, I., Snåre, M., Eränen, K., Mäki/Arvela, P., Hydrocarbons for diesel fuel via decarboxylation of vegatable oils, *Catal. Today*, **2005**, 106, 197.

71 Ashida, R., Painter, P., Larsen, J.W., Kerogen chemistry 4. Thermal decarboxylation of kerogens, *Energy Fuels*, **2005**, 19, 1954.

72 Elliott, D.C., Schiefelbein, G.F., Liquid hydrocarbon fuels from biomass, *Am. Chem. Soc., Div. Fuel. Chem.*, **1989**, 34, 1160.

73 Samolada, M.C., Baldauf, W., Vasalos, I.A., Production of a bio gasoline by upgrading of biomass flash pyrolysis liquids via hydrogen processing and catalytic cracking, *Fuel*, **1998**, 77, 1667.

74 Chantal, P.D., Kaliaguine, S., Grandmaissen, J.L., Mahay, A., Production of hydrocarbons from Aspen Poplar pyrolytic oils over HZMS-5, *Appl. Catal.*, **1984**, 10, 317.

75 Evans, R.J., Milne, T.A., Molecular characterization of the pyrolysis of biomass. 2. Applications, *Energy Fuels*, **1987**, 1, 311.

76 Chantal, P.D., Kaliaguine, S., Grandmaissen, J.L., Reactions of phenolic compounds over HZMS-5, *Appl. Catal.*, 18, **1985**, 133.

77 Grandmaissen, J.L., Chantal, P.D., Kaliaguine, S., Conversion of furanic compounds over HZSM-5, *Fuel*, **1990**, 69, 1058.

78 Gayubo, A.G., Aguayo, A.T., Atutxa, A., Aguado, R., Bilbao, J., Transformation of oxygenate components of biomass pyrolysis oil on a HZSM-5 zeolite. I. Alcohols and phenols, *Ind. Eng. Chem. Res.*, **2004**, 43, 2610.

79 Gayubo, A.G., Aguayo, A.T., Atutxa, A., Aguado, R., Olozar, M., Bilbao, J., Transformation of oxygenate components of biomass pyrolysis oil on a HZSM-5 zeolite. II. Aldehydes, ketones, and acids, *Ind. Eng. Chem. Res.*, **2004**, 43, 2619.

80 Gayubo, A.G., Tarrío, A.M., Aguayo, A.T., Olozar, M., Bilbao, J., Kinetic modelling of the transformation of aqueous ethanol into hydrocarbons on a HZSM-5 zeolite, *Ind. Eng. Chem. Res.*, **2001**, 40, 3467.

81 Horne, P.A., Williams, P.T., The effect of zeolite ZSM-5 catalyst deactivation during the upgrading of biomass-derived pyrolysis vapours, *J. Anal. Appl. Pyrolysis*, **1995**, 34, 65.

82 Sharma, R.K., Bakhshi, N.N., Catalytic upgrading of pyrolysis oil, *Energy Fuels*, **1993**, 7, 306.

83 Vitelo, S., Seggiana, M., Frediani, P., Ambrosini, G., Poloti, L., Catalytic upgrading of pyrolytic oils over different zeolites, *Fuel*, **1999**, 78, 1147.

84 Gayubo, A.G., Aguayo, A.T., Atutxa, A., Prieto, R., Bilbao, J., Deactivation of a HZSM-5 zeolite catalyst in the transformation of the aqueous fraction of biomass pyrolysis oil into hydrocarbons, *Energy Fuels*, **2004**, 18, 1640.

85 Seshan, K., results to be published.

86 Mosier, N.S., Ladisch, C.M., Ladisch, M.R., Characterization of acid catalytic domains for cellulose hydrolysis and glucose degradation, *Biotechnol. Bioeng.*, **2002**, 79, 618.

87 Nivarthy, G.S., Feller, A., Seshan, K., Lercher, J.A., Alkylation of isobutane with light olefins catalyzed by zeolite beta, *Microporous Mesoporous Mater.*, **2000**, 75, 35–36.

88 US Department of Energy, Top Value Added Chemicals From Biomass, Vol. 1: *Results of Screening for Potential Candidates from Sugars and Synthesis Gas*, **2004**, available online at: http://www.osti.gov/bridge.

89 Huber, G.W., Chheda, J.N., Barret, C.J., Dumesic, J.A., Production of liquid alkanes by aqueous-phase processing of biomass-derived carbohydrates, *Science*, **2005**, 308, 1446.

90 Heeres, R.H., Jende, J.F., Agterberg, F., Droescher, M.J., *Green Chem.*, **2004**, 6, 544–556.

91 Ragauskas, A.J., Nagy, M., Kim, D.H., *Ind. Biotechnol.*, **2006**, 2, 55.

92 Yoshida, T., Oshima, Y., Matsumura, Y., Gasification of biomass model compounds and real biomass in supercritical water, *Biomass Bioenenergy*, **2004**, 26, 71.

7
Thermal Biomass Conversion

Simone Albertazzi, Francesco Basile, Giuseppe Fornasari,
Ferruccio Trifirò, and Angelo Vaccari

7.1
Introduction

Fossil fuels (coal, oil and natural gas) are used to cover almost the 80% of the power demand in the world via combustion processes. However, firing with these fuels gives rise to carbon dioxide, which is a "greenhouse" gas discharged to the atmosphere. Owing to the growing interest in reducing the CO_2 emissions from sources such as fossil fuels, biomass thermal conversion has attracted a great attention in the last few years. Indeed, the carbon dioxide generated in the combustion of biomass is not considered to give any net contribution to the CO_2 content of the atmosphere, since it is absorbed by photosynthesis when new biomass is growing [1, 2]. However, the use of biomass for power generation is more limited, although several smaller combined heat and power (CHP) plants have been built in recent years [3]. The electrical efficiency of the plants based on conventional technology (i.e., a boiler unit and a steam turbine cycle) is around 30% and the α value (ratio of electrical energy to thermal energy generated) is around 0.5 or below [4]. Although there is potential development for these plants, the electrical efficiency and the α value cannot be expected to increase to any significant extent. A technology for achieving higher electrical efficiencies is based on the gasification of solid fuels and combustion of the gas thus produced in a gas turbine and in a following steam cycle (integrated gasification combined cycle, IGCC). Several studies [5] have shown that well-optimized generation plants, rated at 30–60 MW_e, based on pressurized gasification of wood and integrated into a combined cycle can achieve net electrical efficiencies of 40–50% and an overall efficiency of 85–90% with competitive generation costs and low emission levels. An interesting alternative to the production of power is gas upgrading to obtain a mixture of CO and H_2, thus having a high added value. Hydrogen is an important raw material for the chemical industry and is a clean fuel that can be used in fuel cells and internal combustion engines [6, 7]. The main process for hydrogen production is currently the catalytic steam reforming of methane, light hydrocarbons and naphtha. Partial oxidation of heavy oil residues and coal gasifi-

Catalysis for Renewables: From Feedstock to Energy Production
Edited by Gabriele Centi and Rutger A. van Santen
Copyright © 2007 WILEY-VCH Verlag GmbH & Co. KGaA, Weinheim
ISBN: 978-3-527-31788-2

cation are also alternative processes for producing hydrogen [8]. Two innovative technologies using biomass as feedstock are steam gasification [9–13] and catalytic steam reforming of pyrolysis oils [14, 15]. The latter route begins with fast pyrolysis of biomass to produce bio-oil, which can be converted into hydrogen via catalytic steam reforming followed by a shift conversion step. Moreover, the economy of scale can make possible a more efficient but more complex utilization of the pyrolysis slurry. Since the costs per kg of transporting low-density biomass (such as straw and bagasse) are much higher than those of transporting high-density liquor, the slurries can be transported from a wide number of small and local pyrolysis plants to a large central gasification facility [16], to produce the most valuable products such as syngas and H_2. However, this technology is still under development, due to several problems. Tar formation (heavy hydrocarbons, refractory aromatics etc.) is one of the major problems, since tar condenses at reduced temperature, thus blocking and fouling process equipment. Other species initially present in the biomass (sulfur and nitrogen compounds, alkali and heavy metals) act as poisons for the gas upgrading units. Therefore, an optimal gas cleaning stage and the development of more tolerant catalysts are key points for commercialization of the process.

7.2
Biomass Resources and Biomass Pre-treatment

Biomass resources include various natural and derived materials, such as woody and herbaceous species, wood wastes, bagasse, agricultural and industrial residues, waste paper, municipal solid waste, sawdust, biosolids, grass, waste from food processing, animal wastes, aquatic plants and algae etc. The major organic components of biomass can be classified as cellulose, hemicellulose and lignin [17]. Cellulose is a polysaccharide having the general formula $(C_6H_{10}O_5)_n$ and an average molecular weight range of 300 000–500 000 Da. Hemicellulose is a complex polysaccharide that consists of branched structures, which have the general formula $(C_5H_8O_4)_n$ and a degree of polymerization (number of monomeric units) of less than 200. Lignin consists of highly branched, substituted, mononuclear aromatic polymers, especially present in woody biomass species, and it is responsible for the heating value of the biomass [18].

Ensuring a constant feed supply is very important, because most of the biomass is only available on a seasonal basis. In such cases continuous operation of the conversion facility will require either extensive long-term storage of the feedstock or a feed reactor that is flexible enough to accommodate multiple feedstocks. Most thermal conversion processes demand a finely divided, substantially dry feed and therefore some pre-treatment is required to match the feedstock to the process. The main pre-treatment operations are [19]:

(a) Screening of the feedstock to remove metal or rocks.

(b) Drying of the feedstock to a moisture content suitable for the conversion technology. Drying is generally the most important pre-treatment operation, and is necessary for high efficiency. Drying reduces the moisture content to 10–15%. Drying can either be done with flue gas or with steam. Steam drying results in very low emissions and it is safer with respect to risks for dust explosion. However, using flue gas is the cheapest way to dry the feedstock.

(c) Tailoring the feedstock to an appropriate particle size. To avoid a pressure drop, sawdust and other small particles must be pelletized, while smaller particles can be used in circulating fluidized bed (CFB) gasifiers.

7.3
Biomass Combustion

Firing biomass was the first civilization step in human history. Consequently, nowadays combustion with a Rankine cycle for power generation from biomass is well established, with hundreds of plants operating world wide [5, 20, 21]. Most are based on conventional technology using either fixed beds or grates, or, more recently, fluidized and circulating fluid bed systems. In grate-fired combustors the feed burns as it passes through the furnace by means of a moving grate. Fluid bed designs burn the feed in a turbulent bed of inert material that is fluidized by the air flow from underneath. The inert material is required to maintain fluidization and to provide a large heat transfer area (see below). A catalyst (e.g., Ni on olivine) can eventually be put inside the furnace to improve combustion of tars.

Grate-fired combustors are in use for old biomass-fired plant, while fluid bed combustors are rapidly becoming the preferred technology for biomass combustion because of their low NO_x emissions.

Electricity generation is carried out by internal and external combustion engines or turbines. The basic steam turbine Rankine cycle is bound by thermodynamic and materials limitations to modest efficiencies of around 35%. Such cycles are optimized through the use of high pressure, highly superheated steam combined with complex steam generation, reheat and regeneration options, which, however, boost capital costs, especially on a small scale. As a result, most small-scale steam cycles are relatively simple and, consequently, inefficient. However, excess heat or steam can be used as process steam or in district heating, thus increasing the overall system efficiency.

7.4
Biomass Gasification

Gasification is the conversion by partial oxidation at elevated temperature of a carbonaceous feedstock into a gaseous energy carrier consisting of permanent, non-condensable gases. Ideally, the process produces only a non-condensable gas and an ash residue. However, since gasification processes are carried out far from equilibrium, tars (condensable organic material) are produced and the ash resi-

Table 7.1 Main reactions occurring in a gasifier.

Gasifying reactions	$\Delta H°_{298}$ (kJ mol^{-1})
Organic compounds $\leftrightarrows$ CH$_4$ + C	<0
C + 1/2O$_2$ $\leftrightarrows$ CO	−111
CO + 1/2O$_2$ $\leftrightarrows$ CO$_2$	−254
H$_2$ + 1/2O$_2$ $\leftrightarrows$ H$_2$O	−242
C + H$_2$O $\leftrightarrows$ CO + H$_2$	+131
C + CO$_2$ $\leftrightarrows$ 2CO	+172
C + 2H$_2$ $\leftrightarrows$ CH$_4$	−75
CO + 3H$_2$ $\leftrightarrows$ CH$_4$ + H$_2$O	−206
CO + H$_2$O $\leftrightarrows$ CO$_2$ + H$_2$	−41
CO$_2$ + 4H$_2$ $\leftrightarrows$ CH$_4$ + 2H$_2$O	−165

due often contains some char. Table 7.1 lists the main reactions occurring in a gasifier [22]. In the oxygen-deficient zone (far from the injection of the oxidant), pyrolysis of the organic compounds into methane and carbon occurs. In the oxygenated zone these compounds are then converted mainly into CO, CO$_2$, H$_2$O and H$_2$ by reforming and partial oxidation reactions.

Development of gasification technology dates back to the end of the 18th century when hot gases from coal and coke furnaces were used in boiler and lighting applications. Gasification of coal is now well established, and biomass gasification benefited from that technology.

The main parameters of a gasifying process are [23]:

(a) Temperature: gasification is generally carried out at 750–900 °C; higher temperatures allow a higher conversion of tars into gaseous compounds but increase the overall cost of the process.

(b) Pressure: all gasifier types can be operated at either atmospheric or elevated pressure. With increasing pressure, the gasifier and other process equipments can be made smaller. However, less energy is consumed by gas compression in atmospheric gasification.

(c) Gasifying media: air or oxygen. The presence of nitrogen decreases the heating value of the produced gas and increases the size of downstream equipments. If the produced gas is intended for synthesis gas, oxygen should be used. Steam would have to be used to dilute the oxygen, when maintaining enough gas flow is necessary (i.e., in fluidized beds) but avoiding combustion of the biomass. Obviously, operating in an oxygen/steam flow is more expensive than operating in air.

Gasifiers have been designed in various configurations [4, 5, 24], each having its own advantages and disadvantages, in terms of product gas composition and operating parameters.

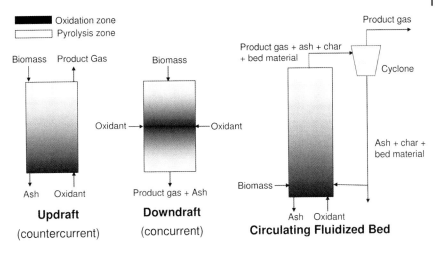

Fig. 7.1 Main gasifying reactor configurations.

Fixed bed gasifiers are generally used for small-scale operation with gas engines, having an electrical output of about 80–500 kW$_e$. They divide into updraft and downdraft gasifiers.

In an updraft gasifier (Fig. 7.1) the downward-moving biomass is first dried by the upflowing hot product gas. After drying, the solid fuel is pyrolyzed, giving char that continues to move down to be gasified, and pyrolysis vapors that are carried upward by the upflowing hot product gas. The organics in the vapor either condense on the cool descending fuel or are carried out of the reactor with the product gas, thus forming a high amount of tars (higher than any other gasifier type). For heat applications this is no problem as long as blocking of pipes can be overcome. However, it makes this gasifier unsuitable for power applications because of the extensive tar cleaning required. Since the gas leaves this gasifier at relatively low temperatures, this process has a high thermal efficiency, and biomass up to 50% moist can still be gasified without pre-drying of the feed. Moreover, size specifications of the fuel are not very critical for this gasifier. As a result, for heat applications at capacities below 10 MW$_{th}$ updraft gasifiers are the most popular.

The downdraft gasifier (Fig. 7.1) features concurrent flow of gases and solids through a descending packed bed that is supported across a throat, where most of the gasification reactions occur. The reaction products are intimately mixed in the turbulent high-temperature region around the throat, which aids tar cracking, thus leading to a relatively clean gas and making this configuration generally favored for small-scale electricity generation with an internal combustion engine. There is a practical upper limit to the capacity of this configuration of up to 500 kWe.

Fluidized beds provide many features not available in the fixed-bed types, including high rates of heat and mass transfer and good mixing of the solid phase, which means that reaction rates are high and the temperature is more or less constant in the bed. A relatively small particle size compared with that in fixed bed gasifiers is required. In these gasifiers a bed material is used to maintain fluidization and to provide a large heat transfer area. Sand, magnesite (MgO) or magnesium carbonate ($MgCO_3$) are usually chosen as bed material. The latter are basic and do not react with alkali from the fuel, have high strength so that they will not be worn down, are cheap even if more expensive than sand. Loss of fluidization due to bed sintering is one of the commonly encountered problems.

The most common design among fluidized bed reactors is the circulating type, due to its highest carbon conversion obtained by means of the recycling system and to the confidence in scaling up such a reactor in applications that generate over 1 MW_e (Fig. 7.1). Thus, circulating fluidized bed (CFB) gasifiers are claimed to be available from several manufacturers in thermal capacities ranging from 2.5 to 150 MW_{th} for operations at atmospheric or elevated pressures, using air or oxygen as gasifying agent.

A new technology under development is the entrained bed gasifier. In entrained flow gasifiers no inert material is present but a finely reduced feedstock is required. Entrained bed gasifiers operate at much higher temperatures of 1200–1500 °C, depending on whether air or oxygen is employed, and hence the product gas has low concentrations of tars and condensable gases. However, this high-temperature operation creates problems of materials selection and ash melting. Conversion in entrained beds effectively approaches 100%; however, there are some problems coupled with the pulverizing of the biomass to be fed entrained with air.

To prevent damage to downstream equipments and poisoning of downstream catalysts, the following contaminants have to be removed from the flue gas: particulate (ash, char, and fluid bed material) causing erosion, alkali metals (sodium and potassium compounds) responsible for hot corrosion, tars (high molecular weight hydrocarbons and refractory aromatics), and catalyst poisoning species (H_2S, HCl, NH_3, and HCN).

Tar formation is one of the major problems to deal with during biomass gasification [3]. Tar condenses at reduced temperature, thus blocking and fouling process equipments such as engines and turbines. Tar removal technologies can broadly be divided into two approaches (Fig. 7.2): hot gas cleaning after the gasifier (secondary methods), and treatments inside the gasifier (primary methods). Although secondary methods are proven to be effective, treatments inside the gasifier are gaining much attention as the gasifier is optimized to produce a gas with low tar concentration. There is a potential to use some active bed additives such as dolomite or olivine inside the gasifier [25–27]. Nickel-based catalysts are reported to be very effective not only for tar reduction but also for reforming methane [28–30]. Since the cracking of tars inside the gasifier can significantly improve the heating value of the produced gas, catalytic cracking is preferred to removal methods. However, in the gasifier environment, the lifetime of Ni catalysts is very short, due to the high amount of sulfur compounds and other contaminants.

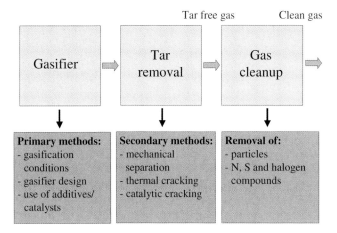

Fig. 7.2 Tar removal and gas cleaning approach for biomass gasification.

Gas clean up methods can be classified into two distinct routes: "wet" low temperature cleaning and "dry" high temperature cleaning.

Conventional "wet" low-temperature syngas cleaning (Fig. 7.3) is the preferred technology in the short term. This technology has the highest removal efficiency and it is well established, but requires additional wastewater and solid treatments. Firstly, a gas filter collects particulate matter. Bags work at low temperature, while candles work at higher temperatures. Ceramic candles are commonly used for atmospheric gasification, while metal candles are suitable for pressurized systems. In new configurations, a Ni based catalyst is loaded on the filter to, additionally, crack tars. Then a scrubber operating with a NaOH solution removes the acid contaminants and another scrubber with a H_2SO_4 solution removes the basic ones. Finally, a ZnO guard bed is a very efficient trap at low temperature, even for a few ppm of H_2S. However, the bed has to be constantly replaced and the ZnS so-produced has to be discharged and treated.

Hot gas cleaning consists of several filters and separation units in which the high temperature of the product gas (≈ 800 °C) can be maintained, achieving efficiency benefits and lower operational costs. Hot gas cleaning is specifically advantageous when preceding gas upgrading stages, because these process steps have high inlet temperatures. Hot gas cleaning after atmospheric gasification

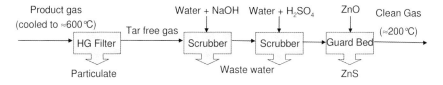

Fig. 7.3 Schematic view of "wet" low-temperature cleaning.

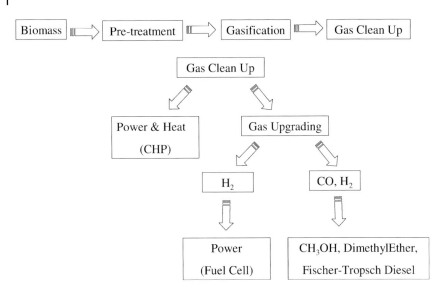

Fig. 7.4 Basic pathway of producing power, hydrogen and fuels from biomass gasification.

does not improve efficiency, because the subsequent essential compression requires syngas cooling anyway. Hot gas cleaning is not a commercial process yet, since it is still in the experimental phase. However, within a few years hot gas cleaning will become established and commercially available, improving the economic and technical feasibility of producing fuels from biomass feedstocks (Fig. 7.4).

7.5
Pyrolysis

Pyrolysis is the thermal degradation of biomass in the absence of an oxidizing agent, whereby the volatile components of a solid carbonaceous feedstock are vaporized in primary reactions by heating, leaving a residue consisting of char and ash. Pyrolysis always produces a gas mainly containing CO_2, CO, CH_4, H_2, C_2H_6, C_2H_4, a vapor that can be collected as a liquid, and a solid char (bituminous and anthracitic coal). The char may be sold as solid fuel or even as precursor of active carbon, but often it is burned internally to provide heat for the process. The gas has a medium heating value and it is used to provide process heat, re-circulated as an inert carrier gas or exported for feed drying. The liquid product has many denominations, such as pyrolysis liquid, pyrolysis oil, bio-oil, bio-crude-oil, bio-fuel-oil, wood liquids, wood oil, liquid smoke, wood distillates and so on. Bio-oil (Table 7.2 [19]) is a dark brown liquid with a distinctly smoky smell and can irritate the eyes on prolonged exposure. It contains over 100 different chemicals with a wide range of molecular weight distribution and functional groups (most are

Table 7.2 Composition of pyrolysis oil and comparison with other fuels.

Property	Pyrolysis oil	Diesel	Heavy fuel oil
Density at 15 °C (kg m^{-3})	1220	854	963
C (wt.%)	48.5	86.3	86.1
H (wt.%)	6.4	12.8	11.8
O (wt.%)	42.5	0.0	0.0
S (wt.%)	0.0	0.9	2.1
Viscosity at 50 °C (cSt)	13	2.5	351
Flash point (°C)	66	70	100
Pour point (°C)	−27	−20	21
Ash (wt.%)	0.13	<0.01	0.03
Water (wt.%)	20.5	0.1	0.1
LHV (MJ kg^{-1})	17.5	42.9	40.7
Acidity (pH)	3	–	–

oxygenated, such as carbonyl, carboxyl and phenolic). Bio-oil has a heating value of 16–18 MJ kg^{-1}. This is about 40% on weight basis of what conventional fuel oil has and about 60% if you compare the same volumes due to the higher density of bio-oil (1.2 kg L^{-1} compared with 0.85 kg L^{-1} for conventional fuel oil). Bio-oil is a homogenous mixture of organic compounds and water. Owing to its high water content, the pyrolysis oil is not miscible with fossil fuels and, therefore, cannot be mixed to pure fuel blends, even at low rates, as happens instead for some residual streams in the refinery system (i.e., light cycle oils coming from fluid catalytic cracking). The high water content is also detrimental for ignition. Moreover, organic acids in the oils are highly corrosive to common construction materials. Char in the liquid can block injectors or erode turbine blades. In addition, over time, the reactivity of some components in the oils affords larger molecules that result in high viscosity [31]. As a result, the pyrolysis oil as it stands can be used as a fuel only in slightly modified boilers and engines, where the start-up is carried out with fossil fuel. Some upgrading methods, such as catalytic hydrotreating [32] or steam reforming [8, 33], are used to improve the stability, to remove oxygen and water, thus increasing the burning properties of the pyrolysis liquid to convert it into economic and environmentally acceptable liquid fuel. Reactive pyrolysis indicates that the pyrolysis processes is carried out in a reactive atmosphere, i.e., hydrogen, to directly obtain an upgraded bio-oil.

Besides fuel and power production, there is the opportunity of recovering chemicals from pyrolysis liquids (Fig. 7.5). Even if there is a wide range of specialties that can be extracted or derived, including food flavorings, resins, agrochemicals, fertilizers and emissions control agents, this application is likely to lie in niche markets.

Since the costs per kg of transporting low-density biomass are much higher than those of transporting high-density liquor, the bio-oil can be trans-

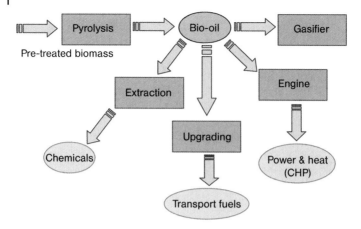

Fig. 7.5 Main utilizations of pyrolysis oil.

ported from a wide number of small and local pyrolysis plants to a large central gasification–upgrading facility, to take advantage of the gasifier scale economy. Moreover, since pyrolysis oil is relatively free of contaminants (especially inorganic), excessive gas cleaning may be avoided after the gasifier [34].

All the different pyrolysis processes work in the same T range (500–600 °C), the best compromise between gas and char yields, but they can be discriminated according to the heating rate, thus leading to a different ratio between gas, liquid and solid fractions [35, 36].

The pyrolysis processes may be, therefore, divided into:

(a) Flash (or ultra fast) pyrolysis: a heating rate of 10 000 °C s^{-1} and a rapid quenching of the vapors (vapor residence time of a few hundred milliseconds) maximize the liquid fraction (up to 80% respect to the dry feed).

(b) Fast pyrolysis: heating rate of 300 °C min^{-1}; gaseous fraction 20–30%; liquid fraction 50–60%; solid fraction 20–30%.

(c) Slow pyrolysis: heating rate of 30 °C min^{-1}; liquid fraction is below 50%; driven to equilibrium, cracking reactions maximize the production of char and incondensable compounds.

The most interesting process is, therefore, the flash pyrolysis, because it leads to the maximum yield of the most valuable product, the oil. For this process, the key parameters are the char separation and the vapor residence time (determined by the quenching method).

Some fine char is inevitably carried in the liquid and can only be removed by liquid filtration using, for example, cartridge or rotary filters. Almost all of the ash in the biomass is retained in the char, so successful char removal gives successful ash removal. Char separation, however, is difficult and may not be necessary for all applications. Char contributes to secondary cracking by catalyzing secondary cracking in the vapor phase. Even char in the cooled collected liquid product contributes to instability problems, accelerating the slow polymerization processes, which are manifested as increasing viscosity. Rapid and complete char separation is therefore desirable.

The time–temperature profile between formation of pyrolysis vapors and their quenching influences the composition and quality of the liquid product. High temperatures will continue to crack the vapors, and the longer the vapors are at higher temperatures the greater the extent of cracking. Although secondary reactions become slow below around 350 °C, some will continue down to room temperature, contributing to the instability of the bio-oil. Long residence times result in significant reductions in organic yields from cracking reactions. Thus, the collection of liquids has long been very important in the operation of flash pyrolysis processes. Quenching, i.e., contact with cooled liquid, is effective but careful design and temperature control is needed to avoid blockage from differential condensation of heavy ends. Light ends collection is important in reducing liquid viscosity. Electrostatic precipitation is very effective in recovering the aerosols.

Various reactor configurations for flash pyrolysis have been developed [37–39]:

(a) Ablative pyrolysis, in which biomass is pressed against a heated surface and rapidly moved. The biomass melts at the heated surface and leaves an oil film behind that evaporates. This process may use large particles and is typically limited by the rate of heat supply to the reactor. It leads to compact and intensive reactors that do not need a carrier gas, but with the penalty of a surface area controlled system and moving parts at high temperature.

(b) Fluidized bed and circulating fluidized bed pyrolysis, which transfers heat to the biomass by a mixture of convection and conduction. The heat transfer limitation is within the particle; consequently, very small particles are required to obtain good liquid yields. Substantial carrier gas is needed for fluidization.

(c) Vacuum pyrolysis, which has slow heating rates but removes pyrolysis products as rapidly as in the previous methods, thus simulating flash pyrolysis. Larger particles are needed and the vacuum leads to larger equipment and higher costs. Total liquid yields are slightly lower (60–65 wt.%) than for previous technologies (75–80 wt.%).

(d) Rotating cone reactor, in which biomass particles are mixed intensively with an excess of hot sand particles. The circulating hot sand provides the heat for the pyrolysis process and prevents fouling of the cone wall. New versions [40] are equipped with an external circulation loop of char particles, thus combining the pyrolysis reactor with a section for char combustion. Indeed, the combustion of char may provide the energy necessary for the endothermic pyrolysis process, which would enable an overall autothermal operation.

7.6
Fuels via Thermal Biomass Conversion

The exit gas from the gasifier needs to be improved to a rather clean syngas to produce fuels or other products. The predominant commercial technology for syngas generation has been and continues to be steam reforming (SRM, Reaction 1), in which methane and steam are catalytically and endothermically converted into hydrogen and carbon monoxide [41]. An alternative approach is partial oxidation (POX, Reaction 2), the exothermic, non-catalytic reaction of methane and oxygen to produce a syngas mixture:

$$\text{SRM: } CH_4 + H_2O \leftrightarrows CO + 3H_2 \; (\Delta H°_{298} = 250.1 \text{ kJ mol}^{-1}) \tag{1}$$

$$\text{POX: } CH_4 + 1/2 O_2 \leftrightarrows CO + 2H_2 \; (\Delta H°_{298} = -35.7 \text{ kJ mol}^{-1}) \tag{2}$$

SRM and POX inherently produce syngas mixtures having appreciably different compositions. In particular, SRM produces a syngas having a much higher H_2/CO ratio. This, of course, represents a distinct advantage for SRM in hydrogen-production applications and, in large measure, accounts for its overall dominance among syngas production technologies to date. A further innovation is catalytic partial oxidation (CPO), in which the oxidation reactions and the reforming ones occur on the catalytic bed. This allows working at low residence time (few ms), with the advantages of small reactor dimension and high productivity. This process is particularly interesting for small-medium size applications, and it is under development by many companies (Air Liquide, Shell, Amoco; currently, it is at a demonstrative step with Snamprogetti. Catalyst containing Ni but also Ni/Rh or just Rh dispersed on high temperature (above 1000 °C) resistant supports are suitable for this reaction. A different approach is autothermal reforming (ATR), which combines POX and SRM in one reactor. The process is "autothermal" in that the endothermic reforming reactions proceed with the assistance of the internal combustion (or oxidation) of a portion of the feed hydrocarbons – in contrast to the external combustion of fuel characteristic of conventional tubular reforming.

Commercial plants commonly use supported nickel catalysts [42, 43]. The catalyst contains 15–25 wt.% nickel oxide on a mineral carrier (α-Al_2O_3, aluminosilicates, magnesia and MgAl spinel). Before start up, nickel oxide must be reduced to metallic nickel. This is done, preferably, with hydrogen but also with natural gas or even with the feed gas itself at high temperature (above 600 °C, depending on the reducing stream). Catalyst carriers require a relatively high specific surface area, low pressure drop and high mechanical resistance at up to 1000 °C. The main catalyst poison in SRM plants is sulfur. Concentrations as low as 50 ppm deactivate the catalyst. During biomass thermal conversion and SRM, all sulfur compounds are converted into H_2S, which chemisorbs on metal sites, forming NiS, in accordance with the equilibrium:

$$Ni + H_2S \leftrightarrows NiS + H_2 \tag{3}$$

The low melting point and high surface mobility of NiS also accelerate the sintering process of Ni crystallites. Since the formation of NiS is exothermic, activity loss can be partially recovered by raising the reaction temperature, which, however, also accelerates thermal degradation of the catalyst and increases carbon formation through cracking reactions.

In the specific case of biomass gasification, several alkaline salts and heavy metals and metal oxides particles may act as additional poisons by enhancing the sintering of the Ni crystallites or by being adsorbed on the Ni sites [44]. While acid supports such as alumina react with alkali to form crystalline phases, basic supports (like MgO) do not react directly with them; however, alkali causes coverage of the surface and plugging of the pores.

Another cause of activity loss is carbon deposition, which can be avoided if a high steam to carbon (S/C) ratio is employed [45, 46]. However, economic evaluations indicate that the optimum S/C ratio tends to be low. The presence of tars in the reforming reactor enhances coking and it is the main cause of carbon formation in reforming a gas from biomass thermal conversion [29].

Depending on the reason for converting the produced gas from biomass gasification into synthesis gas, for applications requiring different H_2/CO ratios, the reformed gas may be ducted to the water-gas shift (WGS, Reaction 4) and preferential oxidation (PROX, Reaction 5) unit to obtain the H_2 purity required for fuel cells, or directly to applications requiring a H_2/CO ratio close to 2, i.e., the production of dimethyl ether (DME), methanol, Fischer–Tropsch (F-T) Diesel (Reaction 6) (Fig. 7.6).

$$\text{WGS: } CO + H_2O \leftrightarrows CO_2 + H_2 \ (\Delta H°_{298} = -41 \text{ kJ mol}^{-1}) \quad (4)$$

$$\text{PROX: } CO + 1/2 O_2 \leftrightarrows CO_2 \ (\Delta H°_{298} = -254 \text{ kJ mol}^{-1}) \quad (5)$$

$$\text{F-T: } nCO + 2(n+1)H_2 \leftrightarrows C_nH_{2n+2} + nH_2O \ (\Delta H°_{298} = -167 \text{ kJ mol}^{-1}) \quad (6)$$

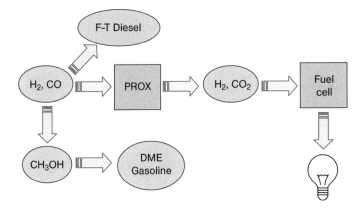

Fig. 7.6 Possible applications of the reformed product gas from a gasifier.

The WGS reaction [47] is a critical step in fuel processors for preliminary CO clean up and additional hydrogen generation prior to the CO preferential oxidation. The reaction is moderately exothermic, with low CO levels resulting at low temperatures but with favorable kinetics at higher temperatures. Since the flow contains CO, CO_2, H_2O, H_2, additional reactions can occur, depending on the H_2O/CO ratio, that are favored at high temperatures: methanation, CO disproportionation or decomposition. In industrial applications, the classical catalyst formulations employed are Fe–Cr oxide for the first stage (high-temperature shift HTS, typically in the range 360–400 °C) and Cu–ZnO–Al_2O_3 for subsequent stages (low-temperature shift LTS, operating just above the dew point, the lowest possible inlet temperature, i.e., about 200 °C), for good performance under steady state conditions (CO exit concentration in the range 0.1–0.3%). However, these catalysts are very sensitive to the contaminants from biomass gasification. WGS of gases containing appreciable amounts of sulfur and tar requires catalysts consisting mainly of Co and Mo oxides. Moreover, their activity is at a maximum only when Co and Mo are in the sulfided forms.

Since CO acts as a poison to the proton exchange membrane fuel cell in the 50 ppm range, it has to be removed before feeding the H_2 enriched gas to the fuel cell. This CO removal occurs in the PROX reactor, where Pt/Al_2O_3 catalysts are common, even though some interest in Au-based catalysts is growing due the lower cost of the active phase [48].

The H_2 with low CO content can than be fed into the fuel cell: an electrochemical device in which the chemical energy stored in a fuel is converted directly into electricity. A fuel cell consists of an anode (negatively charged electrode), to which a fuel, commonly hydrogen, is supplied, and a cathode (positively charged electrode) to which an oxidant, commonly oxygen, is supplied. The oxygen needed by a fuel cell is generally supplied by air. The two electrodes of a fuel cell are separated by an ion-conducting electrolyte. The input fuel is catalytically reacted (electrons removed from the fuel elements) in the fuel cell to create an electric current [49] (Reaction 7):

$$H_2 + 1/2O_2 \leftrightarrows H_2O + e^- \qquad (7)$$

There are five established types of fuel cells, on the basis of the electrolyte employed. They are alkaline fuel cells (AFC), phosphoric acid fuel cells (PAFC), proton-exchange membrane fuel cells (PEMFC), molten carbonate fuel cells (MCFC), and solid oxide fuel cells (SOFC). In all types there are separate reactions at the anode and the cathode, and charged ions move through the electrolyte, while electrons move round an external circuit. Another common feature is that the electrodes must be porous, because the gasses must be in contact with the electrode and the electrolyte at the same time. The advantages of fuel cells include operation without combusting fuel and few moving parts, but the main one is that they can be two to three times more efficient than an engine in converting fuel into electricity. Many types of fuel cell can be used coupled with fuel produced from biomass: syngas using MCFC, SOFC with internal reforming, and direct methanol fuel cell (DMFC) are examples of processes under development.

F-T synthesis produces hydrocarbons of different length from syngas. The large hydrocarbons can be hydrocracked to form mainly diesel of excellent quality [2]. Commercial catalysts contain Fe and Co on different oxides (silica, alumina, titania). These catalysts are even less tolerant to contaminants than those of the gas processing stages; therefore much care must be taken while scaling up the integrated process of producing fuels from biomass. Methanol (MeOH) and dimethyl ether (DME) are other possible liquid fuels produced by CO and H_2 conversion from gasified biomass. Especially, DME for its low toxicity and high cetane number is a promising fuel for diesel engine. It can be obtained from MeOH using high synthetic alumina zeolite as catalyst. New one-step processes from syngas have been claimed but no commercial plants are available at present. The Värnamo Biomass Gasification Centre (VBGC) in Sweden is a unique plant and an important site and, together with the CHRISGAS Project, is acting to demonstrate the development of an innovative technology that produces liquid fuel from biomass. It aims to convert the gas produced from biomass gasification into syngas in an oxygen/steam circulating fluid bed followed by hot gas filtration and ATR. The system should be able to work with streams having high contents of sulfur, tars and alkali metals. The synthesis gas can than be transformed into liquid fuels (DME or F-T diesel) or hydrogen with high efficiency [23, 50, 51].

7.7
Conclusions

The thermoconversion of lignocellulosic biomass for energy, syngas and fuel production is a field in continuous evolution and of growing interest, because biomass is a renewable and CO_2 neutral source. The ability to produce biomass-derived syngas on a large scale will help to reduce greenhouse gas emission and pollution, increase the security of energy supplies, and enhance the use of renewable energy. This chapter has attempted to overview the processes and problems in obtaining syngas from the pyrolysis and gasification of biomass.

References

1 D. Sutton, B. Kelleher, J.H.R. Ross, *Fuel Process. Technol.* 73 (**2001**) 155.
2 M.J.A. Tijmensen, A.P.C. Faaij, C.N. Hamelinck, M.R.M. van Hardeveld, *Biomass Bioenergy* 23 (**2002**) 129.
3 L. Devi, K.J. Ptasinski, F.J.J.G. Janssen, *Biomass Bioenergy* 24 (**2003**) 125.
4 A.A.C.M. Beenackers, *Renewable Energy* 16 (**1999**) 1180.
5 A.V. Bridgewater, *Fuel* 74 (**1995**) 631.
6 J.L. Cox, A.Y. Tonkovich, D.C. Elliott, E.G., Baker, E.J. Hoffman, in: *Proceedings, Second Biomass Conference of the Americas: Energy, Environment, Argriculture, and Industry.* Meeting held August 21–24, **1995**, Portland, Oregon; published by National Renewable Energy Laboratory, Golden, Colorado, p. 657.
7 E.D. Larson, R.E. Katofsky, in: A.V. Bridgwater (Ed.), *Advances in Thermochemical Biomass Conversion*, Blackie, London, **1994**, p. 495.
8 L. Garcia, R. French, S. Czernik, E. Chornet, *Appl. Catal. A: General* 201 (**2000**) 225.
9 M.P. Aznar, J. Corella, J. Delgado, J. Lahoz, *Ind. Eng. Chem. Res.* 32 (**1993**) 1.

10 M.P. Aznar, MA. Caballero, J. Gil, A. Olivares, J. Corella, in: R.P. Overend, E. Chornet (Eds.), *Making a Business from Biomass*, Pergamon Press, New York, **1997**, p. 859.
11 S. Rapagna, N. Jand, P.U. Foscolo, *Int. J. Hydrogen Energy* 23 (**1998**) 551.
12 S. Turn, C. Kinoshita, Z. Zhang, D. Ishimura, J. Zhou, *Int. J. Hydrogen Energy* 23 (**1998**) 641.
13 L. Garcia, M.L. Salvador, J. Arauzo, R. Bilbao, *Energy Fuels* 13 (**1999**) 851.
14 D. Wang, S. Czernik, D. Montané, M. Mann, E. Chornet, *Ind. Eng. Chem. Res.* 36 (**1997**) 1507.
15 D. Wang, S. Czernik, E. Chornet, *Energy Fuels* 12 (**1998**) 19.
16 E. Henrich, F. Weirich, *Environ. Eng. Sci.* 21 (**2004**) 53.
17 A.M.C. Janse, R.W.J. Westerhout, W. Prins, *Chem. Eng. Process.* 39 (**2000**) 239.
18 A. Demirbaş, *Energy Conversion Manage.* 42 (**2001**) 183.
19 A.V. Bridgewater, A.J. Toft, J.G. Brammer, *Renewable Sustainable Energy Rev.* 6 (**2002**) 181.
20 J.L. Easterley, M. Burnham, *Biomass Bioenergy* 10 (**1996**) 79.
21 D.A. Tillman, *Biomass Bioenenergy* 19 (**2000**) 365.
22 S. Albertazzi, F. Basile, F. Trifirò, *Chim. Ind.* 2 (**2006**) 62.
23 S. Albertazzi, F. Basile, J. Brandin, C. Hulteberg, J. Einvall, G. Fornasari, V. Rosetti, M. Sanati, F. Trifirò, A. Vaccari, *Catal. Today* 106 (**2005**) 297.
24 T. Kivisaari, P. Björnbom, C. Sylwan, *J. Power Sources* 104 (**2002**) 115.
25 L. Devi, K.J. Ptasinski, F.J.J.G. Janssen, S.V.B. van Paasen, P.C.A. Bergman, J.H.A. Kiel, *Renewable Energy* 30 (**2005**) 565.
26 L. Devi, K.J. Ptasinski, F.J.J.G. Janssen, *Fuel Process. Technol.* 86 (**2005**) 707.
27 S. Rapagnà, N. Jand, A. Kiennemann, P.U. Foscolo, *Biomass Bioenergy* 19 (**2000**) 187.
28 R. Zhang, R.C. Brown, A. Suby, K. Cummer, *Energy Conversion Manage.* 45 (**2004**) 995.
29 R. Coll, J. Salvadò, X. Farriol, D. Montanè, *Fuel Process. Technol.* 74 (**2001**) 19.
30 T.J. Wang, J. Chang, C.Z. Wu, Y. Fu, Y. Chen, *Biomass Bioenergy* 28 (**2005**) 508.
31 S. Yaman, *Energy Conversion Manage.* 45 (**2004**) 651.
32 M.N. Islam, F.N. Ani, *Bioresource Technol.* 73 (**2000**) 67.
33 D. Wang, S. Czernik, D. Montané, M. Mann, E. Chorlet, in: Hydrogen production via catalytic steam reforming of fast pyrolysis oil fractions. *Proceedings of the 3rd Biomass Conference of the Americas*, Pergamon, Oxford, **1997**, p. 845.
34 A. Demirbaş, *Energy Conversion Manage.* 43 (**2002**) 877.
35 A.V. Bridgwater, G.V.C. Peacocke, *Renewable Sustainable Energy Rev.*, 4 (**2000**) 1.
36 O. Onay, O.M. Kockar, *Renewable Energy* 28 (**2003**) 2417.
37 A.V. Bridgwater, D. Meier, D. Radlein, *Org. Geochem.* 30 (**1999**) 1479.
38 D. Meier, O. Faix, *Bioresource Technol.* 68 (**1999**) 71.
39 A.V. Bridgwater, *J. Anal. Appl. Pyrolysis* 51 (**1999**) 3.
40 A.M.C. Janse, P.M. Biesheuvel, W. Prins, W.P.M. van Swaaij, *Chem. Eng. J.* 76 (**2000**) 77.
41 D.J. Wilhelm, D.R. Simbeck, A.D. Karp, R.L. Dickenson, *Fuel Process. Technol.* 71 (**2001**) 139.
42 H.J. Arpe (Ed.), *Ullmann's Encyclopedia of Industrial Chemistry*, 5th ed., Vol. A12, **1989**, p. 238.
43 M.V. Twigg (Ed.), *Catalyst Handbook*, 2nd ed., Wolfe Publ., London, **1989**.
44 J.S. Choi, H.H. Kwon, T.H. Lim, S.A. Hong, H.I. Lee, *Catal. Today* 95 (**2004**) 553.
45 J.R. Rostrup-Nielsen, *Catal. Today* 37 (**1997**) 225.
46 D.L. Trimm, *Catal. Today* 37 (**1997**) 233.
47 A.F. Ghenciu, *Curr. Opin. Solid State Mater. Sci.* 6 (**2002**) 389.
48 A.N. Fatsikostas, D.I. Kondarides, X.E. Verykios, *Catal. Today* 75 (**2002**) 145.
49 C. Song, *Catal. Today* 77 (**2002**) 17.
50 *Värnamo Demonstration Plant*, Berling Skogs, Trelleborg, **2001**.
51 K. Ståhl, L. Waldheim, M. Morris, U. Johnsson, L. Gårdmark, GCEP Energy Workshop, April 27, 2004.

8
Thermal Biomass Conversion and NO$_x$ Emissions in Grate Furnaces

Rob J. M. Bastiaans, Hans A. J. A. van Kuijk, Bogdan A. Albrecht, Jeroen A. van Oijen and L. Philip H. de Goey

8.1
Introduction

A popular small-scale thermal biomass conversion method is combustion in grate furnaces. To meet the emission regulations for such a furnace, the operating conditions and design of the furnace have to be chosen carefully. Numerical models, known as computational fluid dynamics (CFD), can support the making of these choices, provided that accurate sub-models for the phenomena occurring in the oven are available.

In a grate furnace, the conversion processes take place in two zones. In the primary conversion zone, the solid fuel is gasified on a moving grate. In the secondary combustion zone there is burnout of the resulting gas mixture. This chapter deals with the development of sub-models for both the primary and secondary combustion zones, focusing on the formation of NO$_x$, especially by conversion of fuel nitrogen. The fuel nitrogen is mainly released as HCN and NH$_3$, of which the ammonia is the most important component under grate furnace conditions. In the gas phase the ammonia is converted into NO$_x$ at higher temperatures and in the presence of radicals.

With respect to the considerations above, research is split into three parts. The first is connected to the kinetic description of the release of ammonia from the biomass as function of temperature. This research employs infrared spectroscopy using a tunable diode laser. Here very small biomass particles are used that are heated up very rapidly in a small reactor, which ensures that transport effects are virtually excluded from the kinetic release effects. Since ammonia is released in very small quantities it is quite hard to detect. Therefore, we first measure CO release, which is easier. In the second part we investigate the propagation of a conversion front in biomass layers. Here we perform experiments and try to establish a modeling approach for the propagation by analytical and numerical approaches. In the third part the gas-phase conversion processes are described in terms of

8.2
Tunable Diode Laser Measurements of Biomass Kinetics

8.2.1
Introduction

Species release rates of heated biomass can be measured by means of absorption of coherent light at a specified frequency. We intend to measure the release rate of ammonia as an important precursor for NO_x emissions. However, the amount of ammonia is very low and hard to detect. Therefore initial measurements are performed to determine the apparent rate of CO release during pyrolysis [1]. The CO measurements were conducted using bark and Medium Density Fiberboard (MDF). The material was heated by a pre-heated wire grid technique. Experiments were performed under ambient nitrogen and at atmospheric pressure conditions over the temperature range 620–940 K. The temperature is almost constant for each pyrolysis experiment. The CO release was monitored as a function of the grid temperature using a tunable diode laser set to a single infrared absorption line at 2082 cm^{-1}. The release followed a typical growth curve in time to a steady-state level at each temperature. Curves were fitted using a simple first-order kinetic analysis and Arrhenius parameters were extracted. Both uncertainties in applied temperature and release rate were taken into account. The calculated apparent activation energies E_a were 64 ± 6 kJ mol^{-1} for bark and 71 ± 6 kJ mol^{-1} for MDF. A single apparent rate seems to be a fair description although two time scales can be identified.

8.2.2
Tunable Diode Laser Grid Reactor Experiments

The development of accurate, predictive models for thermal decomposition rely not only on a good understanding of heat and mass transfer processes but also on the complicated decomposition and product formation kinetics involved. There are large deviations in the pyrolysis kinetics reported in the literature. This is symptomatic of the difficulties involved in measurements of this type. Most kinetic pyrolysis experiments on bark and other cellulose-containing materials have used a slow pyrolysis technique, of which the most widely used are Thermo-Gravimetric Analysis (TGA) and the fluidized bed technique [2, 3]. Data collected are based on weight loss as a function of time and temperature. In addition, volatile component release, including CO, has been investigated by coupling the technique with Gas Chromatography (GC) and Fourier-transform Infrared Spectroscopy (FTIR) [4–6]. The advantage of this method is that at low heating rates the particle temperature can be measured accurately. The disadvantage is that kinetics determined at low heating rates are generally not applicable under conditions where the heating rate is significantly higher, such as in a furnace plant.

Real kinetic data, not influenced by the relatively slow processes of convection and diffusion, are therefore needed to model the behavior of thermal degradation in fuel beds accurately. The present chapter reports on CO release kinetics at high heating rates and small particle sizes, both diminishing transport effects. The heated grid technique, as developed in Refs. [7, 8] for the gasification and pyrolysis of coal and char, was applied to measure the fast pyrolysis kinetics of CO release. Furthermore, the temperature measurements are more accurate. No reports have been made of the application of this method to wood pyrolysis studies although the technique has also been used in the study of coal pyrolysis [9]. Additionally, correlations were derived that include the uncertainty in the measurements by fluctuating temperatures and the uncertainty in the time-constants of the release rate.

8.2.3
Experimental Setup

Pyrolysis experiments were performed using a heated wire grid technique. Reactor details have been reported previously [7, 8]. The technique has been successfully applied in both pyrolysis and gasification studies [8, 10, 11]. In brief, the mesh or grid is housed in a stainless steel chamber known as the grid reactor (Fig. 8.1). The reactor is 227 mm long and has an inner diameter of 15 mm. The grid (9 × 4 mm) is constructed of interwoven wires (platinum/rhodium 10%)

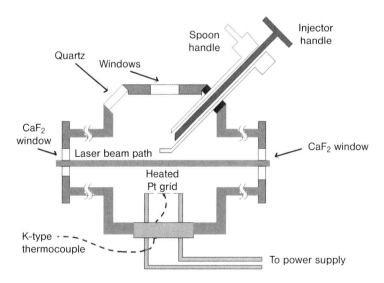

Fig. 8.1 Grid reactor setup.

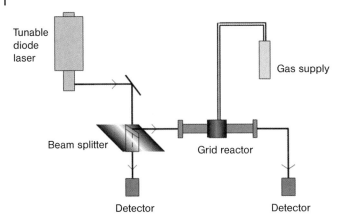

Fig. 8.2 Laser system and optical path.

with an individual wire diameter of 76 μm. The grid is mounted in the center of a chamber clamped between two copper electrodes. Power is supplied to the grid and its temperature is adjusted by varying the applied current. The maximum temperature that the grid can reach is ± 2000 K. A K-type thermocouple, having a wire 270 μm in diameter, is attached to the grid and is used to measure temperatures below 1000 K. Quartz windows situated on top of the reactor allow the use of a manual optical pyrometer for grid temperature determinations over 1000 K. Windows made of CaF_2 are mounted on both sides of the reactor, giving a spectral transmission range of 1200–58 000 cm^{-1}.

A tunable diode laser system and optics (Laser Photonics, L5736) was used to monitor the species release rates during an experiment. It consists of a liquid-nitrogen cooled diode emitter. For CO absorption measurements the laser emits a beam tuned to a wavenumber of 2082 cm^{-1}. This line is chosen as it exhibits strong absorption and is not subject to interference from other species likely to be present in the pyrolysate.

Figure 8.2 illustrates the optical path followed by the beam. The beam is split with one component directed through a reference cell, containing a sample of the gas of interest, in this case CO, to a liquid-nitrogen cooled detector (Fermionics HgCdTe pn). Using a reference signal aids the initial tuning of the laser and also makes it possible to monitor and correct any drift in the signal. The second component travels through the reactor and is directed to a second detector. Both detector signals are amplified and processed through a data acquisition system.

8.2.4
Results

Figure 8.3 shows the typical time-dependent CO release curve for a bark sample at an initial nominal grid temperature of 915 K and a pressure of 1 atm. At this

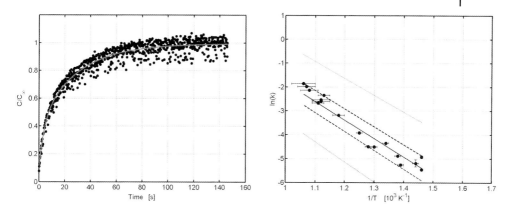

Fig. 8.3 CO release from bark pyrolyzed at 915 K, and Arrhenius plot for CO release from bark pyrolyzed at 680–940 K.

low temperature the release rate is relatively slow and the CO concentration history has a characteristic time, τ, of 17 s.

The CO release for MDF has also been measured [1], at an initial grid temperature of 652 K. At this lower temperature CO release is slower with a τ of about 21 s. In addition, the grid temperature is seen to gradually fall off during the run. The gradual CO release under these conditions can be taken as indicative of overall gas release from the particle. Both slow gas release rate and gradual fall off in particle temperature could be indicative of ongoing internal molecular rearrangements within the particle, such as bridge breaking and crosslinking mechanisms. This process of graphitization affords a char material of increasing porosity and therefore increasing surface area for heat loss. At higher temperatures this process is not evident but may be masked by the higher grid temperature applied to the particle.

As already mentioned, the CO release exhibited an exponential growth to a steady-state level. The curves were fitted using a first-order kinetic scheme. Several sets of pyrolysis investigations were performed between 620 and 940 K. The lower temperature is a limit due to a large scatter in data received at lower temperatures and the upper temperature is limited by the temporal resolution of the concentration measurements.

For the Arrhenius kinetics, a mean temperature had to be extracted from each run and plotted with its standard deviation. The approach was to use the time constant $\tau = 1/k$ to define a time limit within which CO evolution and temperature show a strong interdependency. Based on the first-order kinetic model at $t = \tau$, $C = 0.63 C_\infty$, a temporal range was defined from which the average temperature is extracted. The Arrhenius plot (Fig. 8.3, right-hand side) shows that the error in temperature increases with increasing temperature. At lower grid temperatures the impact of a slower CO release rate results in a mean temperature that has a significant contribution from post-trough temperatures.

The CO evolution rate shows a strong dependence on the initial temperature setting of the grid – fast at high temperatures and slower at lower settings. A linear fit to the data points of the Arrhenius plots allows the calculation of an activation energy and pre-exponential factor for CO release. Here the uncertainties in time constant and temperature had to be taken into account. In the figures, the uncertainty in the activation energies and pre-exponential factors are displayed in the dashed and dotted lines, respectively, whereas the drawn line gives the actual fit. Here the dashed line gives the range in uncertainty of the fit when using the raw data including the measurement uncertainties. The dotted lines give the uncertainties at amplified measurement errors in such a way that the linear model is within acceptability.

As already indicated, a physical description in terms of a process with a single time constant is fair but an assessment on the basis of two time-scales gives even improved results. Therefore, more research has to be carried out to determine the characteristics of the most important additional phenomena. As an example a two time-scale model is applied to the previously reported measurements of Fig. 8.3 and displayed in Fig. 8.4. Clearly, the release is governed by two rates, typically a smaller and a larger time scale appear compared with the single rate case. However, the single rate results are still very valuable because they describe the apparent rate very well and this would be the only thing that can be described in coarse scale models of devolatilization, e.g., in CFD of biomass conversion.

An example of a fuel layer model is discussed in the next section. In a first approximation no detailed description of the release of volatiles need to be included. Later, the results of the tunable diode laser experiments can be implemented in such a fuel layer model.

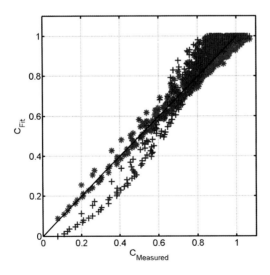

Fig. 8.4 CO release ($C_{Measured}$) from bark pyrolyzed at 915 K, correlated with a single rate fit (+) and a fit for a two-time scale process (*).

8.3
Propagation of Thermal Conversion Fronts

8.3.1
Introduction

Solid fuel is transported by a moving grate in the primary conversion zone of a grate furnace (Fig. 8.5). Air is supplied through the grate. Due to the hot environment in the combustion chamber, ignition will take place at the top of the fuel layer. After ignition, a thermal conversion front starts to propagate downwards through the bed. Initially, the front consists of different layers, in which drying, pyrolysis, gasification and combustion reactions take place. In the case of incomplete combustion, the top layer consists of char. When the conversion front has reached the grate, the hot char is directly exposed to oxygen, which means that char burnout will take place. The conversion front is now propagated upwards, because the hot char closest to the grate will ignite first. When all the char is burnt out, only ashes are left on the grate.

The conditions in the reaction zone determine the release rate of the N-precursors HCN and NH_3 [11]. Among these conditions are the properties of the fuel (e.g., N-content and particle size), parameters related to the combustion front (temperature and propagation velocity) and the gas composition in and directly above the combustion front. As the prediction of the mass fractions of the N-precursors is important for the final goal of this research, i.e., the prediction of NO formation of the complete furnace, a model is needed in which all these conditions are represented.

To develop such a model, a further simplification can be made. In a reference frame attached to the grate, the process can be regarded as a one-dimensional sys-

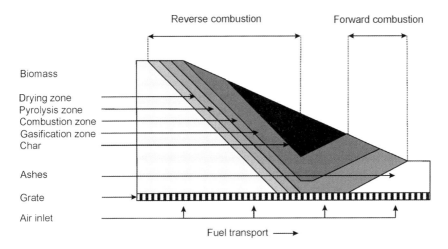

Fig. 8.5 Sketch of the processes occurring during solid fuel combustion in a grate furnace.

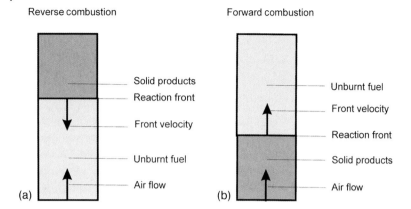

Fig. 8.6 Reverse combustion (a) and forward combustion (b) in a fixed bed. Note that the unburnt fuel in (b) corresponds to the char layer of Fig. 8.5.

tem. In this frame, the combustion front propagates in the vertical direction. In the first part of the furnace, reverse combustion takes place, whereas in the second part of the furnace (char burnout), forward combustion takes place (Fig. 8.5). This has the advantage that the combustion process in a moving grate furnace can be studied experimentally in a fixed bed reactor and theoretically by means of a one-dimensional (1D) model (Fig. 8.6).

8.3.2
Modeling Approach

The model used here is based on a 1D model [12] for reverse combustion in a bed of coal particles following the reaction:

$$C + O_2 \rightarrow CO_2 \tag{1}$$

and is implemented in the 1D flame code CHEM1D [13]. The model consists of three equations [Eqs. (2–4)]: an equation for the temperature T, for the oxygen mass fraction Y and for the porosity of the solid fuel, ε:

$$\frac{\partial}{\partial t}\{[(1-\varepsilon)\rho_s c_{ps} + \varepsilon \rho_g c_{ps}]T\} + \frac{\partial}{\partial x}(\varepsilon u_g \rho_g c_{pg} T) - \frac{\partial}{\partial x}\left(\Lambda \frac{\partial T}{\partial x}\right) = -\Delta H_r R \tag{2}$$

$$\frac{\partial}{\partial t}(\varepsilon \rho Y) + \frac{\partial}{\partial x}(\varepsilon u_g \rho_g Y) - \frac{\partial}{\partial x}\left(\rho D \frac{\partial Y}{\partial x}\right) = -\upsilon R \tag{3}$$

$$\frac{\partial}{\partial t}[(1-\varepsilon)\rho_s] = -R \tag{4}$$

Table 8.1 Details of the reverse combustion model.

In the differential equations of the reverse combustion model, the reaction source term, dispersion coefficient, and the conductivity coefficient need to be defined. This is done based on data presented in Ref. [12]. At present, the model operates with a source term based on the combustion of coal particles. The source term is given by Eq. (5):

$$R = \frac{1}{k_r^{-1} + k_m^{-1}} S_u \left(\frac{1-\varepsilon}{1-\varepsilon_u} \right)^{2/3} \rho_g Y \qquad (5)$$

where k_r is the chemical reaction rate constant, k_m a mass transfer coefficient and S_u is the surface area per unit volume of the coal particles under unburnt conditions. For the rate constant, an Arrhenius expression is used:

$$k_r = A_f T \exp\left(-\frac{E_a}{RT}\right) \qquad (6)$$

The mass transfer coefficient describes the effect of mass transfer resistance of the reactants flowing from the gas phase to the surface of the individual particles in the bed. The mass transfer coefficient can be obtained from a correlation for the Sherwood number (or dimensionless mass transfer coefficient) given by Eq. (7):

$$Sh = \frac{0.81}{\sqrt{\varepsilon}} Re_p^{1/2} Sc^{1/3} \quad (5 < Re_p < 500) \qquad (7)$$

in which the Sherwood number, particle Reynolds number and Schmidt number are defined as:

$$Sh = \frac{k_m d_p}{\delta_g}, \quad Re_p = \frac{u_g \varepsilon d_p}{\eta_g}, \quad Sc = \frac{v_g}{\delta_g} \qquad (8)$$

where δ_g and η_g are the diffusion coefficient and the viscosity of the gas, respectively, for which correlations from Ref. [14] are used.

For the dispersion coefficient, the expression:

$$D = 0.5 \varepsilon u_g d_p \qquad (9)$$

is used. The conductivity is given by:

$$\Lambda = \lambda_{s0} + 0.8 \varepsilon u_g \rho_g c_{pg} d_p \qquad (10)$$

where λ_{s0} is the conductivity of the bed without fluid flow, which has a value of $\lambda_{s0} = 9.0 \times 10^{-3}$ J cm^{-1} s^{-1} K^{-1}. Data for all other constant parameters are given in Table 8.2.

Table 8.2 Model parameters.

Parameter	Unit	Value	Parameter	Unit	Value
c_{ps}	J g^{-1} K^{-1}	1.8	ε	–	0.615
c_{pg}	J g^{-1} K^{-1}	1.13	d_p	cm	2.5
ρ_{s0}	g cm^{-3}	1.22	υ	–	2.67
$\rho_{gu}(T_u)$	g cm^{-3}	1.23×10^{-23}	ΔH_r	J g^{-1}	28.9×10^3
T_u	K	288	E_a	J mol^{-1}	1.5×10^5
Y_u	–	0.23	A_f	cm K^{-1} s^{-1}	6×10^4

Here, the densities of the gaseous and solid fuels are denoted by ρ_g and ρ_s respectively and their specific heats by c_{pg} and c_{ps}. D and Λ are the dispersion coefficient and the effective heat conductivity of the bed, respectively. The gas velocity in the pores is indicated by u_g. The reaction source term is indicated with R, the enthalpy of reaction with ΔH_r and the mass based stoichiometric coefficient with υ. In Ref. [12] an asymptotic solution is found for high activation energies. Since this approximation is not always valid we solved the equations numerically without further approximations. Tables 8.1 and 8.2 give details of the model.

Stationary, traveling wave solutions are expected to exist in a reference frame attached to the combustion front. In such a frame, the time derivatives in the set of equations disappear. Instead, convective terms appear for transport of the solid fuel, containing the unknown front velocity, u_s. The solutions of the transformed set of equations exist as spatial profiles for the temperature, porosity and mass fraction of oxygen for a given gas velocity. In addition, the front velocity (which can be regarded as an eigenvalue of the set of equations) is a result from the calculation. The front velocity and the gas velocity can be used to calculate the solid mass flux and gas mass flux into the reaction zone, i.e., $m_{su} = \rho_s(1-\varepsilon)u_s$ and $m_{gu} = \rho_g(1-\varepsilon)u_g$.

Results of the model for two parameters, i.e., the spatial temperature profile and the mass flux into the reaction zone as a function of gas mass flux are presented in Fig. 8.7. The temperature profile of the solid fuel flame (Fig. 8.7, left) is similar to that of a premixed laminar flame: it consists of a preheat zone and a reaction zone. (The spatial profile of the reaction source term, which is not depicted here, further supports this conclusion.) The temperature in the burnt region (i.e., for large x) increases with the gas mass flux. The solid mass flux (Fig. 8.7, right) initially increases with an increase of the air flow, until a maximum is reached. For higher air flows, it decreases again until the flame is extinguished.

Currently, the 1D model describes the essential features of the propagation of a combustion front in the reverse combustion mode. With an adapted version of the model, the combustion of biomass could be modeled accurately. To obtain

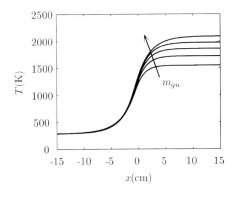 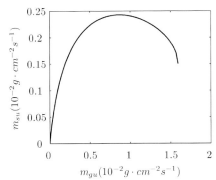

Fig. 8.7 Results of the fixed bed model. Spatial profiles of the temperature, $T(x)$, for $m_{gu} = 4.9$–14.6×10^{-3} g cm^{-2} s^{-1} in steps of 2.4×10^{-3} g cm^{-2} s^{-1} (left), and the solid mass flux, $m_{su} = \rho_s(1-\varepsilon)u_s$, into the reaction zone as a function of gas mass flux (right).

such a model, the reaction source term and the number of species have to be extended to include the formation, gasification and combustion of char. In addition, the pyrolysis process of wood has to be included in the model.

8.3.3
Experiments

To validate the numerical work and to study the phenomena that play a role in fixed bed combustion, experiments with a fixed bed reactor (Fig. 8.8) were performed [15]. Essentially, the reactor consists of an insulated metal tube filled with biomass. The biomass is ignited at the top, while air is supplied at the bottom of the reactor. The conversion front can be tracked with thermocouples. A mass balance is used to record the conversion of the biomass. As the results of the previous section are for coal conversion, the two sets of results can not yet be compared directly.

The raw data of the thermocouples consist of the temperature as a function of time (Fig. 8.9, left). In the raw data, the passing of the conversion front can be observed by a rapid increase in temperature. Because the distance between the thermocouples is known, the velocity of the conversion front can be determined. The front velocity can be used to transform the time domain in Fig. 8.9 (left) to the spatial domain. The resulting spatial flame profiles can be compared with the spatial profiles resulting from the model. The solid mass flux can also be plotted as a function of gas mass flow rate. The trend of this curve is similar to the model results (Fig. 8.9, right).

The experimental setup will be tested and improved. In the raw thermocouple data, the temperature of the conversion font decreases strongly when the conversion front has passed a thermocouple. This is caused by heat losses through the

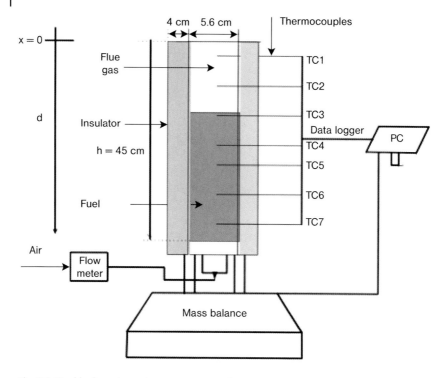

Fig. 8.8 Fixed bed reactor and measurement equipment.

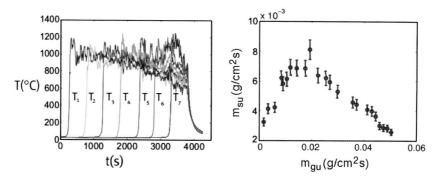

Fig. 8.9 Typical measurement result. Example of raw data of the thermocouples (left) and the solid mass flux as a function of gas mass flux (right).

reactor walls to the environment. The effect of these heat losses will be investigated further with a two-dimensional (2D) model. Furthermore, the setup has been used to study the effect of parameters affecting the combustion process. These parameters include the moisture content and particle size of the fuel.

If gas release is included in the measurements and in the model, these results can be used as boundary conditions for the gas phase combustion, as discussed in the next section.

8.4 Gas-phase CFD Modeling of Grate Furnaces

8.4.1 Introduction

To optimize the design of biomass combustion furnaces for NO_x emission reduction, numerical models can be used. The requirements of these models are twofold: the model has to be sufficiently detailed to predict the NOx production accurately and the calculation time of the model must be within reasonable limits to be able to calculate the emission levels for different design parameters and operating conditions. To that end a flamelet-PDF approach is used to derive a model for the combustion and NO_x formation in the biomass grate furnace (see also [16, 17]). The model reduces the known detailed gas-phase combustion chemistry using the flamelet generated manifolds method. The species mass fractions and temperature become functions of the mixture fraction, Z, and a reaction progress variable, c. They are tabulated in look-up tables in a pre-processing step to speed up the numerical calculations. The turbulence–chemistry interaction is described by an assumed shape PDF approach. Transport equations are solved for mean and variance of mixture fraction and progress variable and the mean NO mass fraction. The model is validated by comparing model predictions with measurements for a partially premixed flame: Sandia Flame D. Good agreement between predictions and experimental data is found. The model has also been applied to a 2D biomass grate furnace. Boundary conditions are taken from experiments, with data provided by TU-Graz. Later, the fuel layer model from the previous section can be coupled to the gas phase modeling. The first results are presented here.

8.4.2 Description of the Model

Some basic results of a new flamelet-PDF model are also presented here. The model is intended for the simulation of the gas-phase combustion in a biomass grid furnace. The model works as follows. A database is calculated using the flamelet generated manifold (FGM) chemical reduction method [18], developed in the Combustion Technology Group at Eindhoven University of Technology. The chemistry is described by a detailed reaction mechanism. Several premixed flamelets covering the flammability range are computed using the CHEM1D code [13], developed in the same research group as above. A progress variable is defined, to monitor the advance of the chemical reaction in the gaseous system. The species mass fractions, temperature, density and species source terms are tabulated as functions of the mixture fraction and the progress variable. The resulting laminar database is PDF-averaged using an assumed shape PDF function. The calculation of the turbulent database, which serves as a look-up table, in a pre-processing step, speeds up the simulations. During a simulation, turbulent transport equations for the mean and variance of mixture fraction and progress variable are solved, making use of the tabulated density and progress variable source term. A transport equation for the NO mass fraction is also solved to predict the NO formation. The extra transport equation for NO is needed since production is slow and there is no resolution of the table outside the flame zone.

This combustion model benefits from the low calculation time and memory usage of a flamelet model and is thought to improve the accuracy of gas-phase species and temperature predictions due to the use of a reaction progress variable. Moreover, turbulence–chemistry interactions are taken into account in a physical way.

8.4.3
Construction of Look-up Tables for Numerical Simulations and Validation

Following the description given in the previous section, the detailed chemical system is mapped on two controlling variables: the mixture fraction (Z) and a reaction progress variable (c). The mixture fraction describes the mixing of the species and enthalpy, while the progress variable follows the advance of the chemical reaction. Hence, the species mass fractions, temperature, density and species chemical source terms become functions of Z and c. The mixture fraction is defined according to [19] as:

$$Z = \frac{Y^* - Y^*_{ox}}{Y^*_{fu} - Y^*_{ox}}, \quad Y^* = 0.5 \frac{Y_H}{M_H} + 2.0 \frac{Y_C}{M_C} \tag{11}$$

where Y_H and Y_C are the element mass fractions for H and C, respectively and M_H, M_C are their molar masses. Y^* is a combined element mass fraction. Y^*_{ox} and Y^*_{fu} are the values of Y^* corresponding to the oxidizer and fuel streams, respectively. Following Eq. (11), Z varies between 0 in the oxidizer stream and 1 in the fuel stream. The progress variable is defined as:

$$c = \frac{Y_{CO_2} - Y_{CO_2}^{min}}{Y_{CO_2}^{max} - Y_{CO_2}^{min}} \tag{12}$$

where Y_{CO_2} is the mass fraction of CO_2, which varies between $Y_{CO_2}^{min}$ in the unburned mixture and $Y_{CO_2}^{max}$ in the burned (equilibrium) gas. The species mass fractions, temperature, density and chemical source terms present in the laminar database correspond to laminar premixed flamelets with mixture fraction values spanning the flammability range (depending on the fuel type) and progress variable values from 0 (reactants) to 1 (products).

In turbulent reactive flows, the chemical species and temperature fluctuate in time and space. As a result, any variable can be decomposed in its mean and fluctuation. In Reynolds-averaged Navier–Stokes (RANS) simulations, only the means of the variables are computed. Therefore, a method to obtain a turbulent database (containing the means of species, temperature, etc.) from the laminar data is needed. In this work, the mean variables are calculated by PDF-averaging their laminar values with an assumed shape PDF function. For details the reader is referred to Refs. [16, 17]. In the combustion model, transport equations for the mean and variances of the mixture fraction and the progress variable and the mean mass fraction of NO are solved. More details about this turbulent implementation of the flamelet combustion model can also be found in Ref. [20].

The model was validated using a well-documented flame: Sandia Flame D [21]. This is a turbulent piloted flame, the fuel consisting of a mixture of CH_4 (25% vol.) and air (75% vol.). The flame is surrounded by a pilot that produces the same equilibrium composition as a CH_4/air flame with an equivalence ratio of 0.77. The bulk velocities of the fuel jet, pilot and air co-flow are 49.6, 11.4 and 0.9 m s^{-1}, respectively. The inlet profiles of velocity, turbulent kinetic energy and dissipation rate are taken from Ref. [21]. To construct the database, the GRI-Mech3.0 reaction mechanism [22] was used. The realizable k-ε model was employed for turbulence modeling. The predictions are compared with measurements reported in Ref. [23]. Results of this validation are presented in Ref. [17]. The axial velocity and temperature are generally well predicted. The prediction of the mixture fraction and progress variable agree quite well with the measurements, and the shape of the mixture fraction variance is also close to experiments, but the position of its maximum value is moved to the left. In addition, the shape of the axial NO profile is well predicted by the present combustion model. The maximum value is only slightly under-predicted, within 10% of the measured data. It could be concluded that the present model performs like other state-of-the-art models, with low computational cost.

8.4.4
Application of the Combustion Model on a 2D Grate Furnace

In cooperation with Graz University of Technology (TU-Graz), the model has been applied to a biomass combustion grate furnace for the prediction of combustion and NO_x emissions [24]. A 2D geometry together with appropriate boundary conditions for such a furnace was supplied by TU-Graz. This was used to test the present flamelet model by TU-Eindhoven. Two inlets have been set: one at the grate, where the primary air is mixed with the biomass gasification/combustion products, and the other one corresponding to the secondary air that is added to complete the combustion process. However, the recirculation gases

178 | 8 Thermal Biomass Conversion and NO$_x$ Emissions

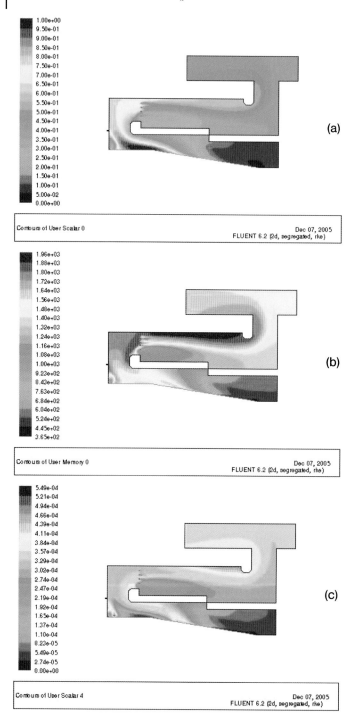

Fig. 8.10 (legend see p. 179)

that are usually introduced for temperature control in the furnace to prevent ash slagging or sintering and to reduce the NO_x emissions have not been considered in the present stage of the simulations. This is because including recirculation gases at low temperature (200 °C) in the simulation requires the use of an enthalpy variable. At the fuel bed, profiles of velocity, species mass fractions and temperature have been supplied by TU-Graz. They result from the gasification/combustion of a biomass fuel on the grate when blowing air from below the grate. A fuel and an oxidizer are defined, which are mixed by using the mixture fraction to generate an approximation of the species profiles provided by TU-Graz.

A detailed reaction scheme dedicated to biomass combustion was used: the SKG03 reaction mechanism [25]. The simulations were performed with the FLUENT flow solver using the realizable k-ε model for turbulence modeling.

Figure 8.10(a–c) shows the 2D profiles of mixture fraction, temperature and NO mass fraction, as predicted by the current flamelet model. In Fig. 8.10(a), at the bottom line of the furnace, the distribution of the mixture fraction on the grate can be seen. The mixture fraction is 1 somewhere in the middle of the grate, where only gaseous fuel is present, and 0 at the sides of the grate, where only air is present. In between, there are mixtures of fuel and air with mixture fractions between 0 and 1. This distribution of the mixture fraction on the grate is a result of different degrees of biomass conversion into gasification and combustion products along the grate when blowing air from below the grate, through the fuel bed. Later, the streams with different mixture fractions coming from the grate mix with each other and start to burn, leading to a more homogeneous mixture with rising temperature (see also Fig. 8.3b). After the first turn in the furnace, the main flow meets the secondary air flows (four tiny flows in a row in Fig. 8.10a), which dilute the mixture and insure complete combustion of the fuel still present in the mixture. This results in an additional temperature increase, as indicated by the red areas in Fig. 8.10(b). Figure 8.10(c) shows that the NO is formed where the combustion takes place. This is indicated by the high temperature zones in the furnace (Fig. 8.10b).

From the results it can be concluded that the requirements of the model are met: the model is sufficiently detailed to predict the NOx production accurately and the calculation time of the model is within reasonable limits to be able to calculate the emission levels for different design parameters and operating conditions.

Fig. 8.10 2D predicted profiles of (a) mean mixture fraction, (b) mean temperature and (c) mean NO mass fraction for the biomass grate furnace.

8.5
Conclusions

The present research has treated important parts of the modeling of combustion and NO_x formation in a biomass grate furnace. All parts resulted in useful approaches. For all these approaches successful first steps were taken. Currently, more research is underway to obtain improved results: NH_3 production is measured in the grid reactor with the tunable diode laser, detailed kinetics will be attached to the front propagation model, including the measured NH_3 release functionalities, and for the turbulent combustion model heat losses are taken into account. In addition, the fuel layer model has to be coupled to the turbulent combustion model in the furnace.

Acknowledgments

The EC funded the present work within the European project OPTICOMB, NNE5–2001–000639. The authors thank TU-Graz for their valuable contribution.

References

1 Toland, A., Bastiaans, R.J.M., Holten, A.P.C., de Goey, L.P.H. Kinetics of CO release from bark and medium density fibreboard pyrolysis, *Biomass Bioenergy*, submitted (**2005**).
2 Tran, D.Q., Rai, C. A kinetic model for pyrolysis of Douglas fir bark, *Fuel*, 57, 293–298, (**1978**).
3 Thurner, F., Mann, U. Kinetic investigation of wood pyrolysis, *Ind. Eng. Chem. Process Des. Dev.*, 20, 482–488, (**1981**).
4 Chen, W.Y. Pyrolysis and combustion kinetics of pine bark, *AIChE Symp. Ser.* 307, 91, 143–153, (**1995**).
5 de Jong, W., Pirone, A., Wojtowicz, M.A. Pyrolysis of Miscanthus Giganteus and wood pellets: TG-FTIR analysis and reaction kinetics, *Fuel*, 82, 1139–1147, (**2003**).
6 Simmons, G.M., Lee, W.H. Kinetics of gas formation from cellulose and wood pyrolysis, in: *Fundamentals of Thermochemical Biomass Conversion*, Overend, R.P. et al. (Eds.), Elsevier Applied Science, London, pp. 385–395, (**1985**).
7 Moors, J.H.J. Pulverised char combustion and gasification at high temperatures and pressures, Ph.D. thesis, Eindhoven University of Technology, The Netherlands, (**1999**).
8 Guo, J. Pyrolysis of wood powder and gasification of wood-derived char, Ph.D. thesis, Eindhoven University of Technology, The Netherlands, (**2004**).
9 Solomon, P.R., Serio, M.A., Suuberg, E.M. Coal pyrolysis: experiments, kinetic rates and mechanisms, *Prog. Energy Combust. Sci.*, 18, 133–220, (**1992**).
10 Moors, J.H.J. Pulverised Char Combustion and Gasification at High Temperatures and Pressures, Ph.D. Thesis, Eidhoven University of Technology, The Netherlands, (**1999**).
11 Glarborg, P., Jensen, A.D., Johnson, J.E. Fuel nitrogen conversion in solid fuel fired systems, *Prog. Energy Combust. Sci.*, 29, 89–113, (**2003**).
12 Gort, R. On the propagation of a reaction front in a packed bed, Ph.D. thesis, Twente University, The Netherlands, (**1995**).

13. CHEM1D, A one-dimensional laminar flame code, Eindhoven University of Technology. http://www.combustion.tue.nl/chem1d.
14. Reid, R.C., Prousnitz, J.M., Poling, B.E. *The Properties of Gases and Liquids*, McGraw-Hill Book Company, London, (**1987**).
15. Katunzi, M. Biomass conversion in fixed bed experiments, Msc. thesis, Eindhoven University of Technology, Report number WVT 2006.18, (**2006**).
16. Albrecht, B.A., van Kuijk, H.A.J.A., Bastiaans, R.J.M., van Oijen, J.A., de Goey, L.P.H., Prins, M.J. A flamelet-PDF approach for the modeling of combustion, and NOx formation in a biomass grate furnace, in *Proceedings of the 14th European Biomass Conference and Exhibition*, Sjunnesson, L., Carrasco, J.E., Helm, P., Grassi, A., (Eds), Paris, France, V2BVI19, (**2005**).
17. Albrecht, B.A., Bastiaans, R.J.M., van Oijen, J.A., de Goey, L.P.H. NOx emissions modelling in biomass combustion grate furnaces, in *Proceedings of the 7th European Conference on Industial Furnaces and Boilers (INFUB)*, Reis, A., Ward, J., Leuckel, W., (Eds.), Porto, Portugal, ISBN 972–99309–1-0, (**2006**).
18. Van Oijen, J. Flamelet-generated manifolds: development and application to premixed laminar flames, PhD thesis, Eindhoven University of Technology, Eindhoven, The Netherlands, (**2002**).
19. Bilger, R.W. On reduced mechanisms for methane-air combustion in non-premixed flames, *Combustion Flame*, 80, 135–149, (**1990**).
20. Ramaekers, W.J.S. The application of flamelet generated manifolds in modeling of turbulent partially-premixed flames, MSc thesis, Eindhoven University of Technology, Eindhoven, The Netherlands, (**2005**).
21. Barlow, R.S., Frank, J. Piloted CH4/air flames C, D, E and F–release 2.0, Livermore: Sandia National Laboratories, http://www.ca.sandia.gov/TNF, (**2003**).
22. Bowman, G., Frenklach, M., Gardiner, B., Smith, G., Serauskas, B. GRI-Mech, Gas Research Institute, Chicago, Ilinois. http://www.me.berkeley.edu/gri-mech/index.html
23. Schneider, Ch., Dreizler, A., Janicka, J., Hassel, E.P. Flow field measurements of stable and locally extinguished hydrocarbon-fuelled jet flames, *Combustion Flame*, 135, 185–190, (**2003**).
24. Scharler, R., Fleckl, T., Obernberger, I. Modification of a Magnussen constant of the Eddy Dissipation Model for biomass grate furnaces by means of hot gas in-situ FT-IR absorption spectroscopy, *Progr. Comp. Fluid Dyn.*, 3, 102–111, (**2003**).
25. Skreiberg, O., Kilpinen, P., Glarborg, P. Ammonia chemistry below 1400 K under fuel-rich conditions in a flow reactor, *Combustion Flame*, 136, 501–518, (**2004**).

9
Bioethanol: Production and Pathways for Upgrading and Valorization

*Stephane Pariente, Nathalie Tanchoux, François Fajula,
Gabriele Centi, and Siglinda Perathoner*

9.1
Introduction

Bioethanol is produced from sugars (particularly sugar cane) and starch by fermentation with yeasts and can be used pure or as a gasoline extender. In 2004, around 49 billion liters of ethanol were produced world wide, most of it being for use in cars. Brazil produced around 18 billion liters and used 2.7 million hectares of land for this production, representing 4.5% of Brazilian land area used for crop production in 2005, and roughly 50% of the total production of sugar is further transformed into ethanol. The Brazilian vehicle pool consisted of 16 million vehicles using a blend of 25% anhydrous ethanol in gasoline (E25), and 2.3 million vehicles using pure ethanol (92% volume ethanol and water), with no vehicle using only gasoline [1a].

The other large-scale ethanol user is the USA, where ethanol has been used to increase the octane rating of gasoline, to decrease carbon monoxide emissions, and, more recently, to replace MTBE (methyl *tert*-butyl ether) in reformulated gasoline. Ethanol production in the USA grew from about 0.6 billion liters in 1980 to 16.1 billion liters in 2004. The major raw material is corn grain (starch) with a potential production capacity of over 50 billion liters, to which may also added agricultural residues, wood, municipal solid waste and dedicated energy crops for a global potential production capacity of about of about 300 billion liters, e.g., close to 30% displacement of current fossil fuel usage, as estimated from the US Departments of Agriculture and Energy.

In the European Union, ethanol is consumed in Spain, France, Sweden and Germany, especially after conversion into ETBE (ethyl *tert*-butyl ether), except in Sweden, but its use is increasing in all the other countries. New uses of bioethanol, e.g., in ethanol-direct fuel cells or as raw material for other chemicals, will further expand bioethanol use and production. Table 9.1 summarizes bioethanol production in different countries by 2004 [1]. Owing to political decisions (EU directive setting at 5.75% the proportion of biofuels in fuels) and incentive taxation

Table 9.1 World ethanol production by 2004. (Adapted from [1a, 1b]).

Country	Billion liters
Brazil	18.1
United States	16.1
China	4.4
EU	2.4
India	2.1
Russia	0.9
South Africa	0.5
Saudi Arabia	0.4
Thailand	0.4
Total	49.0

rates, the production volumes in Europe are growing very rapidly, particularly in Germany and the UK, but the main producers in 2004 were Spain, France and Sweden [1b].

Both in the USA and the EU, the introduction of renewable fuels standards is likely to increase considerably the consumption of bioethanol. Lignocelluloses from agricultural and forest industry residues and/or the carbohydrate fraction of municipal solid waste (MSW) will be the future source of biomass, but starch-rich sources such as corn grain (the major raw material for ethanol in USA) and sugar cane (in Brazil) are currently used. Although land devoted to fuel could reduce land available for food production, this is at present not a serious problem, but could become progressively more important with increasing use of bioethanol. For this reason, it is important to utilize other crops that could be cultivated in unused land (an important social factor to preserve rural populations) and, especially, start to use cellulose-based feedstocks and waste materials as raw material.

Table 9.2 shows the yields of common crops associated with bioethanol production [2]. Fast growing crops such as Miscanthus and Switchgrass increase by a factor 3–4 the bioethanol yield (L ha^{-1}). These crops can also be grown on land not suitable for other cash crops. In addition, research on improving ethanol yields from each unit of corn or other crops is underway using biotechnology. By utilizing hybrids designed specifically with higher extractable starch levels, the energy balance is improved significantly. As yields improve or different feedstocks are introduced, bioethanol production may become more economically feasible.

Figure 9.1 reports the prospective average biofuel yield from different crops in EU-15 over 2005–2010 (GJ ha^{-1}) [3]. Bioethanol yield in EU-15 normally is higher than the biodiesel yield, e.g., a smaller land area will be needed to produce

Table 9.2 Yields of bioethanol from the most common crops. (Adapted from [2]).

Crop	Bioethanol (L ha^{-1})
Miscanthus	14031
Switchgrass	10757
Sweet potatoes	10000
Poplar wood (hybrid)	9354
Sweet sorghum	8419
Sugar beet	6679
Sugar cane	6192
Corn (maize)	3461
Cassava	3835
Wheat	2591

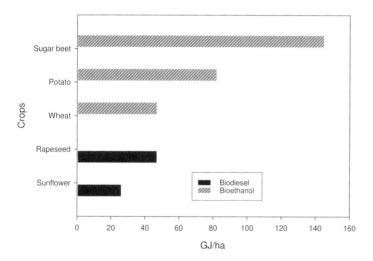

Fig. 9.1 Prospective average biofuel yield from different crops in EU-15 over 2005–2010 (GJ/ha). (Adapted from [5]).

the same amount of transport biofuel, if this biofuel is bioethanol, rather than biodiesel.

Notably, however, any comparison of biodiesel vs. bioethanol should be done with great caution, because analysis of an industry such as that related to biofuels is a very complex task and all conclusions are country dependent. It may be interesting, however, to compare the energy balance and environmental impact in producing biodiesel from oilseed rape and bioethanol from wheat crops [4]. Table 9.3 reports this comparison. The energy balance for bioethanol is more positive than for biodiesel, in particular when straw is utilized, mainly due to the higher yield

Table 9.3 Energy values, costs and emissions per hectare in the comparison of the production of biodiesel by oilseed rape and bioethanol by wheat crops. (Adapted from [4]).

	Biodiesel from oilseed rape; straw utilized	Bioethanol from wheat; straw utilized
Seed/grain yield (t)	4.08	8.96
Biodiesel/bioethanol yield (t)	1.51	3.47
Energy value of biofuel (MJ)	54346	74189
Energy value of straw (MJ)	60000	97500
Energy value of cake (MJ)	1316	0
Energy cost (MJ)	31162	68257
Energy balance (MJ)	+84500	+103432
Emissions, to air		
NOx (kg)	8.58	9.08
N_2O (kg)	6.66	7.41
CO_2 (kg)	751	799
Emissions, to water		
N compounds (kg)	2.16	3.39
P compounds (kg)	0.02	0.03

per hectare. However, the impact of bioethanol production on the environment, particularly on water, is higher than that of biodiesel.

Data shown in Table 9.3 evidence a net energy return of both biofuels, but whether bioethanol can contribute to energy and environmental goals, e.g., the answer to the question "Doesn't it take more energy to make ethanol than is contained in the ethanol?", is an endless debate [5]. Over 300 articles published in the period 2000–2004 discussed the energy balance of ethanol, arriving at quite contradictory conclusions. The debate started in 1981 with the Pimentel report of the advisory committee of the US Department of Energy on the viability of fuel ethanol (and coal-derived methanol) [6]. The conclusion of this report, backed up by Pimentel and coworkers in over 20 consecutive papers, was that more fossil fuel energy is needed to grow corn and convert it into ethanol than is contained in ethanol. In a more recent paper [7] the criticism was extended from corn-derived ethanol to ethanol derived from cellulosic materials like wood or switchgrass and to diesel fuel substitutes derived from sunflowers and soybeans. In addition, ethanol from sugar cane was indicated as a net energy loser. However, the arguments were not fully sustained from a full critical analysis of the data.

A more recent assessment [8] has re-evaluated six representative analyses of fuel ethanol, evidencing that the studies reporting negative net energy incorrectly

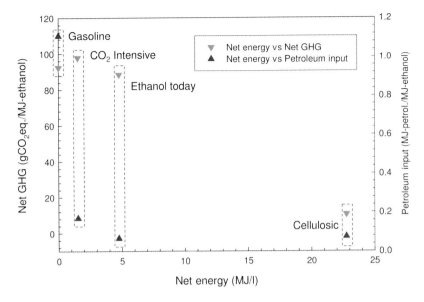

Fig. 9.2 Net energy and net greenhouse gases (GHG) or petroleum inputs for three case studies (see text) with respect to gasoline. (Adapted from [8]).

ignored co-products and used some obsolete data. Three case studies were considered by Farrell et al. [8] by applying a biofuel analysis meta-model (EBAMM) in comparison with gasoline as the reference: *Ethanol Today*, which includes typical values for the current US corn ethanol industry; CO_2 *Intensive*, which considers the shipment of Nebraska corn to a lignite-powered ethanol plant in North Dakota; and *Cellulosic*, which assumes that production of cellulosic ethanol from switchgrass becomes economic. The *Cellulosic* case is a preliminary estimate of a rapidly evolving technology, but highlights the dramatic reductions in greenhouse gas (GHG) emissions that could be achieved.

Figure 9.2 shows the results. For all three cases, producing one MJ of ethanol requires far less petroleum than is required to produce one MJ of gasoline. However, the GHG metric illustrates that the environmental performance of ethanol varies greatly, depending on production processes. With the use of the petroleum intensity metric, the *Ethanol Today* case would be slightly preferred over the *Cellulosic* case (a petroleum input ratio of 0.06 compared with 0.08); however, on the GHG metric, the *Ethanol Today* case is far worse than *Cellulosic* (83 compared to 11).

It may, therefore, be indicated that, when using the appropriate assumptions (e.g., the inclusion of co-product credits calculated by a displacement method) and accurate data, there are clear benefits in expanding the use of bioethanol, but progress toward attaining these goals will require new technologies and practices, such as sustainable agriculture and cellulosic ethanol production. Bioethanol is not only already a chemical produced in large volumes, but its use will expand over the world thanks to progress in its production technology and consequent reduction of the price. Wooley et al. [9] have estimated a current cost of bioethanol production of about $ 40–60 barrel^{-1}, depending on the technology and the availability of low-cost feedstocks for conversion into ethanol. Therefore, it may already be competitive with gasoline, especially considering that the expected progress in its production may further reduce costs [10].

There are several motivations to increase the use of bioethanol [11]: (a) energy security, e.g., widen the energy raw material portfolio; (b) air pollution, because compared with reformulated gasoline a 95% ethanol/5% gasoline blend (E95) reduces sulfur and volatile organic emissions (VOC) [12]; and (c) reduce GHG emissions (Fig. 9.2). Although the expansion of its market will be driven by the direct use as biofuel, there are various reasons (as discussed in the following sections) that point to the need for its further conversion, either for a better blending into gasoline or diesel or to produce chemicals: (a) the expected reduction in bioethanol costs and, at the same time, the increase in ethylene production costs will make bioethanol a competitive raw material; (b) progressively, with the increase of ethanol production capacity, it will be necessary to diversify its uses to other applications to reduce the dependence on the fluctuations in prices for a single production line; and (c) ethanol upgrading is necessary for a better blending with diesel, as noted below. However, studies and analyses on the further conversion of bioethanol are quite limited, apart from its use for the production of ETBE [13–18].

This chapter, after an introduction on the production methods for bioethanol and its use as biofuel, discusses the catalytic upgrading and valorization of bioethanol.

9.2
Production, a Short Overview

Bioethanol is currently produced from the fermentation of sugar by enzymes produced from specific varieties of yeast. The five major sugars are the five-carbon xylose and arabinose and the six-carbon glucose, galactose, and mannose [19]. Traditional fermentation processes rely on yeasts that convert six-carbon sugars into ethanol. Glucose, the preferred form of sugar for fermentation, is contained in both carbohydrates and cellulose. Also, the organisms and enzymes for carbohydrate conversion and glucose fermentation on a commercial scale are readily available.

Two main types of crops are used in the production of bioethanol today: sugar-producing crops (sugar cane, sugar beets) and amylaceous plants (wheat, corn). The production process for each requires a fermentation stage to convert the sugar into ethanol, as well as an advanced distillation stage to separate the alcohol from water. Large volumes of co-products are also formed.

Native sugars found in sugar cane and sugar beet can be easily derived from these plants, and refined in facilities that require the lowest level of capital input. Starch, consisting of glucose linked via α-1,4 and α-1,6 glycosidic linkages (amylose and amylopectin), is the dominant component of cereal crops such as corn and wheat. Starch should be first processed by an acid- or enzyme catalyzed hydrolysis step to liberate glucose, but a single family of enzymes may be used, the amylases, which makes bioconversion relatively simple. Downstream processing of sugars includes traditional fermentation, which uses yeast (*Saccharomyces* yeasts) to produce ethanol; other types of fermentation, including bacterial fermentation under aerobic and anaerobic conditions, can produce various other products from the sugar stream.

Starch processing is a fairly mature technology, but recent advances in starch processing have improved the economics and efficiency of the process. For example, a simpler low-pH α-amylases process has been developed, improving the ethanol yield and reducing costs [20]. The other major advance is the development of enzymes that function on raw, uncooked starch, thereby improving overall process economics [21, 22].

Most of the biomass available, however, does not contain starch as the major carbohydrate, but is typically composed of cellulose (40–50%), hemicellulose (25–35%) and lignin (15–20%). The predominant polysaccharide in the primary cell wall of biomass is cellulose, the second most abundant is hemicellulose, and the third is pectin. The secondary cell wall, produced after the cell has stopped growing, also contains polysaccharides and is strengthened through polymeric lignin covalently crosslinked to hemicellulose. Cellulose is composed of glucose linked via β-1,4 glycosidic bonds. Because of the β-1,4 linkage, cellulose is highly crystalline and compact, making it very resistant to biological attack. In general, hemicellulose consists of a main chain xylan backbone (β-1,4 linkages) with various branches of mannose, arabinose, galactose, glucuronic acid, etc. The degree of branching and identity of the minor sugars in hemicellulose tends to vary, depending upon the type of plant. Furthermore, lignin can be covalently linked to hemicellulose via ferulic acid ester linkages.

The compactness and complexity of (ligno)cellulose makes it much more difficult to attack by enzymes with respect to starch. Therefore, the cost of bioethanol production is higher [23]. To be cost competitive with grain-derived ethanol, the enzymes used for biomass hydrolysis must become more efficient and far less expensive. In addition, the presence of non-glucose sugars in the feedstock complicates the fermentation process, because conversion of pentose sugars into ethanol is less efficient than conversion of the hexose sugars.

The passage from the 1st-generation of biofuels, which use agricultural products, to 2nd-generation biofuels that use all the structural components of plants and trees, e.g., able to process (ligno)cellulose, requires the following developments:

1. A new generation of cheap enzymes for hydrolysis of cellulose and lignocellulose to fermentable sugars (able to complete the biomass hydrolysis during fermentation).

2. Biocatalysts (of chemo/bio-catalysts working in tandem) able to fully convert the carbohydrates into ethanol and other fuels, and be robust and tolerant of the toxic compounds formed during the pretreatment (hydrolysis) process.

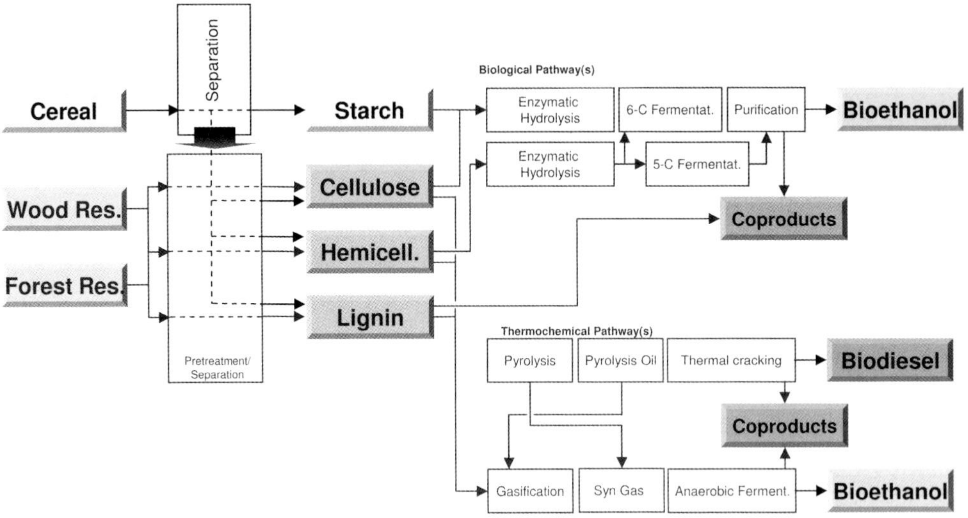

Fig. 9.3 Simplified flowchart of the biomass-to-bioethanol pathways. (Adapted from [24]).

They must be also able to withstand the stress of high ethanol and substrate concentrations, low pH, etc.

3. Biocatalysts able to expand the substrate usage spectrum of microorganisms (e.g., C_5-sugars) and to increase tolerance to industrial conditions (e.g., high product tolerance, fast growth, high yield and productivity).

Figure 9.3 shows a simplified flowchart of the biomass-to-ethanol pathways for 1st- and 2nd-generation biofuels [24]. The figure evidences that every biofuel pathway links fuel production to the generation of co-products. In 1st-generation biofuel production, co-products play an essential role in process economics, adding value and profit to production. The pathways for biofuel production become more complex as one moves from starch to wood residues, but at the same time new opportunities arise for co-product valorization, if the bioethanol production is effectively integrated in a biorefinery. A research challenge, and also a key step towards decreasing the cost of production of bioethanol, is to optimize this integration of bioethanol processes into the general scheme of biorefineries.

Figure 9.3 also points out that two main complementary technological platforms may be identified in processing biomass to ethanol: a bioconversion platform and a thermochemical platform. The first uses biological agents to carry out a structured deconstruction of lignocellulose components. This platform combines process elements of pretreatment with enzymatic hydrolysis to release carbohydrates and lignin from the wood. The thermochemical platform uses processes to gasify wood, producing synthesis gases. This platform combines process elements of pretreatment, pyrolysis, gasification, cleanup and conditioning to generate a mixture of hydrogen, carbon monoxide, carbon dioxide, and other gases. The products of this platform may be viewed as intermediate products, which can then be assembled into chemical building blocks and, eventually, end products.

An overview of the bioconversion platform is given in Fig. 9.4(a), while Fig. 9.4(b) shows a simplified thermochemical platform flowchart.

The first step is a pretreatment stage, which must optimize the biomass feedstock for further processing by increasing the surface area of the substrate for enzymatic action. As in traditional pulping, lignin is either softened or removed, and individual cellulosic fibers are released creating pulp, but the effectiveness of the subsequent enzymatic hydrolysis step depend considerably on this stage. New pretreatment technologies under study include water-based systems, such as (a) steam-explosion pulping; (b) acidic treatments, using concentrated or dilute acids such as H_2SO_4; (c) alkaline treatments that utilize recirculated ammonia or modified steam explosion; and (d) organic solvent pulping systems, such as acetic acid or ethanol. However, further research is needed in this area, particularly on understanding pretreatment technologies, including improved chemical hydrolysis of lignocellulose, and in designing new, more reliable reactors and equipment, and minimizing the energy and chemical (acid) input.

After pretreatment, the cellulose and hemicellulose can be hydrolyzed, typically by enzymatic hydrolysis using cellulases, usually produced by fungi such as *Trichoderma*, *Penicillum*, and *Aspergillus* [25]. A cocktail of cellulases is required to break down efficiently the cellulosic microfibril structure into its carbohydrate components, unlike the bioconversion of starch, which has a simpler chemical structure. The enzymatic hydrolysis step may be completely separate from the other stages of the bioconversion process, or it may be combined with the fermentation of carbohydrate intermediates to end-products. Separating hydrolysis and fermentation allows greater flexibility, and the ability to process different end products. However, a separate process implies higher costs.

Further research is also needed in this area. Particularly, (a) to create a new generation of cheap enzymes for hydrolysis of cellulose and lignocellulose to fermentable sugars (able to complete the biomass hydrolysis during fermentation); (b) to develop improved biocatalysts that allow us to simplify the process and reduce energy input; and (c) to improve separation and recovery.

The bioconversion platform allows the production of a range of intermediate products, including glucose, galactose, mannose, xylose, and arabinose, which can be relatively easily processed into value-added bioproducts. In addition, relevant amounts of lignin or lignin components are also produced. Depending upon the pretreatment, lignin components may be found in the hydrolysate after enzymatic hydrolysis, or in the wash from the pretreatment stage. Finally, a relatively small amount of extractives may be retrieved from the process. These extractives are highly variable, depending upon the feedstock employed, but may include resins, terpenes, or fatty acids.

After hydrolysis, six-carbon sugars can be fermented to give ethanol using age-old yeasts and processes. Five-carbon sugars, however, are more difficult to ferment. New yeast strains are being developed that can process these sugars, but

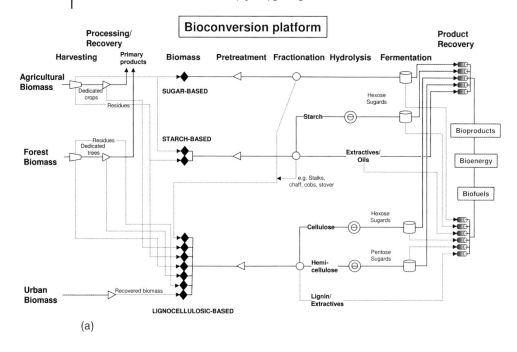
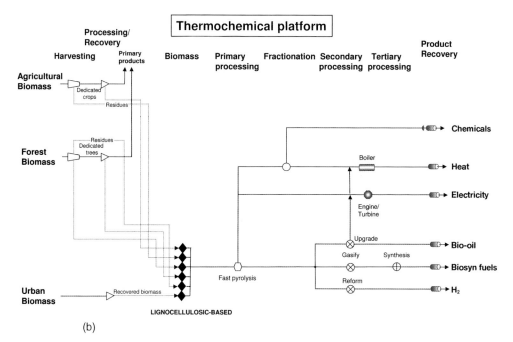

Fig. 9.4 Bioconversion (a) and thermochemical (b) platforms flowcharts for 2nd-generation biofuel production. (Adapted from [24]).

issues remain with process efficiency and the length of fermentation. Relevant research targets are, thus, to develop (a) novel technologies for the physical/chemical fractionation of biomass in separate (C5/C6) sugar and lignin fractions and (b) effective biocatalysts (enzymes or chemo- and bio-catalysts working in tandem) that are active in five-carbon sugar conversion under industrial conditions (e.g., high product tolerance, fast growth, high yield and productivity).

It may be estimated that ethanol yields from lignocellulosics will range between 0.12 and 0.32 L kg^{-1} undried feedstock, depending upon the efficiency of five-carbon sugar conversion [26]. Other types of fermentation, including bacterial fermentation under aerobic and anaerobic conditions, can produce various other products from the sugar stream, including lactic acid.

Bioconversion platforms for lignocellulosics-to-ethanol are beginning to become commercially viable, but the effectiveness of the pretreatment stage should still be improved, the cost of the enzymatic hydrolysis stage decreased, and overall process efficiencies improved by better synergies between various process stages. There is also a need to improve process economics by creating co-products that can add revenue to the process.

The thermochemical platform (Fig. 9.4b) is not finalized to produce bioethanol, apart from the possible catalytic conversion of syngas into higher alcohols, including ethanol. Ethanol and other higher alcohols are also formed as by-products of both Fischer–Tropsch and methanol synthesis, and modified catalysts have been shown to provide better yields. The thermochemical platform provides the opportunity for several additional co-products, as well as energy in the form of heat or electricity and biofuels. Thermochemical transformations are discussed in detail in other chapters of this book and, thus, are not discussed further here. However, notably, Fig. 9.3 evidences that the concept that both bio and thermochemical conversions should be integrated in a biorefinery and the cost of the products, including bioethanol, depends on the effectiveness of this integration and valorization of all the products of reaction.

9.3
Uses as Biofuel

9.3.1
Bioethanol as Fuel Additive

9.3.1.1 Gasoline/Bioethanol Blends

Bioethanol can be combined with gasoline in any concentration up to pure ethanol (E100), but a problem is the presence of water. Conventional gasoline, depending on its aromatics content, can dissolve up to 150 parts per million (ppm) water at 21 °C. By adding ethers to the gasoline, water solubility increases to 600 ppm. Cooling water-saturated blends turns them hazy because some of the water becomes insoluble, without causing any performance problems.

The situation is different for gasoline oxygenated with 10 vol.% ethanol. The gasoline–alcohol blend can dissolve more water (6000–7000 ppm at 21 °C) and, upon cooling, both the water and some of the ethanol become insoluble. As a result, two layers of liquid form: an upper ethanol-deficient gasoline layer and a lower ethanol-rich (up to 75% ethanol) water layer. The same demixing problem happens when water exceeds the solubility limit. After phase separation, the gasoline layer has a lower octane number and may knock in an engine. The fuel also is less volatile. The engine will not run on the water/ethanol layer.

Owing to this problem of phase separation, gasoline/ethanol blends should be not exposed to water during distribution or use in a vehicle. For this reason, they cannot be transported in pipelines, which sometimes contain water. This water sensitivity also means that extra care should be taken when gasoline/ethanol blends are used as a fuel for boat engines.

Another problem lies in material compatibility. Some metal components in the fuel system will rust or corrode in the presence of water/ethanol deriving from the phase separation of gasoline/ethanol blends. In addition, ethanol can swell and soften natural and some synthetic rubbers (elastomers). The elastomeric materials used in today's vehicles have been selected to be compatible with oxygenated gasolines, not for gasoline/ethanol blends above a certain ethanol concentration. Bioethanol-fueled engines should thus be modified with respect to conventional gasoline engines to operate reliably. Typically, it is necessary to at least substitute the various seals. Vehicles that may use these gasoline/ethanol blends are typically indicated as "Flex-Fuel" in the marketplace.

Another problem is volatility. Adding ethanol to a conventional gasoline not designed for alcohol blending can produce a blend that is too volatile. Blend volatility is increased only slightly by other oxygenated molecules (ethers) such as MTBE (derived from methanol and isobutylene) and not at all by ETBE (derived from ethanol and isobutylene) and TAME (derived from methanol and isoamylenes). Ethanol volatility increase also causes problems in meeting the very tight vapor pressure limits of this gasoline, especially when used in high temperature regions. Furthermore, ethanol has a higher heat of vaporization than ethers. Some of the degradation in drivability of gasoline/ethanol blends can be attributed to the additional heat needed to vaporize the fuel.

Another volatility situation can occur if ethanol-blended fuel is mixed with hydrocarbon-only fuel in a vehicle fuel tank. This event is called "commingling". In effect, the ethanol in the blend increases the vapor pressure of the hydrocarbon-only gasoline. The increase in vapor pressure is dependent on the ratio of the two components and the amount of ethanol in the blend.

Ethanol consumption in an engine is ca. 34% higher than that of gasoline, but higher compression ratios in an ethanol-only engine allows for increased power output. Pure ethanol has a much higher octane rating (129 RON) than ordinary gasoline (91/92 RON), allowing a higher compression ratio and different spark timing for improved performance. To change a pure-gasoline-fueled car into a pure-ethanol-fueled car, larger carburetor jets (about 30–40% larger by area) or fuel injectors are needed. With the current electronic injections malfunctioning sometimes occurs.

9.3.1.2 Diesel/Bioethanol Blends

There has been a recent revival in interest in the use of ethanol–diesel fuel blends (E-diesel) in heavy-duty vehicles as a means to reduce petroleum dependency, increase renewable fuels use, and reduce vehicle emissions [27]. E-diesel blends containing 10–15% ethanol could be prepared via the use of additives. However, several fuel properties that are essential to the proper operation of a diesel engine are affected by the addition of ethanol to diesel fuel – in particular, blend stability, viscosity and lubricity, energy content and cetane number (increasing concentrations of ethanol in diesel lower the cetane number proportionately) [28]. Materials compatibility and corrosiveness are also important factors that need to be considered.

Ethanol solubility in diesel is affected mainly by two factors, temperature and water content of the blend. At warm ambient temperatures, dry ethanol blends readily with diesel fuel. However, below about 10 °C the two fuels separate. This separation can be prevented by adding an emulsifier that acts to suspend small droplets of ethanol within the diesel fuel, or by adding a co-solvent that acts as a bridging agent through molecular compatibility and bonding to produce a homogenous blend. Co-solvents allow fuels to be "splash-blended", thus simplifying the blending process. The aromatic content of diesel fuel also affects the solubility of ethanol in diesel and, therefore, the effectiveness of emulsifiers and co-solvents.

The addition of ethanol to diesel lowers fuel viscosity and lubricity. Lower fuel viscosities lead to greater pump and injector leakage, reducing maximum fuel delivery and hence power output. Hot start problems may also be encountered as insufficient fuel may be injected at cranking speed when fuel leakage in the high-pressure pump is amplified because of the reduced viscosity of the hot fuel.

The quality of the ethanol has a strong influence on its corrosive effects. Three categories of problems have been identified: general corrosion (caused by ionic impurities, mainly chloride ions and acetic acid), dry corrosion due to ethanol polarity, and wet corrosion caused by azeotropic water. Corrosion inhibitors should thus be incorporated in ethanol–diesel blends.

A critical safety issue of using diesel–ethanol blends relates to flashpoint and flammability. E-diesel blends containing 10–15% ethanol have the vapor pressure and flammability limits of ethanol. This means that ethanol concentrations in enclosed spaces such as fuel storage and vehicle fuel tanks are flammable over the temperature range 13–42 °C. Thus, there are higher risks of fire and explosion than with diesel fuel, or even gasoline. Other vehicle performance-related concerns are (a) a decreased maximum power; (b) an increased incidence of fuel pump vapor lock; and (c) a reduced fuel pump and fuel injector life due to the decreased lubricity of ethanol.

On the other hand, biodiesel is a good additive to diesel fuel, reducing smoke, particulate matter (PM) and CO emissions. Engines using E-diesel with 10%

ethanol have been reported to produce up to 27% less soot and 20% less carbon monoxide, although the effect is much less relevant in the latest generation of "common-rail" diesel engines.

9.3.2
Bioethanol and Hydrogen

The social, more than technological, push towards an hydrogen economy, e.g., to use H_2 as the primary energy vector, has progressively increased the search for alternative ways of producing H_2, including from biomass. About 95% of the hydrogen used today comes from reforming natural gas. The remaining, high-purity hydrogen from water electrolysis is primarily produced using electricity generated by burning fossil fuels. Therefore, to realize the full benefits of a hydrogen economy, hydrogen must be produced cleanly, efficiently, and affordably from renewable resources.

There are two main possible routes to H_2 from biomass. They are schematically shown in Fig. 9.5 together with other possible alternative routes of producing fuels from biomass [29].

As discussed in previous sections, sugars, starch and (ligno)cellulose can be converted into ethanol by fermentation, the latter via preliminary chemical and physical pretreatment followed by enzymatic breakdown of the biopolymers. Pure ethanol can be added to gasoline or diesel. However, this requires an energy-intensive distillation step. This and the energy used in fertilizers, transportation

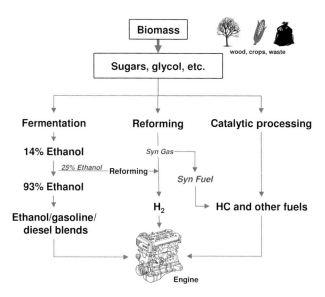

Fig. 9.5 Main possible alternative routes of producing fuels from biomass. (Adapted from [29]).

of biomass, etc. should be subtracted from the energy gained from the biofuel to assess the net energy output. The distillation problem can be reduced if ethanol is converted into hydrogen. Ethanol may thus serve as an energy carrier for a future hydrogen society. The technology for this conversion is available [30], but there also exists the alternative that transportable fuels (liquid alkanes with the number of carbon atoms ranging from 7 to 15) may also be formed directly from the carbohydrates by reaction with water (steam reforming) in the liquid phase, hence saving the energy and costs associated with distillation, but adding the energy cost of the endothermic reaction [31], exemplified by the conversion of sorbitol into hexane. It takes 1.5 mol of sorbitol to produce 1 mol of hexane by combining the catalytic hydrogenation of sorbitol in aqueous solution to the sorbitol reforming to produce the H_2 necessary for first reaction. The net process is thus:

$$\frac{19}{13} C_6O_6H_{14} \rightarrow C_6H_{14} + \frac{36}{13} CO_2 + \frac{42}{13} H_2O \qquad (1)$$

The advantage of producing liquid hydrocarbons instead of ethanol is that it avoids the need to change the complex and costly infrastructure for fuel distribution and use (engines), and on-board (on car) or on-site (at the fuel pump) reforming to H_2. With respect to this alternative it would be preferable to use directly ethanol fuel cells (see next section).

The other alternative in H_2 production is steam-reforming to *syn gas* (a mixture of CO and H_2). Carbon monoxide may be further converted into hydrogen and carbon dioxide by the catalytic water-gas shift, with eventual CO_2 sequestration by injection in oil fields or cavities. H_2 can be then used directly as fuel (solving the problem of storage). Alternatively, the syn gas can be transformed into synthetic liquid fuels (*synfuels*) in the form of synthetic diesel by Fischer–Tropsch reaction or gasoline components (by methanol synthesis and further conversion into higher alcohols, see later), again avoiding the need to change transport infrastructure.

Ethanol is the most heavily researched renewable liquid for H_2 production, but other renewable liquid options exists: sorbitol, glucose, glycerol, bio-oil, methanol, propylene glycol, and less-refined sugar streams (cellulose, hemicellulose). Ethanol has the highest theoretical hydrogen yield per kg of feedstock, but its cost is higher than that of most other alternatives. Figure 9.6 compares the costs of H_2 produced by reforming from different potential bio-liquids [32]. This cost is only a theoretical cost, based on the current bio-liquid cost (plant-gate) and the theoretical H_2 yield. Therefore, they do not consider the production costs because, for several of these bio-liquids, catalysts and/or processes for their use in H_2 production by reforming may not be available. Nevertheless, the comparison indicates that there are other preferable alternatives to ethanol for H_2 production, even if the catalysts and/or process are developed and/or improved. The use of bio-oil shows possible problems in the formation of carbonaceous deposits and/or products difficult to convert into H_2, while the direct use of glucose solutions

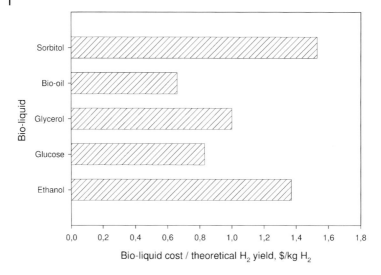

Fig. 9.6 Comparison of the ratio between bio-liquid cost (plant-gate) and the theoretical H_2 yield for different bio-liquid raw materials to be used in H_2 production. (Adapted from [32]).

appear quite interesting. With respect to ethanol, the potential cost of H_2 production is lower because fermentation and separation costs may be avoided. In addition, the solution has lower volatility, it is non-toxic and non-flammable. Therefore, it may be preferable to use bioethanol either directly as fuel additives, as fuel in direct ethanol fuel cells, or as raw material for upgrading to more valuable chemicals than H_2.

There are several technical challenges in reforming bioethanol and other bio-liquids. Various catalyst systems have been investigated for the steam reforming of ethanol, bio-oil, sugar alcohols, and other bio-liquids [30, 33–40]. A common problem with the catalysts is deactivation due to coking, which occurs when side-reaction products (e.g., acetaldehyde, ethylene) deposit on the catalyst [41]. Process parameters such as the steam-to-carbon (water-to-ethanol) ratio can limit carbon deposits, but at the cost of greater process energy requirements [42, 43]. The water-to-ethanol feed ratio and operating temperature also influence the selectivity to hydrogen. Water-to-ethanol molar ratios of three or more, and temperatures above 500 °C, favor the production of hydrogen over methane and other reaction intermediates. However, higher water-to-ethanol ratios require increased energy inputs.

Low-temperature (<500 °C) reforming technologies are also under investigation. The advantages of low-temperature technologies are reduced energy intensity, compatibility with membrane separation, favorable conditions for the water-gas shift reaction, and minimization of undesirable decomposition reactions typically encountered when carbohydrates are heated to high temperatures [44].

Aqueous-phase reforming is a promising technology that has been applied to glucose, ethylene glycol, sorbitol, glycerol, methanol [31, 35, 45]. Studies have shown that the following factors promote selectivity to hydrogen rather than alkanes [35]:

- catalysts made of platinum, palladium, and nickel-tin (nickel catalysts favor alkane production),
- more basic catalyst support materials (e.g., alumina),
- neutral and basic aqueous solutions,
- feedstock type (in descending order of hydrogen selectivity) – polyols (selectivity decreases with increasing carbon number), glucose (selectivity decreases as the wt.% increases from 1 to 10).

Catalyst coking and deactivation are not significant problems as they are in steam reforming. While hydrogen yields are highest from the aqueous phase reforming of sorbitol, glycerol, and ethylene glycol, glucose reforming may be more practical. Improvements in catalyst performance, reactor design, and reaction conditions may help increase hydrogen selectivity [45]. Low-temperature gas phase reforming of ethanol is also being investigated, but there is a tradeoff between catalyst activity and resistance to deactivation (due to coking) and research is ongoing [44].

9.3.3
Bioethanol for Fuel Cells

There is a great interest in using bioethanol directly for fuel cells (DEFC) [46–50]. Electrochemical fuel cells convert the chemical energy of bioethanol directly into electrical energy to provide a clean and highly efficient energy source. In DEFC there is no need for a prereformer to produce on-site H_2, and it is much easier to store ethanol than hydrogen. The energy density of ethanol (the amount of energy released by using a given volume of ethanol) is largely greater than even highly compressed hydrogen. With respect to direct methanol fuel cells the advantage is the much lower toxicity, but there is the drawback of lower reactivity and easy cross-over of the membrane. Ethanol is a hydrogen-rich liquid and it has a higher energy density (8.0 kWh kg^{-1}) than methanol (6.1 kWh kg^{-1}).

There are several issues to be solved for practical commercialization of DEFC even though for applications such as portable batteries for notebooks they will be probably commercialized in a few years:

- The need to develop alternative electrocatalysts with low or, even better, without noble metals (platinum), to decrease the costs due to Pt shortage. New nanostructured electrocatalysts (HYPERMEC™ by ACTA SpA for example, http://www.acta-nanotech.com) [51, 52] have been developed, which are based on non-noble metals, preferentially mixtures of Fe, Co, Ni at the anode, and Ni, Fe or Co alone at the cathode. With ethanol, power densities as high as 140 mW cm^{-2} at 0.5 V have been obtained at 25 °C with self-breathing cells containing commercial anion-exchange membranes.

- Also, using noble metals, the performances are still not satisfactory. There is not a total oxidation of ethanol and byproducts such as acetaldehyde are formed [48].

- There is a high anode overpotential, which should be reduced by a better electrode-membrane assembling design.

- There is still significant fuel cross-over from the anode to the cathode through the electrolyte membrane. In general, the membrane stability characteristics should also be improved [49].

- The stability of the DEFC should be increased.

Therefore, DEFC presents various chemical and catalytic challenges not previously encountered in PEM type cells. This cell has the potential to provide good energy density, if electrocatalysts are found that efficiently carry out the 12-electron oxidation of ethanol to carbon dioxide and water. Unlike electrocatalysts presently utilized in fuel cells, an ethanol catalyst must efficiently cleave carbon–carbon bonds, while adding oxygen atoms to the substrate. Presently available catalysts tend to stop at acetaldehyde or acetic acid products, leaving the C–C bond intact. The ethanol cross-over and the necessity of operations at higher temperatures than H_2 fuel cells, e.g., 120–150 °C vs. 80 °C for the latter, also impose the need to develop new improved membranes. However, significant progress has been made in all these areas and, therefore, DEFC may be commercialized in a relatively few years, opening up new opportunities for the use of bioethanol.

9.4
Bioethanol Upgrading and Valorization

9.4.1
Conversion into Fuel Components

As commented in Section 9.3, there are various problems in adding bioethanol directly to gasoline and, especially, diesel. Owing to better fuel economy, the market for diesel cars is fast expanding and in Europe it is already the main type of light-duty vehicle commercialized. Therefore, to increase the share of biofuels in the transportation sector up to the target levels in Europe (at least 5.75% by 2010; European directive) it is necessary to consider in addition to the blending with biodiesel also the use of bioethanol or its derivates. To use this blending in all actual vehicles and avoid some of the costly additives necessary in the case of bioethanol (Section 9.3), it is preferable to upgrade bioethanol to other oxygenated molecules. Owing to the large increase in the availability of glycerol as a byproduct in the trans-esterification process to produce biodiesel (see Chapters 10 and 11), a quite interesting opportunity is the production of oxygenate additives from bioethanol and glycerol to be used in diesel fuel blends.

The incorporation of oxygenated molecules in diesel fuel blends have been shown for several years to induce an increase in the cetane numbers as well as a reduction of soot and particles emissions [53, 54]. More recently, several technical papers investigating the use of oxygenates as additives in diesel fuel have showed the growing interest of the automobile industry for such additives [55–61]. Recent patents [62] have shown that the addition of glycerol acetals to diesel blends reduces particulate emissions. In general, most of the cited studies seem to agree that the reduction of particles emissions is proportional to the oxygen content of the additives.

An extensive review has been conducted to identify potential oxygenates for blending into diesel fuels [55]. Over 70 molecules were identified and tested, taking into account numerous physical properties such as oxygen content, flash point, viscosity, cetane number, corrosivity, toxicity, and miscibility with diesel blends. Five key aspects were considered critical to develop commercially valuable diesel additives:

1. The amount of oxygenate added to diesel fuel to give 7 wt.% oxygen should not exceed 20 vol.%.
2. The flash point should not be less than 52 °C.
3. The oxygenate should be soluble in low aromatic diesel fuel at temperatures down to 6 °C.
4. The oxygenate must tolerate at least 1000 ppm of water and should not separate into two phases.
5. The oxygenate must not decompose into corrosive products.

Biological, toxicological and environmental issues were also taken into account. Based on these aspects, bioethanol itself does not possess the necessary characteristics as diesel additives, and its conversion into ETBE suffers from limits in the availability of isobutene. There is thus the need to develop alternative oxygenated fuel additives.

Only 10 out of the 70 molecules analyzed met all the requirements to blend successfully into diesel fuels. Of these molecules the most interesting ones seem to be acetals, ethers and esters.

Diacetals can be synthesized from bioethanol only and from bioethanol and glycerol (the co-product of the diester production) for the ether-type molecules, which could be even more interesting.

Two possible interesting acetals are diethoxy ethane or butane. They can be synthesized by the catalyzed reaction of acetaldehyde (obtained by ethanol catalytic oxidation) with two molecules of ethanol, or by the catalyzed reaction of butanal (obtained by catalytic conversion of two molecules of acetaldehyde) with two molecules of ethanol. To achieve a "one-pot" synthesis, a key aspect for a possible commercial development, it is necessary to develop suitable multifunctional catalysts. Research on these aspects is in progress [63].

Nord and Haupt [61] using diethoxy ethane as diesel fuel additive in an unmodified diesel engine observed a reduction of 34% in particulate mass emissions, a decrease of 6.4% in CO_2 emissions in the atmosphere, and a reduction of 3.4% in NOx emissions, although a large increase in VOC emissions, particularly acetaldehyde, was found. The need to introduce catalytic converters also in diesel cars can solve this problem of VOC emissions, and they can probably be beneficial in enhancing NOx catalytic conversion. In fact, various evidence indicates that acetaldehyde is a good reductant to catalytically convert NOx in autoexhaust emissions [64–66]. In addition, the formation of acetaldehyde is one of the primary motivations for the low-temperature conversion of NOx over catalysts in the presence of non-thermal plasma [64], a research line of significant interest recently.

Finally, notably, the cetane number for diethoxybutane is 96, which is excellent considering that the cetane number for a usual diesel fuel is set around 45.

Interesting ethers of glycerol with ethanol are (a) triethoxypropane, obtained by catalytic reaction of three molecules of ethanol with one of glycerol; and (b) 1,3-dioxolane, -4(ethoxymethyl)-2-methyl-, obtained by catalytic reaction of glycerol first with ethanol to form 3-ethoxy-1,2-propanediol, which then reacts with acetaldehyde (formed again by catalytic ethanol oxidation) to form the target compound. Also in this case, research is in progress [63] and the characteristics of the target products are well suited for a blending with diesel.

In conclusion, there is interest for bioethanol upgrading to fuel additives and some research is active in this direction, but much more effort is needed to demonstrate the validity and the viability of the concept of preparing oxygenated diesel fuel additives from bioethanol and glycerol. The key to success is to develop selective multifunctional solid catalysts, in which interest is more general, because similar multifunctionality is necessary in the catalytic synthesis of fine chemicals [67]. There is, thus, the possibility of cross-fertilization between the two research areas.

Other possibilities for the transformation of ethanol into gasoline are:

- Decomposition and aromatization of ethanol on Fe-ZSM-5 [68] or Mo_2C/ZSM type catalysts [69, 70]. This reaction, however, produces mainly aromatics; there are continuous more stringent regulations on the use of aromatics in fuels due to their carcinogenicity (benzene) and toxicity.

- Dehydration into ethylene (over acid catalysts) [71–73] followed by oligomerization (via coordination or multifunctional catalysis) [74–77]. The main problem is the need to remove water from the feed and the vaporization of ethanol, making the process costly.

- Transformation of ethanol into butadiene over basic catalysts [78, 79] and further conversion of the latter, but there are problems of deactivation of the catalysts due to the reactivity of butadiene.

As shown in Fig. 9.7, bioethanol may be also used in a biorefinery for the production of higher-alcohols, to integrate ethanol and methanol (produced via fermen-

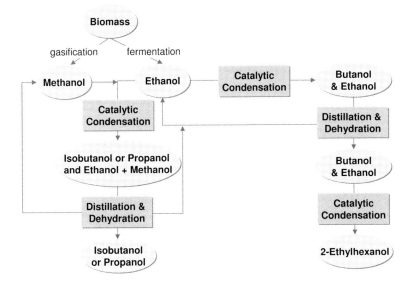

Fig. 9.7 Bioethanol catalytic transformation in a bio-refinery based on Guerbet catalytic condensation.

tation and biomass gasification, respectively) through the Guerbet reaction [80, 81]. 1-Butanol results from the self-condensation of ethanol in this multistep reaction, occurring on a single catalytic bed. Combining methanol with ethanol gives a mixture of propanol, isobutanol, and 2-methyl-1-butanol. All of these higher alcohols are useful as solvents, chemical intermediates, and fuel additives and, consequently, have higher market values than the simple alcohol intermediates. The most successful catalysts are multifunctional with basic and hydrogen transfer components. Also in this case, the catalytic performances need to be improved further.

9.4.2
Conversion into Chemicals

Ethanol itself finds use as chemical in various applications, particularly as solvent (toiletries and cosmetics, paints, lacquer thinners, printing inks, dyes, detergents, disinfectants and pharmaceuticals), besides its uses in the food and beverage industry, and as a disinfectant.

As a chemical raw material, ethanol is used for the production of a range of monomers and solvents, and is essential in pharmaceutical purification. In the presence of an acid catalyst ethanol reacts with carboxylic acids to produce ethyl esters. The two largest-volume ethyl esters are ethyl acrylate (from ethanol and acrylic acid) and ethyl acetate (from ethanol and acetic acid). Ethyl acrylate is a monomer used to prepare acrylate polymers for use in coatings and adhesives. Ethyl acetate is a common solvent used in paints, coatings, and in the pharmaceutical industry. Various other ethyl esters are used in much smaller volumes as artificial fruit flavoring.

Another large class of chemicals produced starting from ethanol are ethylamines. When heated to 150–220 °C over a silica- or alumina-supported nickel catalyst, ethanol and ammonia react to produce ethylamine. Further reaction leads to diethylamine and triethylamine. The ethylamines find use in the synthesis of pharmaceuticals, agricultural chemicals, and surfactants.

Ethanol in the past has been used commercially to synthesize dozens of other high-volume chemical commodities. However, at present, it has been substituted in many applications by less costly petrochemical feedstocks, e.g., ethylene. The availability of low-cost ethanol and the rising cost of ethylene, however, may change this scenario. For example, there is interest in producing ethylene from ethanol [71–73], while the opposite reaction is commercially current. Already, in markets with abundant agricultural products, but a less developed petrochemical infrastructure, such as the People's Republic of China, Pakistan, India, and Brazil, ethanol can be used to produce chemicals, including ethylene and butadiene, that would be produced from petroleum in the West. For example, ethanol may substitute alkenes for the alkylation of aromatics [82].

Figure 9.8 summarizes the different possible conversion pathways and/or uses of bioethanol [83]. Notably, however, a decrease in ethanol cost may open the door to a large range of other possible applications, and there is interest in biorefineries to diversify the possible uses of ethanol so as to decrease dependence on market fluctuations. There is thus the need to rediscover, and improve as well, the catalytic chemistry of ethanol.

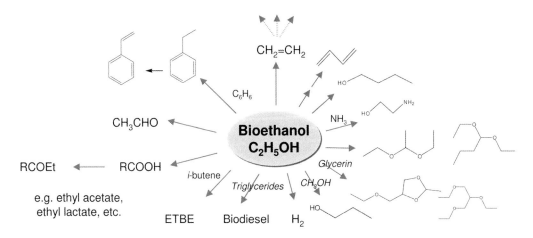

Fig. 9.8 Schematic overview of the possible transformations of bioethanol into chemicals or fuel components. (Adapted from [79]).

9.5
Conclusions

Bioethanol is already a produced world-wide in large amounts (over 50 million tons), mainly by fermentation of sugars and crops. Its market is expected to grow largely in the next 5–10 years, mainly due to its use as biofuel, because of various socio-economic and strategic motivations, as discussed in this chapter and elsewhere in this book.

The expansion of the market, however, will depend considerably on the possibility of an efficient use of other biomass sources, particularly lignocellulosic-based materials, fast growing dedicated crops, and waste resources. Effective integration of bioethanol production into biorefineries will also be a key aspect in decreasing the price by a better use of all the components of biomass.

There are significant challenges for catalysis to realize these objectives, as detailed in this chapter. There is the need for better integration between bio-, homo- and heterogeneous catalysis, and to foster cross-fertilization between these and contiguous areas (reaction engineering, membrane, etc.). New tools for the synthesis and understanding of these catalysts are need to be developed.

Bioethanol upgrading and valorization is another area in which catalysis will be a key player. The decrease in ethanol production cost and the need to realize a transition to a society based more on renewables are two driving forces to develop new catalytic processes for bioethanol conversion. However, without improvements in the efficiency and selectivity of the various processes for ethanol, and in general biomass conversion, which are possible only by the introduction of better and/or new catalysts, this transition to a bio-based economy will probably not be possible. Research on catalysis will thus be the enabling factor for this change towards a more sustainable society.

References

1 (a) D. Ballerini, *Les Biocarburants*, IFP Publications, (Technip Ed., Paris, **2006**), pp. 19 and 28; (b) F.O. Licht, *Ethanol Industry Outlook 2005*, Renewable Fuel Association (Washington, DC: **2005**), p. 14 (http://www.ethanolRFA.org).

2 http://en.wikipedia.org/wiki/Ethanol_fuel.

3 B. Kavalov, *Biofuel Potential in the EU*, EU Commission Jan. **2004**, Report EUR 21012 EN.

4 I.R. Richards, Energy balances in the growth of oilseed rape for biodiesel and of wheat for bioethanol, *Levington Agriculture Report*, British Association for Bio Fuels and Oils (BABFO) Jan. **2000** (http://www.biodiesel.co.uk).

5 D. Morris, *The Carbohydrate Economy, Biofuels and the Net Energy Debate*, Institute for Local Self-Reliance Pub. Minneapolis, Aug. **2005** (http://www.newrules.org/agri/netenergyresponse.pdf).

6 D. Pimental (chair), *Report on Biomass Energy*. The Biomass Panel, Energy Research Advisory Board. U.S. Department of Energy. Washington, D.C., **1981**.

7 D. Pimentel, T.W. Patzek, Ethanol production using corn, switchgrass and wood: Biodiesel production using soybean and sunflower, *Nat. Resources Res.* March **2005**.

8 A.E. Farrell, R.J. Plevin, B.T. Turner, A.D. Jones, M.O'Hare, D.M. Kammen, *Science*, 311 (**2006**) 506.

9 R. Wooley, M. Ruth, D. Glassner, J. Sheehan, *Biotechnol. Prog.*, 15 (**1999**) 794.

10 J. DiPardo, *Outlook for Biomass Ethanol Production and Demand*, US Energy Information Administration, Washington D.C., April (**2000**), http://www.eia.doe.gov/oiaf/analysispaper/biomass.html.

11 J. Sheehan, M. Himmel, *Biotechnol. Prog.*, 15 (**1999**) 817.

12 K.S. Tyson, C.J. Riley, K.K. Humphreys, *Fuel Cycle Evaluations of Biomass-Ethanol and Reformulated Gasoline*, National Renewable Energy Lab., Golden CO (**1993**), Vol. 1, Report NREL/TP-463–4950.

13 W. Kaminsky, D. Meier, J. Puls, *Chem. Tech.*, 5 (**2005**) 1215.

14 N.K. Rohatgi, J.D. Ingham, *Appl. Biochem. Biotechnol.*, 34–35 (**1992**) 515.

15 F. Rosillo-Calle, *Biomass*, 11 (**1986**) 19.

16 W. Swodenk, *Chem. Ing. Techn.*, 55 (**1983**) 683.

17 F. Aiouache, S. Goto, *Chemical Industries (Dekker)*, 101 (Handbook of MTBE and Other Gasoline Oxygenates) (**2004**) 129.

18 J.-P. Leroudier, *Zuckerindustrie (Berlin)*, 127 (**2002**) 614.

19 M. McCoy, *Chem. Eng. News*, (Dec. 7, **1998**), p. 29.

20 T.H. Richardson, X. Tan, G. Frey, W. Callen, M. Cabell, D. Lam, J. Macomber, J.M. Short, D.E. Robertson, C. Miller, *J. Biol. Chem.*, 277 (**2002**) 26501.

21 J.K. Shetty, O.J. Lantero, N. Dunn-Coleman, *Int. Sugar J.*, 107 (**2005**) 605.

22 S. Bhargava, H. Frisner, H. Bisgard-Frantzen, J.W. Tams, US Patent 2005113785 (**2005**), assigned to Novozymes A/S.

23 C.E. Wyman, *Biotechnol. Prog.*, 19 (**2003**) 254.

24 W. Mabee, *Economic, Environmental and Social Benefits of 2nd Generation Biofuels in Canada*, BIOCAP Canada March **2006**. (http://www.biocap.ca/rif/report/Mabee_W.pdf).

25 M. Galbe, G. Zacchi, *Microbiol. Biotechnol.*, 59 (**2002**) 618.

26 A. Wingren, M. Galbe, G. Zacchi, *Biotechnol. Prog.*, 19 (**2003**) 1109.

27 L.R. Waterland, S. Venkatesh, S. Unnasch, *Safety and Performance Assessment of Ethanol/Diesel Blends (E-Diesel)*, NREL Report SR-540–34817 Sept, **2003**, Cupertino, California (http://www.nrel.gov/docs/fy03osti/34817.pdf).

28 A.C. Hansen, P.W.L. Lyne, Q. Zhang, Ethanol-diesel blends: a step towards a bio-based fuel for diesel engines, 2001 ASAE Annual International Meeting, Sacramento, CA, July **2001**, Paper 01–6048 (http://www.ag-bioeng.uiuc.edu/faculty/ach/ediesel/Publications/infopub.pdf).

29 J.R. Rostrup-Nielsen, *Science*, 308 (**2005**) 1421.

30 G.A. Deluga, J.R. Salge, L.D. Schmidt, X.E. Verykios, *Science*, 303 (**2004**) 993.

31 G.W. Huber, J.W. Shabaker, J.A. Dumesic, *Science*, 300 (**2003**) 2075.

32 A. Anderson, *Bio-Derived Liquids to Hydrogen Distributed Reforming Working Group Background Paper*, U.S. Dept. Energy, Sept. **2006** (http://www1.eere.energy.gov/hydrogenandfuelcells/pdfs/biliwg_nov06_background_ paper.pdf).

33 A. Haryanto, S. Fernando, N. Murali, S. Adhikari, *Energy Fuels* 19 (**2005**) 2098.

34 P.D. Vaidya, A.E. Rodrigues, *Chem. Eng. J.* 117 (**2006**) 39.

35 R.R. Davda, J.W. Shabaker, G.W. Huber, R.D. Cortright, J.A. Dumesic, *Appl. Catal. B: Environ.*, 56 (**2005**) 171.

36 H.-S. Roh, A. Platon, Y. Wang, D.L. King, *Catal. Lett.*, 110 (**2006**) 1.

37 A. Casanovas, J. Llorca, N. Homs, J.L.G. Fierro, P. Ramirez de la Piscina, *J. Mol. Catal. A: Chem.*, 250 (**2006**) 44.

38 E.C. Wanat, K. Venkataraman, L.D. Schmidt, *Appl. Catal., A: General*, 276 (**2004**) 155.

39 S. Cavallaro, V. Chiodo, S. Freni, N. Mondello, F. Frusteri, *Appl. Catal., A: General*, 249 (**2003**) 119.

40 D.K. Liguras, D.I. Kondarides, X.E. Verykios, *Appl. Catal., B: Environ.*, 43 (**2003**) 345.

41 M. Benito, J.L. Sanz, R. Isabel, R. Padilla, R. Arjona, L. Daza, *J. Power Sources*, 151 (**2005**) 11.

42 J. Comas, F. Marino, M. Laborde, N. Amadeo, *Chem. Eng. J.*, 98 (**2004**) 61.

43 Y. Yang, J. Ma, F. Wu, *Int. J. Hydrogen Energy*, 31 (**2006**) 877.

44 DOE Hydrogen Program, FY **2005** Progress Report, IV.A.6 *Production of Hydrogen by Biomass Reforming*, pp. 98–105.
45 R.D. Cortright, R.R. Davda, J.A. Dumesic, *Nature*, 418 (**2002**) 964.
46 S. Song, P. Tsiakaras, *Appl. Catal., B: Environ.*, 63 (**2006**) 187.
47 F. Vigier, S. Rousseau, C. Coutanceau, J.-M. Leger, C. Lamy, *Top. Catal.*, 40 (**2006**) 111.
48 K. Taneda, Y. Yamazaki, *Electrochim. Acta*, 52 (**2006**) 1627.
49 G. Andreadis, P. Tsiakaras, *Chem. Eng. Sci.*, 61 (**2006**) 7497.
50 J. Mann, N. Yao, A.B. Bocarsly, *Langmuir*, 22 (**2006**) 10432.
51 P. Bert, C. Bianchini, PCT Int. Appl. (**2004**) WO 2004036674, assigned to Idea Lab S.R.L., Italy.
52 P. Barbaro, P. Bert, C. Bianchini, G. Giambastiani, S. Moneti, A. Scaffidi, A. Tampucci, F. Vizza, Francesco PCT Int. Appl. (**2006**) WO 2006063992, assigned to Acta S.p.A., Italy.
53 B. Delfort, I. Durand, A. Jaecker, T. Lacombe, X. Montagne, F. Paille, FR Patent 2 833 606 (**1983**).
54 D.M. Dillon, R.Y. Iwamoto, U.S. Patent 4,891,049 (**1990**).
55 M. Natarajan, E.A. Frame, T. Asmus, W. Clark, J. Garbak, M.A. Gonzalez, E. Liney, W. Piel, J.P. Wallace III, *SAE Tech. Pap. Ser.* 2001–01–3631 (**2001**).
56 B.E. Hallgren, J.B. Heywood, *SAE Tech. Pap. Ser.* 2001–01–0648 (**2001**).
57 M.J. Murphy, *SAE Tech. Pap. Ser.*, 2002–01–2848 (**2002**).
58 A.S. Cheng, R.W. Dibble, B.A. Buchholz, *SAE Tech. Pap. Ser.*, 2002–01–1705 (**2002**).
59 T.C. Zannis, D.T. Hountalas, D.A. Kouremenos, *SAE Tech. Pap. Ser.*, 2004–01–0097 (**2004**).
60 C.D. Rakopoulos, D.T. Houtalas, T.C. Zannis, Y.A. Levendis, *SAE Tech. Pap. Ser.*, 2004–01–2924 (**2004**).
61 K.E. Nord, D. Haupt, *Environ. Sci. Technol.* 39 (**2005**) 6260 and references therein.
62 B. Delfort, I. Durand, A. Jaecker, T. Lacome, X. Montagne, F. Paille, FR Patent FR2833606 and FR2833607 (**2003**), assigned to Institut Francais du Petrole, France.
63 S. Pariente, G. Centi, S. Perathoner, N. Tanchoux, F. Fajula, Bioethanol valorization for incorporation in diesel fuel blends, *Prepr. Symp. – Am. Chem. Soc., Div. Fuel Chem.*, 51 (**2006**) 434.
64 Y. Itoh, M. Ueda, H. Shinjoh, M. Sugiura, M. Arakawa, *J. Chem. Technol. Biotechnol.*, 81 (**2006**) 544.
65 Y.H. Yeom, M. Li, W.M.H. Sachtler, E. Weitz, *J. Catal.*, 238 (**2006**) 100.
66 J.-H. Lee, S.J. Schmieg, Se H. Oh, *Ind. Eng. Chem. Res.*, 43 (**2004**) 6343.
67 G. Centi, S. Perathoner, *Microporous Mesoporous Mater.* (**2007**), in press.
68 N.R.C.F. Machado, V. Calsavara, N.G.C. Astrath, A. Medina Neto, M.L. Baesso, *Appl. Catal. A: General*, 311 (**2006**) 193.
69 R. Barthos, A. Szechenyi, F. Solymosi, *J. Phys. Chem. B*, 110 (**2006**) 21816.
70 A. Szechenyi, R. Barthos, F. Solymosi, *Catal. Lett.*, 110 (**2006**) 85.
71 I. Takahara, M. Saito, M. Inaba, K. Murata, *Catal. Lett.*, 105 (**2005**) 249.
72 S. Golay, L. Kiwi-Minsker, R. Doepper, A. Renken, *Chem. Eng. Sci.*, 54 (**1999**) 3593.
73 R. Le Van Mao, T.M. Nguyen, G.P. McLaughlin, *Appl. Catal.*, 48 (**1989**) 265.
74 H. Olivier-Bourbigou, J.A. Chodorge, P. Travers, Olefin oligomerization with homogeneous catalysis, *Petrol. Technol. Quarter (PTQ)*, (**1999**) 141.
75 J. Skupinska, *Chem. Rev.*, 91 (**1991**) 613.
76 J.R. Sohn, W.C. Park, *Appl. Catal. A*, 239 (**2003**) 269.
77 V. Hulea, F. Fajula, *J. Catal.*, 225 (**2004**) 213.
78 R. Ohnishi, T. Akimoto, K. Tanabe, *J. Chem. Soc., Chem. Commun.*, (**1985**) 1613.
79 Y. Kitayama, A. Michishita, *J. Chem. Soc., Chem. Commun.*, (**1981**) 401.
80 E.S. Olson, R.K. Sharma, T.R. Aulich, *Appl. Biochem. Biotechnol.*, 113–116 (**2004**) 913.
81 C. Carlini, M. Di Girolamo, A. Macinai, M. Marchionna, M. Noviello, A.M. Raspolli Galletti, G. Sbrana, *J. Mol. Catal. A: Chem.*, 200 (**2003**) 137.
82 J.-J. Yuan, B.S. Gevert, *Indian J. Chem. Technol.*, 11 (**2004**) 337.
83 H. van Bekkum, A.C. Besemer, *Chem. Sustainable Dev.*, 11 (**2003**) 11.

# 10
Conversion of Glycerol into Traffic Fuels

Tiia S. Viinikainen, Reetta S. Karinen, and A. Outi I. Krause

10.1
Introduction

The role of biocomponents in traffic fuel is increasing. The European Union Directive [1] on the promotion of the use of biofuels for transport purposes states that by the end 2005 traffic fuels should have contained 2% of components produced from renewables. The figure rises to 5.75% by the end of 2010 and up to 20% by the end of 2020. This directive defines biofuel as a liquid or gaseous fuel for transport produced from biomass, biodiesel as a methyl ester produced from vegetable or animal oil, of diesel quality, to be used as biofuel and synthetic biofuel as synthetic hydrocarbons or mixtures of synthetic hydrocarbons, which have been produced from biomass. The European Commission also encourages member states to lower tax rates on pure and/or blended biofuels, to the offset cost premium over petroleum-based fuels [1, 2].

Glycerol is a major by-product in biodiesel process, for every 1000 kg of biodiesel around 100 kg of glycerol is formed. Glycerol cannot be directly added to fuel due to its low solubility in hydrocarbons. Further, it has a high viscosity, and decomposition, polymerization and consequential engine problems would be encountered at high temperatures [3]. New uses for glycerol need to be found; although glycerol could be burnt as heating fuel, it could be processed into more valuable components. One alternative is to etherify glycerol with either alcohols (e.g., methanol or ethanol) or alkenes (e.g., isobutene, 2-methylpropene) and produce branched oxygen-containing compounds, which could have suitable properties for use, for example, in fuel or solvents. Another interesting approach is to reform glycerol in an aqueous phase into hydrogen and carbon dioxide, which can be further converted into liquid fuels by the Fischer–Tropsch process or to methanol (raw material of biodiesel). Also, steam reforming of glycerol has been studied in the production of hydrogen. An advantage of using glycerol as a starting material for traffic fuel components is that, being a biocomponent, it could be included in the renewables category and help to meet the biocomponent target of the EU directive.

10.2
Glycerol

10.2.1
Properties, Production and Use of Glycerol

Glycerol or glycerin (1,2,3-propanetriol) has a particular combination of chemical and physical properties and it is physiologically harmless. Glycerol is a colorless, odorless and sweet-tasting hygroscopic liquid. At room temperature glycerol is viscous (1.495 Pa s). The boiling point of glycerol is 290 °C (101.3 kPa), its freezing point is 18 °C and its density is 1.249 g cm^{-3} (20 °C). It is water soluble and almost insoluble in hydrocarbons [4]. The heating value of glycerol is -1499 kJ mol^{-1} (heat of combustion) [5].

Glycerol is a reactive molecule that undergoes various reactions. It is easily oxidized, yielding glyceric acid, tartronic acid, ketomalonic (or mesoxalic) acid and dihydroxyacetone. These are useful compounds as such and as intermediates [6]. At 180 °C alkaline glycerol begins to dehydrate, forming polyglycerols [4]. Propylene glycol (1,2-propanediol) can be produced from glycerol with a catalytic [7] or microbial conversion. The possible by-products are 1,3-propanediol and ethylene glycol [7]. Glycerol can be etherified with alkenes, e.g., with isobutene to tertiary ethers of glycerol, which are valuable components for traffic fuels [8]. Glycerol monoesters (monoglycerides) can be obtained from glycerol esterification with fatty acids. Naturally, di- and triesters (di- and triglycerides) are by-products of this reaction [9]. Other ways of converting glycerol lead to, for example, the production of hydrogen, dichloropropanol, polyglycerols and polyglycerol esters.

The splitting of fats by saponification (soap preparation) with caustic alkali or alkali carbonates is a traditional method for obtaining glycerol. The most important industrial process for synthetic glycerol uses propene as starting material. Three processes are known for the production of glycerol from propene, but only one of them is industrially important. The intermediate stages involve allyl chloride, which is converted into dicholorohydrin; this is then saponified to glycerol with caustic solution. Glycerol can also be derived from acrolein and hydrogen peroxide as well as from the hydrogenation of carbohydrates with a nickel catalyst [4].

Today, glycerol has over 2000 different applications, in cosmetics, pharmaceutics, foods and drinks, tobacco, paper, inks and printing colors, the production of phthalic and maleic alkyd resins and crosslinked polyesters, and as a hydraulic agent. Polyglycerols have a wide range of applications as emulsifiers, and technical esters of glycerol with fatty acids are used as synthetic lubricants [4].

World-wide demand for glycerol is projected to be 0.2 Mt in 2007 and is forecasted to grow by 2.2% per year, driven mainly by personal care and food products [10]. If the target of the European Union directive is achieved, European biodiesel demand could increase to 10 Mt per year by 2010, resulting in 1 Mt of glycerol [2]. The price of glycerol has already fallen by 50% over the past ten years [10, 11]. Thus, the application of glycerol, obtained from the biodiesel process, should be increased, and new applications must be invented to improve the economics of the biodiesel processes [12]. USP grade (United States Pharmaceutical grade) glycerol has substantial value and many uses, but the glycerol from the biodiesel processes is not suitable for these uses as such [13], because of many impurities, resulting in expensive and challenging purifying steps [14].

10.2.2
Glycerol from Biodiesel Production

Biodiesel can be produced from various vegetable oils, such as rape seed, soy bean, sun flower, peanut and corn or from used oils like recycled cooking grease [15]. Biodiesel can also be produced from animal fats, but animal fats have not been studied to the same extent as vegetable oils [15]. One hundred years ago, Rudolf Diesel tested vegetable oils as a fuel for his engine. When cheap petroleum occurred, appropriate crude oil fractions were refined to serve as fuel and diesel fuels and diesel engines evolved simultaneously. In the 1930s and 1940s, vegetable oils were used as diesel fuels from time to time, but usually only in emergency situations. Recent increases in crude oil price, limited resources of fossil oil and environmental concerns have renewed the focus on vegetable oils and animal fats to make biodiesel fuels. Natural vegetable oils (and animal fats) are extracted or pressed to obtain crude oil (or fat). The crude oil (or fat) usually contains free fatty acids, phospholipids, sterols, water, odorants and other impurities. The free fatty acids and water have significant effects in biodiesel production [15].

Biodiesel is a mixture of methyl esters of fatty acids and is produced from vegetable oils by transesterification with methanol (Fig. 10.1). For every three moles of methyl esters one mole of glycerol is produced as a by-product, which is roughly 10 wt.% of the total product. Transesterification is usually catalyzed with base catalysts but there are also processes with acid catalysts. The base catalysts are the hydroxides and alkoxides of alkaline and alkaline earth metals. The acid catalysts are hydrochloride, sulfuric or sulfonic acid. Some metal-based catalysts can also be exploited, such as titanium alcoholates or oxides of tin, magnesium and zinc. All these catalyst acts as homogeneous catalysts and need to be removed from the product [16, 17]. The advantages of biodiesel as fuel are transportability, heat content (80% of diesel fuel), ready availability and renewability. The

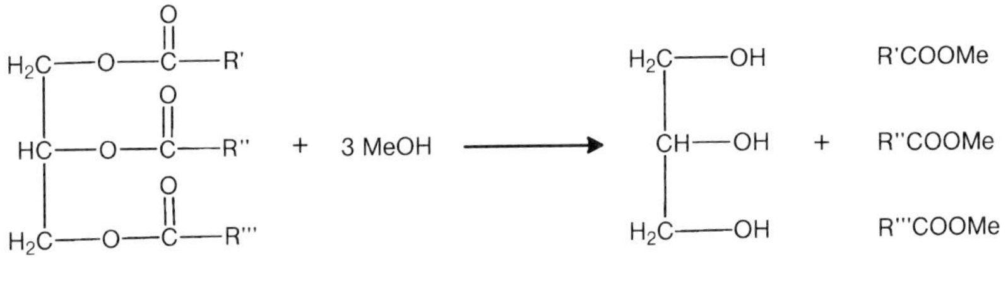

Fig. 10.1 Transesterification of vegetable oils to produce biodiesel (methyl esters).

disadvantages are a higher viscosity and lower volatility than conventional diesel, and the reactivity of unsaturated hydrocarbon chains [2, 15].

The glycerol obtained from the transesterification is separated from the biodiesel by gravity. Owing to the low solubility of glycerol in the esters, this separation generally occurs quickly and may be accomplished with either a settling tank or a centrifuge. The glycerol stream from the separator contains only about 50% glycerol, including some of the excess alcohol, soap and most of the catalyst. In this form glycerol has little value and disposal may be difficult, because the methanol content requires glycerol to be treated as hazardous waste [13, 14].

The first step in the purification of the raw glycerol is to split the soaps with acids into free fatty acids and salts. The free fatty acids are not soluble in glycerol and will rise to the top, where they can be recycled; the salts remain mainly with glycerol (some may precipitate out). The glycerol stream is then neutralized with caustic soda. A vacuum flash process or another type of evaporator removes the excess methanol. At this point glycerol should have a purity of 80–85%. It is often most cost effective to purify the raw glycerol and sell the so-called crude glycerol to industrial glycerol refiners. The refining of the crude glycerol raises the purity to 99.5–99.7% by vacuum distillation [13, 14].

10.3
Etherification of Glycerol with Isobutene

10.3.1
Reaction Scheme

Etherification of glycerol with isobutene (2-methylpropene) is an acid-catalyzed reaction and produces five ethers: two monoethers, two diethers and one triether (Fig. 10.2). The IUPAC names for the two monosubstituted ethers are 3-*tertiary*-butoxypropane-1,2-diol and 2-*tertiary*-butoxypropane-1,3-diol, the two disubsti-

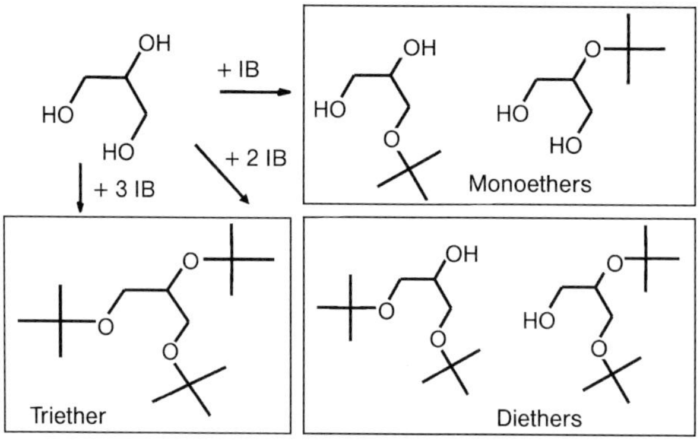

Fig. 10.2 Reaction scheme leading to ethers.

tuted ethers are 2,3-di-*tertiary*-butoxypropan-1-ol and 1,3-di-*tertiary*-butoxypropan-2-ol, and the one trisubstituted ether is 1,2,3-tri-*tertiary*-butoxypropane. The etherification of glycerol is preferred on primary hydroxyl groups, forming of 3-*tertiary*-butoxypropane-1,2-diol and 1,3-di-*tertiary*-butoxypropan-2-ol [8]. The product ethers are not 100% biocomponents if the isobutene is of fossil origin. In that case the percentage of bio-origin is 62% for the monoethers, 45% for the diethers and 35% for the triether.

The main side reaction in the etherification of glycerol with isobutene is the dimerization of isobutene to C_8 hydrocarbons; trimers (C_{12}) and tetramers (C_{16}) of isobutene are also formed [8]. The polymerization of glycerol to polyglycerols is not a side reaction of glycerol etherification, since it requires a basic catalyst. This polymerization of glycerol is also referred as a selective etherification of glycerol because the glycerol molecules join together with ether linkage [18].

10.3.2
Etherification Catalysts

Some studies and patents related to the etherification reaction between glycerol and isobutene are available in the literature, which report the production of the ethers with various catalysts, such as zeolites [19], ion-exchange resins [20, 21] and some homogeneous catalysts, e.g., *p*-toluene sulfonic acid [21, 22] and methane sulfonic acid [22]. Recent studies, however, have concentrated on the etherification of glycerol with isobutene on ion-exchange resins [8, 23].

The ion-exchange resins used as etherification catalysts are strongly acidic cation-exchange resins. These materials consist typically of polystyrene chains that have been linked with divinylbenzene (DVB), the amount of which determines the degree of crosslinking and regulates the rigidity of the structure schematically presented in Fig. 10.3 [24].

Resins with a DVB content of less than 8 wt.% are of the gel-type without permanent porosity. Such resins function only in the presence of polar components that swell the resin structure. Resins with a DVB content of 12 wt.% or more have permanent macroporosity. These materials also have a microporous gel phase consisting of gel-type microspheres [25].

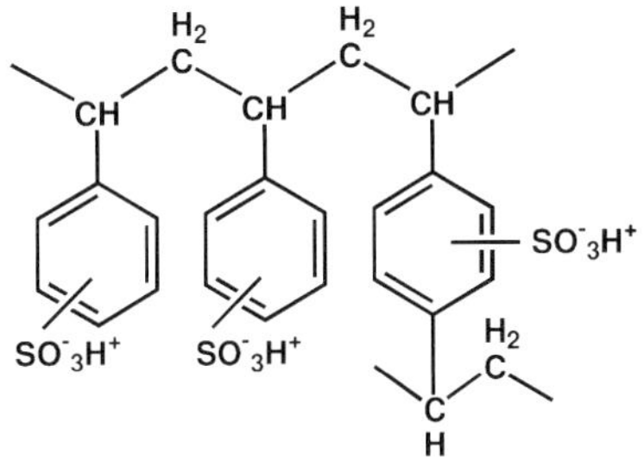

Fig. 10.3 Acidic ion-exchange resin.

Acid groups are introduced by sulfonation, e.g., with sulfuric acid. If all the benzene rings are monosulfonated, an acid capacity of the order of 5 mmol g_{cat}^{-1} should be obtained. The sulfonic acid sites are situated in the easily accessible macropores, and also inside the gel-type microspheres [26].

The strongly acidic cation-exchange resins are thermally stable up to 117–127 °C. Stability up to 147 °C, though, has been reported for some resins [27]. Above these temperatures, a loss of catalytic activity is observed, owing to the hydrolysis of active sulfonic acid groups. Thermal stability generally decreases as crosslinking increases [28]. Water has an inhibition effect because it adsorbs strongly on these catalysts. Various cations are harmful because they neutralize acid groups (ion exchange occurs). In the case of glycerol, only pure waterless and salt-free glycerol is suitable for etherification with ion-exchange resins [24, 29].

Cation-exchange resins are used as catalysts in the production of MTBE (methyl *tertiary*-butyl ether, 2-methoxy-2-methylpropane) and various other oxygenates and, lately, also in the dimerization of isobutene [30]. Other commercial applications of the cation-exchange resins include dehydration of alcohols, alkylation of phenols, condensation reactions, alkene hydration, purification of phenol, ester hydrolysis and other reactions [31]. The major producers of ion-exchange resins are Sybron Chemicals Incorporated [32] (Lewatit® resins), Dow Chemical Company [33] (DOWEX™ resins), Purolite [28] (Purolite® resins), and Rohm and Haas Company [27] (Amberlyst™ resins).

Klepáčová et al. [23] have tested several commercial strongly acidic ion-exchange resins with a macroreticular structure (Amberlyst A15, A35, A36 and A39) and with a gel structure (Amberlyst A31 and A119) and two commercial large-pore zeolites (H-Y and H-Beta). Ion exchange resin catalysts A15 and A35 with a highly crosslinked structure were very active in the dry and also in the wet form [23].

10.3.3
Process Conditions

Isobutene is only sparingly soluble in glycerol and, consequently, in the initial state before the reaction starts the reactants are in separate liquid phases. At stoichiometric initial conditions (molar ratio of isobutene to glycerol = 3:1) in the equilibrium mixture the glycerol phase contains less than 3 mol.% isobutene and the isobutene phase is almost free of glycerol (based on phase equilibrium calculations). Proper mixing in the reactor is required for a thorough mixing of the phases and to ensure that both reactants reach the catalyst and the reaction can occur. As the reaction proceeds the products accumulate in one of the phases, according to their solubilities. The monoethers of glycerol accumulate in the polar glycerol phase and diethers, triether and C_8–C_{16} hydrocarbons in the non-polar hydrocarbon phase. Notably, all of the components are present to some degree in both phases. At some stage the two phases dissolve in each other and only one phase remains. This phase combination occurs at a glycerol conversion of about 60–70% [8].

10.3 Etherification of Glycerol with Isobutene

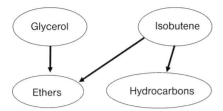

Fig. 10.4 The two parallel reactions in the etherification system.

The etherification reaction was studied in the temperature range 45–90 °C. The initial molar ratio of isobutene to glycerol varied between 1.5 and 4.5 [8, 23].

In a system of two parallel reactions (Fig. 10.4), selectivity is an important parameter to monitor. The measure of selectivity is defined here as the ratio of "isobutene reacted to ethers"/"total amount of isobutene reacted". Glycerol is consumed only in the etherification reaction, so the selectivity is calculated with respect to isobutene. Depending on the extent of the reaction, the formation of an ether molecule consumes from one to three isobutene molecules, and oligomerization consumes from two to four isobutene molecules [8].

The distribution of ethers shifts towards more substituted species at constant temperature (80 °C) with the initial isobutene to glycerol ratio (Fig. 10.5). Changing the initial isobutene to glycerol ratio does not affect the by-product (C_8, C_{12} and C_{16} hydrocarbons) distribution. Temperature, however, has a clear effect on the hydrocarbon distribution; the higher the temperature the smaller the fraction of C_{12} and C_{16} hydrocarbons of the total amount of hydrocarbons [8].

The higher the initial isobutene to glycerol ratio the more substituted species occur; so the optimal condition to obtain a high amount of triether is a high initial isobutene to glycerol ratio. Up to a certain point the fraction of diethers increases with increasing initial isobutene to glycerol ratio, but thereafter it starts to decrease at the expense of the fraction of the triether. The highest diether fraction was observed when the initial isobutene to glycerol ratio was 3 (stoichiomet-

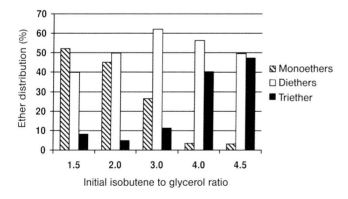

Fig. 10.5 Ether product distribution at 80 °C.

Table 10.1 Optimal conditions for the production of ethers.

Product	Optimal conditions
Triether	High isobutene to glycerol ratio
Diether	Stoichiometric isobutene to glycerol near 80 °C
Monoether	Low isobutene to glycerol, low conversion

ric ratio) and the conversion of glycerol was complete or almost complete; 80 °C was the optimal temperature for the formation of diethers. The highest monoether fractions were observed when the conversion of glycerol was low. With the lowest initial isobutene to glycerol ratio (1.5) at 80 °C the conversion of glycerol was higher. From this it was concluded that, when stoichiometry limits the reaction, monoether is the main product even though the conversion of glycerol is relatively high [8]. Table 10.1 presents the optimal conditions for mono-, di- and triethers.

In the dimerization of isobutene, *tertiary*-butyl alcohol (TBA, 2-methyl-2-propanol) has a strong role in modifying the selectivity of the reaction to C_8 hydrocarbons and limits further oligomerization to C_{12} and C_{16} hydrocarbons [34]. Also, in the etherification of glycerol with isobutene the addition of TBA has a clear effect on the selectivity and on hydrocarbon distribution. The selectivity to ethers increased and the fraction of the C_{12} and C_{16} hydrocarbons decreased while the concentration of TBA was increased from 0 to 2.6 mol.%. As a conclusion, the formation of C_{12} and C_{16} hydrocarbons can be prevented in two ways: either TBA should be added to the reaction mixture or the reaction should be carried out at high temperatures [8].

TBA and isobutene have been compared as the etherifying agent at 60 °C. The initial molar ratio of isobutene to glycerol was 4.0 with Amberlyst A35 as the catalyst. The conversion of glycerol is lower when etherified with TBA than when etherified with isobutene. More hydrocarbons are formed with isobutene than with TBA. But, with TBA, mainly monoethers are formed and valuable triethers are formed only in small amounts. In addition, TBA dehydrates to water, which has an inhibition effect on ion-exchange resin catalysts [23].

10.3.4
Etherification Kinetics

Behr and Obendorf [21] proposed a step-wise reaction model, according to which diethers are formed from monoether and isobutene and triether is formed from diethers and isobutene. In the simplified kinetic model no difference was considered between the two monoethers and the two diethers, and disproportion reactions and all side reactions were neglected (Fig. 10.6). The conversion rate was modeled without taking into account any mass transfer processes and phase glycerol (G) + isobutene (IB) $\underset{k_{-1}}{\overset{k_1}{\rightleftarrows}}$ monoether (M)

monoether (M) + isobutene (IB) $\underset{k_{-2}}{\overset{k_2}{\rightleftarrows}}$ diether (D)

diether (D) + isobutene (IB) $\underset{k_{-3}}{\overset{k_3}{\rightleftarrows}}$ triether (T)

Fig. 10.6 Alternative reaction scheme.

equilibrium separately. A power law approach on the basis of the overall molar concentrations (the sum of a component's concentrations in the polar and the non-polar phase) was used to model the observed reaction rates. Good agreement between simulation and experimental data was obtained on simplifying the complex reaction system by using first-order concentration dependencies [21].

10.3.5
Ethers of Glycerol as Fuel Components

Some possibilities to improve the combustion characteristics of diesel fuels with oxygenate additives are known. *tertiary*-Butyl ethers of glycerol with a high content of diethers and, especially, triether have been seen as potential oxygenates for diesel fuels (diesel, biodiesel and their mixtures) since 1994 [19, 20, 22]. A mixture of di- and triethers has infinite solubility in diesel fuels but monoethers have a low solubility in diesel [23]. Ethers of glycerol have mainly been studied as diesel fuel additives, but ethers are potential components for use in gasoline as well and they offer an alternative to oxygenates such as MTBE and ETBE (ethyl *tertiary*-butyl ether, 2-ethoxy-2-methylpropane), which are currently used in gasoline [8]. Di- and triethers are potential oxygenates in gasoline because of their low water solubility and suitable boiling points (169 and 148 °C, respectively) [35].

When ethers of glycerol are blended with biodiesel, the resulting fuel has lower viscosity, cloud and pour points than standard biodiesel, with the flash point remaining in the acceptable range, the mixture closely resembles a petroleum-based diesel. No increase in the cetane number was observed with oxygenated diesel fuel. In the emission testing of mixture of ethers and diesel fuel, the NOx emissions slightly increased but the addition of chemical cetane improvers can be utilized to reduce NOx emission increase. The most significant reduction observed in emission testing was for particulate matter (PM); the lowest amount of PM was observed with oxygenated diesel containing 5% di- and triethers of glycerol; the second lowest PM value was achieved with a mixture of diesel (90%) and oxygenated biodiesel (10%) (which contained 80% biodiesel and 20% ethers) [20, 36].

218 | *10 Conversion of Glycerol into Traffic Fuels*

Blending octane numbers of 112–128 (BRON) and 91–99 (BMON) have been reported for an ether product mixture from the etherification of glycerol with isobutene [35]. These octane numbers are very suitable for gasoline components. The BRON was observed to be higher for mixtures where monoethers were the main product and lower for mixtures containing more substituted ethers. If the product ethers were to be used in gasoline, the C_8 hydrocarbon fraction that is produced in a side reaction in the system would be a very valuable gasoline component. The main isomers of the C_8 fraction are 2,4,4-trimethyl-1-pentene and 2,4,4-trimethyl-2-pentene, which have high octane numbers (RON 100, MON 89) [30]. However, the formation of higher hydrocarbons, e.g., C_{12} and C_{16} alkenes should be avoided.

10.4
Improvements to Biodiesel Process

Modifications in the production of biodiesel can result in valuable glycerol as a by-product and in fewer separation steps. The modifications studied or considered include combining etherification of glycerol into the biodiesel production process, etherification *in situ* within the biodiesel process and a biodiesel process with heterogeneous catalyst.

10.4.1
Etherification with Biodiesel Process

Noureddini [36] proposed a two-step process consisting of the transesterification of soybean oil using a homogeneous base catalyst and the separate etherification of glycerol by-product with isobutene or isoamylene using Amberlyst A15 as catalyst. Subsequently, the obtained mono-, di- and tributyl ethers of glycerol were blended with the previously prepared biodiesel. This approach not only makes a good use of the reaction by-product glycerol but potentially also could increase the fuel yield by ca. 20 vol.%. Table 10.2 gives the properties of ethers blended in diesel and in biodiesel–diesel mixtures.

Table 10.2 Properties of ethers blended in diesel and in mixtures of biodiesel and diesel.

Mixture	Oxygen content (wt.%)	PM reduction (%)	Cetane number
100% diesel (EPA base fuel)	0	–	43
5% ethers and 95% diesel	2.80	26	44
2% ethers, 8% biodiesel and 90% diesel	2.06	21	Not applicable

10.4.2
Heterogeneous Biodiesel Process

Much effort has been spent searching for a solid acid or base catalyst that could be used in a heterogeneous biodiesel production process. Some solid metal oxides, such as oxides of tin, magnesium and zinc, are known catalysts but they act according to a homogeneous mechanism and end up as metal soaps or metal glycerates [17]. In a heterogeneous biodiesel process, transesterification is promoted by a completely heterogeneous catalyst [17]. The catalyst consists of a mixed oxide of zinc and aluminum, which promotes the transesterification reaction without any catalyst lost. The reaction is performed at a higher temperature and pressure than the homogeneous biodiesel process. The excess methanol that is used is removed by vaporization and recycled to the process with fresh methanol. The chemical conversion required to produce biodiesel at the European specifications is reached with two successive stages of reaction and glycerol separation in order to shift the equilibrium of reverse transesterification (methanolysis). The process includes two fixed-bed reactors and the excess methanol is removed after each reactor by partial evaporation. Then, esters and glycerol are separated in a settler, and residual methanol is removed by evaporation. Clear, colorless glycerol is obtained with a purity of at least 98% [17].

10.5
Reforming of Glycerol

Hydrogen can be generated from glycerol by liquid phase reforming (aqueous reforming) or by gas phase reforming (steam reforming). The reforming reaction of glycerol is:

$$C_3H_8O_3 + 3H_2O \leftrightarrow 3CO_2 + 7H_2 \tag{1}$$

Hydrogen and carbon dioxide can be converted into methanol over a copper and zinc catalyst [37]. The product stream from the reforming can also be utilized in traffic fuels through carbon dioxide hydrogenation. The two-step reaction pathway consists of a reverse water-gas shift reaction (RWGS) and a Fischer–Tropsch (FT) reaction [38, 39]. First, the RWGS reaction is used to convert carbon dioxide and hydrogen into carbon monoxide and water. Then, liquid fuels are obtained from the FT reaction, where carbon monoxide and hydrogen react to give hydrocarbons and water. The advantages of FT hydrocarbon products are the suitability to present infrastructure, high cleanliness (no sulfur, no aromatics and high quality diesel) and applicability as organic bulk chemicals as such or as feedstocks [38, 39]. Carbon dioxide could also be available from fossil fuel fired power plants. Thermodynamic calculations did not indicate equilibrium constraints for CO_2 hydrogenation. Iron is typically used as the active metal on the catalyst in the utilization of carbon dioxide. Potassium and copper are important promoters [40].

10.5.1
Aqueous Phase Reforming

Virent Energy System Incorporated (USA) have developed a novel APR (aqueous phase reforming) process, where purified hydrogen can be generated from oxygenated hydrocarbons, such as glucose, sorbitol, ethylene glycol and glycerol via liquid phase reforming [37, 41]. The APR process occurs at low temperatures (225–265 °C) and pressures (above the bubble point of water) where the water-gas shift reaction is favorable and undesirable decomposition reactions are at minimum. Low-value raw glycerol streams can be fed to APR process, where the reforming takes place over a 3 wt.% Pt/Al_2O_3 catalyst in a single reactor. The selectivity to hydrogen from glycerol was 75% at 225 °C and 51% at 265 °C, when the feed concentration of glycerol was 1 wt.%. Hydrogen is purified from the product stream (H_2 and CO_2) utilizing pressure swing adsorption (PSA) technology [37, 41]. Huber et al. [42] have produced hydrogen by APR of glycerol over a tin-promoted Raney-nickel catalyst. This catalyst exhibited higher selectivity for the production of H_2 than Pt/Al_2O_3, especially at 265 °C when selectivity to hydrogen was 76%.

10.5.2
Steam Reforming

Steam reforming of crude glycerol has been studied in a fluidized-bed reactor over a commercial nickel-based naphtha reforming catalyst at 850 °C. The hydrogen yield was 77% of the stoichiometric potential. The promising results suggest that a low-value by-product from biodiesel production could be viable renewable raw material for producing hydrogen. An integration of transesterification and steam reforming technologies could be beneficial for the economics of both processes [43]. Hirai et al. [44] have developed a novel 0.5 wt.% Ru/Y_2O_3 catalyst for glycerol steam reforming. Activity tests were performed in a fixed-bed reactor operated at 600 °C under atmospheric pressure. The activity order for the metals tested was Ru ≈ Rh > Ni > Ir > Co > Pt > Pd > Fe. The screened supports for ruthenium were Y_2O_3, ZrO_2, CeO_2, La_2O_3, SiO_2, MgO and Al_2O_3. Yttria- and zirconia-supported catalysts exhibited highest glycerol conversions and highest H_2 yields (82.8 and 81.6%, respectively).

10.6
Future Aspects

Demand for biodiesel is growing, which increases the overproduction of glycerol. Processes are being changed so that glycerol can be used as a raw material. For example, a plant has been started to produce propylene glycol from glycerol instead of propylene oxide [45]. The use of glycerol for chemicals is not enough if the oversupply remains. However, burning the CO_2 neutral glycerol for energy is always a possibility. Although glycerol as such is not suitable as a fuel component, it can be converted into potential fuel components such as ethers of glycerol, or glycerol can be reformed to hydrogen. Improving existing biodiesel processes to produce more valuable glycerol and/or integrating etherification or reforming units to the biodiesel production will have a significant effect on the glycerol market.

However, one future alternative might be a synthetic biofuel process without glycerol production. An example of a glycerol-free biofuel production technology is NExBTL, which is a refinery-type process of synthetic biofuel by Neste Oil Corporation (Finland). It is due for start-up in the summer of 2007 and the capacity of the new plant will be 170 000 t a^{-1}. NExBTL biodiesel is basically a mixture of n- and isoalkanes, contains no sulfur, oxygen, nitrogen or aromatics and has a very high cetane number. When blended in diesel, both regulated and unregulated exhaust emission components are reduced. The production process consists of hydrodeoxygenation (HDO) and isomerization steps. In the HDO step, the structure of the biological component is decomposed and the alkene bonds are hydrogenated. In the next step, isomerization is carried out to branch the hydrocarbon chain. Vegetable oils, animal fats and similar materials can be used as raw materials. No glycerol is produced in this type of synthetic biofuel process [46, 47].

References

1 EU Commission: *The Promotion of the Use of Biofuels or Other Renewable Fuels for Transport*, EU Directive 2003/30/EC, May 8, **2003**.
2 D. de Guzman, *Chem. Market Rep.*, **2005**, 267(6), 19.
3 H. Noureddini, W.R. Dailey, B.A. Hunt, *Adv. Environ. Res.* **1998**, 2, 232.
4 R. Lewis, *Hawley's Condensed Chemical Dictionary (14th Edition)*, John Wiley & Sons, New York, **2001**.
5 C. Yaws, *Yaws' Handbook of Thermodynamic and Physical Properties of Chemical Compounds*, Knovel, Norwich (NY), **2003**.
6 R. Ciriminna, M. Pagliaro, *Adv. Synth. Catal.* **2003**, 3, 345.
7 M.A. Dasari, P-P. Kiatsimkul, W.R. Sutterlin, G.J. Suppes, *Appl. Catal. A: General* **2005**, 281, 225.
8 R.S. Karinen, A.O.I. Krause, *Appl. Catal. A: General* **2006**, 306, 128.
9 J. Barrault, Y. Pouilloux, J.M. Clacens, C. Vanhove, S. Bancquart, *Catal. Today* **2002**, 75, 177.
10 Anon., *Chem. Market Rep.* **2005**, 267(4), 31.
11 Anon., *Chem. Market Rep.* **2001**, 260(23), 35.
12 D. de Guzman, *Chem. Market Rep.* **2005**, 268(3), 29.
13 M.J. Haas, A.J. McAloon, W.C. Yee, T.A. Foglia, *Bioresource Technol.* **2006**, 97, 671.
14 J. van Gerpen, *Fuel Process Technol.* **2005**, 86, 1097.
15 F. Ma, A. Hanna, *Bioresource Technol.* **1999**, 70, 1.
16 http://www.biodiesel.org, April 19, 2006.
17 L. Bornay, D. Casanave, B. Delfort, G. Hillion, J.A. Chodorge, *Catal. Today* **2005**, 106, 190.
18 J.-M. Clacens, Y. Pouilloux, J. Barrault, *Appl. Catal. A: General* **2002**, 227, 181.
19 C. Dewattines, H. Hinnekens, EP Patent 0,649,829, assigned to FINA Research, **1994**.
20 H.S. Kesling Jr., L.J. Karas, F.J. Liotta Jr., US Patent 5,308,365, assigned to ARCO Chemical Technology, **1994**.
21 A. Behr, L. Obendorf, *Chem. Ing. Tech.* **2001**, 73, 1463.

22 V.P. Gupta, US Patent 5,476,971, assigned to ARCO Chemical Technology, **1995**.
23 K. Klepáčová, D. Mravec, M. Bajus, *Appl. Catal. A: General* **2005**, 294, 141.
24 A.O.I. Krause. K.I. Keskinen, Etherification, in: *Handbook of Heterogeneous Catalysis*, 2nd edn, G. Ertll, H. Knözinger, F. Schmith, J. Weitkamp (Eds), Wiley-VCH, Weinheim, **2007**.
25 A. Chakrabarti, M.M. Sharma, *React. Polym.* **1993**, 20, 1.
26 K. Jěrábek, *Collect. Czech. Chem. Commun.* **1979**, 44, 2612.
27 Rohm and Haas Company, Amberlyst™ ion exchange resins product data sheet, http://www.rohmhaas.com/ionexchange/IP/sac.htm, 20 August **2006**.
28 Purolite Company, Purolite® ion exchange resins product data sheet, http://www.puroliteusa.com/ftp/ProductData/Catalysts/CTSERIES.pdf, 20 August **2006**.
29 F. Cunill, M. Vila, J.F. Izquierdo, M. Iborra, J. Tejero, *Ind. Eng. Chem. Res.* **1993**, 32, 564.
30 M. Di Girolamo, M. Lami, M. Marchionna, E. Pescarollo, L. Tagliabue, F. Ancillotti, *Ind. Eng. Chem. Res.* **1997**, 36, 4452.
31 M.A. Harmer, Q. Sun, *Appl. Catal. A: General* **2001**, 221, 45.
32 Sydron Chemicals Inc., Chemical and Environmental Ion Exchange Products (Lewatit® ion exchange resins product data sheet), http://www.ion-exchange.com/service/pdf/chem._env.pdf, 20 August **2006**.
33 Dow Chemical Company, Dowex™ ion exchange resins product data sheet, http://www.dow.com/liquidseps/prod/prd_dowx.htm, 20 August **2006**.
34 M.L. Honkela, A.O.I. Krause, *Catal. Lett.* **2003**, 87, 113.
35 R. Wessendorf, *Erdöl, Kohle, Erdgas, Petrochem.* **1995**, 48, 138.
36 H. Noureddini, US Patent 6,174,501, assigned to The Board of Regents of the University of Nebraska, **2001**.
37 http://www.virent.com, April 19, **2006**.
38 T. Riedel, M. Claeys, H. Schulz, G. Schaub, S.-S. Nam, K.-W. Jun, M.-J. Choi, G. Kishan, K.-W. Lee, *Appl. Catal. A: General* **1999**, 186, 201.
39 T. Riedel, G. Schaub, K.-W. Jun, K.-W. Lee, *Ind. Eng. Chem. Res.* **2001**, 40, 1355.
40 M. Niemelä, M. Nokkosmäki, *Catal. Today* **2005**, 100, 269.
41 R.D. Cortright, R.R. Davda, J.A. Dumesic, *Nature* **2002**, 418, 964.
42 G.W. Huber, J.W. Shabaker, J.A. Dumesic, *Science* **2003**, 300, 2075–2077.
43 S. Czernik, R. French, C. Feik, E. Chornet, *Ind. Eng. Chem. Res.* **2002**, 41, 4209–4215.
44 T. Hirai, N. Ikenaga, T. Miyake, T. Suzuki, *Energy Fuels* **2005**, 19, 1761–1762.
45 M. McCoy, *Chem. Eng. News* **2006**, 84(6), 7.
46 J. Jakkula, V. Niemi, J. Nikkonen, V.-M. Purola, J. Myllyoja, P. Aalto, J. Lehtonen, V. Alopaeus, EP Patent 1,396,531, assigned to Fortum OYJ, **2004**.
47 http://www.nesteoil.fi, April 19, **2006**.

11
Catalytic Transformation of Glycerol

Bert Sels, Els D'Hondt, and Pierre Jacobs

11.1
Introduction and Scope

The decreasing reserves of fossil carbon and environmental problems related to their use have led to a continuously growing interest in the use of renewable feedstocks from easily grown crops such as soy bean, rape seed, and sunflower. Glycerine, also denoted as glycerol, is a co-product of primary products such as fatty acids, fatty alcohols, soap and biodiesel, all using fats and oils as feedstock. Although its production amounts to only about 10% of these primary products, glycerine is often the focus of more attention than its stoichiometry would suggest because of the price volatility of this commodity [1].

Nowadays, glycerine receives particular attention as by-product from biodiesel production [2]. Extracted triglycerides can undergo transesterification with methanol (methanolysis) or (bio)ethanol (ethanolysis), yielding the corresponding methyl or ethyl esters (biodiesel) and 14% glycerol as by-product on a stoichiometric basis. With current technology 1 kg of crude glycerol by-product is obtained for every 9 kg of biodiesel produced [3]. A recent process model for continuous production of biodiesel from soybean oil containing 96% triglycerides at a rate of 100 kg h^{-1}, yields 10.3 kg h^{-1} of 80% aqueous glycerol. After the methanol recovery train, a glycerol refining section consists of an acidulation reactor to neutralize with mineral acid the base catalyst and to remove the soaps formed as well, resulting in less than 1% of free fatty acids [4]. With an estimated production capacity of biodiesel of 12 million metric tons per year in 2010, glycerol is becoming a bulk renewable feedstock [5]. Independent forecasts of global glycerol production by 2010 amount to significantly lower quantities [6]. Costs of raw glycerol of around 15 dollar cent per pound ($\approx$ 26 eurocent per kg) are figures currently handled in economic models [7].

Glycerol is a colorless, odorless, viscous (syrupy), nontoxic and hygroscopic liquid with a very sweet taste; its etymologic roots stem from the Greek *glykys*, sweet. Pure glycerol (1,2,3-propanetriol) is a trihydric alcohol with a specific gravity of 265 at 15 °C. Below 0 °C it solidifies to a white crystalline mass, which melts at

Catalysis for Renewables: From Feedstock to Energy Production
Edited by Gabriele Centi and Rutger A. van Santen
Copyright © 2007 WILEY-VCH Verlag GmbH & Co. KGaA, Weinheim
ISBN: 978-3-527-31788-2

17 °C. When heated it partially volatilizes, though a greater part decomposes. At a pressure of 1.6 kPa it boils at 170 °C. In an atmosphere of steam it distils without decomposition at atmospheric pressure. It is soluble in water and alcohol in all proportions, but not in ether [8]. Compared with the hydrocarbon feedstocks currently used in classic petrochemistry, it is already a highly functionalized molecule. Obviously, new (catalytic) transformations of glycerol will require controlled defunctionalization.

Traditional uses of glycerol are among the following [9]. Its use in medical and pharmaceutical applications is related to its properties as a humectant and lubricating agent. Glycerol is an essential component in (glycerol) soaps, hair and skin care products, mouthwashes and toothpaste because of its properties as an emollient, humectant, lubricant and solvent. It has a similar role as a food additive (with number E422). Moreover, it helps food preservation, operates as a softening agent and as an emulsifier. It is also used as a raw material for polyols. The latter are required for the manufacture of polyether polyols for flexible foams and as plasticizer in alkyd resins and cellophane. It is needless to emphasize its long-standing use as basic chemical for nitroglycerine production.

In view of the use of glycerol as a chemical commodity for the production of chemical intermediates, an overview will be made of existing catalytic knowledge. More specifically, glycerol oxidation, dehydration, hydrogenolysis, oligomerization/polymerization, polyol formation, and formation of a few miscellaneous products will be dealt with.

The formation of mono- and diacylglycerols via transesterification processes of triacylglycerols (triglycerides) with glycerol (glycerolysis) with mostly a homogeneous mineral base or (mineral) acid are established technologies that only economically (via the glycerol price) and not technically will be affected by biodiesel production. Recent catalytic developments in this area are outside the scope of the present chapter. Also, no explicit attention will be paid to the role of biocatalysts in the area of glycerol transformation. Notably, recently, attempts have been made to produce biodiesel with a chemo- or biocatalyst without the parallel production of glycerol [10, 11] (see also Section 10.6 of Chapter 10). This approach is also not further considered here.

11.2
Catalytic Dehydration of Glycerol and Acrolein Formation

Acrolein is traditionally manufactured from propylene in a catalyzed gas-phase oxidation using air or dioxygen. Acrolein is used in the production of acrylic acid and methionine, while many fine chemical synthesis procedures also rely on its use. The conversion of glycerol into acrolein has long been known, but for economic reasons the reaction has not been applied industrially. However, with possible cheap glycerol prices ahead, acrolein production from glycerol (Scheme 11.1) might be an elegant green alternative to the petrochemical route.

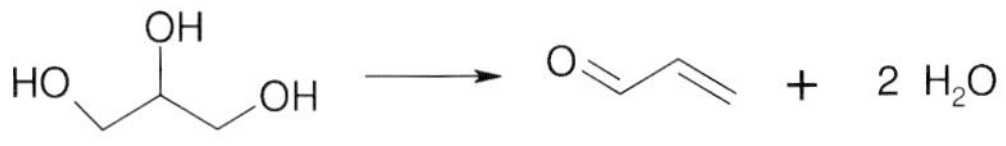

HO−CH(OH)−CH₂OH ⟶ CH₂=CH−CHO + 2 H₂O

Scheme 11.1

Distillation of glycerol at temperatures below 623 K in the presence of a metal or salts ($MgSO_4$, alkali phosphates and silica) forms acrolein [12, 13]. With potassium bisulfate as dehydrating agent for instance, glycerol is converted into acrolein in 33–48% yield [14]. Phosphoric acid salts, either as such or impregnated on diatomaceous earth, produce acrolein from gaseous glycerol [15, 16]. However, the exceptional acrolein yields around 80% reported for 1% lithium or iron phosphate on pumice could not be reproduced by others [17]. Instead, acidic catalysts having an H_0 between −8.2 and −3, such as phosphoric acid on alumina, gave satisfactory acrolein formation, preferably in the gas (573 K) but also in the liquid (533 K) phase [17, 18]. At 300 °C for instance, gaseous glycerol (20 wt.% water) is fully converted, yielding 70.5% acrolein and ca. 10% hydroxyacetone. Na zeolites and γ-alumina led to practically no glycerol conversion, while acrolein yields formed over H-ZSM-5 never exceeded 14%. Despite the high yields obtained with alumina-supported phosphoric acid, its service time was very limited. In contrast, stable high acrolein yields were recently reported in the gas phase conversion of glycerol using solid catalysts such as zirconium tungstate (90.7% ZrO_2–9.3% WO_3) with ultrahigh acid strength ($-9 < H_o < -18$). As an example, a full steady-state glycerol conversion is achieved during 78 hours, yielding around 70% acrolein, with hydroxypropanone and propanal as the main side products [19].

Some homogeneous metal catalysts have been examined in the production of acrolein from glycerol [20]. However, considering all reaction components present, it is more likely that soluble acids, such as HCl and CF_3SO_3H, are responsible for glycerol dehydration instead of Pt and Pd phosphine complexes.

Non-catalyzed reactions have also been performed. For instance, the pyrolysis of glycerol in steam was studied in a laminar flow reactor in the temperature range 923–973 K. Acrolein is the principal product along with acetaldehyde and CO [21]. Reported yields were as high as 52% when pyrolysis was carried in steam at 923 K and atmospheric pressure.

Sub- and supercritical conditions have also been applied to selectively degrade glycerol. Water in these conditions experiences an enhanced self-dissociation, forming H^+ and OH^-. The highly ionic water is capable of converting glycerol without the need for acids or bases. Besides ionic reaction steps, radical pyrolysis of the alcohol also occurs. Allyl alcohol and methanol are the main products under conditions (<25 MPa, $T > 700$ K) where the radical pathways predominate, whereas glycerol is mainly converted into acetaldehyde and formaldehyde according to ionic routes at higher pressure and lower temperature. Acrolein is formed in 10–15% yield irrespective of the pressure applied, while lower yields are obtained at higher T [22]. The addition of acid catalysts such as sulfuric acid in supercritical water improved the acrolein yield up to 47% at 55% glycerol conversion (55 s residence time, 34.5 MPa and 623 K, ~4 wt.% glycerol), hinting that acrolein formation is acid catalyzed. There is ongoing debate as to whether 3-hydroxypropanal is a direct intermediate of acrolein, or whether the cross-aldol condensation of acetaldehyde and formaldehyde mainly contributes to acrolein formation [23]. Ott et al. recently promoted the usage of non-mineral acid catalyst for the formation of acrolein from glycerol in sub- and supercritical water to avoid the disadvantage of intensifying corrosion. Their highest acrolein selectivity was 75% at ca. 50% glycerol conversion, when operating at 633 K and 25 MPa with a residence time of 60 s in the presence of $ZnSO_4$ [6].

11.3
Etherification of Glycerol via Catalytic Dehydration

11.3.1
Glycerol Oligomerization

Polyglycerol has been used typically for the preparation of polyglycerolesters of fatty acids, the latter compounds being used in tensioactive applications, as lubricants, cosmetics and food additives such as in cake batters, confectionaries, sugar-icing, creams and ointments, or as emulsifiers in vegetable fats and margarines. To minimize "graininess" of the latter products and to maximize "butter-like" properties, the glycerol polymer distribution should be specific [24], and consist of 50–100% of di- to tetraglycerol, and less than 10% hexaglycerols or larger polymers. Moreover, the iodine value (degree of unsaturation) of the fatty acid residues should be low and the degree of esterification high. When the selectivity in the glycerol oligomerization is not controlled, upon subsequent esterification with fatty acids the polyglycerol ester product does not have a well-defined hydrophilic–lipophilic balance.

Glycerol polymerization in basic medium is assumed to occur via an S_N2-type mechanism, in which the basic site (agent) weakens the glycerol O–H bond, thus increasing the nucleophilicity of the oxygen atom involved, allowing it to exert a nucleophilic attack on an accessible carbon atom of a second glycerol molecule (Scheme 11.2) [25].

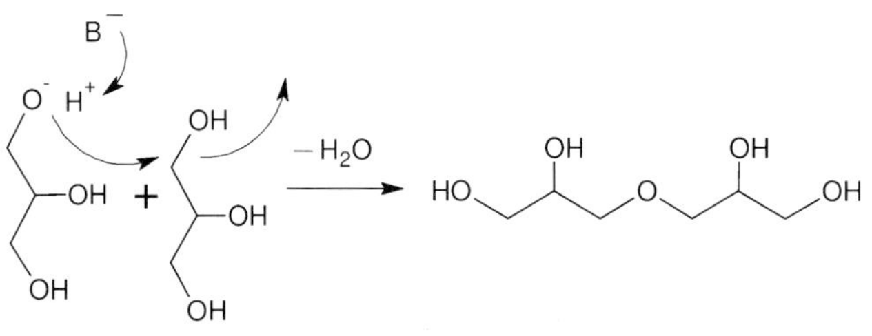

Scheme 11.2

11.3 Etherification of Glycerol via Catalytic Dehydration

Acid-catalyzed glycerol etherification probably occurs via an S_N1 mechanism, consisting of the consecutive formation of an oxonium and carbenium ion, and its electrophilic attack on a glycerol O atom (Scheme 11.3).

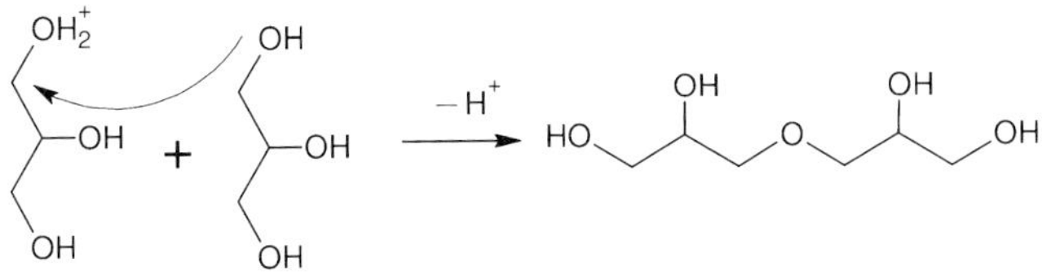

Scheme 11.3

By subjecting glycerol under vacuum to an alkali-catalyzed dehydration at 240 °C, products ranging from dimers to octamers with exponentially decreasing selectivity are obtained (Fig. 11.1). When Cs_2O via a Cs-acetate precursor is impregnated in the pores of a MCM-41 mesoporous structured silica, significantly more dimer is formed at the expense of the higher oligomers. The restricted formation of the larger oligomers is probably correlated with steric restrictions present in the pores of MCM-41 materials [25, 26]. However, when a systematic data set is considered, encompassing the di- and triglycerol selectivity against glycerol conversion, the consecutive formation of both product groups is evident, though the effect of the catalyst on the respective selectivities at a given conversion is low [27, 28]. At a given glycerol conversion in the range 60–80%, the diglycerol selectivity in MCM-41 pores is not more than 15% higher than with Na_2CO_3 as cata-

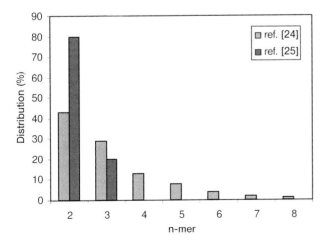

Fig. 11.1 Distribution of the oligomers (n-mer) of glycerol (100% conversion) formed by alkali-catalyzed dehydration [24] and with Cs_2O supported on mesoporous silica (MCM-41) [25].

Table 11.1 Selectivity (%) among the individual isomers in the diglycerol class (s,s: coupling of two secondary alkanols; p,p: coupling of two primary alkanols in glycerol) (after [29]). For the synthesis of authentic standards of pure diglycerol isomers see [30].

	Na_2CO_3	Cs-MCM-41
s,s dimer	4	21
s,p dimer	30	50
p,p dimer	66	29

lyst. When the same comparison is made for triglycerol, its selectivity is less than 10% lower when comparing the same catalysts.

Inspection of the chromatograms of the individual oligomers reveals that the composition within each product group, viz. di- and triglycerols, depends on the nature of the catalyst. Although Cs-MCM-41 catalysts are less active on a weight basis than the homogeneous catalysts, viz. Na_2CO_3, the former are recyclable, provided Cs-ion exchange was performed on an Al-substituted MCM-41.

Clearly, the reactivity of the secondary hydroxyl group of glycerol, viz. of dimer s,s, is enhanced in a mesoporous medium. A parallel reduced reactivity is seen for the primary alkanols, viz. the p,p dimer (Table 11.1) [29]. The data, however, are insufficient to exclude an effect of the catalyst basic strength, resulting in an enhanced selectivity for p,p, for the homogeneous catalyst, which is more basic than the ion-exchanged MCM material. This seems to be confirmed by the significantly higher activity of the homogeneous catalyst.

11.3.2
Reaction of Glycerol with Alkenes

Etherification of glycerol with alkenes has been practiced to obtain fossil fuel adducts. The etherification of glycerol with isobutene, using stoichiometric molar ratios with a strongly acidic macroporous and highly crosslinked sulfonated ion-exchange resin, yielded isobutene selectivities for ethers of 85%, with selectivities among the three ethers given in Scheme 11.4; the remaining 15% of the con-

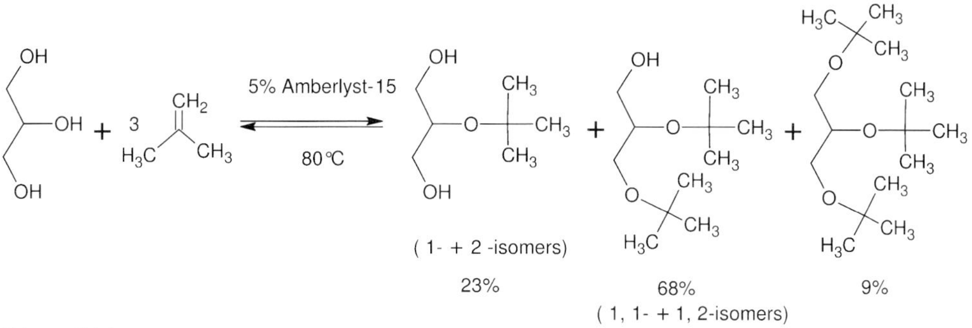

11.3 Etherification of Glycerol via Catalytic Dehydration | 229

Scheme 11.4

verted isobutene was oligomerized on the acid catalyst. A 20% blend of the ether product mixture into biodiesel resulted in a viscosity reduction of 8% and a cloud point decrease of the fuel of 5 °C [31]. With the aim of enhancing the cetane number of fuels and reducing particulate formation from diesel engine emissions, the scope of the solid acid catalysts was further enlarged. Superior results, viz. 100% glycerol conversion and di-+tri-ether selectivity of over 90%, were obtained at 60 °C with strongly acidic macroreticular ion exchange resins [32]. The macroreticular ion exchange resins in dry form are claimed to be superior to other solid acids such as gel-type resins and acid zeolites like HY and HBeta. Differences among the catalysts were attributed to a pore-size effect. However, claims exist as to the use of siliceous large-pore zeolites in the terbutylation of glycerol [33]. It is not clear how they compare with the systems mentioned here that use other acid catalysts.

These selectivities obtained with Amberlyst-15 seem to be rather optimistic in view of recent work, where only with overstoichiometric ratios high di-+tri-ether selectivity is obtained. It is also not settled why addition of *tert*-butanol to the reaction mixture suppresses isobutene oligomerization selectivity [34, 35]. At this stage it should also be stressed that the *tert*-butylglycerol ethers (TBGE) mentioned are excellent substitutes for MTBE (ETBE) as gasoline octane boosting components [36].

However, as far as glycerol conversion and diether selectivity (80–90%) is concerned, the terbutylation of glycerol at 90 °C and 0.2 MPa into TBGE with homogeneous catalysts like sulfonic and heteropoly acids was shown to be superior to that of Amberlyst-15, montmorillonite type K10, and heteropoly acid adsorbed on montmorillonite [37]. Traditionally, homogeneous batch processes use partial conversion to apply phase separation followed by reinjection of the lower glycerol-rich phase. Monoethers and catalyst residue after water washing are then extracted from the ether-rich upper phase [38]. In an improved procedure, which allows high conversion levels, glycerol feed is first fed to the product before entering the reactor, resulting in the extraction and reinjection of catalyst and monoethers.

A typical (industrially applied) example for the synthesis of monoalkylglyceryl ethers [39, 40] consists of the transition metal-catalyzed conversion of glycerol and butadiene (telomerization) to yield glyceryl unsaturated octyl and dioctyl ethers (Scheme 11.5).

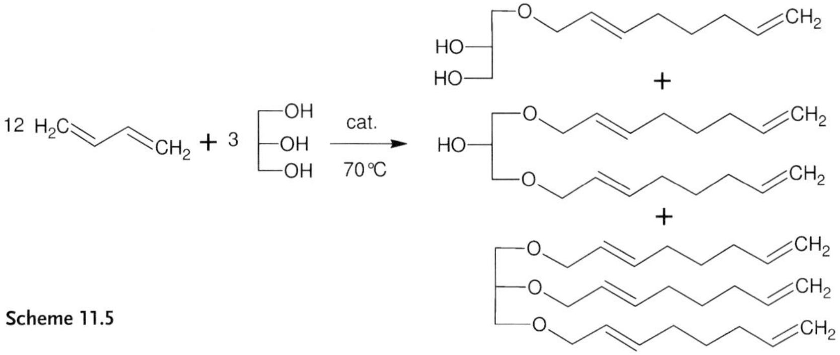

Scheme 11.5

A Pd(acac)$_2$PPh$_3$ complex is used as catalyst for the telomerization of the alkadiene, followed by a nucleophilic addition to glycerol (Fig. 11.2). A high feed conversion is obtained under mild conditions and very low catalyst concentration. The catalyst can perform some 75000 turnovers at a rate of about 7500 h^{-1} – a mixture of three esters being formed that is rich in diester (S$_{DE}$). Selectivity can, to a limited extent, be varied with the ratio of the reactants [41]. After traditional catalytic hydrogen reduction, octylglyceryl ethers are obtained easily. After sulfonation (with SO$_3$) and neutralization with NaOH, the corresponding glycerine ether sulfonates are obtained. They are used as anionic surfactants in detergents and cosmetics. A whole family of industrially very important ether lipids can be made based on the glyceryl ether skeleton. Selectivity control in the Pd-catalyzed

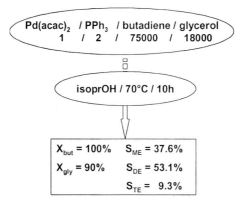

Fig. 11.2 Simultaneous etherification of butadiene oligomers with glycerol. (After [41]).

telomerization of butadiene with glycols (including glycerol), occurs under biphasic conditions [42]. The monoester (ME) could be obtained with a selectivity surpassing 95% using a water-soluble Pd(acac)$_2$TPPTS catalyst [TPPTS = tris(*m*-sulfophenyl)phosphine].

Alkyl glyceryl ethers are now being viewed as C-3 building blocks in lipid chemistry. The presence of two free OH groups allows the introduction of much functionality for specific synthetic surfactants and specialty lipids. Their use in newer synthetic approaches has been reviewed [43–45].

Certain diglyceryl alkyl ethers (*vide supra*) that form, by self-assembly, liquid crystals have a future in nanochemistry [46, 47]. Evidently, therefore, selective catalytic synthesis of specific oligomers of glycerol in a shape selective environment (*vide supra*) will contribute significantly to the generation of new classes of bio- and nano-materials.

11.4
Catalytic Oxidation of Glycerol

Glycerol oxidation products invariably have chelating properties and can be used as synthons in organic synthesis [48]. Dihydroxyacetone and hydropyruvic acid can be starting materials for DL-serine synthesis [49]. Dihydroxyacetone can be used as a tanning agent in cosmetics and as a drug and industrial intermediate [50]. Ketomalonic acid is a valuable compound as an organic synthon, with potential applications as an anti-HIV agent [51].

Traditional industrial oxidation routes of alcohols use stoichiometric amounts of heavy metals or mineral acids [51]. Glycerol is also easily converted into formaldehyde, formic acid and carbon dioxide [52].

11.4.1
Electrochemical Oxidation

Fuels cells are of interest both from energetic and environmental considerations. When methanol is fed directly to an anode, as in "Direct Methanol Fuel Cells", electric power is generated, making the devices suitable for small and lightweight uses [53]. Alternative fuels such as polyhydric alcohols like ethylene glycol and glycerol are much less volatile and toxic, on the one hand, and electrochemically oxidizable on the other [54]. Therefore, the electrochemical oxidation of various polyhydric alcohols has been investigated in acidic as well as in alkaline conditions.

Whereas in alkaline medium gold in particular gives very high current densities, in aqueous acidic medium platinum electrodes present excellent electroactivity [55]. In the latter case, glycerol yielded mainly glyceraldehydes, formic acid, oxalic acid and glycolic acid as products, the selectivity towards glyceraldehydes depending on the applied potential and the pH of the electrolyte [56]. In alkaline medium, gold is the preferred electrocatalyst, in particular for the oxidation of the alcohol and aldehyde functional groups [57]. Electro-oxidation rates of glycerol on Au(111) in aqueous perchlorate is greatest, while the Au(110) plane exhibits very low activity indeed, in contrast with the behavior in CO oxidation. It seems that steps or sites with a low surface atomic coordination number have stereochemical limitations for the adsorption of larger polyfunctional reactants, and are, thus, important for alcohol oxidations [58].

The electrochemical oxidation of polyhydric alcohols, viz. ethylene glycol, glycerol, meso-erythritol, xilitol, on a platinum electrode show high reactivity in alkaline solutions of KOH and K_2CO_3 [53]. This electro-oxidation shows structural effects, Pt(111) being the most active orientation. This results from different adsorption interactions of glycerol with the crystal planes [59].

Prominent co-catalysts for the Pt-on-carbon anode catalyst in the oxidation of polyhydric alcohols are Ru or CeO_2 [54, 60]. Their increased resistance to poisoning with mainly CO during operation is associated with the existence of a bifunctional mechanism (Scheme 11.6).

$$3\,Pt + C_3H_8O_3 + 4\,OH^- \longrightarrow 3(Pt\text{-}CO_{ads}) + 4\,H_2O + 4\,e^-$$

$$CeO_2 + OH^- \longrightarrow CeO_2\text{-}OH_{ads} + e^-$$

$$Pt\text{-}CO_{ads} + CeO_2\text{-}OH_{ads} + OH^- \longrightarrow Pt\,CeO_2 + CO_2 + H_2O$$

Scheme 11.6

Adsorbed CO is produced by polyol dehydrogenation in basic medium. The formation of adsorbed hydroxyl anions at lower potential cleans the metal surface from CO, liberating CO_2 from solution. Sufficient CeO_2 has to be present to efficiently release adsorbed CO, while at too high a concentration the current density decreases. A Pt:Ce atomic ratio of 1.3:1 shows optimal performance for the system. The decrease in electrode conductivity is assumed to be linked to increasing amounts of the semiconductor CeO_2 [54, 61].

Platinum-free electrocatalysts for fuel cells could be designed when Pd on carbon electrocatalysts promoted with nanocrystal oxide particles like Co_3O_4, Mn_3O and NiO, were used [62]. In terms of activity and poison tolerance, the latter were significantly superior.

Alkaline glycerol electrochemical oxidation on Pt-Pd alloys is dependant on the surface composition of the alloys. Compared with the pure metals, enhanced activity is found when the alloy contains about 33% Pd, leading to a synergetic effect [63].

The electro-oxidation of organics and more specifically of alcohols and polyols is also possible on silver electrodes in the following activity sequence: methanol < ethylene glycol < glycerol [64]. With a bulk silver electrode and with a silver-modified glassy carbon electrode, oxidations proceed only in the area of silver oxide formation.

Electrocatalysts using conducting polymers such as films of polyaniline as support have been developed for the electro-oxidation of organics, viz. glycerol [65]. In acidic medium, an electrode consisting of platinum electrodeposited on the polymer substrate showed higher glycerol oxidation activity when the films of polymer exhibited a fibrillar rather than a compact morphology. The advantageous morphology could be obtained on gold rather than on a glassy carbon substrate, pointing to the importance also of Pt-particle size.

11.4.2
Gas-phase Catalytic Oxidation

Gas-phase oxidation of glycerol occurs via a two-step catalyzed reaction between 200 and 400 °C, involving first the dehydration of glycerol into acrolein, followed by its oxidation. The final product depends on the nature and selectivity of the oxidation process. One of few communications in this area reports the formation of 2-oxo-propanal (methylglyoxal) (reaction) on a single catalyst, namely, silver supported on an acid alumino-silicate impregnated with vanadium oxide. Intermediate formation of 1,3-dihydroxyacetone by oxidation of the secondary C atom is assumed to occur (Scheme 11.7). The intermediate is subsequently dehydrated, undergoing a not further defined rearrangement into methylglyoxal [66]. This product is classically obtained from 1,2-propanediol and used for the preparation of pyrethrum insecticides.

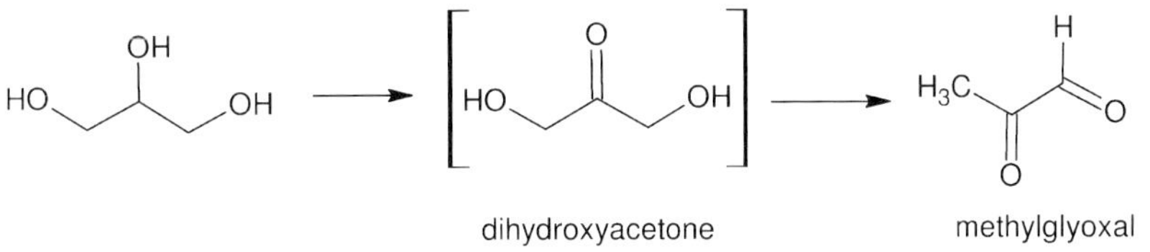

Scheme 11.7

Alternatively, acrylic acid can be obtained in a two-step reactor in which glycerol is catalytically dehydrated with an acid catalyst like H_3PO_4 on α-alumina [67]. The obtained acrolein is then oxidized with a commercially available oxidation catalyst, viz. Mo/V/W/Cu-oxide on α-alumina, yielding up 55% polymerization grade acrylic acid (Scheme 11.8) [68].

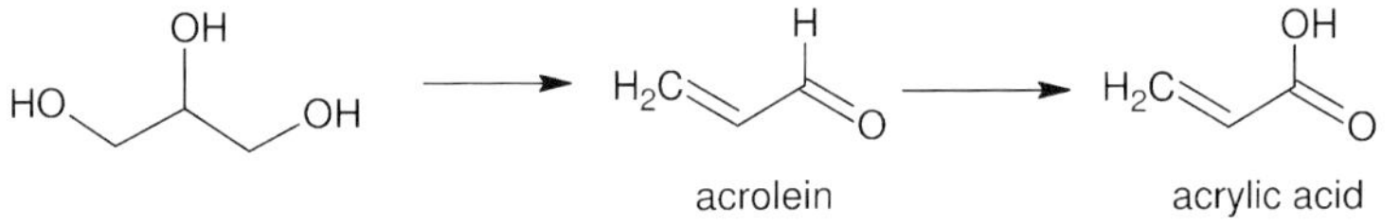

Scheme 11.8

11.4.3
Selective Oxidation with Molecular Oxygen on Pt/Bi Catalysts

The aqueous phase air oxidation of glycerol with supported noble metal catalysts occurs under mild conditions (60 °C), but is very dependant on the pH of the reaction medium. Relevant data are shown in Fig. 11.3 [48]. For Pd, Pt and Bi-promoted Pt the glycerol oxidation rate increases significantly with the pH of the medium, with Pd showing the lower activity.

The reduced activity of Pd was attributed to its lower redox potential, resulting in higher oxygen coverage of the metal and a reduced accessibility for the organic substrate. At more basic pH values, either the first step in the reaction is enhanced, viz. the deprotonation of the hydroxyl group involved [69], or acid desorption from the metal is favored [70]. The selectivity on Pd/C catalysts was high for glyceric acid, being further enhanced at higher pH, up to 77% at 90% conversion at pH 11 [48]. Oxalic acid was definitely a secondary product, whereas dihydroxyacetone and tartronic acid were by-products. The results are in line with the general behavior of alcohol oxidations, as the reactivity of the primary OH group exceeds that of the secondary one. The reaction is also not expected to stop at the aldehyde stage, as this is only observed for substrates with the OH near a double bond or aromatic nucleus [71].

The Pt/C catalyst, compared with Pd/C, showed not only enhanced activity (*vide supra*) but also reduced selectivity for glyceric acid (only 55% at 90% conversion), favoring dihydroxyacetone formation up to 12%, compared with 8% for the Pd case [48]. The Pt/C catalyst promoted with Bi showed superior yields of dihydroxyacetone (up to 33%), at lower pHs. Glyceric and hydroxypyruvic acids, apparently, are formed as by-product and secondary product, respectively [48]. The addition of Bi seems to switch the susceptibility of glycerol oxidation from the primary towards the secondary carbon atoms.

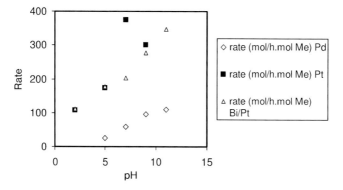

Fig. 11.3 pH dependence of the rate of glycerol conversion at 60 °C using 5 wt.% metal (Me) on carbon catalysts. (After [48]).

Bi/Pt atomic ratios (changing from 0.1 to 0.4), as well as the precise preparation procedure of the BiPt/C catalysts, yielded differences in selectivity between glyceric acid, dihydroxyacetone and hydroxypyruvic acid that were not always easy to rationalize [72, 73]. The catalyst preparation procedures using acid-treated carbon (with enhanced number of surface carboxyl groups) are essentially as follows:

- Pt loading occurs via ion exchange, while Bi loading involves a redox surface reaction of BiO^+ with hydrogen covered platinum particles (1–2 nm), according to the following stoichiometry [74]: $(Pd\text{-}H)_{ads} + BiO^+ \rightarrow (Pd\text{-}Bi)_{surf} + H_3O^+$;
- co-impregnation of Pt and Bi salts, followed by formaldehyde reduction [74];
- Pt impregnation, formaldehyde reduction, Bi loading and reaction of BiO^+ with hydrogen covered platinum particles [74];
- impregnation with acidic Pt and Bi ions, neutralization with NaOH, and reduction with sodium borohydride in NaOH [73].

With the BiPt/C catalyst with a Pt/Bi surface atom ratio of 3, in acidic medium ($2 < pH < 4$) glycerol could be oxidized with a selectivity of 80% into dihydroxyacetone. It is proposed that the Bi adatoms function as blockers of the Pt(111) surface, controlling the glycerol surface orientation.

Practically, this chemo-catalytic batch process operated for dihydroxyacetone selectivity has significantly higher productivity [73] than the fermentation process using *Acetobacter xylinum* [75]. Importantly, although work on glycerol oxidation in a continuous flow reactor in a trickle bed has been reported [76], catalysis is done mostly in batch operation. Although no systematic catalyst deactivation studies have been performed, it is occasionally stated that the presence of acids is at the basis of deactivation, due to their strong sorption on the noble metal. In this way the initial dihydroxyacetone selectivity of 80% in glycerol air oxidation with PtBi/C at acidic pH dropped rapidly towards 37%, due to accumulation of adsorbed acids [74].

The use of glyceric acid as substrate instead of glycerol [77, 78] allowed a network for air oxidation of glycerol to be put together, with different catalysts and pH conditions (Scheme 11.9). In the absence of a Bi promoter or at basic pH, the Pd/C or Pt/C catalyst converts the primary alcohol function of glycerol and consecutive products into the corresponding acid. The transiently formed aldehydes are never observed as intermediates. Secondary alcohol functions at acid pH and with PtBi/C are selectively oxidized into the corresponding carboxylic acid. A complex between glycerate and Bi atoms is assumed to be responsible for this [74]. The role of Bi is dual, preventing overoxidation by these adsorbed acids and by orientating the selectivity towards oxidation of the secondary C atom (Fig. 11.4) [76, 77].

Also on a $PtBiCeO_2/C$ catalyst, ketomalonic acid is formed in a consecutive way from glycerol, followed by its anionic polymerization into polyketomalonate (Scheme 11.10), a polyether polymer with a MW of about 9000 and a polydispersity index of 1.04 [79].

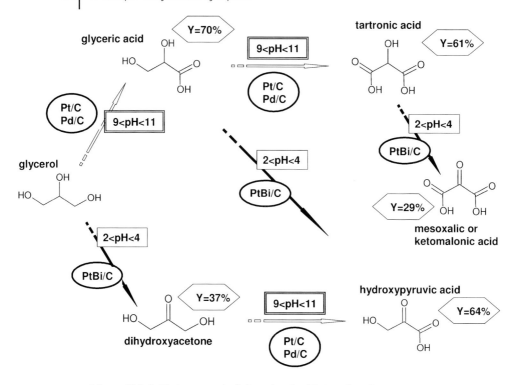

Scheme 11.9 Oxidation network of glycerol and yields (Y) of product at different pH and with discrete catalysts. (After [74, 76]).

However, the precise structure of the catalyst and the precise role of CeO_2 in the present case and of Bi is not completely clear. In general terms, several explanations for the rate and selectivity enhancements by the promoter are possible [80]: (a) geometric blocking of a fraction of sites and generation of specific surface ensembles, viz. formation of an ordered alloy; (b) neighboring atom participation (Fig. 11.4), although the partial oxidized state of the promoter (Bi^{x+}) of the model is not confirmed by surface studies (LEED, XPS, EXAFS); (c) occurrence of bifunctional catalysis, assuming that O or OH radicals formed on the promoter participate in the oxidation.

In a second wave of activity in the area of glycerol oxidation with Pt or Pd-on-carbon catalysts, the high yields for glyceric acid at 60 °C and atmospheric pressure described earlier, were initially no longer obtained by other authors. It seems that there is appreciable formation of compounds other than C_3 and C_2 oxidized

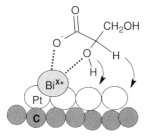

Fig. 11.4 Schematic representation of glycerate dehydrogenation on a PtBi/C catalyst. (After [74, 76]).

products, with mass balances amounting to a maximum of 40%. Only at very high flow rates (C_2+C_3) could mass balances of 80% be obtained, implying that sequential oxidation of C_3 partially oxidized products occurs easily [81]. Significant differences existed among 5% Pd/C catalysts, depending on the exact preparation procedure. Differences were detected among the particle size distribution, the better catalyst having significantly more large particles in the 10–20 nm range. High mass balances for C_2+C_3 and selectivities for glyceric acid over 60% required optimization of catalyst preparation procedure and reaction conditions. Glyceraldehyde and glyceric acid were particularly unstable products, while hydroxypyruvic acid and oxalic acid were much more stable and dihydroxyacetone was particularly inert under these conditions.

At 0.3 MPa of oxygen, the sum of the observed C_2 and C_3 oxidation products amounted to correct mass balances, with glyceraldehyde, glyceric acid and oxalic acid being the dominant products in absence of added NaOH [81]. NaOH addition resulted in the disappearance of oxalic acid formation.

11.4.4
Selective Oxidation with Molecular Oxygen on Gold-based Catalysts

Under the same conditions (enhanced pressure, added base), an Au-on-carbon catalyst easily showed over 90% selectivity for glyceric acid at 90% glycerol conversion [82]. In analogy with oxidations of other alcohol substrates, a particle size effect or the Au(III)/Au(0) ratio were assumed to be important. For Au-on-graphite, 50-nm gold nanoparticles performed optimally with respect to glyceric acid selectivity, while the presence of an oxide species, detected with cyclic voltammetry, seemed essential as well [83]. Gold-on-carbon was confirmed to be a catalyst that allowed full and stable conversion of glycerol into glyceric acid in the following conditions: 30 °C, NaOH/glycerol ratio of 4, glycerol/Au of 500, 0.3-M glycerol in water and 0.3 MPa pressure of oxygen gas [84]. The Au particles were larger than 20 nm, while 6 nm particles were inactive.

In addition, the catalyst preparation method plays an important role; sol immobilization appears superior to incipient wetness or impregnation methods. Gold sol immobilization occurs via citrate reduction of $HAuCl_4$, yielding a narrow particle size distribution around 20 nm [85]. A systematic study established that the liquid-phase glycerol oxidation into glyceric acid using Au/C catalysts, prepared via the sol-method with different particles sizes covering the range 2–42 nm, was a structure-sensitive reaction. At high conversions, the selectivity showed a maximum glyceric acid yield for gold particles of 23 nm [84, 86]. The nature of the support is also influential, though it is not clear whether this effect is indirect, viz. through an influence on particle size. In any case, carbon black as a support yields superior gold catalysts than activated carbon, graphite and inorganic supports like MgO [84].

Furthermore, a base-catalyzed transformation by OH^- from the reaction medium between glycerate and hydroxypyruvate aldehyde (or hydroxypyruvic acid) could be excluded, while hydroxyacetone and glyceraldehyde interconversion was possible (Scheme 11.11). The existence of two major routes, of which hydroxyacetone and glyceric aldehyde are the primary oxidation products and glycolic and oxalic acid are the end-members, respectively, is now firmly established. Clearly, rapid oxidation of glyceraldehydes favors glyceric acid rather than hydroxyacetone formation.

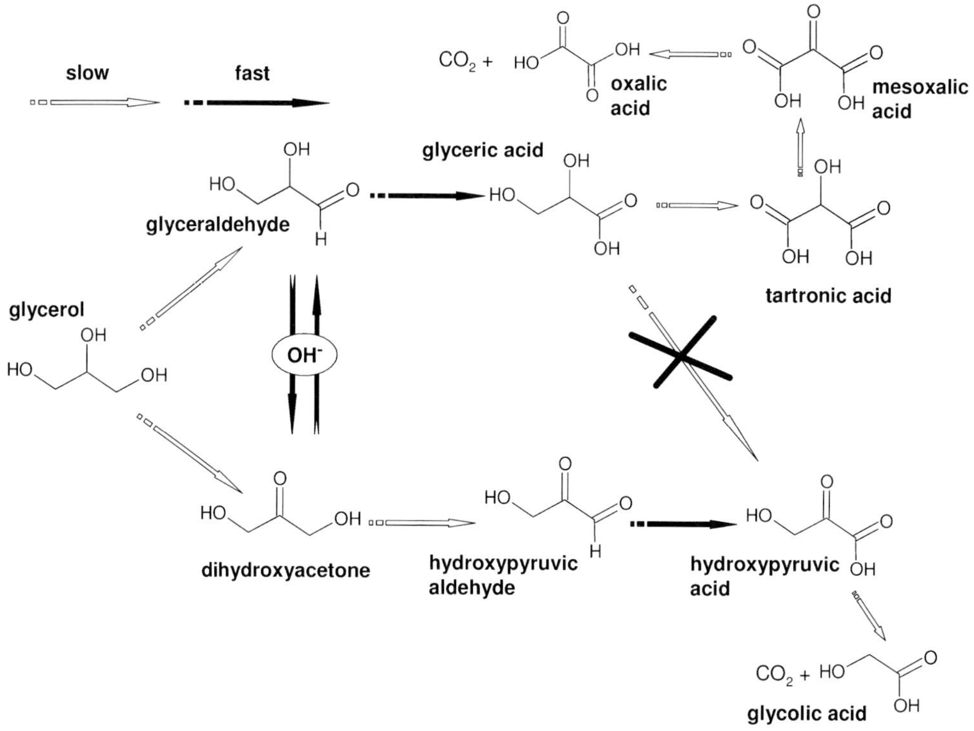

Scheme 11.11 Reaction scheme for glycerol oxidation with dioxygen on Au/C catalysts. (After [84]).

The scheme also indicates that high glyceric acid selectivity is obtainable either via a selective catalyst for glyceraldehydes formation, or with a less selective catalyst in basic medium, where the hydroxyacetone–glyceric aldehyde equilibration is established.

Recently, attention has been devoted to the behavior of bimetallic catalysts consisting of gold and palladium or platinum. Compared with monometallic catalysts such as Au/C, which in turn was more active and stable than Pd/C or Pt/C, the bimetallic catalysts (Pd@Au/C and Pt@Au/C) show synergetic effects on the oxidation rate [87, 88]. In terms of selectivity, Pd in Pd-Au was found to promote tartronic acid formation, in Pt-Au following the channel to glycolic acid. Under standard basic conditions and long contact times, overoxidation of glyceric into tartronic acid (Scheme 11.11) occurred with Pt (single or bimetallic), while this was much less the case with Pd.

The sol-method used to prepare gold-based catalysts has to be applied carefully to avoid segregation of the two metals. A suitable procedure consists in the use of preformed particles of one metal as nucleation center for the other [89]. Usually, a gold sol preformed via BH_4^- reduction is immobilized on the carbon support. A Pd sol is then prepared in the presence of Au/C, using hydrogen gas as reducing agent. The key to avoiding segregation of the two metals resides in the realization of slow reduction of the Pd ions, thus avoiding homogeneous nucleation [87]. Superior 1 wt.% Pd@Au/C catalysts showed a turnover frequency (TOF) [mol converted/(mol Au × h)] in the glycerol oxidation reaction of up to 6500 under standard conditions (0.3-mM aqueous glycerol; glycerol/M ratio of 3000; NaOH/glycerol of 4; 50 °C; 0.3 MPa of dioxygen). Compared with the gold case, the alloyed particle with Au_6Pd_4 phase composition showed unchanged dispersion. The enhanced TOF from 900 to 6500 h^{-1} was significant, while the glyceric acid selectivity at 90% conversion increased from 68 to 77% [89]. Beyond doubt, the alloy phase with diameter of 3.4 nm, as measured by TEM, was responsible for the enhanced performance. Whereas the synergetic effect resulting from alloy formation can be attributed to changed interatomic distances, a distinction between an electronic or geometric effect as the cause of the enhanced catalytic parameters can not be made [89].

All these tendencies, viz. the effect of particle size on mono- and bimetallic catalysts at the level of the carbon-supported gold catalyst alloyed with Pd or Pt, were confirmed recently [90, 91]. An in-depth study of the metal (alloy) phase is needed to allow correlation with general oxidation mechanisms of alcohols, with specific active sites at the metal surface, and with possible support effects. Preliminary work with gold on titania catalyst has revealed similar though not always the same effects [92]. In a standard glycerol oxidation, the better Au/TiO_2 catalysts showed a catalytic performance comparable to that of the better Au/C catalysts. In contrast to Au/C, the best catalyst was prepared via deposition–precipitation of a gold salt, followed by wet chemical reduction. Increases in particle size from 2 to 5 nm, resulting from differences in the preparation procedure, viz. precalcination of precipitated salt and use of different sol protecting agents, allowed suppression of overoxidation and can be used to lever reaction selectivity.

11.4.5
Selective Oxidation with Oxidants Differing from Molecular Oxygen

Stable organic nitroxyl radicals of the TEMPO (2,2,6,6-tetramethylpyperidine-1-oxyl) family find increasing use in the catalytic oxidation of alcohols into carbonyl/carboxyl compounds [93]. In alkaline medium, the oxyammonium ion (TEMPO$^+$) is regioselective for the oxidation of primary versus secondary alcohols and can be regenerated *in situ* by oxidants such as Na$^+$OCl$^-$/Br$^-$ in water, organic solvents or biphasic medium (Scheme 11.12). Encapsulated sol–gel silica-entrapped versions of TEMPO (SG-TEMPO) exist and can be used as heterogeneous, reusable and non-leaching versions [94]. At a pH of 10 and 2 °C, catalytic amounts of TEMPO as well as SC-TEMPO (6.5 mol.%) in combination with NaOCl/Br catalyzed the one-pot oxidation of glycerol into ketomalonic acid [95]. In the absence of TEMPO, only small amounts of dihydroxyacetone and glyceric acid were formed. In the presence of TEMPO, glycerol was converted in the homogeneous as well as the heterogeneous case into, mainly, ketomalonic acid with TOFs of 70 and 4 h^{-1}, respectively. The same was found with tartronic acid as feed. Given the regioselectivity of TEMPO as oxidant of primary alcohols, very rapid formation of glyceric and tartronic acid must be assumed.

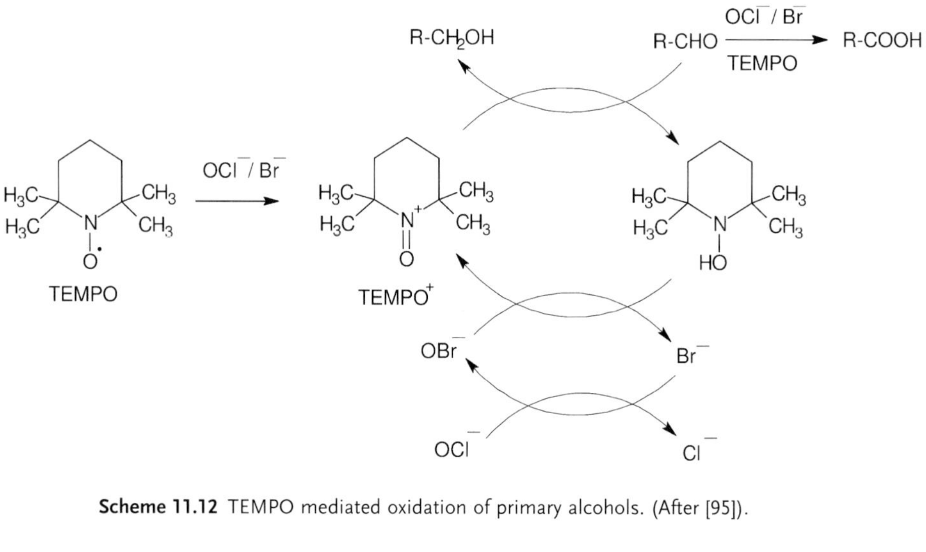

Scheme 11.12 TEMPO mediated oxidation of primary alcohols. (After [95]).

In presence of 3 equivalents of oxidant, TEMPO mediates then the rapid conversion into ketomalonic acid. The pH of the reaction medium is critical: at pH 7 no reaction occurs, while at high alkalinity TEMPO$^+$ undergoes basic dismutation. At pH 10, even the immobilized catalyst is stable and can be used during several subsequent reactions [95].

The Ru(III)-catalyzed oxidation of glycerol by an acidified solution of bromate (BrO_3^-) at 45 °C consumes the required amount of 2 moles of bromate to obtain pure glyceric acid. Traces of Hg(OAc)$_2$ were used as scavenger for potentially formed bromide, thereby eliminating the formation of bromine (formed by reaction of bromide and bromate) as an alternative oxidant [96]. The reaction is first order in Ru(III) (0.58 ms^{-1} at 45 °C) and zero order in substrate and protons. The addition of $RuCl_6^{3-}$ to protonated bromate is assumed to be rate limiting. Similar catalytic chemistry is obtained with Rh(III)Cl$_3$ [97].

With aqueous H$_2$O$_2$, catalyzed glycerol oxidation has also been examined [52]. With Ti, V- and Fe-substituted silicalite (TS-1, VS-1, FeS1) low conversion was obtained from 70 °C on. Formate esters of glycerol were dominant (60–80% selectivity) (Scheme 11.13), along with some hydroxyethanoic acid.

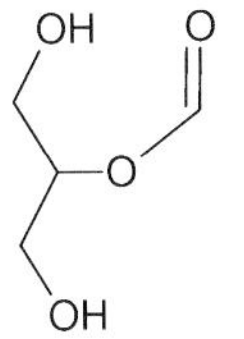

Scheme 11.13 A formate esters of glycerol.

With Cr, Mn and Co-substituted AlPO$_4$-5, less than 40% of the theoretical conversion was achieved, yielding almost exclusively glycerolformate esters. With larger pores (as in Ti-MCM-41 and TiO$_2$-SiO$_2$ cogels), increasing amounts of glyceraldehydes were obtained at the expense of the glycerolformate ester. It seems that in materials were slow glycerol diffusion can be expected, overoxidation easily occurs, resulting in formic acid formation.

11.5
Catalytic Hydrogenolysis of Glycerol

11.5.1
Heterogeneous Catalytic Hydrogenolysis of Glycerol

Glycerol conversion into propanediols offers an elegant and profitable way to valorize glycerol (Fig. 11.5) [98]. Propane-1,2-diol production is attractive because of its use in the manufacture of unsaturated polyester resins. Other uses include functional fluids (e.g., anti-freeze agent), solvent, preservative, cosmetic and paint additive. In particular the market of antifreeze for propane-1,2-diol is growing due to recent toxic concern around glycol. Increasing demand for propylene oxide, the industrial precursor for propane-1,2-diol manufacture, however, has provoked the search for alternative production routes. Propane-1,3-diol is more expensive than the 1,2-diol, and is applied mainly in polymeric applications for the production of carpet and textile fibers. This compound is conventionally produced from acrolein (Degussa–DuPont) in two steps, hydrolysis to 3-hydroxypropanal

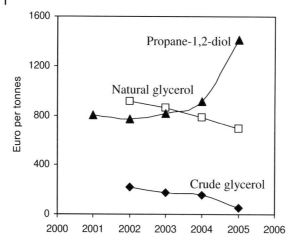

Fig. 11.5 Price evolution for glycerol and propane-1,2-diol. (After [98]).

followed by hydrogenation. Alternatives use, for example, hydroformylation of ethylene oxide (Shell route) or ethylene and formaldehyde, but they are not commercialized for economic reasons. Recently, DuPont and Cargill have exploited biorefinery units to synthesize this compound from sugar feedstock, taking advantage of modified microorganisms.

Conversion of glycerol into propanediols assumes a selective dehydroxylation of the triol. The dehydroxylation usually draws upon the cleavage of carbon–oxygen bonds accompanied by the addition of hydrogen (Scheme 11.14). This type of hydrogenation is often referred to as hydrogenolysis [99]. A key factor in applying this chemical transformation to glycerol conversion relies on the design of a proper catalyst along with its process. Only a catalyst that discriminates between C–O and C–C bond breaking in the presence of other C–O bonds will be successful. Otherwise, large quantities of hydrocarbon gases such as CH_4 and CO_2 will be formed (see glycerol reforming).

Hydrogenolysis of C–O bonds in saturated alcohols is generally very difficult under the usual hydrogenation conditions, and requires fairly harsh conditions [100]. Primary and secondary alcohol groups for instance survive hydrogenolysis in the presence of Raney nickel and copper-chromium oxide up to 533 K and 20 MPa H_2. Glycerol is more reactive due to the presence of neighboring alcoholic groups; the hydroxyl groups are thus removed at lower temperatures.

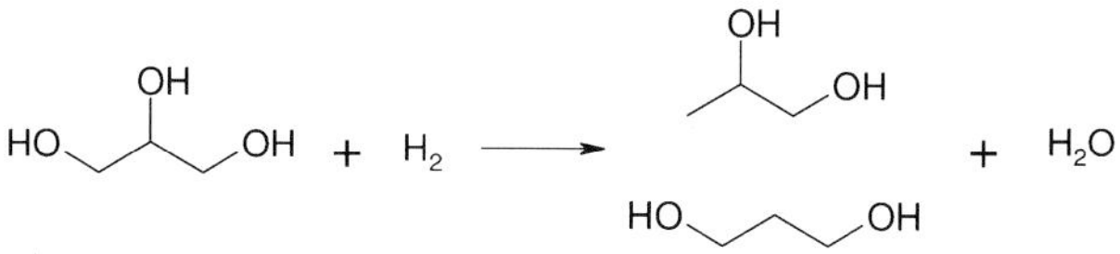

Scheme 11.14

Two reaction routes are generally accepted to explain propanediol formation from glycerol (Scheme 11.15). One route involves a dehydration of glycerol to hydroxyacetone (or acetol) and lactaldehyde, which is further hydrogenated to propanediols [101–103]. This route is closely related to the biocatalytic routes in which propanediol is formed from glycerol and sugars according to acetol-like intermediates. In fact, optically active propane-1,2-diol is currently prepared via biocatalytic reduction of acetol [104]. Several authors have analyzed transiently formed carbonyls at low temperature and H_2 pressure, while the dehydration pathway agrees with the rate acceleration effect of acids (*vide infra*). According to another mechanism, dehydrogenation of glycerol to glyceraldehyde and dihydroxyacetone occurs first on the metal, followed by a selective dehydroxylation to pyruvic aldehyde with subsequent hydrogenation to the diols [105, 106]. This mechanism more closely relates to that commonly accepted for hydrogenolysis of poly-ols. As the intermediate keto-aldehyde is not analyzed during glycerol conversion, its formation at (or desorption from) the metal surface was supposed to be rate limiting [107]. Support for this route is given by the rate enhancement effect of hydroxyl base, which is assumed to help the dehydroxylation (*vide infra*).

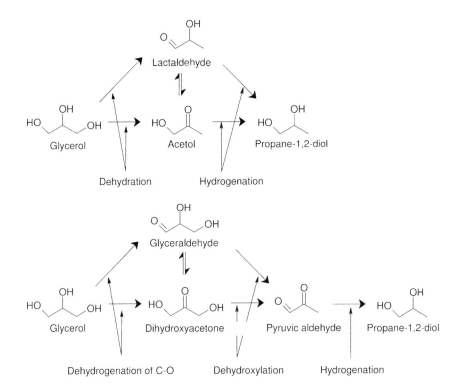

Scheme 11.15 Hydrogenolysis network of glycerol: the two mechanisms.

A detailed study of the reaction network was recently discussed based on hydrogenolysis studies using propane-1,2- and 1,3-diols, and other alcohols (e.g., isopropanol, n-propanol, ethanol) as reactant. Since alcohols are not converted, they can be considered as final products. Propane-1,3-diol and glycerol have comparable reactivity, while 1,2-diol is less reactive. Selective conversion of glycerol into propane-1,3-diol is thus not easy to achieve. Hydrogenolysis of propane-1,3-diol yielded 1-propanol almost quantitatively [100], whereas reaction with 1,2-propanediol gives an almost equimolar mixture of 1-propanol and 2-propanol. As 2-propanol is not observed in the hydrogenolysis of 1,3-propanediol, its formation is linked with that of propane-1,2-diol. Using this information, Miyazawa et al. proposed a selective glycerol hydrogenolysis route towards propane-1,3-diol over Ru on carbon in the absence of acid promoter [103]. Similar to the glucose dehydration mechanism [108], they proposed OH species on Ru that dehydrate glycerol to 3-hydroxypropionaldehyde (Scheme 11.16). However, why these OH species selectively attack H linked to the terminal carbons remains unclear. Unfortunately, the formed propane-1,3-diol is more reactive than glycerol and reacts further, as explained above. Ethylene glycol is not formed from the diols, but originates directly from glycerol [103].

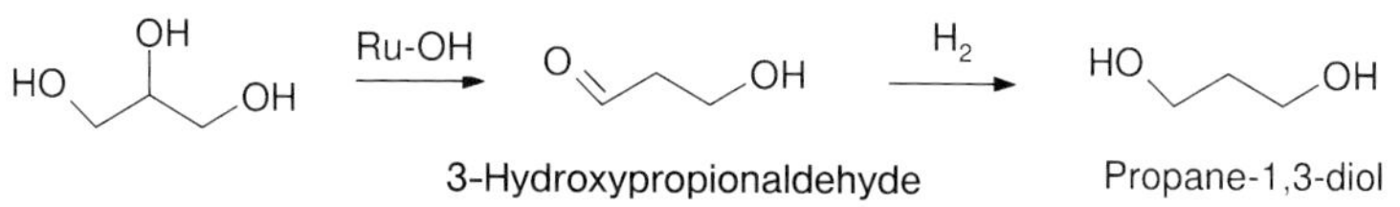

HO⌇OH⌇OH →(Ru-OH) O=⌇⌇OH (3-Hydroxypropionaldehyde) →(H$_2$) HO⌇⌇OH (Propane-1,3-diol)

Scheme 11.16

Table 11.2 lists the main data on the hydrogenolysis of glycerol with H$_2$ using heterogeneous catalysts. The oldest studies, which date back to 1931, describe the use of metal-activated chromium oxide type catalysts. Hydrogenolysis of 1.5 vol.% aqueous glycerol with copper-chromium for instance yields 70% propane-1,2-diol, whereas a 50% yield was obtained over cobalt chromium at 483 K in a dynamic reactor. Glycerol conversion was always incomplete, and 1-propanol is the main byproduct [109]. No information was reported concerning the H$_2$ pressure. Others reported propane-1,2-diol yields of 85% at 523 K and 20 MPa in a batch reactor using comparable catalysts [110–112]. To achieve appropriate glycerol conversion conventional Raney nickel catalysts require higher temperatures, typically above 543 K [113]. At the high temperature, however, considerable quantities of gaseous hydrocarbons, mainly methane, are produced. Unselective hydrogenolysis of glycerol over Raney nickel is confirmed in later studies [105].

In contrast to Raney Ni, no hydrocarbons are formed over pure Raney copper, while propane-1,2-diol is produced with a selectivity of 66% at 85% glycerol conversion at 3 MPa and 513 K. It has been verified that the excellent selectivity is due to the high stability of propane-1,2-diol under the reaction conditions. 1,3-Propanediol is not stable and converts almost quantitatively into 1-propanol

11.5 Catalytic Hydrogenolysis of Glycerol

Table 11.2 Glycerol hydrogenolysis over heterogeneous catalysts in various conditions.[a]

Entry	Catalyst	Temperature (K)	H_2 pressure (MPa)	Vol.% glycerol	Yield (conversion) (%)	Ref.
1	CuCr	483[b]	n.d.	1.5	70 (n.d.)	109
2	CoCr	483[b]	n.d.	1.5	50 (n.d.)	109
3	CuCr	523	20	n.d.	85 (n.d.)	110–112
4	Raney Ni	>543	n.d.	3.2	n.d.	113
5	Raney Ni	533	4	3.2	n.d.	105
6	Raney Cu	513	3	3.2	56 (85)	105
7	Cu/SiO$_2$	533	4	0.8	34 (40)	114
8	ZnO/CuO	543	10	20	84 (99)	115
		523[b]	15	20	88 (97)	115
9	Ru/SiO$_2$	483	4	3.2	12 (99)	105
10	Ru/C	473	8	3.2	8.8 (30)	119
11	Rh/C	453	8	14	2 (2.5)	117
12	Ru/C	543	13	20	10 (82)	118
13	Ru+S/C[c]	543	13	20	5.5 (13)	118
14	Ru+S/C[d]	513	13	20	75 (100)	118
15	CoCuMnMoP	523	25	99.5	96 (100)	123
16	Re+Ni/C[d]	503	9	25	54 (61)	121
17	Ru/C[e]	413	8	20	17.6 (40)	120

[a] n.d. = not determined in the reference. [b] Dynamic reactor. [c] S means sulfided.
[d] With NaOH. [e] With Amberlyst.

(96% yield) [105]. As with Raney Cu, glycerol also degrades selectively to propane-1,2-diol over Cu supported on carbon. The selectivity to 1,2-propanediol increases with H$_2$ pressure, resulting in a maximum value of 84.5% at 40% glycerol conversion, obtained in a three phase dynamic reactor at 533 K and 4 MPa H$_2$ [114]. Although the hydrogenolysis activity of metallic copper is very promising, the catalyst gradually deactivates due to metal sintering and surface oxidation.

Patents describe glycerol (20 vol.% in water) conversion with H$_2$ over copper catalysts stabilized with Zn. The molar percentages of propane-1,2-diol typically range from 27 to 84% at 10 MPa and 543 K using catalysts with varying elemental compositions [115]. An atomic ratio of Cu and Zn of 0.9 is best whenever 1,2-propanediol formation is pursued. The catalyst performs even better in a dynamic reactor, yielding 88% of 1,2-propanediol at close to complete glycerol conversion at 523 K and 15 MPa [115]. Similar data have been reported by others [116, 117]. In particular, CuO/ZnO catalysts are very selective since propane-1,2-diol is the only product formed up to 20% glycerol conversion.

Glycerol hydrogenolysis has also been illustrated over transition metals impregnated on silica. Some metals, such as Co and Pt, appear inactive for glycerol conversion, while others, such as Rh, Pd, Ir and, especially, Ru are active but lack the ability to selectively cleave C–O bonds. For instance, almost no glycerol was converted with Co and Pt, whereas glycerol degrades to various gaseous hydrocarbons, essentially methane, ethane and propane, in addition to ethylene glycol and 1-propanol over Ru on silica at 533 K and 4.0 MPa. Propane-1,2-diol yields never exceed 10% under these conditions [105].

Like Ru on silica, carbon supported Ru also shows unselective hydrogenolysis properties. Ru on charcoal for instance converts 30% of the glycerol feed into 1,2-propanediol with 29.4% selectivity at 473 K and 8 MPa [103], while others found 48% conversion with 24.5% selectivity at 513 K and 13 MPa [118]. The metal particle size affects the hydrogenolysis rate; larger particles are more active, but the propanediol selectivity is not sensitive to structural differences [119]. Other metals, like Rh, Pt and Rh on carbon, are more selective but less active than Ru. Chaminand et al. for instance have found a propanediol selectivity of 80% at 2.5% glycerol conversion over Rh on carbon, when hydrogenolysis was carried out at 453 K and 8 MPa, while Miyazawa et al. reported 58.6% selectivity at 0.3% conversions at shorter reaction times [117, 120]. Interestingly, formation of 1,3-propanediol is recognized over carbon supported with Rh and Ru at low glycerol conversion, but the selectivity is poor (<5%) [103, 117].

The modest C–O hydrogenolysis selectivity of transition metals can be improved by modifying the catalyst. In one example, a sulfided Ru catalyst converts glycerol selectively into propane-1,2-diol, whereas almost no propanediol was observed for unmodified Ru. The highest selectivity (79%) has been found for a catalyst containing equimolar Ru and S surface concentrations. Besides the high selectivity, sulfur addition, unfortunately, induces a 50-fold reduction in hydrogenolysis activity. Accordingly, no significant glycerol conversions were reported for the selective reactions. Sub-equimolar S coverage restores the original hydrogenolysis activity, but the poor propanediol selectivity returns [119]. Analogue effects of sulfur modification on glycerol hydrogenolysis properties have been reported by others [118]. They found that addition of sulfur to Ru on carbon reduces the glycerol conversion from 80 to 13%, while the propane-1,2-diol selectivity doubled from 20.5 to 42%, for reactions operating at 543 K and 13 MPa using 20 vol.% aqueous glycerol. Sulfur also induces the sudden formation of lactic acid. The highest yields reported amounted to 23.5% at complete glycerol conversion. Lactic acid formation from glycerol also appears to be selective over silver catalysts [121]. Although the yields are not yet satisfactory, the idea of producing lactic acid from glycerol is attractive in view of producing biodegradable polylactide plastics.

Modern heterogeneous catalysts increasingly evolve into complex multicomponent systems. In this context, several bimetallic catalysts have been prepared and evaluated for glycerol hydrogenolysis. While in some cases hydrogenolysis is affected beneficially, in other examples combining two metals was not successful. Copper-modified Ru on carbon for instance shows improved C–O hydrogenolysis selectivity, but the presence of Cu poisons the activity of Ru. Similarly, Raney copper modified with Ru is unable to convert glycerol at 493 K under 4 MPa, in contrast to pure Raney copper [122]. A synergistic effect in bimetallic catalysts has been discovered with a Re-promoted Ni catalyst supported on carbon. A yield of

54% propane-1,2-diol at 61% glycerol conversion has been obtained using a catalyst containing 2 wt.% Ni and 0.4 wt.% Re at 503 K and 9 MPa [121]. Here, Re acts as an oxygen acceptor at the surface of the catalyst, which might facilitate glycerol sorption near the active surface sites. Other Ni:Re compositions were either less active or less selective. Whether Re also promotes noble metals such as Rh and Pd could not be properly extracted from the data, due to a lack of reference material. Although no explanation was given, the patent discloses in one particular case the formation of propane-1,3-diol at 503 K when 2.5 wt.%Ni–2.5 wt.%Re was used.

In another patent, very high propane-1,2-diol yields were obtained with a multicomponent oxide catalyst consisting of the metals Co, Cu, Mn and Mo, and P. At 523 K under 25 MPa, glycerol is almost quantitatively converted into propane-1,2-diol. In contrast to most studies, the reactions were performed with pure (99.5%) glycerol instead of aqueous dilutions [123].

Hydrogenolysis of glycerol generally proceeds not so rapidly, and usually requires considerable catalyst amounts or reaction time. The activity is often markedly enhanced by the addition of alkaline additives such as alkali and earth alkali hydroxides without effecting the high 1,2-propanediol selectivity. For instance, a propane-1,2-diol yield as high as 75% is reported at complete substrate conversion for the sulfided Ru catalyst supported on activated coal, when operating at 13 MPa and 513 K in the presence of sodium hydroxide [118]. A similar reaction without base only afforded 3% yield. Close to equimolar quantities of base and glycerol, corresponding to a reaction solution pH of 11–12.5, is apparently ideal. High hydrogenolysis rates in the presence of base were also announced for other metals than sulfided Ru. Poorly active Ni, Pt and Pd catalysts for instance also show improved hydrogenolysis properties after addition of an alkaline promoter.

Like bases, acids also progress dehydroxylation of glycerol. Chaminand et al. have added tungstic acid to noble metals and CuO/ZnO [117]. Both Rh and Pd on carbon catalyzed reactions showed a fourfold increase of propanediol yield upon addition of H_2WO_4. Although the hydrogenolysis rate is not affected in case of Ru on carbon, acidification drastically increased the propanediol selectivity from 18 to almost 80% at isoconversion [120]. Although not understood, the use of H_2WO_4 markedly increases the propane-1,3-diol selectivity from 3.4 to 21% when in combination with Rh, corresponding to a propane-1,3-diol yield of 2.6%, while propane-1,3-diol formation remains marginal with metals like Pd and Ru [117, 120]. Rhodium on alumina shows comparable activity with carbon supported Rh, but lower propanediol selectivity. 1,3-Propanediol was formed slightly more selectively on the alumina support [117].

Kusunoki et al. have used an acidic cation exchange resin (Amberlyst-15) in the hydrogenolysis of glycerol under mild conditions (393–453 K, 4–8 MPa and 20 wt.% aqueous glycerol) in the presence of carbon supported metals like Ru, Pd, Pt and Rh [120]. In their comparative study, the resin provides much better glycerol conversions than other solid acids such as zeolites and sulfated zirconia, H_2WO_4 and liquid H_2SO_4. The addition of Amberlyst is most effective for the metals Ru and Rh. Although the operational temperature is restricted due to the thermostability of the sulfonated organic polymer, high propanediol yields have been reported within a reasonable reaction time. For instance, at 413 K and 8.0 MPa, 40% of the glycerol is converted into propanediols with 44% selectivity after 10 h using Ru on carbon in the presence of the resin. Higher temperatures cause decomposition of the sulfonic acids groups. Repetitive use of the Ru catalyst showed minor changes in the glycerol conversion, but a small increase in hydrogenolysis selectivity in particular, as a result of a decrease of ethylene glycol. As no sintering of the metal (TEM, TPR) was found, the enhanced selectivity is not due to changes of the surface state of the metal particles [103]. Addition of H_2SO_4 in equimolar amounts is less successful than the resin, suggesting that physical separation of the two reactive sites (acid and metal) is crucial.

The rates and selectivity of the hydrogenolysis of glycerol over noble metal with acid promoter also greatly depend on solvent. Activities were generally higher in organic solvents such as sulfolane. While 1,2-propanediol is the dominant product in water, 1-propanol and 1,3-propanediol are mainly formed in sulfolane. The highest propane-1,3-diol yield hitherto reported for heterogeneous hydrogenation systems is ca. 4% (at 32% glycerol conversion), and is obtained in sulfolane over Rh on carbon with H_2WO_4 (8.0 MPa and 453 K) [117].

11.5.2
Homogeneous Catalytic Hydrogenolysis of Glycerol

Soluble group VIII transition metal complexes have been tested in the presence of tungstic acid for the hydrogenation of glycerol to propanediols. Depending on the catalyst combination, yields ranging from 0 to 36% were reported at temperature and pressure ranges of 373–473 K and 30–60 MPa [124]. Reactions are generally performed in weakly basic solvents such as amides and in an CO/H_2 (1:2) atmosphere to stabilize and activate the metal catalyst. As an example, in the presence of $Rh(CO)_2(Acac)$ and H_2WO_4, glycerol is converted in N-methylpyrrolidone (NMP) into propanediols (94% selectivity) and 1-propanol (6% selectivity) at 39% conversion, while only 4% of glycerol is converted in the absence of W promoter. Tungstic acid is more than a simple proton donor, since its replacement with sulfuric acid leads to an unreactive system. Also not understood, but highly interesting, is the equimolar production of 1,2- and 1,3-diol isomers with the coupled W/Rh catalytic system. With a yield of 18%, this system gives by far the best propane-1,3-diol production. As with heterogeneous catalysis, the dehydroxylation selectivity is solvent dependent. When γ-butyrolactone, for instance, is used instead of NMP, 1-propanol predominates, corresponding to a yield of 40%.

Selective glycerol dehydroxylation is also possible with unimetallic catalytic systems using, for example, $H[Ru(CO)_3I_3]$ [125]. Glycerol is converted mainly into n-propanol and its ethers in the presence of $CO:H_2$ (10:3) at 473 K and 13 MPa, while propane-1,3-diol is formed with 10% selectivity. A total n-propanol (plus derivatives) selectivity of over 90% is easily obtained for glycerol conversions below 75%. At higher conversion, products are further hydrogenated to propane, homologated to n-butanol, or carbonylated to butyric esters. The products obtained were related to the bifunctional character of the iodocarbonyl complex, viz. its strong Brønsted acidity and its hydrogenating power in aqueous conditions; the proposed mechanism is via dehydration of glycerol to acrolein (and heavier condensation products), which is further hydrogenated to n-propyl alcohol (and its derivatives). Most interestingly, C–C rupture products are not formed, even though they are the main products when heterogeneous Ru catalysts are used.

Another homogeneous Ru complex, [{Cp*Ru(CO)$_2$}$_2$(μ-H)]$^+$OTf$^-$, was developed recently for the selective deoxygenation of diols to alcohols. The reaction almost exclusively deoxygenates secondary over primary OH groups. Interestingly, preliminary results with glycerol under 5.2 MPa H$_2$ at 383 K in sulfolane gave no detectable propane-1,2-diol, 4.6% propane-1,3-diol and 5.7% n-propanol [126, 127]. Despite the poor yields the Ru system is promising since high propane-1,3-diol selectivity is obtained under mild reaction conditions free of CO/H$_2$. Excess HOTf is, however, required to ensure high propane-1,3-diol selectivity.

A palladium phosphine complex [e.g., BCPE = 1,2-bis(1,5-cyclooctylenephosphino)ethane] was also reported to produce propanediols and n-propanol from glycerol at 443 K under 6 MPa CO/H$_2$ atmosphere in acidic conditions. n-Propanol is the dominant product, while a slight preference for the formation of propane-1,3-diol is seen in the diol fraction. Reactions were performed at different temperatures in the range 413–448 K. Since acrolein was monitored at high temperature, a reaction network was proposed following a sequential dehydration/hydrogenation pathway [20].

11.6
Glycerol Reforming and Hydrogen Production

Current biodiesel can not be considered as a 100% biomass-based fuel as long as methanol is derived from petrochemical resources. A clean way to solve the bio-related problem is the conversion of glycerol waste from the transesterification process into syngas. In this context, glycerol reforming is a suitable target reaction worthy of study.

The first reports concerning glycerol reforming appeared in the pioneering work of the group of Dumesic [128–131]. They considered H$_2$ production, for example, from glycerol at temperatures near 500 K in a single-reactor, aqueous liquid-phase reforming process. According to Scheme 11.17, glycerol is almost completely converted into CO$_2$ and H$_2$. Major steps in the reaction include the dehydrogenation of glycerol on the metal surface with the formation of adsorbed intermediates before cleavage of C–C and C–O bonds. Subsequent C–C bond breaking leads to H$_2$ and CO, while CO reacts with water according to the water-gas shift reaction. Despite the chemical value of CO production, CO must be removed to ascertain high catalytic activity.

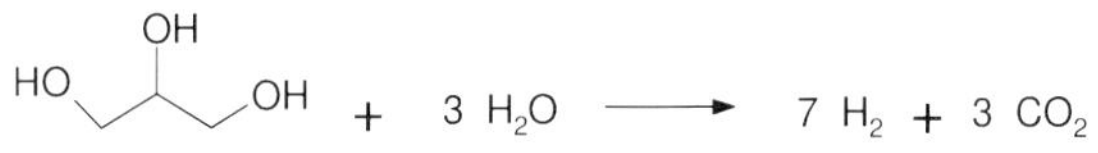

$$\text{HOCH}_2\text{CH(OH)CH}_2\text{OH} + 3\,\text{H}_2\text{O} \longrightarrow 7\,\text{H}_2 + 3\,\text{CO}_2$$

Scheme 11.17

An effective catalyst thus breaks C–C, O–H, and C–H bonds in the glycerol reactant without cleaving C–O bonds or hydrogenating CO or CO_2 to form alkanes. Many metals on various supports have been evaluated in high-throughput mode. Pt on alumina and Sn-doped Ni on alumina have been found most valuable [132, 133]. Catalytic reforming of glycerol gives almost complete conversion of carbon in the liquid phase into gaseous compounds at 538 K, with CO_2 being the major constituent of the gas phase composition in addition to small amounts of methane and ethane. The liquid phase is composed of a pool of dehydroxylated and cleavage derivatives such as ethanol, propane-1,2-diol, methanol, propan-1-ol, acetic acid, ethylene glycol, acetol, propan-2-ol, propionic acid, acetone, propionaldehyde, and lactic acid.

A major challenge for glycerol reforming is the maintenance of high H_2 selectivity – defined as the amount of H_2 produced, normalized by the amount of H_2 that would be expected based on the converted glycerol, i.e., seven molecules of H_2 formed when three molecules of CO_2 are formed. Several H_2 consuming reactions, such as methanation, Fischer–Tropsch and hydrogenation, must thus be avoided as much as possible. H_2 selectivities of 75 and 51% were reported, for instance, at 498 and 538 K, respectively, for full conversion of 1 wt.% glycerol solutions using 3 wt.% Pt on alumina [128]. Somewhat higher H_2 selectivities were reported when reforming catalysis is carried with Sn-modified Ni [132]. Even though high hydrogen yields are obtained, processing such dilute solutions is economically not feasible. Solutions with higher concentrations, though, are more subject to undesirable side-reactions.

Production of H_2 from glycerol has also been attempted in the gas phase. The advantage of this approach is that atmospheric pressure can be maintained in the reactor, while the previous liquid phase process is probably less energy consuming due to the lower operating temperature. Chornet and coworkers were the first to report steam reforming of crude glycerol [134]. They used commercial Ni catalysts. Novel efficient catalysts for steam reforming of glycerol have been studied recently [135]. Ruthenium impregnated on Y_2O_3 showed very high activity in a prolonged time on stream run at 600 °C. A H_2 selectivity of 90% was observed, while almost no methane was formed.

In addition, the catalyst appeared very stable under the reaction conditions; little carbon was deposited on the spent catalyst. Other supported metals were less active. The activity order, Ru ∼ Rh > Ni > Ir > Co > Pt > Pd > Fe, is very comparable to that measured for the steam reforming of methane. Of all the supports tested, Y_2O_3 and ZrO_2 gave the best results for the Ru-catalyzed steam reforming of glycerol.

11.7
Miscellaneous Oxidation Reactions

Glycerol can be converted into glycerine carbonate via an oxidative carbonylation and, subsequently, into glycidol (Scheme 11.18) [136].

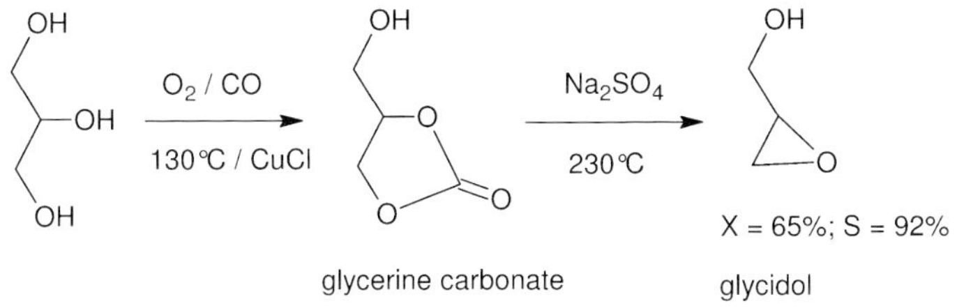

Scheme 11.18

The analogous chlorinated oxirane epichlorohydrin has also been prepared from glycerol. Glycerol is, therefore, first reacted with anhydrous HCl to form dichlorohydrins, which are further contacted with lime water to form epichlorohydrin (Scheme 11.19) [137–139].

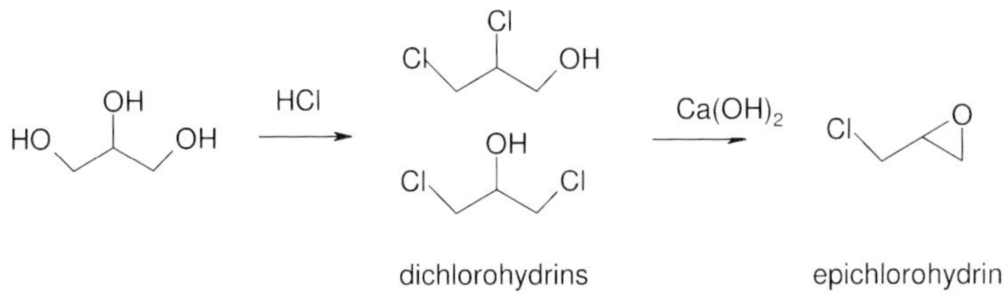

Scheme 11.19

11.8
Conclusions

Glycerol as a by-product from biodiesel production can be considered as a green chemical feedstock for subsequent catalytic transformation. In contrast to traditional petrochemical feedstocks, the present one is highly functionalized, its transformation requiring selective defunctionalization.

In the present economic context, the old gas-phase transformation of glycerol into acrolein may become of interest. Oligo-glycerol formation via catalytic dehydration has been studied by a few groups. It seems that the oligomerization degree of the glycerol oligomers as well as the nature of the individual dimers can be influenced by steric restrictions around the acid sites, though the data sets could still be improved.

Catalytic oxidation of glycerol with dioxygen is receiving considerable attention again. The Pt-Bi on carbon catalysts have been studied in the past in detail. Many consecutive products can be achieved in good yields, although it does not seem easy to come up with 100% C balances. Newer systems such as Pt/Pd on gold colloids have recently been claimed to be superior systems. The catalyst preparation procedure seems crucial to obtain good yields and catalyst stability. However, the fundamental link between surface architecture of the active catalyst and the catalytic selectivity and stability is still missing.

Many systems have been described for the hydrogenolysis of glycerol, yielding mainly 1,2-propanediol. Copper chromite-based catalysts seem to be the better performing ones. Unfortunately, most of the data available is of a descriptive nature and a fundamental understanding of the catalyst is often missing.

With the price evolution of glycerol ahead, old and forgotten reactions will be excavated and reinvestigated to meet current needs, while many unique and inventive catalytic reactions are to be expected in the years ahead.

Acknowledgments

The authors acknowledge sponsorship from the following programmes: IAP (Supramolecular catalysis) from the Federal Government, GOA and VIRCAT (high throughput catalysis) from the Flanders region and K.U. Leuven, CECat from K.U. Leuven. E.D. is grateful to IWT for a fellowship.

References

1 http://www.aocs.org/archives/am2004/session.asp?session=IOP+3.1%3A++New+Uses+of+Glycerin
2 G. Knothe, J. Krahl, J. Van Gerpen, eds., *The Biodiesel Handbook*, AOCS Press, Urbana, IL, **2005**, pp. 302.
3 M.A. Dasari, P.P. Kiatsimkul, W.R. Sutterlin, G.J. Suppes, *Appl. Catal. A* **2005**, 281, 225–231.
4 D. Tapasvi, D. Wiesenborn, C. Gustafson, *Trans. Am. Soc. Agric. Eng.* **2005**, 48, 2215–2221.
5 http://www.bharatbook.com/pr.asp?id=18047
6 L. Ott, M. Bicker, H. Vogel, *Green Chem.* **2006**, 8, 214–220.
7 http://www.virent.com/whitepapers/Biodiesel%20Whitepaper.pdf#search=%22price%20of%20glycerol%20%22
8 http://www.1911encyclopedia.org/Glycerin
9 http://en.wikipedia.org/wiki/Glycerol#Purification
10 Y. Xu, W. Du, D. Liu, J. Zeng, *Biotechnol. Lett.* **2003**, 25, 1239–1241.
11 http://asae.frymulti.com/abstract.asp?aid=16951&t=2
12 E.J. Witzemann, *J. Am. Chem. Soc.*, **1914**, 36, 1766–1770.
13 A.A. Newman, *Glycerol*, CRC Press, Cleveland, OH, **1968**.
14 http://www.orgsyn.org/orgsyn/orgsyn/prepContent.asp?prep=cv1p0015
15 FR 695931, **1930**.
16 H.E. Hoyt, T.H. Manninen, US5258520, **1951**.
17 T. Haas, N. Armin, A. Dietrich, K. Herbert, G. Walter, US 5426249, **1994**.
18 H. Klenk, W. Girke, A. Neher, T. Haas, D. Arntz, US 5387720, **1994**.
19 J.J. Dubois, C. Duquenne, W. Holderich, J. Kervennal, FR 2882053, **2006**.
20 E. Drent, W.W. Jager, US 6080898, **2000**.
21 Y.S. Stein, M.J. Antal, Jr., J. Maitland Jr., *J. Anal. Appl. Pyrolysis* **1983**, 4, 283–296.
22 W. Bühler, E. Dinjus, H.J. Ederer, A. Kruse, C. Mas *J. Supercrit. Fluids* **2002**, 22, 37–53.
23 S. Ramayya, A. Brittain, C. DeAlmeida, W. Mok, M.J. Antal, Jr. *Fuel* **1987**, 66, 1364–1371.
24 J. Van Heteren, C. Poot, F. Rechweg, M.F. Stewart, EPA 0070080, **1982**.
25 J. Barrault, Y. Pouilloux, J.M. Clacens, C. Vanhove, S. Bancquart, *Catal. Today* **2002**, 75, 177–181.
26 J. Barrault, J.M. Clacens, Y. Pouilloux, WO 01/98243 A1.

27 J.M. Clacens, Y. Poullioux, J. Barrault, C. Linares, M. Goldwasser, *Stud. Surf. Sci. Catal.* **1998**, 118, 895–902.
28 J.M. Clacens, Y. Poullioux, J. Barrault, *Stud. Surf. Sci. Catal.* **2002**, 143, 678–695.
29 J. Barrault, J.M. Clacens, Y. Pouilloux, *Top. Catal.* **2004**, 27, 137–142.
30 S. Cassel, C. Debaig, T. Benvegnu, P. Chaimbault, M. Lafosse, D. Plusquellec, P. Rollin, *Eur. J. Org. Chem.* **2001**, 875–896.
31 H. Noureddini, W.R. Dailey, B.A. Hunt, *Adv. Environ. Res.* **1998**, 2, 232–244.
32 K. Klepacova, D. Mravec, M. Bajus, *Appl. Catal. A* **2005**, 294, 141–147.
33 A. Behr, C. Dewattines, H. Hinnekens, EP 0649829, **1995**.
34 G. Hillion, B. Delfort, I. Durand, Fr P 2866653, **2004**.
35 R.S. Karinen, A.O.I. Krause, *Appl. Catal. A* **2006**, 306, 128–133.
36 R. Wesendorf, *Erdöl, Erdgas, Petrochem.* **1995**, 48, 138.
37 A. Behr, *Chem. Ing. Techn.* **2001**, 73, 1463–1467.
38 V.A. Gupta, US P 5476971, **1995**.
39 B. Gruber, B. Fabry, B. Giesen, R. Müller, F. Wangemann, *Tenside Surf. Det.* **1993**, 30, 422–426.
40 K. Urata, N. Takaishi, *J. Am. Oil Chem. Soc.* **1996**, 73, 819–830.
41 B. Gruber, B. Fabry, B. Giesen, R. Müller, F. Wangemann, *Tenside Surf. Det.* **1993**, 30, 422–426.
42 A. Behr, M. Urschey, *Adv. Synth. Catal.* **2003**, 345, 1242–1246.
43 K. Urata, N. Takaishi, *J. Am. Oil Chem. Soc.* **1994**, 71, 1027–1033.
44 K. Urata, N. Takaishi, *J. Surfact. Deterg.* **1999**, 2, 91–103.
45 K. Urata, N. Takaishi, *J. Surfact. Deterg.* **2002**, 5, 287–294, and 403–409.
46 K. Urata, *Eur. J. Lipid Sci. Technol.* **2003**, 105, 542–556.
47 K. Urata, *Chim. Oggi* **2003**, 21/5, 36–42.
48 R. Garcia, M. Besson, P. Gallezot, *Appl. Catal. A* **1995**, 127, 165–176.
49 H. Kimura, K. Tsuto, *J. Am. Oil Chem. Soc.* **1993**, 70, 1027–1030.
50 W.R. Davis, J. Tomsho, S. Nikam, E.M. Cook, D. Somand, J.A. Peliska, *Biochemistry* **2000**, 39, 14279–14291.
51 R.A. Sheldon, I.W.C.E. Arends, A. Dijksman, *Catal. Today* **2000**, 57, 157–166.
52 P. McMorn, G. Roberts, G.J. Hutchings, *Catal. Lett.* **1999**, 63, 193–197.
53 K. Matsuoka, M. Inaba, Y. Iriyama, T. Abe, Z. Ogumi, M. Matsuoka, *Fuel Cells* **2002**, 2, 35–39.
54 C. Xu, R. Zeng, P. Kang Chen, Z. Wei, *Electrochim. Acta* **2005**, 51, 1031–1035.
55 A. Kahyaoglu, B. Beden, C. Lamy, *Electrochim. Acta* **1984**, 29, 1489–1492.
56 L. Roquet, E.M. Belgsir, J.M. Leger, C. Lamy, *Electrochim. Acta* **1994**, 39, 2387–2394.
57 B. Beden, I. Cetin, A. Kahyaoglu, D. Takky, C. Lamy, *J. Catal.* **1987**, 104, 37–46.
58 A. Hamelin, Y.H. Ho, X.P. Gao, M.J. Weaver, *Langmuir* **1992**, 8, 975–981.
59 M. Avramov-Ivic, J.M. Leger, B. Beden, F. Hahn, C. Lamy, *J. Electroanal. Chem.* **1993**, 351, 285–297.
60 T. Kawaguchi, W. Sugimoto, Y. Murakami, Y. Takasu, *Electrochem. Commun.* **2004**, 6, 480–483.
61 C. Xu, P. Kang Chen, *Chem. Commun.* **2004**, 2238–2239.
62 P. Kang Shen, C. Xu, *Electrochem. Commun.* **2006**, 8, 184–188.
63 G. Yildiz, F. Kadirgan, *Annal. Chim.* **1994**, 84, 455–466.
64 M. Avramov-Ivic, V. Jovanovic, G. Vlajnic, J. Popic, *J. Electroanal. Chem.* **1997**, 423, 119–124.
65 E.C. Venancio, W.T. Napporn, A.J. Motheo, *Electrochim. Acta* **2002**, 47, 1495–1501.
66 H. Baltes, E.I. Leupold, *Angew. Chem.* **1982**, 94, 544–545.
67 G. Bub, J. Mosler, A. Sabbach, F.F. Kuppinger, S. Nordhoff, G. Stochniol, J. Sauer, U. Knippenberg, WO 2006/092272 A2.
68 S. Masahide, T. Tsukasa, Jpn. KokaiTokyo Koho **2005** 213225 A2.
69 P. Vinke, D. de Wit, A.T.J.W. de Goede, H. van Bekkum, *Stud. Surf. Sci. Catal.* **1992**, 72, 1–20.
70 A. Abbadi, M. Makkee, W. Visscher, J.A.R. van Veen, H. van Bekkum, *J. Carbohydr. Chem.* **1993**, 12, 573–587.
71 T. Mallat, A. Baiker, *Catal. Today* **1994**, 19, 247–283.

72 G.H. Kimura, K. Tsuto, T. Wakisaka, Y. Kazumi, Y. Inaya, *Appl. Catal. A* **1993**, 96, 217–228.
73 H. Kimura, *Appl. Catal. A* **1993**, 105, 147–158.
74 R. Garcia, P. Gallezot, *Catal. Today* **1997**, 37, 405–418.
75 G. Bertland, *C.R. Acad. Sci. Ser. 2* **1986**, 122, 900.
76 P. Fordham, R. Garcia, M. Besson, P. Gallezot, *Stud. Surf. Sci. Catal.* **1996**, 91, 161–170.
77 P. Fordham, M. Besson, P. Gallezot, *Appl. Catal. A* **1995**, 133, L179–L184.
78 A. Abbadi, H. van Bekkum, *Appl. Catal. A* **1996**, 148, 113–122.
79 H. Kimura, *Polym. Adv. Technol.* **2001**, 12, 697–710.
80 T. Mallat, A. Baiker, *Chem. Rev.* **2004**, 104, 3037–3058.
81 S. Carrettin, P. McMorn, P. Johnston, K. Griffin, C.J. Kiely, G.J. Hutchings, *Phys. Chem. Chem. Phys.* **2003**, 5, 1329–1336.
82 S. Carrettin, P. McMorn, P. Johnston, K. Griffin, G.J. Hutchings, *Chem. Commun.* **2002**, 696–697.
83 S. Carrettin, P. McMorn, P. Johnston, K. Griffin, C.J. Kiely, G.A. Attard, G.J. Hutchings, *Top. Catal.* **2004**, 27, 131–136.
84 F. Porta, L. Prati, *J. Catal.* **2004**, 224, 397–403.
85 M.K. Chow, C.F. Zukowski, *J. Colloid Interface Sci.* **1994**, 165, 97–109.
86 S. Demirel-Gülen, M. Lucas, P. Claus, *Catal. Today* **2005**, 102–103, 166–172.
87 C.L. Bianchi, P. Canton, N. Dimitratos, F. Porta, L. Prati, *Catal. Today* **2005**, 102–103, 203–212.
88 N. Dimitratos, F. Porta, L. Prati, *Appl. Catal. A* **2005**, 291, 210–214.
89 D. Wang, A. Villa, F. Porta, D. Su, L. Prati, *Chem. Commun.* **2006**, 1956–1958.
90 N. Dimitratos, J.A. Lopez-Sanchez, D. Lennon, F. Porta, L. Prati, A. Villa, *Catal. Lett.* **2006**, 108, 147–153.
91 N. Dimitratos, C. Messi, F. Porta, L. Prati, A. Villa, *J. Mol. Catal. A* **2006**, 256, 21–28.
92 N. Dimitratos, A. Villa, C.L. Bianchi, L. Prati, M. Makkee, *Appl. Catal. A* **2006**, 311, 185–192.
93 A.E.J. de Nooy, A.C. Besemer, H. van Bekkum, *Synthesis* **1996**, 1153–1174.
94 R. Ciriminna, D. Avnir, J. Blum, M. Pagliaro, *Chem. Commun.* **2000**, 1441–1442.
95 R. Ciriminna, M. Pagliaro, *Adv. Synth. Catal.* **2003**, 345, 383–388.
96 S. Srivastava, H. Tripathi, K. Singh, *Transition Met. Chem.* **2001**, 26, 727–729.
97 S. Srivastava, *Transition Met. Chem.* **1999**, 24, 683–685.
98 G. Suppes, *Inform* **2006**, 17, 553.
99 C. Ellis, *Hydrogenation of Organic Compounds*, 3rd edn., D. Van Nostrand Company, New York, **1930**, p. 564.
100 S. Nishimura, *Handbook of Heterogeneous Catalytic Hydrogenation for Organic Synthesis*, Wiley-VCH, Weinheim, **2001**.
101 M.A. Dasari, P.P. Kiatsimkul, W.R. Sutterlin, G.J. Suppes, *Appl. Catal. A* **2005**, 281, 225–231.
102 G.J. Suppes, W.R. Sutterlin, M.A. Dasari, WO 2005/095536.
103 T. Miyazawa, Y. Kusunoki, K. Kunimori, K. Tomishige, *J. Catal.* **2006**, 240, 213–221.
104 K. Yamada-Onodera, N. Kawahara, Y. Tani, H. Yamamoto, *Eng. Life Sci.* **2004**, 4, 413–417.
105 C. Montassier, D. Giraud, J. Barbier, in: *Heterogeneous Catalysis and Fine Chemicals*, Eds. M. Guisnet et al., Elsevier Science Publisher B.V., Amsterdam, **1988**, pp. 165–170.
106 D.G. Lahr, B.H. Shanks, *J. Catal.* **2005**, 32, 386–394.
107 D.G. Lahr, B.H. Shanks, *Ind. Eng. Chem. Res.* **2003**, 42, 5467–5472.
108 S.K. Tyrlik, D. Szerszen, M. Olejnik, W. Danikiewics, *J. Mol. Catal. A* **1996**, 106, 223–233.
109 O. Schmidt, DE 524 101, **1931**.
110 R. Connor, K. Folkers, H. Adkins, *J. Am. Chem. Soc.* **1932**, 54, 1138–1145.
111 H. Adkins, *Reactions of Hydrogen with Organic Compounds over Copper-Chromium Oxide and Nickel Catalysts*, University Wisconsin Press, Madison, WI, **1937**, 69.
112 R. Conner, H. Adkins, *J. Am. Chem. Soc.* **1932**, 54, 4678–4690.
113 C.L. Lautenschläger, M. Bockmühl, G. Ehrhart, W. Kross, DE 541 362, **1931**.
114 C. Montassier, J.M. Dumas, P. Granger, J. Barbier, *Appl. Catal. A* **1995**, 121, 231–244.

115 B. Casale, A.M. Gomez, US 5214219, **1993**.
116 T. Fleckenstein, G. Gerd, F.-J. Carduck, DE4302464, **1994**.
117 J. Chaminand, L. Djakovitch, P. Gallezot, P. Marion, C. Pinel, C. Rosier, *Green Chem.* **2004**, 6, 359–361.
118 B. Casale, A.M. Gomez, US 5276181, **1994**.
119 C. Montassier, J.C. Ménézo, L.C. Hoang, C. Renaud, J. Barbier, *J. Mol. Catal.* **1991**, 70, 99–110.
120 Y. Kusunoki, T. Miyazawa, K. Kunimori, K. Tomishige, *Catal. Commun.* **2005**, 6, 645–649.
121 T.A. Werpy, J.G. Frye Jr., A.H. Zacher, D.J. Miller, US2003/119952, **2003**.
122 C. Montassier, J.C. Ménézo, L. Moukolo, J. Naja, L.C. Hoang, J. Barbier, *J. Mol. Catal.* **1991**, 70, 65–84.
123 L. Schuster, M. Eggersdorfer, US 5616817, **1997**.
124 T. Che, US 4642394, **1987**.
125 G. Braca, A.M. Raspolli Galletti, G. Sbrana, *J. Organomet. Chem.* **1991**, 417, 41–49.
126 M. Schlaf, P. Ghosh, P.J. Fagan, E. Hauptman, R.M. Bullock, *Angew. Chem.* **2001**, 113, 4005–4008.
127 R.M. Bullock, P.J. Fagan, E. Hauptman, M. Schlaf, WO 0198241, **2004**.
128 R.D. Cortright, R.R. Davda, J.A. Dumesic, *Nature* **2002**, 419, 964–967.
129 R.R. Davda, J.W. Shabaker, G.W. Huber, R.D. Cortright, J.A. Dumesic, *Appl. Catal. B* **2005**, 56, 171–186.
130 G.W. Huber, J.A. Dumesic, *Catal. Today* **2006**, 111, 119–132.
131 R.D. Cortright, J.A. Dumesic, US 2003/0220531, **2003**.
132 J.W. Shabaker, G.W. Huber, J.A. Dumesic, *J. Catal.* **2004**, 222, 180–191.
133 J.W. Shabaker, G.W. Huber, J.A. Dumesic, *Catal. Lett.* **2003**, 88, 1–8.
134 S. Czernik, R. French, C. Feik, E. Chornet, *Ind. Eng. Chem. Res.* **2002**, 41, 4209–4215.
135 T. Hirai, N. Ikenaga, T. Miyake, T. Suzuki, *Energy Fuels* **2005**, 19, 1761–1762.
136 J.H. Teles, N. Rieber, W. Harder, Eur. Patent 0582201 A2.
137 F. Andreas, H. Berthold, H. Slowak, DE1075103, **1960**.
138 A.M. Kurishko, V.T. Vdovichenko, P.E. Bratchanskii, V.A. Stankevich, G.S. Shatalova, N.F. Vorona, V.V. Tsutsarin, M.S. Dyaminov, SU144838, **1962**.
139 P. Kraft, P. Gilbeau, B. Gosselin, S. Classens, FR2862644 A1, **2006**.

12
Catalytic Processes for the Selective Epoxidation of Fatty Acids: More Environmentally Benign Routes

Matteo Guidotti, Rinaldo Psaro, Maila Sgobba, and Nicoletta Ravasio

12.1
Introduction

Unsaturated vegetable oils and fats (Scheme 12.1) are among the important starting materials directly obtained from renewable sources for the production of intermediate and speciality chemicals. In the framework of a sustainable exploitation of triglycerides and fatty acid derivatives as a source of long-chain aliphatic acids, the presence of unsaturated bonds on the carbon framework of these molecules is a basic requirement to allow the selective functionalization of the alkyl chain and the production of a wide range of compounds with different physical and chemical features. For this reason, epoxidation is one of the most important reactions occurring at the C=C double bonds in unsaturated fatty compounds, as these epoxidized oleochemicals can react readily with many nucleophilic agents, leading to the opening of the strained epoxide ring. By reaction of the oxirane moiety with water, alcohols, carboxylic acids, amines or amides, it is possible to obtain diols, alkoxyalcohols, hydroxyesters, aminoalcohols or hydroxyalkylamides derivatives, respectively (Scheme 12.2); these compounds are of increasing interest for industrial applications in several fields [1–4].

In particular, epoxidized triglycerides and the esters derived from them have been used since the 1950s as plasticizers and additives for PVC polymers due to the excellent heat and light stability imparted by the oxiranes [5–7]. Furthermore, they are employed as reactive diluents for paints, as crosslinkers in environmentally friendly solvent-free powder coatings and they can be used as building blocks for the preparation of binders based on renewables, thanks to their good drying and curing properties [8, 9]. Fatty alkoxy alcohols and related polyols are suitable components in polyurethane foams and molding resins [10–13]. Then, other epoxide-related derivatives are utilized in formulations for lubricants, cosmetics, wood impregnants, pressure-sensitive adhesives or for biochemical applications [14–18].

Catalysis for Renewables: From Feedstock to Energy Production
Edited by Gabriele Centi and Rutger A. van Santen
Copyright © 2007 WILEY-VCH Verlag GmbH & Co. KGaA, Weinheim
ISBN: 978-3-527-31788-2

12 Catalytic Processes for the Selective Epoxidation of Fatty Acids

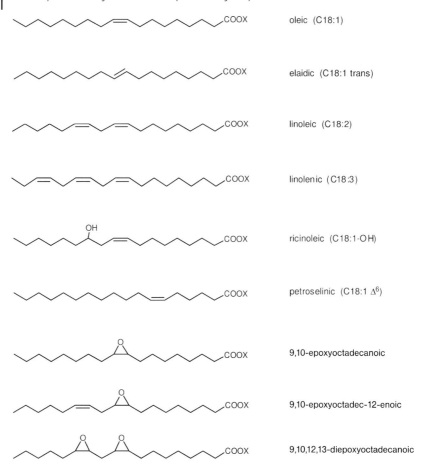

Scheme 12.1 Unsaturated fatty acids (X = H), fatty acid methyl esters (X = CH$_3$) and related epoxy derivatives. Short notation: C xx:y, where xx is the number of carbon atoms of the fatty acid and y is the number of unsaturations; Δ^n, where n is the position of the unsaturation.

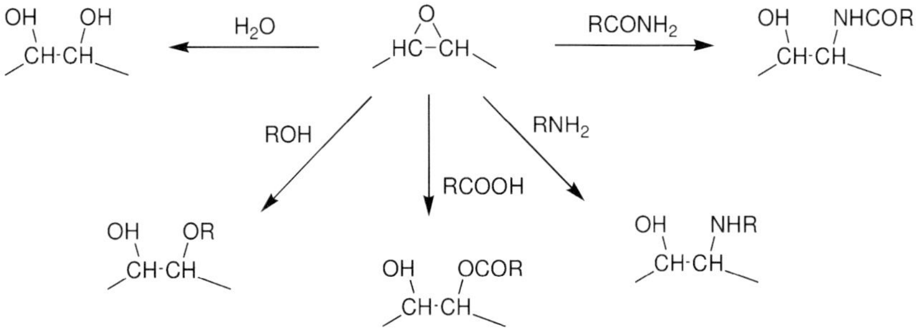

Scheme 12.2 Reactivity of the epoxy group of a fatty-acid derivative.

12.2
Non-catalytic Epoxidation Systems

Currently, the most important fat processed for these purposes is epoxidized soya-bean oil – its total worldwide production is around 2×10^5 t a^{-1} [19]. The epoxidation of such large amounts of raw materials is carried out industrially with a non-catalytic pathway: the so-called Prilezhaev epoxidation process, in which the unsaturated fatty compound reacts with a peroxycarboxylic acid (typically peracetic or performic acid) obtained through the acid-catalyzed oxidation of the corresponding acid with hydrogen peroxide. To avoid the handling of hazardous species, the peracid is generally formed *in situ* for large-scale epoxidations. However, soluble mineral acids (such as sulfuric acid) are essential for peracid formation and the presence of these strongly acidic components leads to several drawbacks: (a) side reactions, due to acid-catalyzed epoxide ring opening; (b) difficult separation of acidic by-products, whose presence may be detrimental for further applications; (c) cumbersome disposal of the salts formed during neutralization; and (d) corrosion problems caused by the strong acids in an oxidizing environment. For these reasons, selectivity to the desired epoxidized species hardly exceeds 80% in industrial plants. Evidently, the conventional Prilezhaev method takes a lot away from the green chemistry credentials of the final products. Indeed, the reagents are dangerous to handle and the process generates large amounts of wastes.

Analogously, the use of *m*-chloroperbenzoic acid (MCPBA) or the *in situ* formation of peroxy acids by means of a sacrificial aldehyde (typically benzaldehyde plus molecular oxygen as a source of perbenzoic acid [20, 21]) are not viable processes, from an industrial point of view, because of the very poor atom efficiency and the huge formation of benzoic acids as side products.

Dioxiranes are practical epoxidizing agents that can be easily obtained at a laboratory scale from peroxomonosulfates and ketones. Dioxiranes can be either prepared directly *in situ* or used as storable ketone solutions (widely known with the trade mark Oxone, i.e., $2KHSO_5 \cdot KHSO_4 \cdot K_2SO_4$); these reagents have been successfully applied to the epoxidation of unsaturated fatty acid methyl esters (FAMEs) or triglycerides [22–25]. Very high yields in epoxidized products (81–96% with triglycerides [23] or 85–99% with methyl esters [24]) have been achieved and the formation of undesired side products is minimized even for conjugated di-unsaturated or unsaturated oxygenated fatty esters. Nevertheless, the very low content of active oxygen in the salt mixture (ca. 5 wt.%) and the need for a remarkable excess of monopersulfate give rise to unsustainable amounts of inorganic wastes to be disposed of (often over six times the weight of the product).

Such environmental concerns and technical problems (having major consequences on the overall cost of the production) prompted many research teams to look for alternative synthetic pathways that rely on catalytic processes and more sustainable technologies.

12.3
Homogeneous Catalytic Systems

After an early attempt to use iron complexes for the conversion of methyl oleate and methyl elaidate [26], the two most notable homogeneous systems reported so far in the epoxidation of fatty derivatives with hydrogen peroxide are based on tungsten and rhenium. The heteropoly acid $H_3PW_{12}O_{40}$ in a two-phase system under phase-transfer conditions showed excellent yields in the epoxidation of both methyl oleate and methyl linoleate, with notable selectivity to the epoxide in the latter case [19]. Analogously, a peroxo polyoxometalate (POM) $[PO_4\{WO(O_2)_2\}_4]^{3-}$ displayed high yields (up to 84%) for the phase-transfer epoxidation of oleic acid, without using halogenated solvents [27].

Likewise, methyltrioxorhenium (CH_3ReO_3) is a widely used catalyst for selective epoxidation; it has been applied mainly to the conversion of fatty methyl esters and triglycerides [19, 28–30]. With oleic acid derivatives, good yields to epoxidized products are obtained (e.g., 70% with methyl oleate, which can be raised up to 88% by addition of bipyridine to reduce acid-catalyzed by-products [19]) and, with polyunsaturated substrates, interesting yields of triepoxides (55% [19]) or diepoxides (ca. 40% [31]) were recorded from linseed oil methyl ester or methyl linoleate, respectively. The reaction proceeds under mild conditions and, over dienic compounds, the formation of di-epoxides occurs in the early stage of the reaction, well before the total consumption of methyl linoleate. So, in optimized studies for epoxidizing soya-bean oil, total double-bond conversion and 95% selectivity were obtained in 2 h at room temperature [30]. Nonetheless, the main drawbacks of CH_3ReO_3-based reactions are the difficult catalyst separation and recovery and use of nitrogen-containing bases in the reaction mixture, which lowers the eco-sustainability of a large-scale process.

Alternatively, organic hydroperoxides (typically, *tert*-butyl hydroperoxide, TBHP, or cumyl hydroperoxide, CHP) can be used as the oxidizing agents – they are efficiently obtained by auto-oxidation of branched alkanes with molecular oxygen. These oxidants need homogeneous catalytic systems based on titanium, vanadium or molybdenum to perform the oxygen transfer from the hydroperoxide to the alkene [25, 32, 33]. By using, for instance, CHP with $MoO_2(acac)_2$, soya-bean oil has been converted into the epoxidized derivative in less than 1 h with a yield of about 60% [34]. Similarly, the epoxidation with TBHP of oleic acid over a molybdenum complex showed 87% selectivity to 9,10-epoxystearic acid at 67% conversion [35]. Also in these cases, however, the difficult recovery of the metal catalyst from the reaction mixture is a major disadvantage to an easy scale-up of the process.

12.4
Chemoenzymatic Epoxidation Systems

With respect to the methods described above, chemoenzymatic epoxidation gained much interest as it does not show any undesired ring opening of the oxiranes. The reaction takes place schematically in two steps: first, the unsaturated fatty acid (or ester) is converted into an unsaturated peroxycarboxylic acid by reaction of the enzyme with hydrogen peroxide; then, it is epoxidized by an essentially intermolecular pathway without the participation of the enzyme (Scheme 12.3). Such a chemoenzymatic method is particularly suitable for the epoxidation of oleochemicals because the necessary carboxyl group and the unsaturation are both in the same reactant molecule. Novozym 435, a *Candida antarctica* lipase B immobilized on a polyacrylic polymer, is one of the most efficient and stable catalyst for this purpose. With this, yields in the range 72–91% for vegetable free fatty acids with internal C=C bonds (including ricinoleic acid) were obtained with excellent selectivity (>98%) in all cases [36]. When triglycerides are used, the addition of 5 mol.% of free fatty acid is necessary to favor the formation of the intermediate peroxy acid and to prevent the side-production of mono- and diglycerides [37, 38]. More recently, methods using immobilized oat seed (*Avena sativa*) peroxygenase were reported and, in this case, TBHP was employed as oxidant instead of hydrogen peroxide, as the latter leads to peroxygenase deactivation, hence precluding high product formation [39, 40].

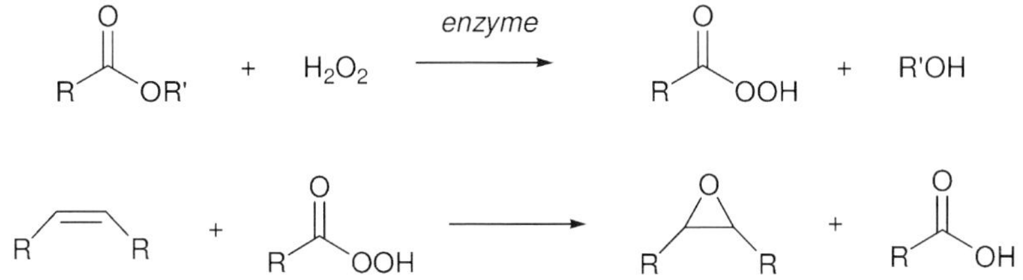

Scheme 12.3 Chemoenzymatic generation of the peroxycarboxylic acid and subsequent epoxidation of the fatty material.

The immobilized enzymes can, therefore, be easily separated by filtration and they are recyclable several times without loss of activity. Actually, total turnovers as high as 2×10^5 moles of product per mole of catalyst have been recorded. Nevertheless, as a drawback, it is worth highlighting that these systems are very sensitive to the kind of substrate employed and hence they can be unsuitable and less versatile for some particular oleochemical applications. For instance, fatty compounds with the C=C in the trans configuration, such as elaidic acid, react very slowly, and the epoxidation of polyunsaturated oleochemicals proceeds step by step. This means that the formation of di-epoxides takes place only after the nearly complete formation of monoepoxides.

12.5
Heterogeneous Catalytic Systems

In the challenge to develop more environmentally benign routes to useful epoxidized fatty derivatives, the exploitation of heterogeneous catalysts is one of the most attractive choices, which could deal with most of the above-mentioned disadvantages. More precisely, the ideal system should possess not only excellent performance but also should be: (a) stable, robust and recyclable, (b) cheap and easy to prepare, (c) versatile, so as to be used with many substrates (also with multifunctionalized fatty derivatives), (d) able to work in a reaction medium free from acidic compounds, and (e) in accordance with the green chemistry guidelines.

The first examples of efficient heterogeneous catalysts were based on molybdenum. Fatty ethyl esters, such as oleate, erucate and linoleate, were epoxidized with 30% aqueous hydrogen peroxide over a molybdenum oxide–tributyltin chloride on charcoal catalyst to give yields of epoxides of 76, 77 and 56%, respectively, after 15 h at 323 K [41]. Several vegetable oils, i.e., rapeseed, olive, soya-bean, corn, cottonseed and linseed oils, were also oxidized, with oxirane contents of 3.5–5.3% [41]. In contrast, a commercial 10% MoO_3-Al_2O_3 catalyst converted various methyl esters (oleate, elaidate, petroselinate, erucate and ricinoleate) into epoxides with yields and selectivities >95% at 388 K after 3–9 h, using TBHP and CHP as oxidants [42]. Unfortunately, these reports did not examine the issue of metal leaching and recyclability of the catalyst.

More recently, solid catalysts have been prepared by heterogenization of well-studied and defined homogeneous systems. A silica gel-supported ruthenium complex prepared by simple impregnation proved to be effective not only in the oxidation of alcohols but also in the epoxidation with TBHP of alkenes, including with methyl oleate (with 67% yield) [43]. Likewise, oleic, linoleic and linolenic acids (as well as their methyl esters) were epoxidized readily with a urea–hydrogen peroxide adduct over CH_3ReO_3 supported on Nb_2O_5 [44, 45]. The yields were very high (>90%) and at 323 K the reaction times were very short (10–30 min) [44]. Over such a catalyst, the reaction rate becomes slower with increasing unsaturation (oleate > linoleate > linolenate) [45]. The main weaknesses of this system are due to the large excess of oxidant used (as high as 4× mol mol^{-1}) and to the extensive use of chloroform as solvent. Nevertheless, in one test, methyl oleate was completely epoxidized under solvent-free conditions in 10 min [44].

As far as porous solids are concerned, the use of redox-active molecular sieves for the epoxidation of oleochemicals has been hindered by the lack, during the 1980s and early 1990s, of stable wide-pore micro- and mesoporous materials that could accommodate the bulky substrate molecules. Then, in 1997, Corma and coworkers tested the epoxidation with both H_2O_2 and TBHP of methyl oleate over titanium-containing beta zeolite (Ti-BEA) and MCM-41 material (Ti-MCM-41) [46]. With hydrogen peroxide, Ti-BEA showed better yields than Ti-MCM-41 (44% vs. 24%) and acetonitrile, thanks to its weak basicity, was a better solvent than methanol, giving rise to less acid-catalyzed by-products. Conversely, with TBHP, excellent epoxide selectivities were obtained and higher yields were recorded over Ti-MCM-41 than over Ti-BEA (59% vs. 48%). In addition, the role of aluminum ions in the zeolite framework and of the hydrophilic character of the solid were also considered and it was concluded that non-acidic (i.e., Al-free) and moderately hydrophobic Ti-BEA and Ti-MCM-41 were the suitable catalysts for obtaining high conversion and selectivity to fatty epoxides [46, 47]. Subsequently, optimization of the experimental conditions, all focused on methyl oleate as substrate, led to improved performances over microporous (Ti-ZSM-5, Ti-BEA) or mesoporous (Ti-MCM-41, Ti-HMS) titanosilicates [48]. For instance, epoxidation of methyl oleate with 35% H_2O_2 over Ti-BEA at 343 K gave, in methanol solvent, the corresponding methyl glycol esters with 98% selectivity and a H_2O_2 conversion of 90%, whereas in acetonitrile it gave the 9,10-epoxy product with 93% selectivity and a H_2O_2 conversion of 96%. In additional examples with mesoporous Ti-MCM-41, very high selectivity to epoxide (up to 100%) was recorded at remarkable TBHP conversion (>90%) [48].

The first case of epoxidation over titanosilicates of FAMEs directly obtained from vegetable oils was reported in 2003; a mixture of high-oleic sunflower oil methyl esters was epoxidized with TBHP over titanium-containing mesoporous silicas of different morphology, giving rise to complete conversions (up to 98%) and high selectivity to monoepoxides (>85%) [49]. In addition, both methyl cis- and trans-octadecenoate (namely methyl oleate and methyl elaidate) were converted in high yields over the ordered Ti-MCM-41 (see next section for details). The epoxidation of soya-bean oil and soya-bean methyl esters was then performed with hydrogen peroxide in dilute solution (6 wt.%) using a non-ordered heterogeneous Ti-SiO$_2$ catalyst in the presence of TBHP [50]. The conversion profiles of both the triglycerides and the derived FAME followed the same behavior, showing that the epoxidation mechanism is not largely affected by the species at the ester group. High yields (up to 88%) of epoxidized derivatives were achieved with moderate excess of oxidant (H_2O_2:substrate molar ratio of 1.1), even if after rather long reaction times (>54 h). It is also worth noting the absence of unwanted side reactions and the negligible decomposition of hydrogen peroxide [50]. Later, another example of epoxidation of methyl oleate over mesoporous titanosilicates with different porosimetric features was reported by Hoelderich and coworkers [51, 52]. All the solids displayed good stability as well as re-usability and, according to the authors, under the conditions tested, the structure of the mesoporous catalyst was not observed to play an important role in the activity. However, from a synthetic point of view, the epoxide was obtained with a 57% yield over the ordered Ti-MCM-41, whereas only 36% yield was achieved over the non-ordered Ti-SiO$_2$ [51]. Finally, titanium-containing silicas were applied to the epoxidation of a series of mixtures of FAMEs derived from vegetable sources, such as high-oleic sunflower, castor, coriander and soya-bean oils [53]. The influence of the nature and the position of functional groups on the C_{18} chain of the FAMEs was studied and very high activity and selectivity were obtained mainly in the epoxidation of castor and soya-bean oil FAMEs (more details are given in the next section). The ordered mesoporous Ti-MCM-41 showed in this case, for the first time, superior performances with respect to non-ordered mesoporous titanosilicates [53].

Taking into account the above-mentioned considerations, a heterogeneous system based on a titanium-containing solid could be considered as an attractive method for the epoxidation of oleochemicals. The need for low oxidant to substrate ratios, the stability and easy recoverability of the catalysts are all advantages shown by these catalysts with respect to conventional non-catalytic processes. Furthermore, the use of TBHP as oxidizing agent could give rise to fewer difficulties for large-scale applications, with respect to the use of hydrogen peroxide. In fact, even though the latter fulfils some green chemistry guidelines in terms of atom efficiency and lack of noxious side products, the former is safer to handle (even in concentrated solutions [32]), rather soluble in many organic solvents as well as in the oleochemicals themselves (hence avoiding the need for phase-transfer co-catalysts) and it can be used under anhydrous and acid-free conditions (minimizing the formation of undesired epoxide hydrolysis products). Notably, if TBHP were produced by auto-oxidation of *iso*-butane with oxygen (as in the Halcon process for the epoxidation of propylene to propylene oxide) and *tert*-butanol were exploited as a valuable side-product, the primary oxidant of the overall process would be molecular oxygen, hence showing an optimal global atom economy.

For such reasons, the following section considers in more detail some of the most significant results obtained by our team on the epoxidation with TBHP of unsaturated FAMEs over mesoporous titanium-grafted silicates. In these examples, the epoxidation tests were carried out either in ethyl acetate, which could be even obtained, in principle, from renewable sources and which is relatively less harmful than other polar non-protic solvents, or under solvent-free conditions.

12.6
Epoxidation of FAMEs Over Titanium-based Catalysts: The Skills in Milan

Various titanium-containing silicas, which had already shown interesting results in the epoxidation of unsaturated alcoholic terpenes [54–59], have been studied, but two solids have displayed peculiar behavior for use over fatty acid esters: Ti-MCM-41 and Ti-SiO$_2$. They are obtained from MCM-41 and commercial Grace Davison SiO$_2$, respectively, by applying and adapting the simple grafting methodology described by Maschmeyer and coworkers [60] to mesoporous silica materials of various origins. Ti-MCM-41 is a mesoporous catalyst (pore diameter ca. 2.5 nm) with an ordered network of cylindrical and parallel pores and with a very high specific surface area (ca. 950 m^2 g^{-1}), whereas Ti-SiO$_2$ is a mesoporous catalyst (pore diameter ca. 13 nm) with a non-ordered pore system and with an intermediate specific surface area (ca. 300 m^2 g^{-1}). Titanium active centers are added onto the siliceous support by grafting titanocene dichloride precursors in the presence of triethylamine (Scheme 12.4). Subsequent calcination at high temperature (773 K) leads to the active catalyst with typical metal loadings of about 2 wt.% [57]. Thanks to the grafting methodology, the titanium sites are all located on the surface of the catalysts and no atoms are buried (at least in principle) within the silicate walls of the porous solids. Such better exposition allows a superior performance of the grafted solids with respect to in-matrix ones (tita-

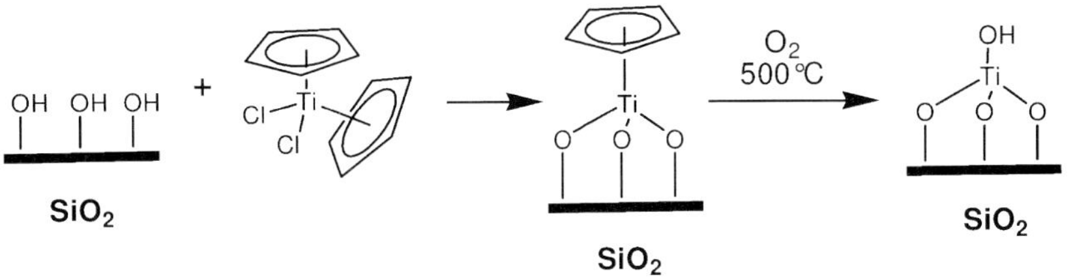

Scheme 12.4 Grafting of titanocene dichloride onto siliceous supports and subsequent formation of active titanium centers.

nium loading being equal) [54]. Additionally, the titanium loading can be easily changed and tuned according to the needs, without changing the original silica support, which can be either specifically synthesized for the purpose or is commercially available. All the titanium-grafted catalysts proved to be mechanically robust and stable towards metal leaching under the epoxidation conditions [53, 57, 59].

12.6.1
Epoxidation of Pure C_{18} Monounsaturated FAMEs

The ability of titanium-grafted silicas in catalyzing the epoxidation with TBHP of fatty compounds was first tested on two pure C_{18} monounsaturated FAMEs: methyl oleate (cis-9-octadecenoate; Scheme 12.1) and methyl elaidate (trans-9-octadecenoate) [49]. In both cases, selectivity to 9,10-epoxystearate was very high (>95%) and the reaction was fully stereospecific, confirming that epoxidation with titanium catalysts and TBHP proceeds via a non-radical mechanism with retention of configuration at the C=C bond. Ti-MCM-41 was more active than Ti-SiO$_2$ (Fig. 12.1). Actually, methyl oleate was almost completely converted after

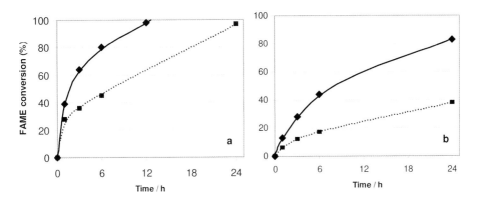

Fig. 12.1 Conversion of fatty acid methyl esters (FAMEs) methyl oleate (♦) and methyl elaidate (■) vs. reaction time over Ti-MCM-41 (a) and over Ti-SiO$_2$ (b). (Adapted from [49]).

Table 12.1 Catalytic performances in the epoxidation of methyl oleate (cis) and methyl elaidate (trans). (Adapted from [49]).

Catalyst	$S^{a)}$ (%)	$A^{b)}$ (h^{-1}) on substrate		A cis to A trans ratio
		cis	trans	
Ti-MCM-41	>98	56	40	1.4
Ti-SiO$_2$	>98	20	9	2.2

Reaction conditions: ethyl acetate solvent; 363 K; TBHP/FAME molar ratio = 1.33; FAME/catalyst molar ratio = 160.
a) Selectivity to monoepoxides.
b) Specific activity ([mol converted C=C] [mol Ti]$^{-1}$ h^{-1}) after 1-h reaction.

12 h and methyl elaidate in 24 h. The better reactivity of oleate is attributable to steric effects, since the approach to the catalyst surface for cis C=C bonds is less hindered than for trans C=C bonds. Nevertheless, it is worth highlighting the good performance in the epoxidation of trans fatty derivatives, as most the promising catalytic systems, such as the immobilized chemoenzymatic ones, are poorly effective in the conversion of unsaturated FAMEs in the trans configuration. Furthermore, by comparing the specific activity (A) values in the epoxidation of methyl oleate and methyl elaidate (Table 12.1), the gap between the two activities is, evidently, minimal for Ti-MCM-41 (cis/trans A ratio = 1.4) with respect to other porous catalysts (the value typically ranges from 2 to 2.5 [41, 42]) or to non-porous catalysts (e.g., pyrogenic titanium-grafted Aerosil silica, with a value of 4.3 [49]). This, the first record of such a small difference in the epoxidation of the two FAMEs, occurs only over Ti-MCM-41, suggesting that the peculiar morphology of this mesoporous material, with its parallel cylindrical-shaped channels, plays a role in influencing the relative reactivity of the two monoenes. Actually, under these conditions, no shape-selectivity effects could be taken into account, because Ti-MCM-41 possesses mesopores that are too wide to act as sieves for FAMEs. However, thanks to a delicate balance of different steric and/or polar interactions involving the substrate molecule and the ordered porous catalyst, the gap between the epoxidation rates of the trans isomer (elaidate) and of the cis isomer (oleate) is noticeably reduced [49].

12.6.2
Epoxidation of a Mixture of FAMEs from Vegetable Sources

12.6.2.1 HO Sunflower-, Coriander-, and Castor-oil FAME Mixtures

The titanium-containing silicas were also employed in the epoxidation of a series of FAME mixtures obtained directly from vegetable oils, namely high-oleic (HO) sunflower oil, coriander oil, castor oil and soya-bean oil, by esterification with so-

Table 12.2 Composition of the fatty acid methyl ester (FAME) mixtures.

Fatty acid		Composition (wt.%)			
		HO sunflower	Coriander	Castor	Soya-bean
Palmitate	C16:0	2	3	2	13
Stearate	C18:0	4	–	1	4
Oleate	C18:1	84	31	6	19
Linoleate	C18:2	10	13	3	56
Linolenate	C18:3	–	–	–	5
Behenate	C22:0	1	–	–	1
Ricinoleate	C18:1–OH	–	–	87	–
Petroselinate	C18:1 Δ^6	–	52	–	–
Others		–	1	1	2

dium ethoxide and subsequent distillation [53]. Since each of the four vegetable oils is rich in one particular C_{18} fatty acid, i.e., high-oleic (HO) sunflower oil in oleic acid, coriander oil in petroselinic acid, castor oil in ricinoleic acid and soyabean oil in linoleic acid (Table 12.2 and Scheme 12.1), the different FAME mixtures were taken as a model to study the molecular interaction between the C=C bond and the titanium active site as well towards understanding how the position of the unsaturations and of other functional groups on the carbon chain could affect the catalytic features.

With the mixture obtained from HO sunflower oil, high yields of methyl 9,10-epoxystearate were obtained after 24 h over both Ti-MCM-41 (70%) and Ti-SiO$_2$ (61%) (Table 12.3). The reaction was carried out with a low excess of oxidizing agent (only 33% molar excess per C=C bond) and the relatively lower selectivity observed over Ti-MCM-41 (85%) is only apparent, as this value refers to the sole monoepoxy-derivatives and does not account for the remarkable amount of diepoxystearate (ca. 8%) detected in the final mixture and derived from the 10 wt.% of methyl linoleate present in the starting mixture. Conversely, epoxidation of the mixture obtained from coriander oil saw a general drastic loss in initial activity (Table 12.3). This is because methyl petroselinate (the major component of coriander methyl ester, with the unsaturation located at the C-6 position) is less easily converted than methyl oleate [42, 53, 61]. However notably, after a reaction time of 24 h, interesting yields in monoepoxy-derivatives can be obtained (i.e., 85% and 57% over Ti-MCM-41 and Ti-SiO$_2$, respectively), notwithstanding the intrinsic low reactivity of the substrate. With the castor oil FAME mixture, whose major component is methyl ricinoleate (87%), Ti-MCM-41 showed an outstandingly high initial activity with respect to Ti-SiO$_2$, and all the unsaturated constituents of the mixture were completely and selectively converted after 24 h (Table 12.3). The dramatic increase in the epoxidation rate over Ti-MCM-41 has to be attributed to the presence of a hydroxyl group in homoallylic position, which improves and assists the oxygen transfer from the oxidant to the C=C bond. This positive

Table 12.3 Catalytic performances of titanium-grafted silicates in the epoxidation of vegetable FAME mixtures. (Adapted from [53]).

FAME mixture	Ti-MCM-41 A (h^{-1})[a]	Conversion (%)[b]	Selectivity (%)[c]	Ti-SiO$_2$ A (h^{-1})[a]	Conversion (%)[b]	Selectivity (%)[c]
HO sunflower	71	98	85	44	76	94
Coriander	26	94	91	12	62	94
Castor	130	97	>98	30	76	97

Reaction conditions: ethyl acetate solvent; 363 K; solvent/FAME volume ratio = 8; TBHP/FAME molar ratio = 1.33; FAME/catalyst molar ratio = 160.
[a] Specific activity ([mol converted C=C] [mol Ti]$^{-1}$ h^{-1}) after 1-h reaction.
[b] Conversion of unsaturated FAMEs after 24 h.
[c] Selectivity to mono-epoxides after 24 h.

effect, already known with homogeneous catalysts [32], has been thoroughly studied by our team in the epoxidation of unsaturated terpenes over heterogeneous titanosilicates [57, 59]; it is particularly enhanced by the peculiar morphology of Ti-MCM-41, which leads to a better interaction between the hydroxyl moiety of methyl ricinoleate and the active centers on the catalyst surface [53].

Because of the relevant results obtained in ethyl acetate, Ti-MCM-41 was tested in the epoxidation of mixture of FAMEs under solvent-free conditions (Table 12.4). High conversion (particularly of castor oil FAMEs) and very high selectivity values were achieved, albeit the oxidizing agent was added in deficit with respect to the unsaturated substrates. Thus, thanks to the use of no solvent, and no acid reactants at all, and to the simple removal of the solid catalyst by filtration,

Table 12.4 Catalytic performances of Ti-MCM-41 in the epoxidation of vegetable FAME mixtures under solvent-free conditions.

FAME mixture	C 1-h (%)[a]	C 24-h (%)[b]	S 24-h (%)[c]
HO sunflower	49	84	90
Coriander	18	62	87
Castor	51	91	98

Reaction conditions: 363 K; TBHP/FAME molar ratio = 0.60; FAME/catalyst molar ratio = 900.
[a] Conversion of unsaturated FAMEs with respect to maximum possible conversion after 1 h.
[b] Conversion of unsaturated FAMEs with respect to maximum possible conversion after 24 h.
[c] Selectivity to monoepoxides after 24 h.

mixtures of epoxidized fatty derivatives with a rather low level of undesired by-products can be easily obtained.

12.6.2.2 Soya-bean Oil FAME Mixture

The epoxidation of mixtures containing large percentages of diunsaturated FAMEs deserves greater attention because of the wide variety of epoxidized compounds that can be obtained. In the epoxidation of the soya-bean FAME mixture, Ti-MCM-41 showed a specific activity that is 2.6× higher than Ti-SiO$_2$ (96 vs. 37 h^{-1}) [53]. Furthermore, with the latter, conversions as high as 90% were recorded after 24 h, whereas, with the former, values not higher than 54% were obtained after the same time. Such remarkable activity of the ordered mesoporous material with respect to the non-porous one in the epoxidation of methyl linoleate is attributed to the concurrence of two main favorable factors: not only the presence of large amounts of accessible and well-defined tetrahedral titanium sites but also the high density of silanol groups surrounding the titanium sites, which accounts for the enhanced formation of the second oxirane ring when the first one is already present on the FAME molecule [53]. This hypothesis has been confirmed by tests carried out over a sample of completely silylated Ti-MCM-41. When all the silanol groups were removed by silylation, over non-silylated Ti-MCM-41 large amounts of di-epoxy derivative were obtained, whereas, at the same conversion values, over the completely silylated Ti-MCM-41 the monoepoxy fatty esters were the major products [62]. So, although excessive hydrophilic character is often detrimental for the epoxidation activity of a porous catalyst on hydrophobic substrates [46, 63], a polar environment surrounding the titanium sites may have positive effects on the conversion of polyfunctionalized substrates such as oleochemicals.

The difference between the two catalysts is even more manifest in the profiles of the yield of epoxidized compounds after 24 h (Fig. 12.2). Only over Ti-MCM-41

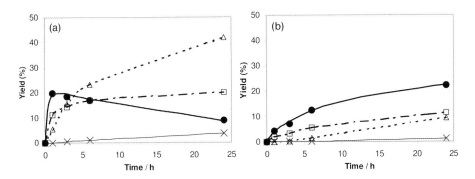

Fig. 12.2 Yield of epoxidized FAMEs vs. time obtained by epoxidation of a soya-bean oil FAME mixture over Ti-MCM-41 (a) and over Ti-SiO$_2$ (b). Methyl monoepoxyoctadecenoate (●), methyl 9,10,12,13-diepoxy-octadecanoate (△), methyl monoepoxyoctadecanoate (□), others (×).

are the monoepoxides, formed at the early stages of the reaction, rapidly converted into di-epoxides and, in fact, methyl 9,10,12,13-diepoxyoctadecanoate is the main component of the reaction mixture after 24 h (42% yield). Conversely, over Ti-SiO$_2$ the formation of doubly epoxidized FAMEs is slower and the methyl monoepoxyoctadecenoate isomers are the major components of the mixture (23% yield). These results are worthy of consideration as, over many catalytic systems, methyl linoleate is reported to be less easily epoxidized than methyl oleate [40, 45]. In contrast, over this kind of Ti-MCM-41, a FAME mixture derived from soya-bean was more readily converted than that obtained from HO sunflower oil (96 vs. 71 h^{-1} specific activity) [53]. Finally, all the catalysts can be reused and recycled, provided they are washed and reactivated in air at 823 K, the calcination being essential to restore catalytic activity and to remove organic deposits adsorbed on the solids.

12.7
Conclusions

Even though no catalytic process has achieved industrial significance in the epoxidation of unsaturated fatty acid derivatives so far, several recent results have opened the way to large-scale exploitation of new catalysis-based sustainable processes to be applied to the transformation of oleochemicals. The chemoenzymatic epoxidation over lipase heterogenized catalysts as well as the catalytic epoxidation over titanium-containing solids seem to be the most promising systems – each of them proved to be suitable for different potential applications and for distinct scale-up volumes. In particular, titanium-containing silicas were shown to be efficient heterogeneous catalysts for the epoxidation of a large variety of oils and fatty derivatives under conditions that are by far more environmentally friendly than those used in the conventional peroxy acid process. Considering indeed that the catalyst can be easily filtered and reused, that less a harmful oxidant such as hydrogen peroxide or TBHP can be utilized, and that various compounds derived from renewable sources can be used as reactants, this process should be judged as a viable and sustainable alternative pathway to long-chain epoxides. For some specific purposes, ordered mesoporous solids, such as Ti-MCM-41, showed unique features in terms of activity and selectivity. Nonetheless, interesting performances were also observed employing non-ordered mesoporous catalysts prepared by means of easy, cheap synthesis methodologies.

References

1 H. Baumann, M. Bühler, H. Fochem, F. Hirsinger, H. Zoebelein, J. Falbe, *Angew. Chem. Int. Ed.*, 27 (**1988**) 41.

2 U. Biermann, W. Friedt, S. Lang, W. Lühs, G. Machmüller, J. O. Metzger, M. Rüsch gen. Klaas, H. J. Schäfer, M. P. Schneider, *Angew. Chem. Int. Ed.*, 39 (**2000**) 2206.

3 S. Warwel, F. Brüse, M. Kunz, *Fresenius Environ. Bull.*, 12 (**2003**) 534.

4 X. Pagès-Xatart-Parès, C. Bonnet, O. Morin, in: *Recent Developments in the Synthesis of Fatty Acid Derivatives*, G. Knothe, J. T. P. Derksen (Eds.), AOCS Press, Champaign, IL, (**1999**), p. 141.

5 F. P. Greenspan, R. J. Gall, *Ind. Eng. Chem.*, 45 (**1953**) 2722.

6 L. H. Gan, K. S. Ooi, S. H. Goh, L. M. Gan, Y. C. Leong, *Eur. Polym. J.*, 31 (**1995**) 719.

7 H. Naito, H. Naito, WO 9927009 (**1999**).

8 G. J. H. Buisman, A. Overeem, F. P. Cuperus, in: *Recent Developments in the Synthesis of Fatty Acid Derivatives*, G. Knothe, J. T. P. Derksen (Eds.), AOCS Press, Champaign, IL, (**1999**), p. 128.

9 J. O. Metzger, *Chemosphere*, 43 (**2001**) 83.

10 M. Skwiercz, C. Priebe, K. Boege, DE 19834048 (**2000**).

11 H. Kluth, P. Daute, J. Klein, R. Gruetzmacher, W. Klauck, DE 4202758 (**1993**).

12 J. V. Crivello, R. Narayan, *Chem. Mater.*, 4 (**1992**) 692.

13 G. Lligadas, J. C. Ronda, M. Galià, U. Biermann, J. O. Metzger, *J. Polym. Sci. A: Polym. Chem.*, 44 (**2006**) 634.

14 G. Bert, DE 4201343 (**1993**).

15 X. Wu, X. Zhang, S. Yang, H. Chen., D. Wang, *J. Am. Oil Chem. Soc.*, 77 (**2000**) 561.

16 M. Niki, K. Otani, JP 02206613 (**1989**).

17 S. P. Bunker, R. P. Wool, *J. Polym. Sci. A: Polym. Chem.*, 40 (**2002**) 451.

18 S. Grinberg, C. Linder, V. Kolot, T. Waner, Z. Wiesman, E. Shaubi, E. Heldman, *Langmuir*, 21 (**2005**) 7638.

19 M. Rüsch gen. Klaas, S. Warwel, in: *Recent Developments in the Synthesis of Fatty Acid Derivatives*, G. Knothe, J. T. P. Derksen (Eds.), AOCS Press, Champaign, IL, (**1999**), p. 157.

20 M. C. Kuo, T. C. Chu, *Ind. Eng. Chem. Res.*, 26 (**1987**) 277.

21 M. C. Kuo, T. C. Chu, *Can. J. Chem. Eng.*, 68 (**1990**) 831.

22 P. E. Sonnet, M. E. Lankin, G. P. McNeill, *J. Am. Oil Chem. Soc.*, 72 (**1995**) 199.

23 P. E. Sonnet, T. A. Foglia, *J. Am. Oil Chem. Soc.*, 73 (**1996**) 461.

24 M. S. F. Lie Kien Jie, M. K. Pasha, *Lipids*, 33 (**1998**) 633.

25 T. A. Foglia, P. E. Sonnet, A. Nunez, R. L. Dudley, *J. Am. Oil Chem. Soc.*, 75 (**1998**) 601.

26 T. Yamamoto, M. Kimura, *J. Chem. Soc., Chem. Commun.*, (**1977**) 948.

27 I. V. Kozhenikov, G. P. Mulder, M. C. Steverink-de Zoete, M. G. Oostwal, *J. Mol. Catal. A: Chem.*, 134 (**1998**) 223.

28 W. A. Herrmann, D. Marz, W. Wagner, J. G. Kuchler, G. Weichselbaumer, R. Fischer, DE 3902357 A1 (**1989**).

29 D. W. Marks, R. C. Larock, *J. Am. Oil Chem. Soc.*, 79 (**2002**) 65.

30 A. E. Gerbase, J. R. Gregorio, M. Martinelli, M. C. Brazil, A. N. F. Mendes, *J. Am. Oil Chem. Soc.*, 79 (**2002**) 179.

31 G. Du, A. Tekin, E. G. Hammond, L. K. Woo, *J. Am. Oil Chem. Soc.*, 81 (**2004**) 477.

32 K. B. Sharpless, T. R. Verhoeven, *Aldrichim. Acta*, 12 (**1979**) 63.

33 M. Nor Bin Omar, R. J. Hamilton, H. A. Moynihan, *Arkivoc*, (7) (**2003**) 190.

34 P. J. Martinez de la Cuesta, E. Rus Martinez, L. M. Cotoluero Minguez, *Grasas Aceites*, 36 (**1985**) 181.

35 J. M. Sobczak, J. J. Ziolkowski, *Appl. Catal. A: General*, 248 (**2003**) 261.

36 S. Warwel, M. Rüsch gen. Klaas, *J. Mol. Catal. B: Enzym.*, 1 (**1995**) 29.

37 M. Rüsch gen. Klaas, S. Warwel, *J. Am. Oil Chem. Soc.*, 73 (**1996**) 1453.

38 M. Rüsch gen. Klaas, S. Warwel, *Ind. Crops Prod.*, 9 (**1999**) 125.

39 G. J. Piazza, T. A. Foglia, A. Nuñez, *J. Am. Oil Chem. Soc.*, 78 (**2001**) 589.

40 G. J. Piazza, A. Nuñez, T. A. Foglia, *J. Mol. Catal. B: Enzym.*, 21 (**2003**) 143.

41 Y. Itoi, M. Inoue, S. Enomoto, *Bull. Chem. Soc. Jpn.*, 59 (**1986**) 3941.

42 A. Debal, G. Rafaralahitsimba, E. Ucciani, *Fat Sci. Technol.*, 95 (**1993**) 236.
43 W. H. Cheung, W. Y. Yu, W. P. Yip, N. Y. Zhu, C. M. Che, *J. Org. Chem.*, 67 (**2002**) 7716.
44 A. O. Bouh, J. H. Espenson, *J. Mol. Catal. A: Chem.*, 200 (**2003**) 43.
45 M. Li, J. H. Espenson, *J. Mol. Catal. A: Chem.*, 208 (**2004**) 123.
46 M. A. Camblor, A. Corma, P. Esteve, A. Martinez, S. Valencia, *Chem. Commun.*, (**1997**) 795.
47 T. Blasco, M. A. Camblor, A. Corma, P. Esteve, J. M. Guil, A. Martínez, J. A. Perdigón-Melón, S. Valencia, *J. Phys. Chem. B*, 102 (**1998**) 75.
48 A. Corma Canos, A. Martinez Feliu, P. Ciudad Esteve, WO 9856780 (**1998**).
49 M. Guidotti, N. Ravasio, R. Psaro, E. Gianotti, L. Marchese, S. Coluccia, *Green Chem.*, 5 (**2003**) 421.
50 A. Campanella, M. A. Baltanàs, M. C. Capel-Sànchez, J. M. Campos-Martìn, J. L. G. Fierro, *Green Chem.*, 6 (**2004**) 330.
51 L. A. Rios, P. Weckes, H. Schuster, W. F. Hoelderich, *J. Catal.*, 232 (**2005**) 19.
52 W. F. Hoelderich, L. A. Rios, P. P. Weckes, H. Schuster, DE 102004010802 (**2006**).
53 M. Guidotti, N. Ravasio, R. Psaro, E. Gianotti, S. Coluccia, L. Marchese, *J. Mol. Catal. A: Chem.*, 250 (**2006**) 218.
54 C. Berlini, M. Guidotti, G. Moretti, R. Psaro, N. Ravasio, *Catal. Today*, 60 (**2000**) 219.
55 M. Guidotti, G. Moretti, R. Psaro, N. Ravasio, *Chem. Commun.*, (**2000**) 1789.
56 C. Berlini, G. Ferraris, M. Guidotti, G. Moretti, R. Psaro, N. Ravasio, *Microporous Mesoporous Mater.*, 44–45 (**2001**) 595.
57 M. Guidotti, L. Conti, A. Fusi, N. Ravasio, R. Psaro, *J. Mol. Catal. A*, 182–183 (**2002**) 149.
58 A. Carati, G. Ferraris, M. Guidotti, G. Moretti, R. Psaro, C. Rizzo, *Catal. Today*, 77 (**2003**) 315.
59 M. Guidotti, N. Ravasio, R. Psaro, G. Ferraris, G. Moretti, *J. Catal.*, 214(2) (**2003**) 247.
60 T. Maschmeyer, F. Rey, G. Sankar, J. M. Thomas, *Nature*, 378 (**1995**) 159.
61 F. D. Gunstone, F. R. Jacobsberg, *Chem. Phys. Lipids*, 9 (**1972**) 26.
62 M. Guidotti, N. Ravasio, R. Psaro, M. Sgobba, *Proceedings of the 1st International IUPAC Conference on Green-Sustainable Chemistry*, Dresden, (**2006**), p. 186. ISBN 3-936028-41-9.
63 J. M. Fraile, J. I. Garcia, J. A. Mayoral, E. Vispe, *Appl. Catal. A: General*, 245 (**2003**) 363.

13
Integration of Biocatalysis with Chemocatalysis: Cascade Catalysis and Multi-step Conversions in Concert

Tom Kieboom

13.1
Overview

The development of cascade conversions, i.e., combined catalytic reactions without intermediate recovery steps, such as those taking place in living cells, is considered as one of the important future directions for carrying out sustainable organic syntheses with inherently safer design. It will drastically reduce operating time and costs as well as consumption of auxiliary chemicals and the use of energy, thus diminishing both feedstock and waste.

The general concept of cascade conversions is demonstrated by a representative selection of illustrative bio-bio, chemo-chemo, and bio-chemo catalytic examples on a laboratory scale as well as on a pilot- or industrial-scale.

Special attention is given to the integration of biocatalysis with chemocatalysis, i.e., the combined use of enzymatic with homogeneous and/or heterogeneous catalysis in cascade conversions. The complementary strength of these forms of catalysis offers novel opportunities for multi-step conversions in concert for the production of speciality chemicals and food ingredients. In particular, multi-catalytic process options for the conversion of renewable feedstock into chemicals will be discussed on the basis of several carbohydrate cascade processes that are beneficial for the environment.

Full exploitation of cascade catalysis and multi-step conversions in concert will require the development of novel, mutually compatible, organic and biosynthetic methods and procedures. Eventually, a full integration of organic synthesis and biosynthesis can be envisaged.

Syntheses routes requiring a smaller range of reaction parameters and reagents than today, but a more intricate array of efficient catalysts, might be expected. Integration with *in situ* product separation techniques will become mandatory, preferably by continuous processing.

13.2
Introduction

Unlike both the petrochemical and fermentation industries [1, 2], stoichiometric synthesis has for many years been a satisfactory tool for the preparation of molecules for application in life science industries. In recent years, however, the timely introduction of bio- and chemocatalysis has allowed organic synthesis in the fine chemical industry to meet both higher demands in molecular complexity of its products and a better process efficiency.

The development of cascade conversions and multi-step conversions in concert, defined as a sequence of chemical transformations where the product of the former reaction is the substrate for the next one, is an important future direction for carrying out sustainable organic syntheses [3, 4]. In particular, for catalytic conversions, using the full range of bio- and chemocatalytic possibilities, it will drastically reduce operating time and costs as well as the consumption of auxiliary chemicals and the use of energy, thus lowering both feedstock use and waste disposal. In this respect, special attention is given to multi-catalytic process options for the conversion of renewables into chemicals on the basis of several cascade processes using carbohydrates as feedstock [5–7].

A major aspect to be overcome in the integration of biocatalysis and chemocatalysis through cascade conversions is the lack of compatibility of the various procedures, both mutually for the many chemocatalytic reactions and between the chemocatalytic and biocatalytic conversions. This is in contrast to biocatalytic reactions, which are, by far, more mutually compatible and can be much more easily combined in a multi-step cascade, as will be shown below.

13.2.1
Human's Chemistry

Organic synthesis, the powerful chemistry developed by humankind, still often uses a simple step-by-step approach to convert a starting material **A** into a final product **D**, in which intermediate products **B** and **C** are isolated and purified for each next conversion step (Fig. 13.1). Catalytic steps are mostly combined with stoichiometric steps in the preparation of precursors or in the further downstream processing. Obvious disadvantages are low space–time yields (kg L^{-1} h^{-1}), laborious recycle loops and large amounts of waste.

A major cause of this problem in organic synthesis is the fact that many of the synthetic tools were developed without knowledge of how nature was performing its chemistry. We now have the heritage of many powerful chemical conversion procedures with a great variety in reaction conditions such as temperature, pressure, solvent, air and moisture sensitivity. In other words, procedures that as such are of great value but lack the possibility to be combined in a cascade mode of conversion for the multi-step syntheses frequently required for specialities.

Many university groups and chemical companies still prepare and produce their molecules by a great diversity of organic synthetic methods that are mutu-

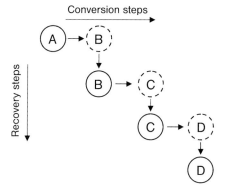

Fig. 13.1 Chemistry as it is carried out in the laboratory or manufacturing plant traditionally involves a recovery step after each conversion step. Dotted circles are compounds in the reaction medium, closed circles are isolated, pure compounds [3, 4].

ally incompatible and, therefore, require an isolation and purification step after each conversion.

In recent years, great advances have been made in the application of new biocatalytic and chemocatalytic procedures, in particular with respect to asymmetric synthesis, but processes are still mainly step-by-step and rather arbitrary combinations of catalytic and stoichiometric conversions. The latter pose increasing problems for eco-efficient and economic manufacturing on an industrial scale.

13.2.2
Nature's Chemistry

Biosynthesis, the powerful chemistry developed by nature, in the cells of living organisms, goes through a multi-step cascade approach to convert a starting material **A** into the final product **D** without separation of intermediates **B** and **C** (Fig. 13.2). Starting materials come in according to fed-batch principles using

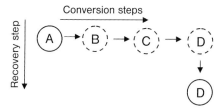

Fig. 13.2 Chemistry as it is performed in the cells of organisms involves coupled conversions without intermediate recovery steps. Dotted circles are compounds in the reaction medium, closed circles are isolated, pure compounds [3, 4].

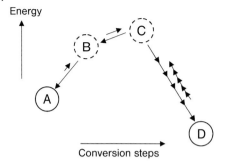

Fig. 13.3 The potential power of cascade conversions to overcome thermodynamic hurdles in multi-step syntheses. Dotted circles are compounds in the reaction medium, closed circles are isolated, pure compounds [3, 4].

controlled membrane transport proteins. End products are removed using the *in situ* product removal (ISPR) principle, again employing controlled transport systems. This allows, in principle, high throughputs and efficiencies. Required activating and protecting groups are kept *in situ* through recyclable energy and redox carriers (cofactors).

A major advantage of biosynthesis is the fact that many enzymatic conversions have been evolved, working under almost the same reaction conditions. With the present knowledge of many biosynthetic pathways we now also have the heritage of many powerful bioconversions that, by far, work at ambient temperature and pressure and in aqueous medium. In other words, it is time to fully exploit biosynthetic procedures and the possibility of combining them with appropriate or new chemocatalytic transformations in a cascade mode of conversion.

Multi-step concerted biosyntheses are also quite common in every-days life to save energy by overcoming thermodynamic hurdles to reach the final product. This, for example, is the case for the well-known glycolysis pathway **A** → **B** → **C** → **D** where **B** and **C** are higher energetic intermediates (Fig. 13.3). The equilibria between fructose 1,6-diphosphate (**A**), dihydroxyacetone phosphate (**B**), and glyceraldehyde 3-phosphate (**C**) are quite unfavorable: [**A**]:[**B**]:[**C**] > 1000:20:1. Nevertheless, complete transformation of **A** into lactic acid (**D**) occurs through a coupled multi-enzymatic conversion of the more highly energetic **B** and **C** intermediates. In addition, the success of this overall cascade conversion is brought about by the very fast aldolase- and isomerase-catalyzed equilibria, **A** ↔ **B** and **B** ↔ **C**, respectively, as the low concentrations of the intermediates **B** and **C** require high reaction rate constants to give acceptable overall **A** → **D** reaction rates [8].

13.2.3
Bio-chemo Integration

Although beyond the scope of this chapter on cascade conversions, metabolic engineering of microorganisms promises to be a powerful means of converting part of the traditional multi-step organic syntheses into one-step fermentation processes. An increasing range of fully fermentative processes for non-natural products might be expected.

As biotechnology, however, cannot replace all multi-step chemical process routes, nature-inspired organic chemical syntheses have to be developed, too. Therefore, for the next generation of organic synthesis, it is the challenge to:

- combine the power of chemical, chemocatalytic, enzymatic and microbial conversions;
- search for multi-step conversions without recovery steps, such as in nature, i.e., to go for simultaneous and/or consecutive (multi-catalytic) clean procedures;
- fine-tune reaction conditions and catalytic systems to allow for the right concerted cooperation without any intermediate isolation and/or purification steps;
- develop alternative conversion procedures that are mutually compatible with respect to the required reaction conditions.

13.3
Types of Cascades

Both literature and patents show an increasing number of cascade reactions based on the concept of one-pot reactions (Fig. 13.4).

Here, bio- and chemocatalysis as well as stoichiometric organic chemistry can work together. The different approaches can be categorized in bio-bio, chemo-chemo and bio-chemo cascades. Their relative development over the years (by number of publications, as shown in Fig. 13.5) reflects the following major features:

- As most enzymes function under compatible ambient conditions, bio-bio cascades had already been successfully developed by the 1970s. By far, most examples have been reported in the field of carbohydrates, using combinations of enzymatic conversions (up to eight enzymes in one-pot), as well as for the *in situ* cofactor regeneration of enzymatic redox reactions towards amino and hydroxy acids.

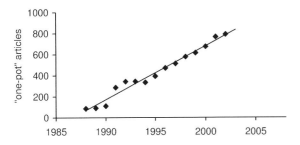

Fig. 13.4 Steadily increasing interest in one-pot cascade-type procedures since the mid-1980s [4].

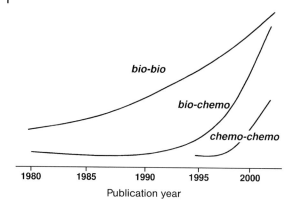

Fig. 13.5 Relative trends of the various types of cascade conversions over the years [4].

- After some early examples of bio-chemo combinations in the 1980s, there was then over a decade of "silence", followed by clearly increasing interest from the mid-1990s in the field of dynamic kinetic resolution processes (i.e., chemocatalyzed racemization combined with enantioselective enzymatic conversion, giving, in principle, 100% yield of an optically pure compound).
- Up to the late 1990s, combined multi-step chemo-chemo conversions were restricted to a few catalytic examples. Apparently, there has been little effort or interest in developing a toolkit of chemocatalytic reactions that are mutually compatible with respect to reaction conditions. Consequently, chemocatalysts have not yet reached the same level of mutual compatibility as biocatalysts. Some recent examples prove, however, the potential power of chemo-chemo catalytic cascades.

Various promising cascade conversions have been described in the literature, involving different combinations of enzymes, homogeneous and heterogeneous catalysts, and of uncatalyzed organic chemical conversions. A representative tabular selection of over 70 examples on the laboratory as well as on the pilot or industrial scale has been reviewed [4]. Compatibility and compartmentalization are key words in developing cascade syntheses. The living cell is furnishing increasing insight as to how to develop new concepts for cascade conversions.

Although the focus here is on the integration of biocatalysis with chemocatalysis (bio-chemo cascades) for carbohydrates as renewable feedstocks, some representative examples (from laboratory to industrial scale) of both bio-bio and chemo-chemo cascades are also given below for comparison of their relative scope and limitations.

13.3.1
Bio-bio Cascades

As enzymes easily work together under comparable reaction conditions, combined enzymatic reactions are by far the most early and abundantly studied. Apart

Four-enzymes mediated synthesis using a *pH switch*

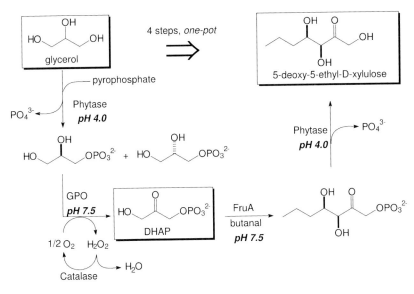

Fig. 13.6 A multi-enzyme one-pot example: cascade conversion of glycerol into a heptose sugar through consecutive phosphorylation, oxidation, aldol reaction and dephosphorylation [11].

from various two-enzyme conversions, including *in situ* redox cofactor regenerations, multi-enzyme carbohydrate conversions on multi-gram scale have been accomplished successfully, using up to eight different enzymes in one pot [9]. The record to date is the 12-step enzymatic conversion in one pot, although on a milligram scale, of the corrin moiety of vitamin B_{12} [10].

Another interesting case is the one-pot four-enzyme cascade conversion of glycerol into a heptose sugar on gram scale [11], in which a pH switch method is applied to temporarily turn off one of the enzymes involved (Fig. 13.6). The four consecutive enzymatic conversion steps in one and the same reactor, without separation of intermediates, consist of:

- Phosphorylation: glycerol is phosphorylated with pyrophosphate by phytase at pH 4.0, 37 °C. Racemic glycerol-3-phosphate is obtained in 100% yield (based on pyrophosphate) in 95% glycerol after 24 h.

- Oxidation: by raising the pH (to 7.5) phytase activity is "switched off"; hydrolysis is prevented. Oxidation of L-glycerol-3-phosphate to DHAP by GPO at 55% glycerol (v/v) is quantitative. Catalase is added to suppress the build-up of hydrogen peroxide. The D-isomer is converted back into glycerol and phosphate in the last step.

- Aldol reaction: Over twenty aldehydes are known to be substrates for the aldolases from *S. carnosus* and *S. aureus*. Stereoselectivity of the aldolases must be looked at for each acceptor substrate, since isomers are formed in different proportions. The oxidation and aldol reaction can be carried out simultaneously.

- Dephosphorylation: lowering the pH back to 4 "switches on" phytase's activity; hydrolysis of the sugar phosphate intermediate is initiated.

An impressive one-pot six-step enzymatic synthesis of riboflavine from glucose on the laboratory scale has been reported with an overall yield of 35–50%. Six different enzymes are involved in the various synthesis steps, while two other enzymes take care for the *in situ* cofactor regenerations [12]. This example again shows that many more multi-enzyme cascade conversions will be developed in the near future, as a much greater variety of enzymes in sufficient amounts for organic synthetic purposes will become available through rapid developments in genomics and proteomics.

On an industrial scale, Codexis reported very recently the development of a three-enzymatic cascade process for making, in high yield and with high selectively, ethyl (R)-4-cyano-3-hydroxyburyrate, the key chiral building block for Pfizer's atorvastatin drug [13]. After reduction of ethyl 4-chloroacetoacetate with both a ketoreductase and a glucosedehydrogenase enzyme, working in tandem with the enzyme's cofactor and glucose as reductant, a halohydrin dehalogenase enzyme subsequently converts the chlorine into a cyanide group. Both enzymatic conversions take place under ambient conditions in water with easy work-up by extraction, as performed at Lonza's production site, thus significantly lowering the costs of the former chemical step-by-step route. It is considered to be the tip of the iceberg of what can be accomplished in biocatalytic process development [13].

13.3.2
Chemo-chemo Cascades

As mentioned above, combined catalytic conversions through combinations of chemocatalytic conversions are not well represented. Although the organic chemist has an extensive synthetic and catalytic toolkit, large differences in reaction conditions often hinder the combined use of tools in one-pot conversions or in a cascade mode without recovery steps.

However, increasing academic interest and success has been demonstrated recently by several multi-step syntheses without isolation of intermediate products. For instance, the *in situ* acid catalyzed three-component synthesis of an oxazole that is compatible with two subsequent cycloaddition reactions is impressive [14]. This one-pot procedure is simple to operate, gives good yields of complicated ring structures, and is also highly adaptable to diversity-oriented parallel syntheses. A very recent triple cascade example [15], using an L-proline-based homogeneous catalyst, gave access to highly-functionalized cyclohexenes with high diastereo- and enantioselectivity from three simple building blocks. This further proves that novel synthetic methodologies based on the cascade concept will still determine the future of organic chemistry.

On an industrial scale, three chemical conversion steps have been combined without isolating intermediates by replacing four different solvents with just one in the synthesis of the drug sertraline by Pfizer [16]. This process change reduced the solvent requirement from 60 000 to 6000 gal ton^{-1} of sertraline and eliminated the use of 440 metric tons of titanium dioxide, 150 metric tons of 35% hydrochloric acid, and 100 metric tons of 50% sodium hydroxide per year, reducing the environmental burden and saving hundreds of thousands of dollars.

Very recently, on a pilot plant scale of over 100 kg, Merck has performed a chemo-chemo cascade, starting from trifluorophenylacetic acid and triazole building blocks, to synthesize their drug sitagliptin [13]. The sequence of reactions was carried out without isolating intermediates, increasing the overall yield by nearly 50% and reducing the amount of waste by over 80%. The benefits to the environment are estimated to be as great as the prevention of 150 000 tons of waste over the lifetime of sitagliptin, which is expected to become a top-selling drug within a few years.

Apart from the impressive recent examples given above, however, there has been too little focus thus far on developing a toolkit of chemocatalytic conversions that are as mutually compatible as enzymatic reactions are in nature (presently there are great differences in solvent, temperature, sensitivity to air and moisture, reactants). This confirms in fact the main difference in approach between organic synthesis and biosynthesis: organic synthesis employs a maximum diversity in reagents and conditions while biosynthesis exploits subtlety and selectivity from a small range of materials and conditions (Fig. 13.7).

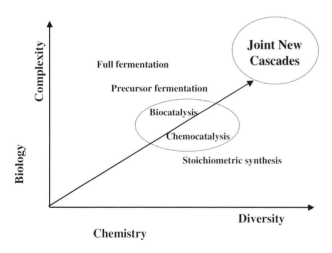

Fig. 13.7 Synthesis as seen from both a chemical and a biological perspective [4].

13.3.3
Bio-chemo Cascades

A very first example involving the combined action of an enzyme and a metal catalyst is the direct one-pot conversion of glucose into mannitol (3× as expensive as

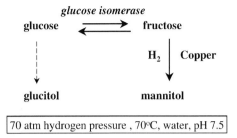

Fig. 13.8 An early example of combined enzyme and metal catalysis: the one-pot cascade conversion of glucose into mannitol [17, 18].

glucitol, the sole product from common glucose hydrogenation) [17] – an attractive alternative to other mannitol processes [18]. In this further optimized cascade process [19], the isomerase enzyme converts glucose into a ~1:1 glucose–fructose mixture and ensures that this mixture remains in equilibrium, while at the same time the copper catalyst hydrogenates preferentially fructose from this equilibrium into mannitol (Fig. 13.8). The high temperature (70 °C) and hydrogen pressure (70 atm) had no negative influence on the enzyme's activity or stability [20].

This combi approach looks, at first sight, quite simple but in practice several fine-tuning measures had to be taken to achieve a balanced cooperation of the two simultaneous catalytic conversion steps:

- Immobilization of the enzyme onto silica to prevent poisoning of the copper metal by protein sulfur moieties.

- Protection of the enzyme by a copper ion complexing agent (EDTA) to avoid inhibition by traces of copper ions from the copper catalyst.

- Right compromise of hydrogen pressure and temperature to fulfill stability and activity requirements for both catalyst systems.

- Slightly basic pH to enhance mutarotation of glucose, i.e., the interconversion of α- and β-glucopyranose forms, which becomes rate limiting as the enzyme only converts the α-form.

- Addition of some sodium borate to increase the fructose → mannitol hydrogenation selectivity due to fructose–borate complex formation [21, 22].

Fig. 13.9 Facsimile of a molecular and kinetic scheme of the one-pot glucose into mannitol bio-chemo cascade, showing the many different species that undergo interconversion during the overall process, involving enzymatic isomerization, homogeneous mutarotation and heterogeneous hydrogenation. For simplicity, the various sugar–borate species have been omitted [23, 24].

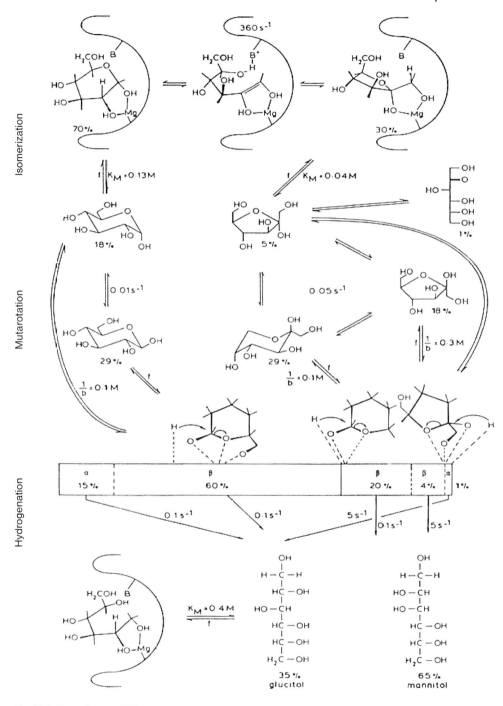

Fig. 13.9 (legend see p. 282)

Table 13.1 Kinetics of three types of catalysis that are in concert in the one-pot glucose-to-mannitol bio-chemo cascade conversion [19].

Type of catalysis	Enzymatic	Heterogeneous	Homogeneous
Reaction	Isomerization	Hydrogenation	Mutarotation
Catalyst	Glucose isomerase	Copper metal	Hydroxide
Activation energy (kJ mol^{-1})	71	67	63
Turnover number (s^{-1})	360	1	0.01
Catalyst concentration (mol L^{-1})	0.00001	0.001	0.3[a]
Reaction rate (mol L^{-1} s^{-1})	0.004	0.001	0.003

[a] Equivalent to the sugar concentration according to the definition of mutarotation kinetics [25].

What is really happening [23, 24] in the glucose → mannitol cascade [19] can be seen from the quite complicated kinetic and molecular picture shown in Fig. 13.9, including three "different" types of kinetics (expressed in turnover numbers in Table 13.1):

- Michaelis–Menten: enzymatic isomerization; only two of the six sugar forms are substrates for the enzyme; K_M values for glucose, fructose, glucitol and mannitol are 0.13, 0.04, 0.4 and >1 M, respectively.

- Langmuir–Hinshelwood: heterogeneous hydrogenation on copper; adsorption constants b vary from 3 to 10 M^{-1}; only ~25% of the copper surface covered with fructose which, fortunately, reacts much faster than glucose adsorbed on the copper catalyst.

- Homogeneous acid–base mutarotation: interconversion of the different glucose and fructose forms; the rate for glucose is 50× slower than that for fructose.

This principle of a dynamic equilibrium between two compounds by one catalyst in combination with a selective conversion of one of those by a second catalyst is of great importance for the so-called 100% e.e.–100% yield synthesis of enantiomerically pure compounds from racemic starting materials. Over ten different examples of such dynamic kinetic resolution on a lab-scale have been reported [4], using the concomitant action of a chemocatalyst and a bio-catalyst (Fig. 13.10). Without such a combination of two catalysts in one reactor, either a maximum yield of only 50% can be obtained or separate recovery and racemization steps are required.

In particular, the combined action of a transition metal catalyst and a lipase in organic solvents for the racemization and esterification steps, respectively, has been applied for the conversion of racemic secondary alcohols into their esters

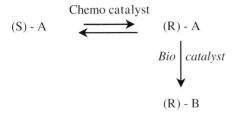

Fig. 13.10 Complete conversion of a racemate A into an optically pure compound B by the cooperation of a bio- and chemocatalyst in one pot [4].

in high yield and high optical purity. The attractiveness of this type of syntheses for enantiomerically pure compounds from relatively inexpensive racemates has already found its way into the fine-chemicals industry. As a first example, an industrial multi-ton per year production has recently been started by DSM Fine Chemicals, using such a one-pot procedure with a homogeneous ruthenium catalyst and the enzyme lipase in an organic solvent [26]. Only one of the enantiomers of the racemic secondary alcohol is esterified by the enzyme, while the homogeneous metal catalyst racemizes both enantiomers of the alcohol. In this way, a nearly quantitative yield of enantiopure ester is obtained.

Additional three- or more-steps bio-chemo cascades are still rare, owing to the incompatibility of many chemocatalytic with enzymatic conversions in terms of reagents, solvent, pH and/or temperature. As an illustrative academic example, the potential power of such a combined catalytic approach, leading to both a 50-fold waste decrease and a five-fold reduction of both operating and space time, has been proven by a one-pot three-step carbohydrate conversion [27]. This cascade consists of a combination of enzymatic, homogeneous and heterogeneous catalysis for, respectively, consecutive oxidation (oxygen, galactose oxidase as enzyme), dehydration (L-proline as homogeneous catalyst) and reduction (hydrogen, palladium metal as heterogeneous catalyst). With water as solvent at neutral pH and mild temperatures, methyl β-D-galactopyranoside is converted quantitatively into its 4-deoxy-6-aldehydo derivative without any intermediate recovery steps, use of stoichiometric amounts of reagents other than oxygen and hydrogen or of protective groups (Fig. 13.11). Extension of this type of cascade, up to five consecutive transformations in one pot, gave rapid access to over 40 uronic acid, amino sugar and deoxy sugar derivatives in water under mild conditions. Workup is avoided by using compatible aqueous reaction conditions for all consecutive reactions involved [28].

This new multi-catalytic method shows a substantial improvement, from an economic as well as an environmental point of view, over traditional organic synthesis, as shown for the synthesis of the rare sugar 4-deoxyglucose (Table 13.2).

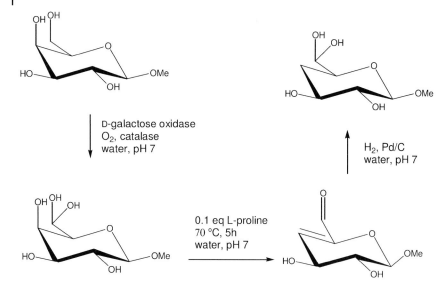

Fig. 13.11 Cascade three-step catalytic conversion in one pot in water, using both bio- and chemocatalysts. This is a quantitative, clean and rapid method compared with traditional organic synthesis [27].

Table 13.2 Benefits of the one-pot bio-chemo catalytic cascade as compared with the traditional step-wise stoichiometric organic synthesis of 4-deoxyglucose [27, 28].

Issue	Step-wise chemistry	Catalytic cascade
Synthesis steps, number of	7	3
Isolation steps, number of	4	1
Solvents, number of	3	1
Reagents, weight ratio to starting sugar	5	0.1
Labor, working days	3	1
Overall yield (% based on starting sugar)	50	100

A similar picture has been obtained very recently for novel multi-step bio-chemo cascade reactions starting from both galactose-derived polyols and aliphatic mono- and diols [29]. Galactose oxidase and alcohol oxidase show complementary synthetic use for this range of alcohols (Table 13.3) [30], allowing *in situ*

Table 13.3 Complementary synthetic use of galactose oxidase and alcohol oxidase for *in situ* aldehyde formation for a range of alcohols, expressed in relative rates under identical reaction conditions [30][a].

Alcohol	Relative rate	
	Galactose oxidase	Alcohol oxidase
Galactose	100	0
2-Deoxygalactose	74	0
Lactitol	14	0
Lactobionic acid	9	0
Lactose	7	0
Threitol	2	4
Glycerol	0.02	0
1,4-Butanediol	0	5
Benzyl alcohol	0.1	3
3-Butenol	0	37
4-Pentenol	0	39
Methanol	0	100

[a] 10 Units of enzyme; 1 mmol of alcohol; 1 atm O_2; 5-mL water; 25 °C; 3 days.

formation and direct use of labile aldehyde intermediates for further transformations, as demonstrated for the simultaneous oxidation and aldol condensation yielding (S)-4-hydroxy-oct-7-en-2-one (Fig. 13.12). In the carbohydrate field, a comparable two-step consecutive approach, although requiring an intermediate filtration step, yielded elegantly the rare sugar L-fructose from glycerol [31].

In the field of food ingredients, the proof of principle for bio-chemo cascade-type conversions has been demonstrated for the modification and/or crosslinking of proteins by reaction of *in situ* generated galactose aldehyde with amines [32, 33]. This approach has been extended to lactose, lactylamine and lactobionic acid [34] by oxidation with galactose oxidase into their 6′-aldehydes (Table 13.3) followed by reaction with butylamine as model compound for protein lysine residues [35]. Application of this bio-chemo cascade allowed the glycosylation of the whey protein β-lactoglobulin, containing 4–10 oxidized lactose-derived moieties per protein molecule [36]. Very recently it was established that enzymatically oxi-

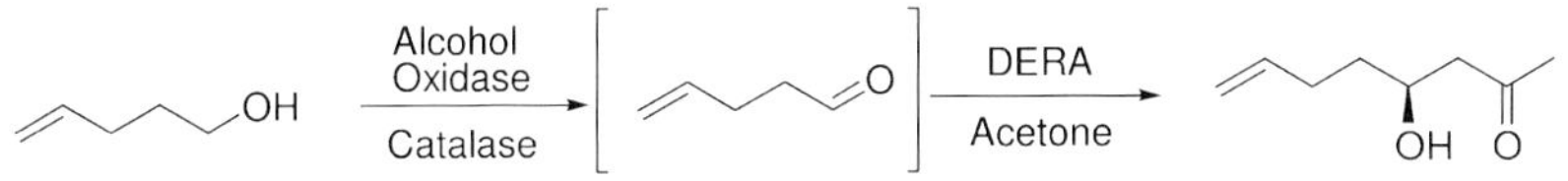

Fig. 13.12 *In situ* aldehyde formation coupled with an aldol condensation as the starting point for further cascade conversions [29, 30].

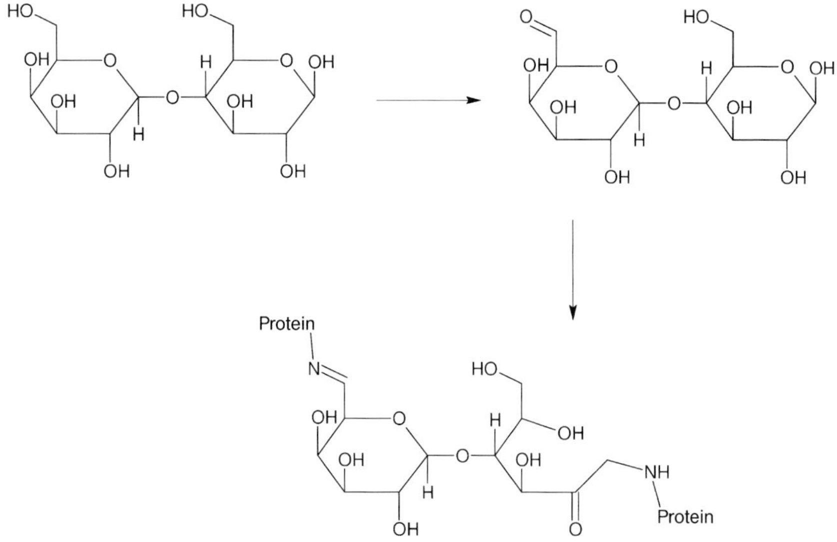

Fig. 13.13 Principle of protein crosslinking by enzymatic oxidation of lactose coupled with the Maillard reaction [35–37].

dized lactose gives crosslinking of proteins through Maillard-type reactions (Fig. 13.13) [37].

In addition to these academic laboratory examples, showing that bio-chemo cascade conversions can be seen as a potentially important tool for novel and more sustainable conversions in the practice of both the fine chemical and the food ingredients industry, the industrial benefits of cascade-type of conversion is demonstrated by the evolution during the past 30 years of the industrial process route for the synthesis of the antibiotic Cephalexin within DSM [24, 38]. Here, the integration of chemistry, enzymatic conversion and fermentation forms the basis of an efficient production method from both an economic and environmental point of view. The 1970s Cephalexin synthesis consisted of one fermentation together with eight organic synthesis steps [39]. This multi-step sequence used

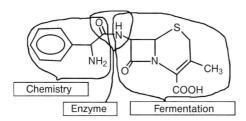

Fig. 13.14 Cephalexin synthesis: cooperation of chemical, enzymatic and fermentative conversions [24].

high levels of energy (such as low temperature conversions, many recovery and solvent recycle steps), of reagents and organic solvents. The 2000s Cephalexin synthesis consists of only three major conversion steps (Fig. 13.14):

- chemo-enzymatic conversion of benzaldehyde into D-phenylglycine amide;
- fermentation of sugar by a metabolic pathway engineered microorganism into the 7-ADCA moiety;
- enzymatic coupling of D-phenylglycine amide with 7-ADCA.

Apart from the energy savings, large amounts of reagents and solvents could be omitted by this chemo-enzymatic-fermentative cephalexin route:

- peracetic acid (as hazardous oxidation agent);
- trimethylsilyl chloride and bis(trimethylsilyl)urea (as protecting group agents);
- pyridine.HBr and trimethylamine.HCl (as catalysts in stoichiometric amounts);
- phosphorpentachloride and butanol (to selectively split the secondary amide);
- dichloromethane (as air- and water-polluting solvent);
- phenylacetic acid (as non-recyclable recovery tool in penicillin fermentation);
- acid chloride HCl salt of D-phenylglycine (for coupling with 7-ADCA).

A substantial part of these savings arises from the novel fermentation process for 7-ADCA that replaces the former multi-step chemical conversion of penicillin (Fig. 13.15).

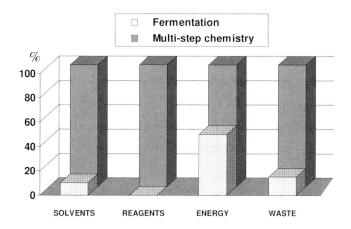

Fig. 13.15 Benefits of fermentative 7-ADCA process as a cascade alternative to a multi-step-by-step organic synthesis [38].

13.4
Technologies for Cascades

As many starting materials and products of the fine-chemicals industry require an organic solvent instead of water, combined conversions in one or two "stan-

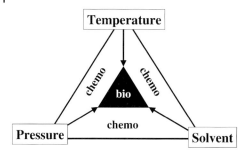

Fig. 13.16 Merging reaction parameters of chemocatalysis (blank area) to that of biocatalysis (black area) to exploit fully the scope of bio-chemo cascade conversions.

dard" organic solvents offers a powerful methodology that yet has to be exploited. In this way, bio-chemo catalytic cascades are also possible since enzymes, although not all [40, 41], are active in organic solvents [42]. This, of course, requires redeveloping part of the synthetic toolkit of organic chemistry to perform multi-step organic syntheses in a combined catalytic mode. At the same time many new synthesis and catalysis methodologies need to be developed to fill the spectrum of possibilities for molecular assembly.

Full exploitation of cascade conversions by the true integration of biocatalytic and chemocatalytic procedures requires merging human's chemistry with nature's reaction conditions; the latter impose a much stricter constraint with respect to reaction temperature, pressure and medium (Fig. 13.16). Consequently, a renaissance in the field of synthetic organic chemistry and catalysis is necessary to develop novel conversion processes that meet biocatalytic conditions.

Apart from new catalytic methods, cascade conversions require new process technologies, such as *in situ* product recovery, reactor design, and compartmentalization. In the long term, part of the present-day stoichiometric chemistry as well as bio- and chemocatalytic conversions in multi-step syntheses will gradually be replaced by cascade catalysis in concert, and full fermentations by cell factory design, or combinations thereof (Fig. 13.17).

Investigations of cascade conversions without isolation of intermediate products require appropriate *in situ* analytical methods to know what is really happening during the consecutive conversions. A quite powerful route to this information is the use of selectively isotope (such as ^{13}C, ^{15}N, ^{17}O) enriched starting materials in combination with NMR [43]. In this way, a sequence of conversions can be well characterized, even *in situ* in complicated matrices of catalysts, reagents and mixed solvent systems [27, 30, 32, 33, 35, 44–48] – as applied to the three-catalytic one-pot reaction of galactose (Fig. 13.11) as well as in the structural investigations of chemo-enzymatic crosslinking phenomena of carbohydrate protein mixtures (Fig. 13.13).

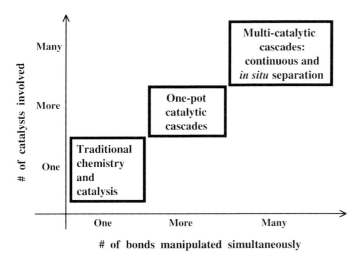

Fig. 13.17 Shift of synthesis methodology: multi-bond manipulation by multi-catalytic transformations [4].

13.4.1
Catalytic Methods

Transformations using biocatalysts are now recognized as attractive alternatives for the synthesis of several important and commercially significant compounds that are complementary to traditional chemical approaches [13]. With rapid access to unique enzymes having improved or novel properties from various sources of biodiversity, and from ingenious DNA recombination techniques along an evolutionary approach, it can be anticipated that critically shorter process development cycles will be achievable – in fact, this may be a precondition to overcome a persisting skepticism about biotechnology and biocatalysis that, so far, has limited the full realization of its potential in the chemical industries. The fast and timely availability of effective new biocatalytic tools is required to tackle current and future challenges presented by the demands of organic synthesis, in a sustainable manner [49].

A marriage of catalysis and synthetic organic chemistry provides a rich wellspring of novel reactivity that expands substantively the scope of transformations that can be realized. Catalysis will continue to play an integral role in shaping the future of sustainable organic synthesis. The way forward is full integration of catalytic methodologies in organic synthesis. There is also a pressing need to integrate the different modes of catalysis: homogeneous, heterogeneous and enzymatic. This would lead to a synergistic cross-fertilization of ideas. The ultimate synergy is the integration of chemo- and biocatalytic steps to afford catalytic cascades that are reminiscent of what occurs in the living cell.

13.4.2
Reactor Design

Successful development and practical implementation of cascade catalysis and multi-step conversions in concert requires a multi-disciplinary approach, ranging from genetic engineering to reactor design. In particular, novel combinations of new reactor concepts, catalytic materials and separation techniques will offer the synergy to afford multi-step sustainable synthesis methodologies that are reminiscent of what occurs in the living cell. Apart from the requisite of very high selectivity for each transformation in cascade conversions, the *in situ* removal of by-products and spent reagents has to be further developed by integrated process design.

Membranes and structured multiphase reactors are a means to achieving compartmentalization of catalysts and reaction media, such as structured multiphase mini-reactors for multi-step cascade conversions through the combined use of enzymes and transition metal catalysts. The advantages of a so-called structured reactor, instead of the traditionally applied stirred batch reactor, are compactness, energy-efficiency, high selectivity, lower amounts of solvent and easy scale-up [50]. For instance, enzymatic bioreactors with both high flow characteristics and mechanical stability on macroporous monoliths have been designed. High catalytic activity of the monolithic reactor is maintained even at a flow velocity of 180 cm min^{-1}, which substantially exceeds those reported for packed bed reactors, for the hydrolysis of both low and high molecular weight substrates such as L-N-benzoylarginine ethyl ester and casein by immobilized trypsin [51].

There is huge potential in the combination of biocatalysis and electrochemistry through reaction engineering as the linker. An example is a continuous electrochemical enzyme membrane reactor that showed a total turnover number of 260 000 for the enantioselective peroxidase catalyzed oxidation of a thioether into its sulfone by *in situ* cathodic generated hydrogen peroxide – much higher than achieved by conventional methods [52].

13.4.3
Compartmentalization

Membrane-Enclosed Enzymatic Catalysis (MEEC) has been developed as a useful, practical new method for the manipulation of enzymes in organic synthesis. The enzyme in soluble form is enclosed in commercially available dialysis membranes. It combines the simplicity of use of soluble enzymes with certain of the advantages of immobilized enzymes. Containment permits separation of the enzyme from the reaction medium, straightforward separation of the product, and recovery of the enzyme for reuse [53].

Enzymes can also be trapped within a nanotubular membrane by capping both faces of the membrane with a thin layer of porous polymer. In effect, arrays of enzyme-filled microcapsules are formed. Small molecules pass through the porous polymer plugs, but not the enzyme, which float freely in the confined space. Loaded with enzyme, the membrane works like a bioreactor [54].

Effective and simple immobilization of enzymes can be obtained by the cross-linking of enzyme aggregates, so-called CLEAs [55]. In this way, essentially any enzyme, including crude preparations, can be transformed into a heterogeneous type of material, insoluble in both water and organic solvents, that is stable and recyclable with high retention of the enzyme's original activity [56]. These enzyme preparations are, therefore, of special value for both bio-bio and bio-chemo cascade processes.

Two types of continuous membrane reactors have been applied for oligomer- or polymer-bound homogeneous catalytic conversions and recycling of the catalysts. In the so-called dead-end-filtration reactor the catalyst is compartmentalized in the reactor and is retained by the horizontally situated nanofiltration membrane. Reactants are continuously pumped into the reactor, whereas products and unreacted materials cross the membrane for further processing [57].

Column chemistry and catalysis is applied to execute a synthetic sequence and deliver a pure product without separate work-up, purification, and catalyst recovery steps. Substrates in solution are converted into products by flowing through solid-phase-bound reagents and catalysts [58]. In this respect, the present-day Merrifield peptide synthesis, by column solid-phase chemistry, is still waiting for a (more) catalytic mode of operation without the need of protecting group reagents. To date, large amounts of reagents and solvents have to be used, as shown for Roche's new industrial process for the 36-peptide HIV drug Fuzeon, which requires 45 tons of raw materials to make 1 ton of product [59]. Wide-pore resins containing long spacer linkages, for instance, could allow both enzymatic and homogeneously catalyzed peptide couplings in a catalytic cascade mode.

13.4.4
Medium Engineering

Catalytic methods that use ecologically benign reaction media and innovative separation schemes are key to future developments; in particular, if the alternative media are sufficiently different from organic solvents to exploit some of their specific features for reaction tuning, or even for new chemistry. Fluorocarbon solvents and carbon dioxide versus water and ionic liquids are at the extremes of the polarity scale [60, 61]. In the same manner, reactions also could be carried out in a mixture of the neat reactants, i.e., without the need of any solvent [62]. As an illustrative example, pure octyl α-D-glucopyranoside.H_2O selectively crystallized out upon cooling a heated suspension of glucose in excess octanol in the presence of a heterogeneous acid catalyst, thus simultaneously removing the water formed during the reaction. Although the yield is just 30%, reheating the supernatant after replenishment of glucose resulted again in the same product yield. Consequently, overall, this recycling procedure with neat reactants gave a quantitative conversion with easy product separation [63, 64].

The fact that enzymes also work in organic solvents [42], ranging from apolar alkanes up to the very polar N,N-dimethylformamide [65], has (Fig. 13.10) and will continue to open up new cascade opportunities for the integration of enzymatic conversions with chemocatalytic methods that require organic solvents. Notably, however, some enzyme classes, for instance the carbohydrate-converting enzymes, do not show activity in non-aqueous media [40, 41].

Precipitation-driven synthesis offers the possibility to obtain high reaction yields using very low volume reactors and is finding increasing applications in biocatalysis. Straightforward prediction of when such a precipitation-driven reaction will be thermodynamically feasible is possible, as demonstrated for a range of enzyme-catalyzed peptide syntheses. The methodology is quite general and is therefore expected to be applicable to a wide range of other (bio)catalyzed reactions [66].

Conventionally, organometallic chemistry and transition-metal catalysis are carried out under an inert gas atmosphere and the exclusion of moisture has been essential. In contrast, the catalytic actions of transition metals under ambient conditions of air and water have played a key role in various enzymatic reactions, which is in sharp contrast to most transition-metal-catalyzed reactions commonly used in the laboratory. Quasi-nature catalysis has now been developed using late transition metals in air and water, for instance copper-, palladium- and rhodium-catalyzed C–C bond formation, and ruthenium-catalyzed olefin isomerization, metathesis and C–H activation. Even a Grignard-type reaction could be realized in water using a bimetallic ruthenium–indium catalytic system [67].

13.4.5
Cell Factory Design

Apart from the chemical technology developments mentioned above, metabolic pathway and flux engineering will have an increasing impact on the way multi-step organic syntheses are carried out in the fine-chemicals industry. For the next generation of microbial conversions, the challenge of molecular biology is to:

- improve productivity in terms of carbon source efficiency of microorganisms, towards a much higher product:biomass ratio;
- broaden the scope of products, from natural products toward the modified products often required for drugs;
- reengineer metabolic pathways as the most efficient way to reach the first two goals.

These steps should free fermentation processes from excessively using renewables that end up in undesired, low-value biomass by-products and minimize additional chemical modification steps to obtain the final product [24].

In particular, the appropriate integration of microbial, enzymatic and chemical catalytic transformations in a cascade mode may be seen as the ultimate tool for sustainable fine chemicals processing. In this respect, the increasing insight in the precise functioning of living cells as high-tech miniature factories will surely be a source of inspiration for the development of novel sophisticated processing devices for the fine-chemicals industry.

13.5 Conclusions

The concepts and various examples given show that we may foresee a renaissance in synthesis methods by the integration of bio- and organic syntheses. Fine chemicals of the future will be produced by cascade multi-step catalytic procedures without intermediate recovery steps.

Whereas we might in principle have a catalyst available today for every synthetic conversion, it will probably take another 10–20 years before enjoying the full potential of cascades of catalysis on an industrial scale (Fig. 13.18).

Integration of reactor and catalyst is a clear trend as well as full-scale application of microsystems technologies. These devices could be used to change large-scale manufacturing from volume-guided to adding numbers of (micro-)reactors. Recent advances in down-scaling rules from batch reactors to laboratory protocols [68, 69] could very well stimulate this scenario, in which pilot plants could also be avoided as well as investments in new manufacturing equipment.

Last but not least, cascade conversions give various paths for safer processing designs. For instance, by reduction of the handling and storage of process intermediates, by replacement of hazardous chemicals, and by more *in situ* creation and use for highly toxic chemicals, i.e., by shifting to just-in-time production [70]. As the intermediate products of multi-step organic syntheses are often highly active and or toxic compounds, cascade processes will prevent the exposure of both operators and the environment to such chemicals.

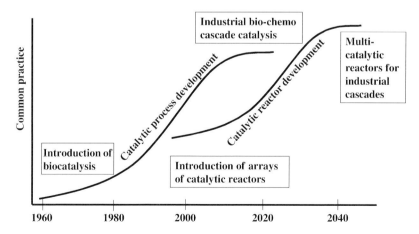

Fig. 13.18 S-curves for developments in catalysis in fine chemicals syntheses [4].

Acknowledgments

Thanks are due to Michiel Makkee, who developed the combi-mannitol process in the 1980s at the Delft University of Technology, as well as to Rob Schoevaart, Arjan Siebum and Arjan van Wijk for their recent research efforts on other cascade conversions at Leiden University, as part of the IBOS Programme (Integration of Biosynthesis and Organic Synthesis) of Advanced Chemical Technologies for Sustainability (ACTS) with industrial support from Friesland Foods and DSM.

Concerning cascade conversions, fruitful discussions with and the interest of Herman van Bekkum (Delft University of Technology), Johan Lugtenburg (Leiden University), Joop Roels and Alle Bruggink (DSM) are gratefully acknowledged.

References

1 A.P.G. Kieboom, J.A. Moulijn, P.W.N.M. van Leeuwen, R.A. van Santen, *Stud. Surf. Sci. Catal.* **1999**, 123, 3.
2 A.P.G. Kieboom, J.A. Moulijn, R.A. Sheldon, P.W.N.M. van Leeuwen, *Stud. Surf. Sci. Catal.* **1999**, 123, 29.
3 R. Schoevaart, T. Kieboom, *Chem. Innovation* **2001**, 12, 33.
4 A. Bruggink, R. Schoevaart, T. Kieboom, *Org. Process Res. Dev.* **2003**, 7, 622.
5 A.P.G. Kieboom, H. van Bekkum, *Recl. Trav. Chim. Pays-Bas*, **1984**, 103, 1.
6 A.P.G. Kieboom, H. van Bekkum, in: *Starch Conversion Technology* (G.M.A. van Beynum, J.A. Roels, eds.), Marcel Dekker, New York and Basel, **1985**, 263.
7 D. de Wit, L. Maat, A.P.G. Kieboom, *Ind. Crops Prod.* **1993**, 2, 1.
8 A.L. Lehninger, *Principles of Biochemistry*, Worth Publishers, New York, **1984**, p. 397.
9 K.M. Koeller, C.-H. Wong, *Chem. Rev.* **2000**, 100, 4465.
10 C.A. Roessner, C.A. Scott, *Annu. Rev. Microbiol.* **1996**, 50, 467.
11 R. Schoevaart, F. van Rantwijk, R.A. Sheldon, *J. Org. Chem.* **2000**, 65, 6940.
12 W. Römisch, W. Eisenreich, G. Richter, A. Bacher, *J. Org. Chem.* **2002**, 67, 8890.
13 S.K. Ritter, *Chem. Eng. News* **2006**, July 10, 24.
14 P. Janvier, J. Zhu, *Angew. Chem. Int. Ed.* **2002**, 41, 4291.
15 D. Enders, M.R.M. Hüttle, C. Grondal, *Nature* **2006**, 441, 861.
16 S.K. Ritter, *Chem. Eng. News* **2002**, July 1, 26.
17 M. Makkee, A.P.G. Kieboom, H. van Bekkum, J.A. Roels, *J. Chem. Soc., Chem. Commun.* **1980**, 930.
18 M. Makkee, A.P.G. Kieboom, H. van Bekkum, *Starch* **1985**, 37, 136.
19 M. Makkee, A.P.G. Kieboom, H. van Bekkum, *Carbohydr. Res.* **1985**, 138, 237.
20 M. Makkee, A.P.G. Kieboom, H. van Bekkum, *Starch* **1985**, 37, 232.
21 M. van Duin, J.A. Peters, A.P.G. Kieboom, H. van Bekkum, *Tetrahedron* **1984**, 40, 2901.
22 M. Makkee, A.P.G. Kieboom, H. van Bekkum, *Recl. Trav. Chim. Pays-Bas* **1985**, 104, 230.
23 Original handicraft drawing (from 1985, found 15 years later in the archives of Delft University of Technology), summarizing the data given in Refs. [18–22].
24 R. Schoevaart, T. Kieboom, Trends in Drug Research III, in: *Pharmacochem. Libr* **2002**, 32, 39.
25 C.E. Grimshaw, *Carbohydr. Res.* **1986**, 148, 345.
26 A.M. Rouhi, *Chem. Eng. News* **2002**, June 10, 43.
27 R. Schoevaart, T. Kieboom, *Tetrahedron Lett.* **2002**, 43, 3399.
28 R. Schoevaart, T. Kieboom, *Top. Catal.* **2004**, 27, 3.
29 A. Siebum, R. Wijtmans, A. van Wijk, H. Schoemaker, F. Rutjes, T. Kieboom, unpublished results.
30 A. Siebum, A. van Wijk, R. Schoevaart, T. Kieboom, *J. Mol. Cat. B* **2006**, 41, 141.

31 D. Franke, T. Machajewski, C.-C. Hsu, C.-H. Wong, *J. Org. Chem.* **2003**, 68, 6828.
32 R. Schoevaart, T. Kieboom, *Carbohydr. Res.* **2001**, 337, 1.
33 R. Schoevaart, T. Kieboom, *Carbohydr. Res.* **2002**, 337, 899.
34 G. de Wit, J.J. de Vlieger, A.C. Kock-van Dalen, A.P.G. Kieboom, H. van Bekkum, *Tetrahedron Lett.* **1978**, 1327.
35 A. van Wijk, A. Siebum, R. Schoevaart, T. Kieboom, *Carbohydr. Res.* **2006**, 341, 2921.
36 A. van Wijk, A. Siebum, T. Kieboom, *Int. Dairy. J.*, submitted.
37 A. van Wijk, A. Siebum, R. Schoevaart, T. Kieboom, *Int. Dairy J.*, to be submitted.
38 DSM Anti-infectives, *DSM Mag.* **1998**, 147, 18.
39 J. Verweij, E. de Vroom, *Recl. Trav. Chim. Pays-Bas* **1993**, 112, 66.
40 A.P.G. Kieboom, *Recl. Trav. Chim. Pays-Bas* **1988**, 107, 347.
41 A.P.G. Kieboom, in: *Biocatalysis* (D. Abramowicz, ed.) Van Nostrand Reinhold, New York, **1990**, 356.
42 A.M. Klibanov, *CHEMTECH* **1986**, 354.
43 J. Lugtenburg, H.J.M. de Groot, *Photosynth. Res.* **1998**, 55, 241.
44 G. de Wit, A.P.G. Kieboom, H. van Bekkum, *Recl. Trav. Chim. Pays-Bas* **1979**, 98, 355.
45 J.A. Peters, A.P.G. Kieboom, *Recl. Trav. Chim. Pays-Bas* **1983**, 102, 381.
46 M. Makkee, A.P.G. Kieboom, H. van Bekkum, *Recl. Trav. Chim. Pays-Bas* **1984**, 103, 361.
47 M. van Duin, J.A. Peters, A.P.G. Kieboom, H. van Bekkum, *J. Chem. Soc., Perkin Trans. II* **1987**, 473.
48 J.M. de Bruin, F. Touwslager, A.P.G. Kieboom, H. van Bekkum, *Starch* **1987**, 39, 49.
49 W.-D. Fessner, *Adv. Synth. Catal.* **2001**, 343, 497.
50 F. Kaptein, J.J. Heiszwolf, T.A. van Nijhuis, J.A. Moulijn, *CATTECH* **1999**, 3, 24.
51 S.F. Xie, F. Svec, J.M. Frechet, *Biotechnol. Bioeng.* **1999**, 62, 30.
52 S. Lütz, S. Steckhan, C. Wandrey, A. Liese, DE PA 10054082.1, **2000**.
53 M.D. Bednarski, H.K. Chenault, E.S. Simon, G.M. Whitesides, *J. Am. Chem. Soc.* **1987**, 109, 1283.
54 B.B. Lakshmi, C.R. Martin, *Nature* **1997**, 388, 758.
55 CLEA Technologies, Delft, The Netherlands.
56 R. Schoevaart, W.W. Wolbers, M. Golubovic, M. Ottens, A.P.G. Kieboom, F. van Rantwijk, L.A.M. van der Wielen, R.A. Sheldon, *Biotech. Bioeng.* **2004**, 87, 754.
57 H.P. Dijkstra, G.P.M. van Klink, G. van Koten, *Acc. Chem. Res.* **2002**, 35, 798.
58 A.M. Hafez, A.E. Taggi, T. Dudding, T. Lectka, *J. Am. Chem. Soc.* **2001**, 123, 10853.
59 L.M. Jarvis, *Chem. Eng. News* **2006**, July 17, 23.
60 W. Leitner, *Acc. Chem. Res.* **2002**, 35, 746.
61 P. Wasserscheid, T. Welton, *Ionic Liquids in Synthesis*, Wiley-VCH, Weinheim, **2002**.
62 M. Wende, R. Meier, J.A. Gladysz, *J. Am. Chem. Soc.* **2001**, 123, 11490.
63 A.J.J. Straathof, H. van Bekkum, A.P.G. Kieboom, *Starch* **1988**, 40, 229.
64 A.J.J. Straathof, H. van Bekkum, A.P.G. Kieboom, *Eur. Pat. Appl.* **1988**, June, 88201204.0.
65 S. Riva, J. Chopineau, A.P.G. Kieboom, A.M. Klibanov, *J. Am. Chem. Soc.* **1988**, 110, 584.
66 R.V. Ulijn, A.E.M. Janssen, B.D. Moore, P.J. Halling, *Chem. Eur. J.* **2001**, 7, 2089.
67 C.-J. Li, *Acc. Chem. Res.* **2002**, 35, 533.
68 J.P.A. Custers, M.C. Hersmis, J. Meuldijk, J.A.J.M. Vekemans, L.A. Hulshof, *Org. Process Res. Dev.* **2002**, 6, 645.
69 M.V. Koch, K.M. Van den Bussche, R.W. Chrisman (eds.), *Micro-instrumentation for High Throughput Experimentation and Process Intensification – A Tool for PAT*, Wiley-VCH, Weinheim, **2007**.
70 J. Johnson, *Chem. Eng. News* **2003**, February 3, 23.

14
Hydrogen Production and Fuel Cells as the Bridging Technologies Towards a Sustainable Energy System

Frank A. de Bruijn, Bert Rietveld, and Ruud W. van den Brink

14.1
Introduction

Today's energy use, based on the conversion of fossil fuels, is not sustainable in the long term. Global warming, security of supply and local air quality are strong driving forces to change the present energy system. Given the huge global demand for energy, no single solution can be imagined to make this energy supply more efficient, less carbon intensive and more sustainable, i.e., using renewable sources to a large extent.

Hydrogen plays a pivotal role in all strategies to lower CO_2 emissions, improve the air quality of urbanized areas and increase the possibilities of covering the energy demand with energy sources other than petroleum.

In the European Union, the Hydrogen and Fuel Cell Platform has set up a Strategic Research Agenda [1] aimed at the development of technologies needed for hydrogen production, storage, transport and application in stationary and mobile systems. In addition, a Deployment Strategy [1] has been made for the market introduction of these technologies. A program of over 2 billion euros is foreseen in the 7th Framework Programme. In the United States, the Freedom Car Initiative [2] and the FutureGEN [3] projects are just two examples of a huge program on hydrogen and fuel cell technologies. The Department of Energy has a well-organized program in which clear technology development targets are set, and progress is frequently assessed. In California, demonstration fleets of fuel cell vehicles are on the road, and state legislation on emissions pose a strong driving force for clean vehicles.

In Japan, being completely dependant on fossil fuels imports, a long tradition exists in developing stationary and mobile fuel cell applications. Programs funded by the NEDO and METI departments have already resulted in hundreds of small scale micro Combined Heat and Power Units as well as fuel cell vehicles running in road demonstrations [4].

14.1.1
The Hydrogen Energy Chain

The energy chain for hydrogen is schematically drawn in Fig. 14.1. For each step in the chain, several options exist or are in development. The energy efficiency and CO_2 emission of the complete hydrogen energy chain should be taken into account when considering the introduction of hydrogen into certain applications. Optimization of each step will be necessary to maximize the benefit of this fuel for the future. At the same time the energy use and CO_2 emissions in society as a whole must be minimized. The consequence is that selection of a primary energy source for, in this case, the hydrogen energy chain alone is not sufficient. The optimal selection of the use of that primary energy source is also as important.

From the perspective of hydrogen in transport for instance, hydrogen obtained by using wind power in combination with an electrolyzer leads to a near zero-emission hydrogen chain. When this wind power can be used directly to substitute fossil fuel powered electricity generation, while the hydrogen for transport is generated by methane steam reforming, the benefit for society in terms of CO_2 reduction and energy savings is two times larger.

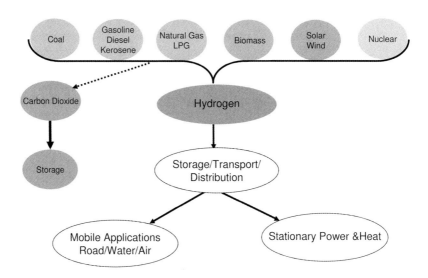

Fig. 14.1 Energy chain for hydrogen.

14.1.2
Hydrogen Sources and Production

Hydrogen can be produced by reforming natural gas and petroleum derived fuels, by gasification of biomass and coal, and by using renewable electricity to power water electrolysers. Longer term options such as thermochemical splitting of water, bacteriological hydrogen production and photo-electrochemical splitting of water are also being investigated.

In this chapter we limit ourselves to several mid-term options that can be used when the large scale use of hydrogen needs to be supplied by cost-effective production routes using available sources, being the reforming of natural gas, with CO_2 capture, and the use of electrolysers that can be powered by renewable electricity.

14.1.3
Use of Hydrogen in Stationary and Mobile Applications

For road transport, fuel cells are the most efficient conversion devices for using hydrogen. For the average drive cycle, which is dominated by a power demand that is only a fraction of the maximum available power, hybrid fuel cell systems offer a clear advantage over internal combustion engines, hybridized or not, when energy use, CO_2 emissions and non-greenhouse pollutants are considered.

For stationary power production, small systems (1–200 kWe) in which heat and power demand can be provided by fuel cells provide an opportunity to save large amounts of energy compared with central electricity production, where waste heat is in general not used. These systems would, preferably, run on natural gas, which is converted on site into hydrogen. For large-scale power production, CO_2-free production of hydrogen from natural gas and coal, with the subsequent burning of this hydrogen in the power plant, can offer the first step to reduce significantly the CO_2 emissions from power production.

We discuss both the Proton Exchange Membrane as well as the Solid Oxide Fuel Cells in this chapter (PEMFC and SOFC). Both types are in full development, the PEMFC for mobile and stationary applications, and the SOFC for stationary applications as well as for auxiliary power generation for transport.

14.2
Hydrogen Production from Natural Gas

The production of syngas and hydrogen is one of the older large-scale industrial catalytic processes. The first steam reforming plant was commissioned by BASF in 1925 [5]. The worldwide annual hydrogen production is 1 billion m^3/day [6]. Hydrogen is mainly used in ammonia synthesis for fertilizer production (700 million m^3 per day) and in hydrotreating of transportation fuels. In recent years, with hydrogen being envisaged to play a very major role as a sustainable transportation fuel and for CO_2-free power generation, new methods for hydrogen production in combination with CO_2 capture and storage (CCS) are being developed. In fact, the production of pure hydrogen already involves CO_2 removal, with the CO_2 being vented or used in the food industry. For bulk production of hydrogen for energy, however, more efficient and cheaper processes are needed. This involves new process steps and new demands on the catalysts used in these processes.

Below, we first briefly describe conventional hydrogen production. Then the combination of hydrogen production and CCS is described. Finally, we elaborate on two of the technologies for more efficient hydrogen production with CO_2 capture that are currently in the R&D phase: hydrogen membrane reactors and CO_2 sorption enhanced reactors.

14.2.1
Conventional Hydrogen Production

14.2.1.1 Hydrogen Production from Natural Gas

Steam methane reforming (SMR) is the most widely practiced commercial process for the production of syngas and hydrogen: almost 50% of the world's hydrogen production comes from natural gas. Two equilibrium reactions, steam reforming and the water-gas shift (WGS) reaction, are at the heart of the hydrogen production process:

$$H_2O + CH_4 \leftrightarrows CO + 3H_2 \ (\Delta H_{298}° = 206 \text{ kJ mol}^{-1}) \tag{1}$$

$$H_2O + CO \leftrightarrows H_2 + CO_2 \ (\Delta H_{298}° = -41 \text{ kJ mol}^{-1}) \tag{2}$$

In addition to SMR, other technologies are used for syngas production from natural gas that involve addition of oxygen or air. The catalytic partial oxidation (CPO) reaction is given in Reaction (3) and in autothermal reforming (ATR) this reaction is combined with Reactions (1) and (2).

$$CH_4 + 1/2O_2 \leftrightarrows 2H_2 + CO \ (\Delta H_{298}° = -38 \text{ kJ mol}^{-1}) \tag{3}$$

ATR is used widely, although costs are relatively high due to the high capital and energy costs related to the cryogenic air separation plant (approx. 40% of the total capital costs). For large-scale plants, however, such as gas-to-liquid plants, but also possibly natural gas based power stations with pre-combustion CO_2 capture, ATR becomes more economic than SMR [7].

Figure 14.2 shows that the production of 99% pure hydrogen requires many catalytic processes. The desulfurization section is used to reduce the sulfur content of the natural gas to 0.01 ppm to protect the SMR and WGS catalysts downstream. A supported cobalt-molybdenum catalyst (CoMoS) converts the sulfur compounds into H_2S, which is removed by a ZnO catalyst [5].

The subsequent steam reforming section is operated at very high temperatures: 850–900 °C. The SMR catalysts themselves are already active below 400 °C, but high temperatures are necessary to drive the strongly endothermic reaction forward [8]. In industry, nickel catalysts are used in high-alloy reaction tubes, which are heated by external burners. This design is expensive and leads to heat losses, although much of the heat is recuperated. Noble metal catalysts such as sup-

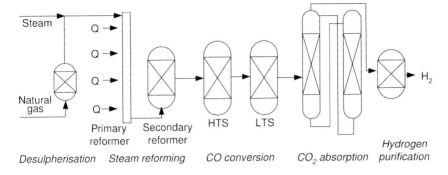

Fig. 14.2 Process diagram for hydrogen production.

ported ruthenium, platinum and rhodium are also very active in SMR [9]. They are only used commercially in special applications, because their price is much higher than that of the Ni-based catalysts. Such applications are, for example, on-board hydrogen production for fuel cell vehicles or auxiliary power units, where compactness of the system is important.

Downstream of the reformer the CO is converted into hydrogen by two subsequent water-gas shift sections: a high-temperature shift (HTS) followed by a low-temperature shift (LTS). This is done because the equilibrium of the WGS reaction lies at the product side at lower temperatures (around 200 °C), but the reaction kinetics are faster at increasing temperature. Therefore, to reach high CO conversions, most of the CO is converted in a HTS section and the remainder is converted within a LTS section.

HTS catalysts have magnetite (Fe_3O_4) as the active metal with chromium oxide (Cr_2O_3) added as a structure stabilizer. They operate between 320 and 500 °C [10], which makes a relatively high CO-slip of a few mole% inevitable due to the thermodynamics of the shift reaction. At higher temperatures and lower steam to CO ratios, some methane may be produced as a by-product. High temperatures combined with low steam-to-CO ratios leads to reduction of the iron oxide, causing sintering and deactivation of the catalyst. LTS catalysts contain copper as the active material, which is supported on zinc oxide (ZnO) and alumina (Al_2O_3). The operation temperature is between 185 and 275 °C. High CO conversions can be reached at these temperatures, up to CO-slip levels of 1000 ppm. As discussed above, the reaction proceeds slowly at low temperatures, so relatively large volumes of catalyst are needed. Above 275 °C, the copper sinters and the catalyst loses its activity. Copper is also very sensitive to sulfur poisoning – for safe operation the sulfur content of the gas should be below 0.1 ppm.

For the final steps, purification of the hydrogen, several technologies are in use. Figure 14.2 also shows CO_2 removal by either physical of chemical absorption followed by conversion of the remaining CO and CO_2 into methane. Alternatively, pressure swing adsorption (PSA) is used to produce pure hydrogen and a purge second stream that contains CO, CO_2, CH_4 and hydrogen. This purge stream is used for underfiring of the reformer. A third option, which is used often in small-scale hydrogen production, is preferential oxidation of CO (PrOx). To use hydrogen in a PEM fuel cell, CO must be removed to below 10 ppm. Some air is added to the gas exiting the LTS catalyst and, at around 150 °C, CO is combusted, while the hydrogen does not react. PrOx catalysts used are supported platinum catalysts and platinum–ruthenium catalysts [11].

14.2.1.2 Hydrogen Production from Other Feedstocks

Approximately 30% of total hydrogen production comes from heavy oil and 20% from coal [6]. Lighter liquid hydrocarbons are – after desulfurization – also converted by steam reforming on nickel catalysts. The downstream treatment of syngas is the same as for natural gas.

Refinery residues, coke as well as coal are converted into syngas by gasification. Several gasification technologies exist, but for hydrogen and power production entrained-flow gasifiers are mainly used [12]. The two major commercially available entrained flow gasifiers are the dry-fed Shell gasifier and the slurry-fed Texaco gasifier (this process is currently marketed by GE). Syngas coming out of the gas clean up section, which still contains up to 1 vol.% of sulfur compounds, can be further converted into hydrogen in two ways. Pathway A in Fig. 14.3 shows that the sulfur-containing, 'raw' gas first enters a sour gas shift catalyst.

The active components of this so-called CoMoS catalyst, cobalt and molybdenum, are only active as sulfides. This implies that sulfur should always be present in the gas to keep the metals in the sulfided state. The operating temperature

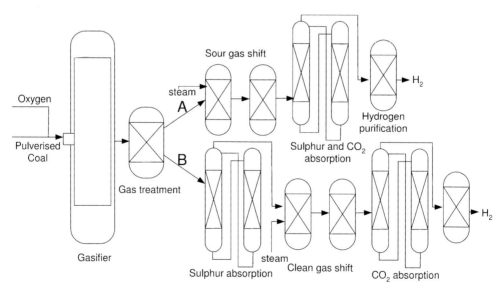

Fig. 14.3 Hydrogen production from coal.

window is 250–500 °C. At higher temperatures, the metal sulfides are more easily converted into metal oxides, so the minimum necessary amount of sulfur in the feed increases with temperature. For operation at 400 °C, the sulfur content of the dry feed should be at least 300 ppm [13].

Subsequently, H_2S is removed from the syngas by a chemical or physical solvent. Upon regeneration of the solvent, the H_2S is fed to a Claus plant to produce elemental sulfur. CO_2 can be removed by the same solvent or in a separate absorber.

In pathway B in Fig. 14.3, H_2S is first removed by absorption and then enters a clean gas shift section. Catalysts used are standard HTS catalysts. The use of LTS catalysts is only possible when sulfur is removed to very low levels (<0.01 ppm), using a Rectisol (methanol) solvent [6]. Downstream the WGS section CO_2 is removed by absorption. When partial pressures are high (>1.5 MPa, [6]) a physical solvent (e.g., methanol) can be used. Physical solvents can be easily regenerated by flashing at reduced pressures. At lower CO_2 partial pressures a chemical solvent [e.g., mono-ethyl amine (MEA)] is used, which requires energy input for steam regeneration of the solvent.

Pathway A, the sour shift case, is especially used for hydrogen production from slurry fed gasifiers. Those gasifiers produce syngas with a relatively high moisture content. Since steam is necessary in the water-gas shift section, this gas can be readily used. In addition, sulfur absorption is generally carried out at low temperature (ambient or even lower), which means that the steam is condensed out of the fuel gas. Therefore, from an efficiency point of view, the need to condense all of the steam from the fuel prior to the sulfur removal unit would undoubtedly make the configuration in pathway B unattractive for slurry fed gasifiers.

Pathway B can be used for dry fed gasifiers, which produce a syngas with very little moisture. After sulfur removal the temperature of the syngas is increased by the addition of superheated steam.

14.2.2
Hydrogen Production with CO_2 Capture

As highlighted elsewhere in this book, CO_2 capture and storage (CCS) has attracted much attention as a possible opportunity to mitigate climate change. The recent publication of the special report on carbon dioxide capture and storage by the Intergovernmental Panel on Climate Change highlights the possibilities for capture of CO_2 from various sources and the storage of CO_2 in geological formations [14].

While the production of hydrogen is a relatively small source of CO_2 nowadays, it is usually produced in large installations with, in many cases, CO_2 separation technology already in place. Currently, the CO_2 produced from hydrogen production is used in the food industry or (in most cases) vented to the atmosphere. In future, hydrogen may well play an important role in two of the largest CO_2 emitting sectors, namely power production and transport.

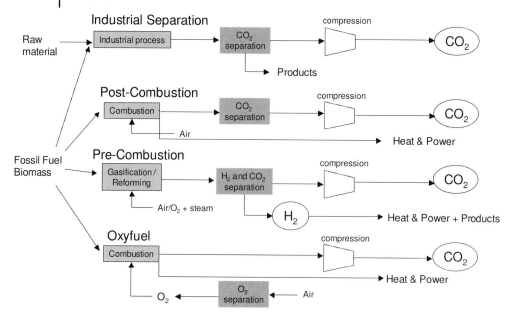

Fig. 14.4 Pathways for CO_2 capture. (Redrawn from IPCC [14]).

14.2.2.1 CO_2 Capture

For the power generation sector several options exist for the capture of CO_2 (Fig. 14.4). Post combustion capture removes the CO_2 from the flue gases of a power plant, using chemical solvents. Much progress is being made in reducing the energy consumption needed for the regeneration of the solvents [14], but the efficiency penalty remains relatively high. Alternatively, combustion of coal or natural gas can be carried out using pure oxygen or oxygen-rich air, to increase the partial pressure of CO_2 in the flue gas. Several industrial oxygen generation technologies exist, such as cryogenic air separation and PSA, but energy demands are relatively high [14]. Chemical looping combustion is an emerging technology in which a solid oxygen carrier is circulated between two beds: a combustor in which a metal oxide (typically NiO) is reduced by natural gas under formation of steam, CO_2, and a hot flue gas and a regenerator in which the depleted oxygen carrier is regenerated using air [15].

The third option, pre-combustion CO_2 capture, involves the conversion of the fuel into a hydrogen and CO_2 mixture and separating the two compounds prior to combustion. The CO_2 is compressed and made ready for transport and storage. The hydrogen is can either be compressed and transported for use as automotive fuel, or be used directly for power generation. The hydrogen production route can in principle be similar to the technologies described above: steam reforming or coal gasification followed by water-gas shift and CO_2 separation. There are, however, a few considerations that make the pre-combustion CO_2 capture different from conventional hydrogen production:

1. CO_2 must be relatively pure. The presence of non-condensables such as nitrogen or methane in the CO_2 stream make compression of the CO_2 to approx. 100 bar very difficult.
2. The efficiency and costs of the complete power production cycle or well to wheel cycle dictate the technology choices. Industrial hydrogen or ammonia synthesis are relatively high-value products compared with electricity and automotive fuel.
3. In a Natural Gas Combined Cycle (NGCC) or an Integrated Gasification Combined Cycle (IGCC), the steam produced from the waste heat boiler is converted into extra electrical power in the steam turbines. Consequently, for hydrogen production, the steam use must be kept as low as possible.

14.3
Novel Processes for Hydrogen Production with CO_2 Capture

For combined hydrogen production and CO_2 capture several novel technologies are in development, most of them for the application in a pre-combustion CO_2 capture combined cycle. The main focus is to reduce the efficiency penalties and other associated costs of CO_2 capture. The most important technologies in the R&D phase, membrane reactors and sorption-enhanced reactors, are described below, with special attention paid to the catalytic aspects.

14.3.1
Hydrogen Membrane Reactors

Hydrogen-selective membranes inside a reactor for reforming remove hydrogen from the reaction zone and, as a consequence of Le Châtelier's Principle, the equilibrium of Reaction (1) is shifted to the product side. (Le Châtelier's Principle states that "any inhomogeneity that somehow develops in a system should induce a process that tends to eradicate the inhomogeneity".) This makes it possible to reach high hydrogen conversions at lower temperatures than ordinary reforming. Hydrogen membranes are also applied in the water-gas shift reaction, which enables the use of just one shift reactor, which can be operated at higher temperatures than the LTS catalyst and still reach high CO conversions.

Hydrogen permeates through the membrane and the gases that do not permeate, the so-called retentate stream, contain a high concentration of CO_2, along with some unconverted methane, CO and non-permeated hydrogen (Fig. 14.5). The driving force of the separation process is provided by the difference in partial hydrogen pressure on both sides of the membrane. This can be achieved by applying a pressure difference over the membrane or by using a flow of a sweep gas (typically nitrogen or steam) on the permeate side [14]. In most cases both a pressure difference and sweep gas are applied.

Membrane materials have to withstand a pressure difference and relatively high temperatures (500 °C and up). Microporous ceramic membranes have been

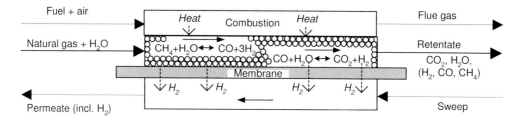

Fig. 14.5 Operating principle of a hydrogen membrane reactor for SMR.

used, but they have a relatively low selectivity for hydrogen [16]. Most commonly studied membranes are dense palladium or palladium-silver alloy membranes [17–22]. Alloying the Pd membrane with silver is necessary to prevent embrittlement of the Pd layer due to the phase change between the α- and β-phase upon cooling the membrane down from, for example, 500 °C to ambient in a hydrogen-containing atmosphere. Generally, the membrane consists of a thin (1–5 μm) layer of Pd-Ag-alloy, supported on a porous metal [18, 22] or ceramic substrate [21]. In addition, high-temperature (700 °C and higher) mixed proton conducting ceramic membranes have been developed for steam reforming applications [23].

Shu et al. [18] provided one of the earlier examples of experimental steam reforming in a membrane reactor. They used a commercial Ni/Al$_2$O$_3$ catalyst and two types of membrane: Pd and Pd-Ag on a stainless steel support. Methane conversion was 51% in the presence of the Pd-Ag membrane at 500 °C and 136 kPa, vs. 37% in the absence of a membrane. Conversion with the pure Pd membrane was somewhat lower. Læsgaard-Jørgensen et al. [19] have used an unsupported 100-μm thick Pd-Ag membrane combined with a commercial Ni/MgO steam reforming catalyst. They observed a strong positive effect of pressure on methane conversion at 500 °C, although the thermodynamics of the SMR reaction are negatively influenced by pressure. At 1 MPa, methane conversion was 61% – to reach this conversion without a membrane a temperature of 707 °C is necessary.

One of the possible problems in a steam reforming membrane reactor is the formation of carbon, either by cracking of methane (Reaction 4) or the Boudouard reaction (Reaction 5).

$$CH_4 \rightarrow C + 2H_2 \quad (4)$$

$$2CO \rightarrow C + CO_2 \quad (5)$$

Whether a driving force for carbon formation exists is dictated by thermodynamics, and so it is dependant on reaction temperature and pressure, and the H/C and O/C ratios in the system (Fig. 14.6). By removing hydrogen from the reaction zone, the H/C ratio decreases and the system moves closer to the thermodynamic area where carbon formation is likely. The arrow in Fig. 14.6 labeled SMR is equal to a reforming mixture with a steam-to-carbon ratio of 3. When hydrogen

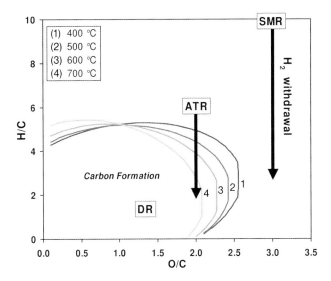

Fig. 14.6 Thermodynamic boundaries for carbon formation at different temperatures and 40 bar.

is removed by a membrane the H/C ratio decreases, but stays out of the thermodynamic carbon formation region, even at 400 °C (in contrast with autothermal reforming, ATR, however). Læsgaard-Jørgensen et al. [19] have calculated that at 500 °C and 7 bar, a steam-to-carbon ratio of higher than 2.5 should be used to avoid carbon formation in a membrane reactor. Notably, however, carbon formation on Ni-catalysts has been observed even outside the thermodynamic region [5]. Carbon formation can be suppressed by adding promoters, such as potassium, to the nickel catalyst [8].

The point labeled DR in Fig. 14.6 represents the conditions for dry reforming (Reaction 6), with a CO_2/CH_4 ratio of 2:

$$CH_4 + CO_2 \leftrightarrows 2CO + 2H_2 \tag{6}$$

For dry reforming, carbon formation is very likely, especially when carried out in a membrane reactor [24]. For this application noble metals are used, which are intrinsically less prone to carbon formation because, unlike nickel, they do not dissolve carbon. Irusta et al. [24] have shown above-equilibrium methane conversion in a reactor equipped with a self-supported Pd-Ag tube. Small amounts of coke were formed on their $Rh/La_2O_3/SiO_2$ catalyst, but this is reported not to have any effect on activity.

Læsgaard-Jørgensen et al. [19] calculated that the rate of reactions involved in SMR is much higher than the rate of penetration of methane through the membrane. Lin et al. [20] have observed that the methane conversion strongly depends on the space velocity and the amount of methane per membrane surface area (load-to-surface ratio). Since the transport of hydrogen through the Pd-layer of a membrane is rate limiting [19], the development of thin, yet stable layers of Pd-Ag with a high permeability is an important goal in membrane R&D. The Energy Research Centre of the Netherlands (ECN) has produced coated ceramic tubes 80 cm long with a smooth 3–5 µm thick Pd-Ag layer on top made by electroless plating [21]. Sintef in Norway have developed a novel method, magnetron sputtering, to make very thin Pd-Ag layers of 1–5 µm, with a very high permeance [22].

14.3.2
Sorption-enhanced Reforming and Water-gas Shift

As with membrane reactors, in sorption-enhanced reforming (SER) or water-gas shift (SEWGS) reactors one of the products is extracted from the reaction zone, thus shifting the reaction equilibrium to the product side. In SER and SEWGS the SMR or WGS catalyst is mixed with a CO_2 sorbent ("acceptor"). The CO_2 produced during the reaction is absorbed and the reverse reaction cannot occur (Fig. 14.7). The hydrogen stream also contains unconverted methane and CO. To produce pure hydrogen an extra hydrogen purification step is needed; for application of SER in pre-combustion CO_2 capture for power generation, the gas can be fed directly to the gas turbine [25].

SER and SEWGS are batch processes: at some point the sorbent is saturated with CO_2 and the equilibrium reaction occurs again. The sorbent subsequently has to be regenerated by either reducing the pressure (pressure swing mode) or increasing the temperature (temperature swing mode). The purge gas is steam, because it can be easily condensed out of the CO_2 stream, yielding pure CO_2 ready for compression and transport (Fig. 14.8). Table 14.1 gives an overview of the most important CO_2 sorbents studied for SER and SEWGS.

Hydrotalcites are layered double hydroxides with the general formula $Mg_6Al_2(OH)_{16}[CO_3] \cdot 4H_2O$. Loading these compounds with potassium carbonate strongly increases their CO_2 uptake [25, 35]. Notably, the hydrotalcite structure already breaks down below 400 °C [26] into a mixed metal oxide.

Air Products and Chemicals pioneered the use of these potassium carbonate promoted hydrotalcite-based materials (K-HTC) for sorption-enhanced reforming of methane [26]. Mixing the K-HTC with a SMR catalyst in a 2:1 ratio gave high (90+%) conversions of methane at temperatures as low as 400 °C. In the first instance Ni-based catalysts were used, but they were not resistant to the environ-

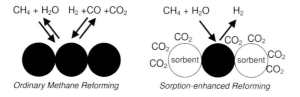

Fig. 14.7 Principle of sorption-enhanced reforming.

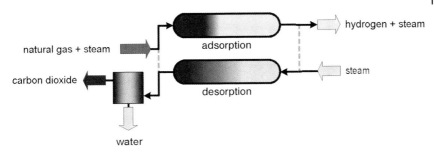

Fig. 14.8 Schematic representation of sorption-enhanced reforming. The topmost reactor is in reforming mode, the bottom one is being regenerated using steam.

Table 14.1 The three most studied classes of CO_2 sorbents in SER and SEWGS.

Sorbent	Mode[a]	T_{ads} (°C)	T_{des} (°C)	Ref.
K-promoted hydrotalcite-based compound	PS	400–500	400–500	25–29
Ca-oxide, dolomite	TS	400–700	>800	30–32
Li-zirconates and silicates	TS	500–650	700–800	33–34

[a] PS = pressure swing, TS = temperature swing.

ment during the regeneration (desorption) step of the cycle. Instead, noble-metal based catalysts were used.

Ding and Alpay also studied sorption-enhanced reforming with K-HTC as sorbent [28], using a commercial Ni-based catalyst. They found that the SER process benefits from higher pressures and that lower steam to methane ratios can be used than in ordinary reforming. Reijers et al. [25] have shown that K-HTC is an effective sorbent between 400 and 500 °C, with an CO_2 uptake of approx. 0.2 mmol g^{-1}. This capacity is low compared with calcium oxides and lithium zirconates. Above 500 °C, the CO_2 sorption capacity of K-HTC decreases rapidly to zero [36].

Reijers et al. [25], using a commercial noble-metal reforming catalyst, have shown that high methane conversions of over 95% are reached at 400 °C, while equilibrium conversion is only 54% under these conditions. Figure 14.9 shows that methane breaks through before CO_2, and that the regeneration part of the cycle is longer than the reaction part. The latter results in a high amount of steam necessary for regeneration, which is very detrimental to the overall efficiency of the system [37]. Using the thermodynamics of the SMR reaction, the conversion of methane can be calculated at a certain equilibrium CO_2 concentration at the exit of the reactor. It follows that for 99% CH_4 conversion at 400 °C, the CO_2 concentration must be well below 1 ppm, which only occurs for a short period

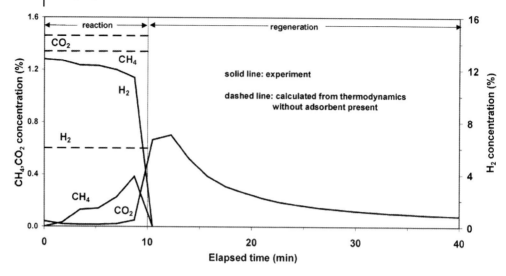

Fig. 14.9 Sorption-enhanced reforming experiment. (Originally published in Reijers et al., *Ind. Eng. Chem. Res.*, 2006 [25]; republished with permission from the American Chemical Society).

in the beginning of an adsorption cycle [37]. At higher temperatures greater CH_4 conversions can be reached at more practical CO_2 output levels but, unfortunately, K-HTC cannot be used above 500 °C.

K-HTC is used more successfully for sorption-enhanced water-gas shift [27]. Breakthrough of CO occurs at the same time as breakthrough of CO_2 [37] and the amount of steam necessary for desorption can be kept relatively low [27]. A commercial high-temperature shift catalyst can be used. During desorption in steam, the catalyst can be oxidized, so it is necessary to add some hydrogen to the purge steam [27].

Calcium oxide, limestone or dolomite (a mineral composed of CaO and MgO) are used as CO_2 acceptors in SER [30] and SEWGS [38]. Calcium oxide acts as a CO_2 acceptor at a wide range of temperatures (450–700 °C). For regeneration, high temperatures are necessary: Balasubramanian et al. have used a regeneration temperature of 975 °C [30]. The energy released by the conversion of CaO into $CaCO_3$ is 170 kJ mol^{-1} [25], which is almost enough to supply the heat for the SMR reaction.

For WGS, commercial catalysts are only operated up to 550 °C and no catalysts are available for higher temperatures, because adverse equilibrium conversion makes the process impractical in the absence of a CO_2 sorbent. Han and Harrison [38] have shown that, at 550 °C, dolomite and limestone have a sufficiently high WGS activity. For SMR a conventional Ni SMR catalyst is used in a 1:1 ratio with CaO [30]. Meyer et al. [32] have also used a Ni-based catalyst in combination with limestone and dolomite, and achieved CH_4 conversions of 95% at 675 °C while the CH_4 conversion at equilibrium was 75%.

A general problem with CaO, limestone and dolomite is the limited lifetime of the CO_2 acceptor material [32]. The capacity for CO_2 is initially very high, but is depleted to almost zero after several cycles. Although the minerals are relatively cheap, this would imply a very considerable stream of waste material coming out of the hydrogen plant. Novel materials are in development with a higher stability [39].

Li-zirconates (Li_2ZrO_3) and Li-orthosilicates (Li_4SiO_4) are also active as high-temperature CO_2 acceptors. The carbonation of Li_2ZrO_3 is accompanied by a heat release of 160 kJ mol^{-1} [34]. The theoretical absorption capacity is huge: 28 wt.% CO_2 for Li_2ZrO_3. For Li_2ZrO_3 the kinetics of CO_2 uptake are slow, and modeling of the SER using this material shows that this limits the applicability of these compounds [34]. Novel specifications of these materials, however, have a better CO_2 uptake kinetics [40]. Lithium orthosilicate has been developed by Toshiba and has more favorable CO_2 uptake kinetics than lithium zirconate [33], as well as reduced volume changes during the reaction–regeneration cycles.

14.4
Conclusions and Catalytic Challenges

To improve the efficiency of combined hydrogen production and CO_2 capture, several technologies are in development that combine catalytic reactions and the separation of either hydrogen or CO_2. Major targeted areas of application are the production of bulk hydrogen as a transport fuel and electricity production with pre-combustion CO_2 capture.

The following catalytic and material challenges can be extracted from an overview of the literature on membrane and sorption-enhanced reforming:

- There is a need for low-cost methane steam reforming catalysts that are active at low temperature and resistant to coke formation under membrane reactor conditions. Low-cost (Ni-based) catalysts are also needed that can withstand regeneration conditions in a sorption-enhanced reformer.

- Although in many WGS applications cases conventional HTS catalysts are used, for some applications water-gas shift catalysts may be needed that operate at temperatures above the current industrial standard. Also, stability in a steam atmosphere, during regeneration of a SEWGS reactor, is an issue.

- CO_2 sorbents should combine a high CO_2 capacity, favorable CO_2 uptake kinetics, low-energy input for regeneration and high stability during cycling. For application of sorption-enhanced water-gas shift in coal gasification gas, sulfur resistance is an issue. Much development work is still needed in this area. Moreover, our understanding of the underlying mechanisms of CO_2 sorption in, especially, promoted hydrotalcite-based materials, and in Li-zirconates and orthosilicates, is still very limited, making development of better sorbents more difficult.

- For Pd-Ag membranes the manufacture of defect-free thin layers, reproducible in production, and stability are major issues. In addition, sulfur resistance and the interaction of the membrane with the catalyst material is important.

14.4.1
Electrochemical Hydrogen Production and Conversion

14.4.1.1 Kinetics of the Electrochemical Hydrogen–Oxygen Processes

Fuel cells and electrolyzers are counterparts for the electrochemical conversion and generation, respectively, of hydrogen and oxygen:

$$H_2 \Leftrightarrow 2H^+ + 2e^-$$

$$1/2 O_2 + 2H^+ + 2e^- \Leftrightarrow H_2O$$

For the overall reaction:

$$H_2 + 1/2 O_2 \Leftrightarrow H_2O$$

$$\Delta G°_{298K, 1atm} = -237.3 \text{ kJ mol}^{-1}$$

The electrochemical rate of an oxidation/reduction reaction is expressed as:

$$I_{ox} = nk^0 FAC_{ox,s} \exp[(1-\alpha)nF/RT)(E - E°)]$$

$$I_{red} = -nk^0 FAC_{red,s} \exp[(-\alpha nF/RT)(E - E°)]$$

in which I = current density of reaction, in A cm^{-2}, n = number of electrons in reaction, k^0 = rate constant (s^{-1}), F = Faraday's constant (96484.56 C mol^{-1}), A = electrode surface area (cm^2), $C_{ox,s}$ = concentration of species to be oxidized (mol cm^{-3}) (at electrode surface), α = transfer coefficient, giving the symmetry of the activation barrier (often taken to be 0.5), R = the gas constant (8.314 J K^{-1} mol^{-1}), E = electrode potential (V), $E°$ = standard potential of electrochemical reaction, 0 V for the H$^+$/H$_2$ couple and 1.229 V for the H$_2$O/O$_2$ couple.

Electrochemistry is in many aspects directly comparable to the concepts known in heterogeneous catalysis. In electrochemistry, the main driving force for the electrochemical reaction is the difference between the electrode potential and the standard potential $(E - E°)$, also called the overpotential. Large overpotentials, however, reduce the efficiency of the electrochemical process. Electrode optimization, therefore, aims to maximize the rate constant k, which is determined by the catalytic properties of the electrode surface, to maximize the surface area A, and, by minimization of transport losses, to result in maximum concentration of the reactants.

Figure 14.10 gives a schematic presentation of the current–voltage characteristics of hydrogen and oxygen on platinum. The theoretical open circuit cell voltage at 25 °C and standard conditions is 1.229 V, for both the fuel cell and electrolyzer.

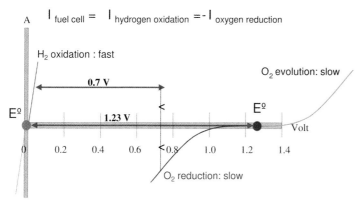

Fig. 14.10 Current–voltage characteristics of hydrogen oxidation and oxygen reduction in a fuel cell, and hydrogen and oxygen evolution in an electrolyzer.

For the reaction of hydrogen and oxygen to generate a current in a fuel cell, the anode needs to be polarized more positive than 0 V vs. NHE (Normal Hydrogen Electrode, the reference potential for all electrochemical reactions) for the oxidation of hydrogen, while the cathode needs to be polarized more negative than 1.229 V vs. NHE for the reduction of oxygen.

As hydrogen oxidation is, on a suitable electrode, a very fast reaction, only a small polarization is needed to generate a significant current density, typically 20–50 mV. The oxygen reduction is a slow reaction, even at the most suitable electrode. To generate a current that is the same as that of the hydrogen oxidation, a much larger polarization is needed, at least 300–500 mV. A fuel cell voltage of 0.5–0.7 V is obtained at normal operation.

For the electrolyzer to generate hydrogen and oxygen by putting electrical energy in the cell the polarization is the mirror-image of the fuel cells: the polarization needs to be negative relative to the standard potential for the facile hydrogen evolution while a polarization to more positive potentials is required for the slow oxygen evolution. An electrolyzer voltage of 1.6–2 V is obtained at the production rate that is often requested for practical purposes.

14.4.1.2 Hydrogen Production by Water Electrolysis

Water electrolysis is a well-known process, dating back to the 19th century, in which DC electricity is used to split water into hydrogen at the cathode and oxygen at the anode. The principle is illustrated for an acid electrolyzer in Fig. 14.11. Electrolysis is the most obvious way to produce hydrogen from wind energy, solar energy and hydroelectricity. This does not necessarily mean that it is the best use of these renewable sources. Direct use of the electricity in fact leads to more avoided CO_2 emissions and primary energy use, as long as the electricity mix in

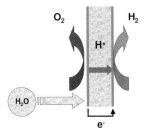

Fig. 14.11 Principle of an acid electrolyzer.

most countries is dominated by fossil fuels. For the long term, however, hydrogen production from renewable electricity is an attractive contribution in the overall mix of options for hydrogen production.

Current electrolyzer technology was originally developed for local hydrogen production to meet industrial demand for high purity hydrogen. Using this technology for hydrogen vehicle fuelling stations leads to an estimated untaxed price of hydrogen of $ 4–8 kg^{-1} [41] and over $ 10 kg^{-1} [42] hydrogen for local hydrogen production on fuelling stations, whereas the target retail cost is $ 2.75 kg^{-1} hydrogen [43]. The cost breakdown shows that both capital cost as well as electricity cost are important contributors to the high overall cost.

The development of electrolyzer technology for the production of hydrogen as a vehicle fuel is aimed at several issues:

1. increase of the electrolyzer efficiency to reduce electricity cost;
2. reduction of capital cost;
3. increase of pressure level of hydrogen output of the electrolyzer.

Three types of electrolysers can be discerned:

1. alkaline,
2. solid polymer,
3. solid oxide.

For the efficiency calculation for the electrolyzer, the operating cell voltage is to be divided by 1.48 V, which is the voltage corresponding to the higher heating value of hydrogen. Other often used numbers for efficiency calculations are the kWh needed for the production of 1 Nm^3 of hydrogen, which is 3.54 kWh, or of 1 kg of hydrogen, which is 39.41 kWh. All electrolyzer efficiencies with liquid water feed quoted below are higher heating value efficiencies.

Alkaline electrolysers are at an industrial stage, especially commercialized for on-site production of ultrapure hydrogen for industrial applications. In general, this hydrogen is needed at low to moderate pressure, and the cost demand is set in comparison to the alternative, which is in general the supply by tube trailers. High purity water is fed to the electrolyzer. State-of-the-art commercial alkaline electrolysers typically operate at HHV systems efficiency of 60–75% [44]. Current densities of alkaline electrolysers lie between 0.2 and 0.4 A cm^{-2} [45, 46]. Hydrogen production capacities are between 10 and 500 Nm^3 h^{-1} [44, 46]. The alkaline electrolysers designs used most often appear to be not very suitable for the variable input power [46] that is to be expected when using renewable electricity.

A clear advantage of alkaline electrolysers is the use of nickel-based electrodes, thus avoiding the use of precious metals. Catalytic research is aimed at the development of more active anodes and cathodes, primarily the development of high surface area, stable structures. Nickel–cobalt spinel electrodes for oxygen evolution and high surface area nickel and nickel cobalt electrodes for hydrogen evolution have been shown at the laboratory scale to lead to a decrease in electrolyzer cell voltage [47]. More active electrodes can lead to more compact electrolysers with lower overall systems cost.

For hydrogen production as an energy carrier, it is beneficial to produce the hydrogen at high pressure. The Solid Polymer Electrolyzer (SPE), or Proton Exchange Membrane Electrolyzer (PEME) offer the best opportunity to generate high pressure hydrogen. The polymer membrane can withstand the pressure differences that can always occur when working at elevated pressure much more easily than the alkaline cells containing liquid electrolyte. By producing hydrogen at high pressure in the electrolyzer, the use of mechanical compression, which poses high maintenance costs and extra capital investment, can be avoided. It is also more suited for use with varying power conditions, in comparison with alkaline electrolysers. Most R&D today is directed to SPE.

As the SPE is an acid electrolyzer, noble metal electrodes are preferred for high current density operation.

SPE electrolysers are at present available at a scale of 0.5–10 Nm^3 hydrogen per hour at an output pressure up to 200 bar. The available systems operate at a rather poor overall efficiency, 50–70% [44, 48]. The projected price level of these systems, at industrial production level, is at least a factor of 3–5 too high.

Platinum on carbon is generally used for the hydrogen-producing cathode, whereas iridium or iridium oxide is used as the oxygen-producing anode [49]. At cell level, current densities of over 1 A cm^{-2} at an electrolyzer voltage of 1.7 V (i.e., at 87% efficiency) are demonstrated, using noble metal loadings of 0.6 mg cm^{-2} at the cathode and 1.5–2 mg cm^{-2} at the anode [49].

Due to the high loading and cost of iridium, binary and ternary mixtures are used in alternative anodes, such as IrO_2/RuO_2 and $IrO_2/RuO_2/SnO_2$ [49]. One should, however, question the long-term stability of these anodes, taking into account the stability domains of ruthenium and tin in acid media [50]. For hydrogen evolution, where corrosion conditions are less severe, nickel alloys are under investigation [51].

A concern for alkaline as well as solid polymer electrolysers should be the sensitivity of the electrolysers to contaminants in the feed water. The minimum required water quality of an industrial solid polymer electrolyzer is ASTM type II water [48], which contains at maximum 5 µg L^{-1} of sodium and chlorine, and 3 µg L^{-1} SiO$_2$ [52]. It should have a maximum conductivity of 0.25 µS cm^{-1}. Clean drinking water meeting Dutch standards contains around 60 mg L^{-1} of sodium, 80 mg L^{-1} of chlorine and has a conductivity of 55 µS cm^{-1} [53]. For both electrolysers, contaminating ions can inhibit the electrochemical reactions, leading to lower power densities and/or efficiencies. In addition, the contaminants can react themselves, leading to poisoning of the electrodes and/or contaminating the product gases. With the SPE electrolyzer, ions are easily exchanged with the protons of the membrane, leading to lower membrane conductivity. For hydrogen production for industrial processes, it is not an issue to process the feed water to meet the water quality requirements. When hydrogen is to be produced as an energy carrier, all systems components that add to system complexity, cost and energy use are highly unwelcome. These new requirements could lead to the need for alternative electrode materials that are less sensitive to specific adsorption of foreign ions. Poisoning of the membrane is a non-catalytic issue but should be addressed as well.

For both low temperature electrolysers, the biggest gain in efficiency is to be expected from an improvement in Balance of Plant components, taking into account the big gap between cell efficiency (80–90%) and system efficiency (50–60%). In the case of SPE electrolysers, catalytic research should therefore be directed to making the catalysts more tolerant to contaminants. For alkaline electrolysers, in addition to this, more active electrodes could lower capital costs.

Solid Oxide Electrolysers (SOE) are in development for steam electrolysis. As electrolysis is an endothermic process, a supply of waste heat can be used beneficially to reduce the electrolyzer voltage, and thus increase its electrical efficiency. Combination with nuclear power generation and geothermal heat sources is often encountered in development programs for SOE.

The simplest SOE electrolyzer splits water into hydrogen and oxygen. Another option is electrolysis of CO_2 and H_2O, which yields syngas, a mixture of CO and H_2 [54]. The syngas can be synthesized further into ethanol or methane. In a conventional steam electrolyzer, air is supplied to the anode side. Increased efficiency can be achieved by supplying CH_4 (natural gas) to the anode, which reduces the required electricity for driving oxygen ions against the concentration gradient across the membrane [55]. In the case of total oxidation of methane, cheap natural gas simply replaces expensive electrical energy. In an advanced device, a catalyst could promote partial oxidation and produce syngas. In this option hydrogen is produced at both the anode and the cathode.

Generally state-of-the-art, zirconia-based fuel cells are being investigated for SOE operation, with LSM (strontium-doped lanthanum manganite) for the anode and nickel zirconia cermets for the cathode [56]. A specific problem of Ni-based cathodes is oxidation at high oxygen (steam) partial pressures. Adding hydrogen to the steam solves this problem but reduces efficiency and increases system complexity. Alternative materials, providing high catalytic activity for the electrolysis reaction and high chemical stability, would simplify the system and reduce costs.

The ultimate catalysis challenge is provided by the reversible SOFC, which can be switched between electrolyzer and fuel cell mode. In this case, to achieve high efficiency in both modes, the electrodes should combine catalytic activity

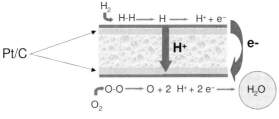

Fig. 14.12 Principle of a proton exchange membrane fuel cell (PEMFC).

for both reactions, as well as chemical stability in reducing and oxidizing gas environments.

14.4.1.3 Proton Exchange Membrane Fuel Cells

The PEMFC (Proton Exchange Membrane Fuel Cell) is a fuel cell with a proton-conducting fluorinated polymer as electrolyte. Figure 14.12 gives a schematic drawing of the PEMFC. At the anode, hydrogen is oxidized to protons. At the cathode, oxygen from air is reduced to water. The PEMFC is in development for various applications.

Transport is the main driving application for the development of the PEMFC, with the toughest specifications with respect to cost, power density and operating conditions. Most car manufacturers aim at fuel cell systems for drivelines operating on pure hydrogen. For auxiliary power systems for passenger cars, trucks, ships and aircrafts, on board hydrogen generation from diesel and kerosene is still the preferred option. In this case, hydrogen is fed to the anode as part of a mixture with carbon dioxide, water, nitrogen and small amounts of carbon monoxide and other contaminants. Such a mixture is generally called "reformate".

For the stationary generation of heat and power the PEMFC is also in development. Fuel cell systems for combined heat and power generation mostly run on natural gas, and sometimes on biogas. Reformate is fed to the anode in these stationary systems. Only for backup power systems, which are designed for only a limited operating time, is pure hydrogen often used as fuel for the anode.

Development for stationary applications is aimed at different specifications, but the materials development mainly coincides with that for transport applications. The operating life time for stationary fuel cells is, however, much longer than for use in cars, 40 000 h or more versus 5000 h for passenger cars. Fuel cells in cars, though, will experience more voltage cycling as well as start–stop cycles.

The PEMFC is technically in quite an advanced status. Fuel cell systems for both transport as well as stationary applications exist in a wide variety and are being operated in demonstration programs under practical conditions [57]. For large-scale market introduction, cost has to be reduced significantly, and durability must be improved. Both items cannot be solved by clever engineering only – new materials are also required.

14 Hydrogen Production and Fuel Cells

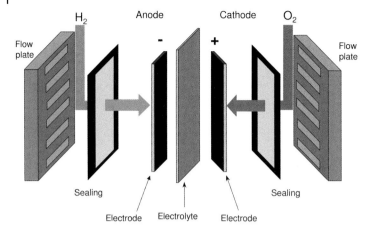

Fig. 14.13 Main components of a PEMFC.

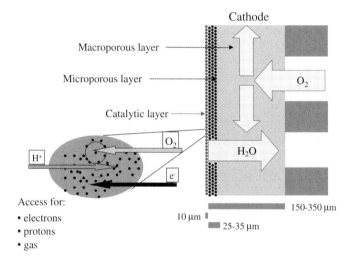

Fig. 14.14 Structural composition of a PEMFC electrode, specifically of a cathode.

Figure 14.13 displays the PEMFC and its components. The electrodes, where the catalytic reactions take place, have a complex structure, as is displayed in Fig. 14.14 for the cathode. Note that, although the reactants and products for the anode differ from those at the cathode, both electrodes have a similar structure.

For a fast catalytic reaction, free access of gas, electrons, protons and water is needed. This leads to a best compromise of the volume fractions of proton-conducting polymer, electron-conducting carbon, active sites and void space.

A current–voltage curve of a PEMFC is displayed in Fig. 14.15. It shows, for a given composition under one set of conditions, the relation between the current

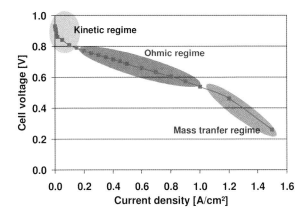

Fig. 14.15 Current density–cell voltage plot for a PEMFC, at 80 °C, 1.5 bar, Pt on carbon electrodes.

density and the cell voltage. The fuel cell electrical efficiency is directly proportional to the fuel cell voltage ΔE_{cell}:

$$\text{Eff}_{FC, LHV} = \Delta E_{cell}/1.23 \text{ V}, \quad \text{or} \quad \text{Eff}_{FC, HHV} = \Delta E_{cell}/1.48 \text{ V}$$

for the lower heating efficiency and the higher heating efficiency respectively.

There is a clear trade-off between power density (voltage × current density) and efficiency. Operating the fuel cell at a high efficiency leads to such a low power density that the investment factor becomes too high. Therefore, most fuel cells are operated at a maximum cell voltage of 0.7 V, giving a cell LHV efficiency of 57%. Contrary to internal combustion engines, operation at low power demands, e.g., when idling, leads to higher fuel cell efficiencies.

In the current–voltage curve in Fig. 14.15, three different regions can be discerned. At low current densities, the performance is kinetically limited. In the linear part, ohmic losses are significant. At high current densities, mass transport losses dominate.

The PEMFC uses platinum on carbon catalysts for the anode and the cathode. State of the art electrodes contain around 0.2–0.4 mg cm^{-2} platinum, generating a power density of 0.5–0.7 W cm^{-2}. Using a total loading of 0.6 mg cm^{-2} and a power output of 0.5 W cm^{-2}, the platinum usage amounts to 1.2 g kWe^{-1}. It has, however, been demonstrated that fuel cells with 0.4 g-Pt kWe^{-1} are achievable when using clean hydrogen and air [58]. The long-term stability of such cells is, though, not known yet, and the use of reformate prescribes higher loadings of PtRu at the anode (0.2 mg$_{PtRu}$ cm^{-2} at minimum). The ultimate goal is to lower the platinum usage to 0.2 g-Pt kW$_e^{-1}$.

The use of noble metals is an important factor in the cost of the fuel cell. Whereas the cost of many components drop when the scale of manufacturing increases, this is not the case for noble metal catalysts. The concern of a real shortage of platinum in the case of large-scale use of fuel cells in vehicles has been proven not to be substantiated [59], but this is based on a significant reduction of its use to 15 g per vehicle, corresponding to the 0.2 g-Pt kWe^{-1} mentioned before. The key issue is to minimize the amount of platinum per kW fuel cell power, while maintaining the power density of the present state of the art. It makes no sense to substitute platinum with another metal that leads to a reduction in power density of several factors.

The catalysts used as a base for electrode manufacturing consist of high loadings of noble metal on carbon (40 wt.% or even higher). These high loadings are used to render a thin electrode with sufficient active sites, typically 10 μm thick. The platinum particle sizes are, even at these high noble metal loadings, in the range 2–3 nm [60].

Catalysts for Hydrogen Oxidation (Anodes) When using pure hydrogen as fuel, low amounts of platinum can be used at the anode, as the oxidation of hydrogen has a high exchange current density. Lowering the loading of platinum at the anode to 0.05 mg cm^{-2} is possible without a significant loss in performance [58]. When reformed fuel is used, platinum cannot be used, as it is too easily poisoned by CO. Even CO_2 has a negative impact, through the reverse water gas shift reaction which leads to CO formation [61].

Platinum alloys offer better tolerance towards CO. Especially, PtRu and PtMo alloys show superior tolerance towards CO [62]. Nonetheless, the performance of presently known catalysts is far from satisfactory.

Figure 14.16 the shows fuel cell stack performance of a 1 kWe atmospheric PEMFC stack using PtRu anodes, operating on various gas compositions. As can be clearly seen, already small concentrations of CO lead to a large decrease of fuel cell performance. An air-bleed of 1.5% air in hydrogen is able to mitigate this ef-

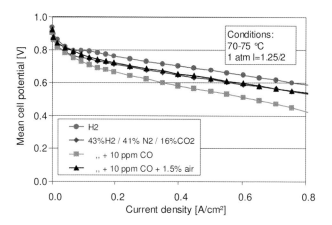

Fig. 14.16 Averaged current density–voltage characteristics of 30 cells for an atmospheric PEMFC stack on various anode feed compositions.

14.4 Conclusions and Catalytic Challenges

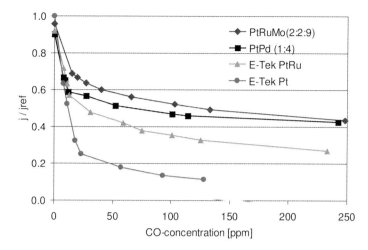

Fig. 14.17 Relative current density for various anode compositions as a function of increasing CO concentrations added to a pure hydrogen feed to the anode; j_{ref} = 840 mA cm^{-2}, at 0.5 V, for platinum/carbon.

fect completely. Such an air-bleed leads to the direct oxidation of CO by O_2 on the noble metal surface. This procedure is frequently used in practice. However, this solution is not ideal. The addition of air needs to be accompanied by careful sensoring of the oxygen concentration, thus leading to increased system complexity. Second, the selectivity of CO oxidation is poor: 1.5% air is needed to oxidize 10 ppm CO. The remainder of the oxygen reacts with hydrogen, leading to lower efficiency and local heat formation. This heat formation is likely to have a negative effect on catalyst and membrane lifetime [63].

Catalyst development has lead to formulations more effective than PtRu, especially at higher CO concentrations. As shown in Fig. 14.17, which gives the drop in performance for different anode formulations when increasing amounts of CO are added to hydrogen, both PtPd [64] as well as PtRuMo [65] lead to a strong improvement in tolerance towards CO.

Another concept, first introduced by Johnson Matthey [66], is the "bilayer" concept. Here, the CO oxidation function is separated from the hydrogen oxidation function. Figure 14.18, measured by using home-made electrodes at ECN, shows the benefit of this bilayer concept. Using bilayer electrodes composed of a PtRu layer closest to the membrane, and a PtMo layer on top of that, a CO concentration of 200 ppm in a gas consisting of 75% H_2 and 25% CO_2 leads to a voltage decline of 50 mV at 350 mA cm^{-2} versus more than 250 mV for PtRu electrodes [67].

Still, the CO tolerance is too low for practical purposes. Ideally, 1000 ppm CO or more should be tolerated without a voltage loss exceeding 20 mV. Moreover, the stability of binary and ternary catalysts under fuel cell operating conditions is an issue.

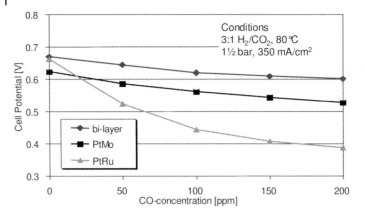

Fig. 14.18 Cell potential at a current density of 350 mA cm^{-2} for a carbon-supported PtRu, a PtMo and a PtRu/PtMo bilayer anode. In the bilayer, PtRu is on the membrane side. (Originally published in G. J. M. Janssen et al., *Fuel Cells*, 2004 [67]; republished with permission from Wiley-VCH).

The strategy towards CO tolerance has therefore been changed, towards the development of proton conducting polymers suitable for high-temperature operation of the PEMFC, i.e., 120 °C and higher. It is already demonstrated [68] that at this temperature, 1000 ppm CO leads to only minor loss of performance. The high temperature operation will be further addressed in the final section of this chapter.

Finally, the anode catalyst can be exposed to very high potentials when a lack of hydrogen occurs in certain parts of the cell. This is known as fuel starvation. The fuel cell, as part of a large series, is intended to generate a high current density anyhow, resulting in electrochemical processes that can generate this current. This often leads the oxidation of the carbon support [63], leading to loss of noble metal. Catalyst formulations that are capable of water oxidation can limit the risk of carbon oxidation.

Catalytic challenges for PEMFC anodes can be summarized as:

- Minimization of the use of platinum to 0.05 mg-Pt cm^{-2}, enabling a fuel cell power density of 1 W cm^{-2}.
- Tolerance towards impurities at the ppm level, even when using hydrogen as fuel.
- For operation on reformate, tolerance towards at least 100–200 ppm CO, as well as to CO_2 in the 10–25% range.
- Stable for 5000 h for automotive and 40 000 h for stationary applications under operating conditions as well as open circuit and shut down conditions.

Catalysts for Oxygen Reduction (Cathodes) At the cathode, the reduction of oxygen requires higher amounts of platinum. Figure 14.10, showing the individual contributions of the hydrogen oxidation and the oxygen reduction to the overall rate, clearly illustrates that optimizing the cathode has the highest priority for hydrogen/air fuel cells. On top of that, mass transfer at the cathode is much more hindered than at the anode. Whereas hydrogen is generally fed as a pure gas, oxygen is fed as part of air. In addition, water is produced at the cathode by the electrochemical reduction of oxygen. Insufficient removal of this product water severely hinders the transport of oxygen to the reaction site.

As a result of poor ionic and electronic conduction the utilization of the platinum surface area at the cathode is much less than 100%.

Strategies towards optimization of the cathode structure consist of improving the removal of water, lowering the ionic and electronic resistance, and preventing the deposition of platinum in inaccessible catalyst pores.

Many researchers have tried to improve the intrinsic activity of platinum. Alloying with chromium has been proven to be very effective [60]. It leads to higher exchange current densities – the challenge is to translate these to higher current densities at the operating voltage [60]. The most promising alloy, reported recently, is a PtCo catalyst [69]. It has a three-fold higher intrinsic activity than platinum, and its stability is promising. Pre-leaching of PtCo prior to the manufacturing of the fuel cell electrode keeps the leaching during fuel cell operation under control. Other alternative active metals are often not stable at the operating conditions, i.e., a rather high potential in combination with a highly acidic environment, leading to leaching of many less noble components from the catalyst surface.

In the search for cheaper materials, an problem often encountered is that the intrinsic activity is so low that it leads to much thicker electrodes, which in turn is counterproductive as it leads to high mass transfer losses.

Catalytic challenges for PEMFC cathodes can be summarized as:

- Minimization of the use of platinum to 0.15 mg-Pt cm^{-2}, enabling a fuel cell power density of 1 W cm^{-2}
- Tolerance towards air impurities at the ppm level.
- Stable for 5000 h for automotive and 40 000 h for stationary applications under operating conditions as well as open circuit and shut down conditions.

High Temperature Operation of the PEMFC The first generation of commercial PEMFCs will use presently known components, consisting of a perfluorosulfonic acid membrane as electrolyte and catalyst compositions as cited above. The electrolyte determines that the fuel cell needs to be operated at fully humidified conditions and limits the operating temperature to 80–90 °C.

There is a strong driving force towards operation at higher temperatures and lower humidity levels: it will make the fuel cell system simpler, heat transfer from the fuel cell will become easier, and tolerance towards impurities will improve [70]. Operation for automotive applications is targeted towards 120 °C, while stationary systems could be operated at even 150 °C and higher. The key component needed to enable this higher operating temperature is the electrolytic membrane. The consequences for the electrocatalysts are, however, also important.

Already in the present set-up, there is concern over the long-term stability of the noble metal particles in the electrodes. Especially during shut-down, high voltages can occur, leading to the corrosion of both the noble metals as well as the carbon support. At higher operating temperatures, these phenomena will be accelerated exponentially. More stable supports (carbon or other), as well as more stable deposited noble metal particles, are needed. The benefit of operation at high temperature could be that water removal at the cathode becomes a less critical issue, as its evaporation will take place much faster. The introduction of new electrolytes, to be used in the electrodes as well for ionic accessibility at high temperature operation, can have a negative impact on electrode kinetics as well. In the present high-temperature fuel cells, using phosphoric acid doped polybenzimidazole, the phosphoric acid leads to poor oxygen reduction kinetics caused by the specific adsorption of phosphoric acid on platinum.

As the state-of-the-art PEMFC electrodes are optimized for operation below 100 °C and the use of perfluorosulfonic acid as electrolyte, significant new R&D will be needed when a new electrolyte emerges with high temperature operation as a consequence.

14.4.1.4 Solid Oxide Fuel Cells (SOFCs)

As the name states, SOFCs consist mainly of metal oxide materials. The electrolyte is a dense oxygen ion conducting ceramic material, typically yttrium-stabilized ZrO_2 (YSZ), which is sandwiched between the porous anode and cathode. The most common type of cathode material is strontium-doped lanthanum manganite (LSM), an ABO_3 type compound having the perovskite crystal structure. The cathode can be mono-phased LSM, but two-phase cathodes, where the LSM is mixed with a good oxygen conductor like YSZ or CeO_2, are also possible. Generally, the anode is nickel based. At the operating conditions the nickel is metallic. YSZ or ceria is mixed with the nickel to prevent sintering during operation, matching of the thermal expansion coefficient of anode and electrolyte and for providing oxygen conductivity of the anode structure.

SOFCs are produced in two geometries, tubular and planar. A further distinction between SOFC types is based on the mechanically supporting structure, which is the thickest component of the cell. The oldest type of SOFC is electrolyte supported, in which case the YSZ electrolyte is, typically, around 0.1 mm thick. Because of the high electrolyte thickness this type of cell must be operated at high temperatures, around 950 °C, for achieving a high oxygen ion conductivity and thereby limiting the losses. Cell types suitable for reduced temperature operation (700–850 °C) are anode-supported (Fig. 14.19) and cathode-supported SOFCs. In these cells the porous anode or cathode structures are relatively thick components while the thickness of the electrolyte layer is reduced to 5–15 µm.

SOFCs are operated with a hydrogen-containing fuel and air. Oxygen from the air is reduced at the cathode. The resulting oxygen ions migrate through the electrolyte to the anode. At the anode the hydrogen is oxidized and reacts with the

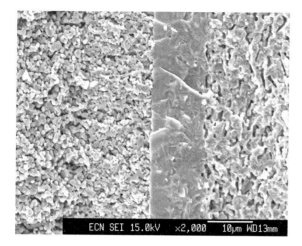

Fig. 14.19 Fracture surface of an anode-supported cell. From left to right, the porous Ni-YSZ anode, the dense 8YSZ electrolyte, and the porous LSM-YSZ cathode.

oxygen ions to water. In contrast to low-temperature fuel cells, a SOFC can also directly oxidize CO, reacting to give CO_2. Whereas low-temperature fuel cells require clean hydrogen as the fuel, SOFC can be fuelled by carbon-containing gases, which may even contain methane, which is steam-reformed internally, either by water added to the fuel stream and/or by the water generated at the anode itself. Figure 14.20 shows a schematic representation of a SOFC cell with two-phase electrodes and the relevant cell reactions.

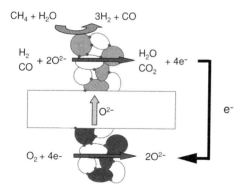

Fig. 14.20 Schematic representation of the relevant SOFC reactions. The steam reform reaction needs nickel as the catalyst. Oxidation of H_2 and CO takes place at the triple phase boundaries (TPBs; represented as smaller dots) in the anode, and is catalyzed by the Ni. At the cathode, O_2 reduction also occurs at TPBs and is catalyzed by LSM.

The supporting structure of tubular cells is commonly manufactured by extrusion of a paste containing the actual electrode ceramic material and, additionally, an organic binder, solvents and additives to achieve the required rheological properties of the paste. After drying, the extrudate is subjected to a two-stage firing process. In the first phase, up to about 300 °C, the organic materials are burned off. In the second phase, at temperatures between 1300 and 1500 °C, the ceramic powder particles are sintered together into a solid structure, still with open porosity in the case of electrode-supported cells, fully dense in the case of electrolyte-supported cells. The thin layers are generally deposited to the tube by dip-coating, after which a further sintering step is applied. Generally, tape-casting of a slurry, containing essentially the same type of components as the paste for extrusion, is applied for manufacturing the mechanically supporting component of planar cells, while the thin layers are applied by screen-printing. The mostly immature processes applied for SOFC manufacturing today, consist of many deposition and sintering steps, involving significant labor. Achieving cost targets for commercial applications will necessitate combining these into a limited number of manufacturing steps (co-extrusion, calandering etc) and co-firing of all the cell components in combination with a high degree of automation. An important aspect in the manufacturing of SOFC cells is matching the sintering shrinkage and thermal expansion coefficients of the components that constitute the cell. Not doing so causes warpage and even cracking of cells. This and other manufacturing requirements put severe restrictions on the freedom of implementing new and improved materials into SOFC cells. Notably, although the nickel in the anode is the only metallic material in the SOFC, it is present as nickel oxide in all manufacturing steps. It is only at the start-up of the SOFC that NiO is reduced to Ni.

Catalysis Issues in SOFC Catalytic processes play in important role in the operation of in particular, the SOFC anode and cathode, and the major challenges for the further development of SOFC towards a commercial project are largely related to improving catalytic activity and robustness. The critical technical issues for SOFC are reliability and lifetime, while economic requirements impose low manufacturing and materials costs.

A general issue addressed by virtually all major SOFC developers is decreasing the operating temperature. Virtually all SOFC cell degradation processes, including the ones involving catalysis, are thermally activated. At operating temperatures around 600 °C, lifetimes in the range of 10–15 years are deemed achievable, which is required for commercialization. In the alternative scenario SOFC stacks will be produced at low cost, which might economically enable stack replacement each five years. Furthermore the reduced temperature will enable the use of cheaper, commercially available materials for the steel interconnects in stacks and system components.

Apart from lifetime, costs requirements demand the use of low amounts of the more expensive materials or the use of low cost materials for cells (preferably both), in addition to low manufacturing costs. As an indication, considering full commercialization of SOFC based systems, cost studies result in required cell production costs around € 100 kW$_e^{-1}$.

Robustness requires a low sensitivity to what might cause permanent damage or degradation of the SOFC system and, hence, the cell – for excursions outside the normal operating window, contaminants in the fuel and the air, thermal and reduction–oxidation (redox) cycling of the anode.

The following subsections address catalysis challenges for each of the cell components separately.

Anode The primary functions of the anode is catalyzing the oxidation of hydrogen and facilitating the reaction with O^{2-} from the electrolyte to H_2O. For hydrogen oxidation Ni appears to be an excellent catalyst, not excelled by any other, relatively cheap, material up to now. In the Ni-YSZ type anode mentioned in the introduction the actual reaction takes place at the triple (gas, YSZ and Ni) phase boundary (TPB) of the anode and, hence, increasing the TPB length by optimizing the anode microstructure is essential for high performances. Nickel also provides electronic conductivity to the anode, which is essential for leading the electrons to the interconnect of the stack. Similarly, YSZ provides oxygen conductivity throughout the anode structure.

Additionally, nickel is a well established steam-reforming catalyst. An ideal SOFC system operated on natural gas applies internal steam reforming, i.e., the reforming of the methane takes place in the anode compartment of the stack. This type of system is favored for system simplicity and costs (no external reformer), and for system efficiency because the heat generated by the cell reaction is directly used by the reform reaction, and hence the cooling requirements of the stack (by air at the cathode side) are significantly reduced.

The main drawbacks of the Ni-based anode are:

- Low tolerance to sulfur compounds that are present in every practical fuel, requiring gas cleanup to below 1 ppm.

- Low redox cycling tolerance. Redox cycling occurs when the fuel flow to the stack is interrupted. Air leaking in from the environment will oxidize the nickel. Upon recovery of the fuel flow the nickel oxide will reduce again. Such a cycle will cause considerable performance loss due to coarsening of the nickel particles in the anode.

- Catalytic promotion of carbon deposition from carbon-containing fuels. Carbon deposited in the anode will generally cause irreversible damage (disintegration) of the anode structure.

Alternative materials potentially capable of solving these issues would be full oxide anodes. Many compounds in this class have been evaluated. The best properties have been achieved by compounds like $La_{0.75}Sr_{0.25}Cr_{0.5}Mn_{0.5}O_3$ [71, 72] and $La_{0.35}Sr_{0.65}Ti_{1-y}Ce_yO_3$ (where $y = 0.05$–0.8) [73]. However, cell performances were still well below those of Ni-cermet anodes, even at high (around 900 °C) operating temperatures, making them unlikely candidates for low-temperature SOFC. The low performance is mainly attributed to the low electronic conductivity of these materials.

Ni-cermet type anodes have been improved by substituting the YSZ by ceria, gadolinium-doped ceria (GDC) and samarium-doped ceria (SDC). Ceria seems to increase the catalytic activity of the cermet for hydrogen oxidation, while SDC and GDC improve the ionic conductivity of the anode. Ni-ceria cermets are considered the main candidate for low-temperature SOFC [74].

Alloying the nickel of the anode to improve tolerance for fuel contaminants has been explored. Gold and copper alloying decreases the catalytic activity for carbon deposition, while dispersing the anode with a heavy transition metal catalyst like tungsten improves sulfur resistance. Furthermore, ceria cermets seem to have a higher sulfur tolerance than Ni-YSZ cermets [75].

It is claimed that Cu-Ceria cermet anodes amply meet targets on sulfur and carbon tolerance and, furthermore, are capable of direct oxidation (no water added to the fuel) of methane and many other higher-carbon fuels, while having high electrochemical performance. Still, a debate is going on as to whether the direct oxidation is not actually internal reforming using the water generated by the cell reaction. Irrespective of this debate, the reported performances with various fuels and high sulfur contaminant levels are impressive [76, 77]. However, because of the low melting point of copper and copper oxide, manufacture of this type of anodes is a very complex and vulnerable process, being the main reason why the vast majority of cell developers disregard this anode option.

The main catalysis challenges related to SOFC anodes can be summarized as the specification of an anode material or mix of materials having the following properties:

- high catalytic activity for hydrogen oxidation at 600 °C;
- controlled catalytic activity for the steam-reforming of methane;
- improved sulfur tolerance related to the catalytic activity for oxidation as well as reforming;
- reduced catalytic activity for carbon deposition;
- sufficient electronic and oxygen-ion conductivity.

These already complex challenges are further complicated by the restrictions imposed by requirements of low cost materials and manufacturing, and chemical and TEC compatibility with the other cell and stack components.

Cathode The primary functions of the cathode are catalyzing the O_2 reduction and transporting the O^{2-} ions to the electrolyte. Furthermore, the cathode should possess sufficient electronic conductivity to lead the electrons from the interconnect to the reaction sites. At practical operating conditions, the ionic conductivity of LSM is several orders of magnitude lower than its electronic conductivity. Therefore, in the single-phase and two-phase LSM type cathodes presented in the introduction the reaction sites are essentially at the TPBs of air, YSZ and LSM. Mixing the LSM with YSZ is one means, apart from refining the microstructure, of increasing the TPB length and extending the reaction sites through a larger part of the cathode structure. In the single-phase cathode the reaction sites are restricted to the electrode–electrolyte interface. LSM type cathodes perform satisfactorily at operating temperatures over 800 °C.

Due to low catalytic activity of the LSM type cathodes at operating temperatures below 800 °C, alternative cathode materials need be used. Many materials have been investigated (Fig. 14.21). For temperatures as low as 700 °C, $La_xSr_{1-x}CoO_3$ (LSC) and $(La_xSr_{1-x})(Co_y,Fe_{1-y})O_3$ (LSCF) are rather established, while $Ba_xSr_{1-x}Co_yFe_{1-y}O_3$ (BSCF) and $Sm_xSr_{1-x}CoO_3$ (SSC) [78] are materials quite recently identified as potentially high-performance cathodes for operation at even lower temperature. The chemical and physical properties of these oxides vary significantly with the values of x and y, and these parameters are selected for the optimum tradeoff between catalytic activity, thermal expansion coefficient (should match with that of the applied electrolyte), electron and ion conductivity and chemical stability.

During the cathode sintering phase at high temperatures, these materials tend to react with zirconia electrolytes to give $La_2Zr_2O_7$, which has low ion conductivity and hence will result in low performance cells. The reaction is generally prevented by sandwiching a thin ceria layer between the electrolyte and the cathode.

The intermediate- and low-temperature materials exhibit both significant oxygen ion and electron conductivity. This means that the oxygen reduction can take place everywhere in the cathode and is not restricted to the TPBs as in the LSM-type cathodes. An option for compensating the relatively low oxygen ion conductivity of LSC and LSCF at operating temperatures around 600 °C is by mixing the material with (doped) ceria [79, 80].

A serious problem of all currently known cathodes is Cr poisoning. SOFC stacks are built with ferritic stainless steel interconnects. This class of steels obtains its high-temperature corrosion resistance by the formation of stable chromium oxide or spinel layers. At the SOFC operating conditions Cr reacts with water in the air to give $CrO_2(OH)_2$, which decomposes to Cr_2O_3 at the reaction sites of the cathode. The Cr_2O_3 reacts locally with the cathode and deactivates the site. Because of the higher number of reaction sites, the mixed conductivity oxide cathodes are less sensitive to Cr poisoning than cathodes that depend on TPBs. However, both cathode types need significant improvement concerning their Cr resistance to achieve lifetime targets.

Hence, catalysis related challenges for SOFC cathode are the development of cathode specifications, i.e., material and microstructure, having high catalytic activity for oxygen reduction at 600 °C, high electron and ion conductivity, and a low sensitivity for poisoning by volatile Cr species. Again, as for the anode, cost and compatibility related requirements have to be considered.

Electrolyte The challenges for electrolyte developments are non-catalytic in nature. However, because anode and cathode development activities have to consider compatibility and interaction with the electrolyte, the most important issues are mentioned here. First, the oxygen ion conductivity of the electrolyte should

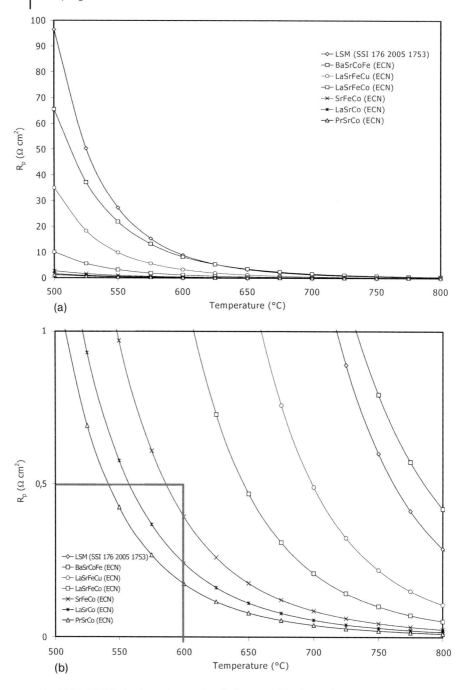

Fig. 14.21 (a) Polarization resistance (partly determined by the catalytic properties of the electrode) of cathode materials, as measured by impedance spectroscopy of symmetrical cells. (b) The area enclosed by the box (bottom left) represents the target area for low-temperature cathode development.

be sufficiently high at 600 °C. This can be achieved by materials selection and by reducing the thickness of the electrolyte layer. Additionally, the electrolyte layer has to be mechanically reliable and chemically stable. Although there are much better oxygen-conducting materials than the zirconias, particularly at 600 °C, mechanical and stability requirements mean that most SOFC developers to base their technology on zirconia. Most common compositions are 3YSZ for electrolyte-supported cells and 8YSZ for electrode-supported cells. Scandium-doped zirconia could offer an attractive combination of conductivity and mechanical properties, but uncertainties on the future price and availability of scandium is a serious drawback. Double layer electrolytes of YSZ and doped ceria, with ceria at the cathode side, might also prove a viable option, particularly for 600 °C operation.

References

1. *Strategic Research Agenda and Deployment Strategy, European Hydrogen and Fuel Cell Platform*, **2005**. Available at http://www.hfpeurope.org/hfp/keydocs
2. *Freedom Car and Vehicle Technologies Multi-year Program Plan*, U.S. Department of Energy, August **2004**. Available at http://www1.eere.doe.gov
3. *FutureGen, Integrated Hydrogen, Electric Power Production and Carbon Sequestration Research Initiative*, U.S. Department of Energy, March **2004**. Available at http://www.fossil.energy.gov
4. Information on Japanese activities can be found at the Japanese Fuel Cell Development Information Center, at http://www.fcdic.com/eng/
5. H. Bartholomew, R.J. Farrauto, *Fundamentals of Industrial Catalytic Processes*, 2nd edn, John Wiley & Sons, Hoboken, NJ, **2006**, Ch. 6, pp. 339–486.
6. A.G. Collot, *Prospects for Hydrogen from Coal*, IEA Clean Coal Centre Report CCC/78, December **2003**, London.
7. J.R. Rostrup-Nielsen, *Catal. Today* 71 (**2002**) 243–247.
8. J.R. Rostrup-Nielsen, J. Sehested, *Adv. Catal.* 47 (**2002**) 65–139.
9. J. Wei, E. Iglesia, *J. Catal.* 225 (**2004**) 116.
10. L. Lloyd, D.E. Ridler, M.V. Twigg, The water gas shift reaction, in: *Catalyst Handbook*, 2nd edn, M.V. Twigg (Ed.), Manson Publishing, London, **1996**.
11. P.J. de Wild, M.J.F.M. Verhaak, D.F. Bakker, Catalysts for the selective oxidation of carbon monoxide in hydrogen-containing gases, Eur. Pat. EP1115651, **1999**.
12. C. Higman, M. van der Brugt, *Gasification*, Elsevier, Amsterdam, **2003**.
13. P. Frank, Sulfur tolerant shift catalyst – Dealing with the bottom of the barrel problem, *Synetix*, Johnson Matthey Group, **2003**.
14. IPCC, **2005**, *IPCC Special Report on Carbon Dioxide Capture and Storage*. Prepared by Working group III of the Intergovernmental Panel on Climate Change (eds. B. Metz, O. Davinson, H.C. de Coninck, M. Loos, L.A. Meyer). Cambridge University Press, Cambridge UK and New York.
15. T. Mattisson, M. Johansson, A. Lyngfelt, *Fuel* 85 (**2006**) 736.
16. M. Bracht, P.T. Alderliesten, R. Kloster, R. Pruschek, G. Haupt, E. Xue, J.H.R. Ross, M.K. Koukou, N. Papayannakos, *Energy Conversion Manage.* 38 (**1994**) S159.
17. E. Kikuchi, *Sekiyu Gakkaishi* 39 (**1996**) 301–313.
18. J. Shu, B.P.A. Grandjean, S. Kaliaguine, *Appl. Catal. A.* 119 (**1994**) 305.
19. S. Læsgaard-Jørgensen, P.E. Højlund Nielsen, P. Lehrman, *Catal. Today* 25 (**1995**) 303.
20. Y.M. Lin, S.L. Liu, Y.T. Chuang, Chu, *Catal. Today*, 82 (**2003**) 127.
21. J.W. Dijkstra, Y.C. van Delft, D. Jansen, P.P.A.C. Pex, Development of a hydrogen membrane reactor for power production with pre-combustion decarbonisation, Proceedings of the 8th International

Conference on Greenhouse Gas Technologies (www.GHGT8.no), 20–23 June **2006**, Trondheim, Norway.

22 H. Klette, T. Peters, A. Mejdell, R. Bredesen, Development of palladium-based hydrogen membranes for water gas shift conditions, Proceedings of the 8th International Conference on Greenhouse Gas Technologies (www.GHGT8.no), 20–23 June **2006**, Trondheim, Norway.

23 B. Vigelund, K. Aasen, Development of a hydrogen membrane for the HMR process concept, Proceedings of the 8th International Conference on Greenhouse Gas Technologies (www.GHGT8.no), 20–23 June **2006**, Trondheim, Norway.

24 S. Irusta, J. Múnera, C. Carrara, E.A. Lombardo, L.M. Cornaglia, *Appl. Catal. A* 287 (**2005**) 147.

25 H.Th.J. Reijers, S.E.A. Valster-Schiermeier, P.D. Cobden, R.W. van den Brink, *Ind. Eng. Chem. Res.* 45 (**2006**) 2522.

26 J.R. Hufton, S. Mayorga, S. Sircar, *AIChE J.* 45 (**1999**) 248.

27 R.J. Allam, R. Chiang, J.R. Hufton, P. Middleton, E.L. Weist, V. White in *Carbon Dioxide Capture for Storage in Deep Geologic Formations–Results from the CO_2 Capture Project*, ed. D.C. Thomas, S.M. Benson, Elsevier, Oxford, **2005**, Vol. 1, Ch. 13.

28 Y. Ding, E. Alpay, *Chem. Eng. J.* 55 (**2000**) 3929.

29 G.H. Xiu, P. Li, A.E. Rodrigues, *Chem. Eng. Sci.* 57 (**2002**) 3893.

30 B. Balasubramanian, A. Lopez Ortiz, S. Kaytakoglu, D.P. Harrison, *Chem. Eng. Sci.* 54 (**1999**) 3543.

31 N. Hildenbrand, J. Readman, I.M. Dahl, R. Blom, *Appl. Catal. A* 303 (**2006**) 131.

32 J. Meyer, D.Ø. Eriksen, R. Glöckner, R. Ørjasæter, Hydrogen production by integrated reforming and CO_2 capture, 1st European Hydrogen Energy Conference, Grenoble, France, September 2003. Association Française de l'Hydrogène, Paris, France (**2003**).

33 M. Kato, S. Yoshikawa, K. Nakagawa, *J. Mater. Sci. Lett.* 21 (**2002**) 485–487.

34 E. Ochoa-Fernández, H.K. Rusten, H.A. Jakobsen, M. Rønning, A. Holmen, D. Chen, *Catal. Today* 106 (**2005**) 41.

35 S. Nataraj, B.T. Carvill, J.R. Hufton, S.G. Mayorga, T.R. Gaffney, J.R. Brzozowski, Materials selectively absorbing CO_2 from CO_2 containing streams, EP patent no. 1006079A1, **2000**.

36 N.D. Hutson, S.A. Speakman, E.A. Payzant, *Chem. Mater.*, 16 (**2004**) 4135–4143.

37 P.D. Cobden, P. van Beurden, H.Th.J. Reijers, G.D. Elzinga, S.C.A. Kluiters, J.W. Dijkstra, D. Jansen, R.W. van den Brink, Sorption-enhanced hydrogen production for pre-combustion CO_2 capture: thermodynamic analysis and experimental results, *Int. J. Greenhouse Gas Control.* Vol. 1 (**2007**) 170–179.

38 C. Han, D.P. Harrison, *Chem. Eng. Sci.* 49 (**1994**) 5875.

39 Z.S. Li, N.S. Cai, Y.Y. Huang, H.J. Han, *Synthesis, Energy Fuels* 19 (**2005**) 1447.

40 E. Ochoa-Fernández, M. Rønning, T. Grande, D. Chen, *Chem. Mater.* 18 (**2006**) 1383.

41 J. Ivy, *Summary of Electrolytic Hydrogen Production*, NREL/MP-560-36734, September **2004**. Available at www.nrel.gov

42 G.J. Kramer, J. Huijsmans, D. Austgen, Clean and green hydrogen, World Hydrogen Energy Conference, 13–16 June 2006, Lyon. Association Française de l'Hydrogène, Paris, France (**2006**).

43 DoE Hydrogen, *Fuel Cells and Infrastructure Technologies Program*, Revision 1, **2005**; available at www.eere.energy.gov/hydrogenandfuelcells

44 E. Breysse-Carabeuf, L. Bocquet, M. Junker, Proceedings of the 1st European Hydrogen Energy Conference, 2–5 September 2003, Grenoble, France. Paper C02/186. Association Française de l'Hydrogène, Paris, France (**2003**).

45 A.F.G. Smith, M. Newborough, *Low Cost Polymer Electrolyses and Electrolysis: Implementation Scenarios for Carbon Abatement*, Report to the Carbon Trust and ITM-Power PLC, November **2004**. Available at www.itm-power.com

46 R. Wurster, J. Schindler, *Handbook of Fuel Cells*, ed. W. Vielstich, A. Lamm, H.A. Gasteiger. Wiley, Chichester, **2003**, Vol. 3, Ch. 5.

47 H.B. Suffredini, J.L. Cerne, F.C. Crnkovic, S.A.S. Machado, L.A. Avaca, *Int. J. Hydrogen Energy*, 25 (**2000**) 415.

48 F. Barbir, *Solar Energy* 78 (**2005**) 661.
49 S.A. Grigoriev, V.I. Porembsky, V.N. Fateev, *Int. J. Hydrogen Energy*, 31 (**2006**) 171.
50 M. Pourbaix, *Atlas of Electrochemical Equilibria in Aqueous Solutions*, 2nd edition, National Association of Corrosion Engineers, Houston, **1974**.
51 E. Navarro-Flores, Z. Chong, S. Omanovic, *J. Mol. Catal. A*, 226 (**2005**) 179.
52 Dossier water, Analis Digest No 4904/02; available at www.analis.be
53 Water quality measurement 2nd quarter **2006**, Location Bergen, NL; downloaded from www.pwn.nl.
54 S.H. Jensen, M. Mogensen, Perspectives of high temperature electrolysis using SOEC. Paper presented at 19th World Energy Congress 2004, Sydney (AU), 5–9 Sep 2004. World Energy Council, London, UK (**2004**).
55 A.Q. Pham, High efficiency steam electrolyzer, NREL/CP-570–26938.28-8-2000, National Renewable Energy Lab., Golden, CO (USA). Proceedings of the 1999 U.S. DOE Hydrogen Program Review. US Department of Energy, Washington DC, USA (**2000**).
56 S.H. Jensen, J.v.T. Høgh, R. Barfod, M. Mogensen, High temperature electrolysis of steam and carbon dioxide, in Energy Technologies For Post Kyoto Targets in the Medium Term. Proceedings Risø International Energy Conference, Risø (DK), 19–21 May 2003. Risø National Laboratory, Risø, Denmark (**2003**).
57 F.A. de Bruijn, *Green Chem.* 7 (**2005**) 132.
58 H.A. Gasteiger, J.E. Panels, S.G. Yan, *J. Power Sources* 127 (**2004**) 162.
59 Platinum Availability and Economics for PEMFC Commercialisation. TIAX report to US Department of Energy, Dec **2003**. DOE report nr. DE-FC04–01AL67601. Available at www1.eere.energy.gov/hydrogenandfuelcells
60 T.R. Ralph, M.P. Hogarth, *Platinum Metals Rev.* 46 (**2002**) 3.
61 F.A. de Bruijn, D.C. Papageorgopoulos, E.F. Sitters, G.J.M. Janssen, *J. Power Sources* 110 (**2002**) 117.
62 T.R. Ralph, M.P. Hogarth, *Platinum Metals Rev.* 46 (**2002**) 117.
63 S.D. Knights, K.M. Colbow, J. St-Pierre, D.P. Wilkinson, *J. Power Sources* 127 (**2004**) 127.
64 D.C. Papageorgopoulos, M. Keijzer, J.B.J. Veldhuis, F.A. de Bruijn, *J. Electrochem. Soc.* 149 (**2002**) A1400.
65 D.C. Papageorgopoulos, M. Keijzer, F.A. de Bruijn, *Electrochim. Acta* 48 (**2002**) 197.
66 S. Ball, S. Cooper, K.Z. Dooley, G.A. Hards, G. Hoogers, ETSU F/02/00160/REP, **2001**.
67 G.J.M. Janssen, M.P. de Heer, D.C. Papageorgopoulos, *Fuel Cells* 4 (**2004**) 169.
68 Q. Li, R. He, J.O. Jensen, N.J. Bjerrum, *J. Electrochem. Soc.* 150 (**2003**) A1599.
69 H.A. Gasteiger, S.S. Kocha, B. Sompalli, F.T. Wagner, *Appl.Catal. B* 56 (**2005**) 9.
70 C. Wieser, *Fuel Cells* 4 (**2004**) 245.
71 S. Tao, J.T.S. Irvine, *Nat. Mater.* 2 (**2003**) 320.
72 A.K. Azad, J.T.S. Irvine, Characterisation of La0.75Sr0.25Mn0.5Cr0.5-xAlxO3 as anode materials for solid oxide fuel cells, Proceedings of 7th European SOFC Forum, Luzern, Switzerland, 2006. European Fuel Cell Forum, Oberrohrdorf, Switzerland (**2006**).
73 O. Marina, M. Walker, J. Stevenson, Development of ceramic composites as SOFC anodes, Proceedings Fuel Cell Seminar, Miami, Florida, 2003. Courtesy Associates, Washington DC, USA (**2003**).
74 S. Zha, W. Rauch, M. Liu, *Solid State Ionics* 166 (**2004**) 241.
75 M. Smith, A.J. McEvoy, Desensitising cermet anodes to sulfur, Proceedings of 7th European SOFC Forum, Luzern, Switzerland, 2006. European Fuell Cell Forum, Oberrohrdorf, Switzerland (**2006**).
76 R.J. Gorte, H. Kim, J.M. Vohs, *J. Power Sources* 4639 (**2002**) 1.
77 R. Remick, O. Spaldon-Stewart, K. Krist, Alternative mechanism for direct oxidation of dry methane on ceria-containing anodes, Proceedings Fuel Cell Seminar, San Antonio, Texas, 2004. Courtesy Associates, Washington DC, USA (**2004**).
78 Z. Shao, S.M. Haile, A high-performance cathode for the next generation of solid-oxide fuel cells, *Nature*, 431 (**2004**) 170.

79 N. Oishi, Y. Yoo, I. Davidson, La0.6Sr0.4Co0.2Fe0.8O3–Ce0.8Sm0.2O2 composite cathode for operation below 600 °C, Proceedings of SOFC IX, 207th ECS Meeting, Quebec City, Canada, 15–20 May 2005. The Electrochemical Society, Pennington NJ, USA (**2005**).

80 F. van Berkel, S. Brussel, M. van Tuel, G. Schoemakers, B. Rietveld, P.V. Aravind, Development of low temperature cathode materials, Proceedings of 7th European SOFC Forum, Luzern, Switzerland, 2006. European Fuel Cell Forum, Oberrohrdorf, Switzerland (**2006**).

15
Pathways to Clean and Green Hydrogen

Gert J. Kramer, Joep P. P. Huijsmans, and Dave M. Austgen

15.1
Introduction

This chapter addresses the issue of the production and distribution of hydrogen for the transportation sector by considering how hydrogen fuel can meet societal expectations within the limits of the anticipated energy resource availability, and under the condition that hydrogen must be a cost-effective alternative fuel. Specifically, we focus on the prospects for "clean hydrogen", defined as hydrogen produced from fossil fuels with carbon capture and sequestration (CCS) and for "green hydrogen", i.e., hydrogen made from renewable electricity or from biomass.

At the highest level of abstraction, hydrogen fuel is one of the options being considered in the World's quest for "sustainable mobility." This, as discussed for instance in a recent report of the World Business Council for Sustainable Development, is a multifaceted subject, encompassing everything from future fuels to future vehicle technology and transportation infrastructure, in all of which major strides forward are needed [1]. In what follows we restrict ourselves to fuels. But the fact that a "sustainable fuel" is just an element of the complex issue of "sustainable mobility" makes clear that such a fuel must meet a complex and sometimes contradictory set of demands. The following three requirements stand out: a sustainable fuel should, in a cost effective way, contribute to

- improving local air quality,
- reducing "well-to-wheel" CO_2 emissions,
- improving the energy supply diversity and thereby the energy security of the transportation sector.

These are the touchstones for all alternative fuels, which – along with hydrogen – include biofuels, compressed natural gas (CNG), and synthetic liquid fuels derived from natural gas (NG) and coal. Although this chapter focuses on hydrogen, it is relevant to recognize that this whole spectrum of future fuels will in the de-

Catalysis for Renewables: From Feedstock to Energy Production
Edited by Gabriele Centi and Rutger A. van Santen
Copyright © 2007 WILEY-VCH Verlag GmbH & Co. KGaA, Weinheim
ISBN 978-3-527-31788-2

cades ahead – in different ways and to different extents – contribute to "sustainable mobility". What makes hydrogen such an attractive fuel is the fact that – if properly manufactured, distributed and used – it can simultaneously deliver on all three of the above-mentioned requirements.

Hydrogen's contribution to local air quality improvement is a well-known attribute for fuel cell vehicles (FCVs), but tailpipe emissions from hydrogen internal combustion engines (H2-ICEs) would also be very low [2]. For hydrogen to properly fulfill its promise in this respect, local emissions from the hydrogen production side must be controlled. Studies attest that if state-of-the-art emission control at hydrogen manufacturing sites is applied, the overall improvements in local air quality are significant [3].

While hydrogen is obviously an energy *carrier*, its contribution to energy supply diversification stems from the unique fact that hydrogen (like electricity but unlike hydrocarbon fuels) can be made from any energy source, either through reforming or gasification of the broad range of fossil feedstocks of NG, (unconventional) oil and coal and from biomass, or from electricity via electrolysis. At present 95% of commercial hydrogen production employs reforming and gasification, 5% is from electrolysis.

This leaves hydrogen's contribution to CO_2 reduction to be assessed. This is arguably a most complicated assessment, and one that is very much interwoven with policy choices and stakeholder preferences. The present chapter aims to support the wider community of stakeholders in the "Hydrogen Economy" in articulating preferences and formulating policy. The main elements we will bring to bear on the topic are:

- Energy resource availability for hydrogen production between now and 2050.
- The scope for CO_2 emissions reduction for the various production and distribution pathways.
- The cost of hydrogen "at the pump," i.e., the sum of production, distribution and retail costs, for the same range of options.

Our focus on "clean" and "green", gives implicit recognition to the often-heard argument that hydrogen cars cannot be justified on the basis of efficiency alone [4]. Stated in a more positive manner: when hydrogen is made from fossil fuels, the opportunity for carbon capture and sequestration is too good to miss, and – as will be demonstrated below – adds relatively little to the total cost. A strong case can be made (to which this chapter hopes to contribute) that "clean hydrogen" will precede "green hydrogen" by a few decades in most regions. This is based on the scope for "clean fossil" in the medium term (10–30 years from now) as well as for the limited availability of "green" energy resources in the same period.

This chapter is split in three main sections, which substantiate the above in three different ways. First we look at energy availability on the basis of Shell's long-term energy scenarios; second, we look at the cost of hydrogen fuel for four NG-based production pathways, including the centralized and decentralized (forecourt) production, and chemical (reforming), and a fifth pathway using electrolysis. Thirdly, we address the scope of carbon capture and come forward with an estimate of the likely level of "remaining" well-to-wheel emissions for "clean hydrogen." As the emphasis in these sections is on NG-based production, we end with two smaller sections that discuss the prospects for coal and biomass.

15.2
Energy Resource Availability

Shell has a long tradition of scenario-based business planning. Scenarios come in pairs and they paint different, but internally consistent, pictures of the future development of the world (Global Scenarios) and of energy supply and demand in these worlds (Long-term Energy Scenarios). The current scenarios are named "Dynamics as Usual" and "The Spirit of the Coming Age". For further background the reader is referred to the Shell scenarios [5]. We use here the energy supply estimate as a proxy to the likely availability of the spectrum of energy resources until 2050.

Figure 15.1 shows how supply and demand matches on a global scale. Though the segmentation is relatively crude, it adequately illustrates the scope by which hydrogen can contribute to CO_2 emissions reduction and energy security. The most striking feature is the shortfall of non-carbon electricity supply in meeting to total energy demand of the power sector. In both scenarios, even in 2050, just

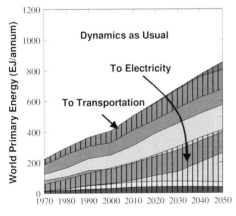
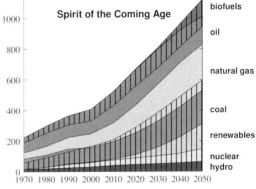

Fig. 15.1 The world energy supply in the two Shell long-term energy scenarios Dynamics as Usual and Spirit of the Coming Age. The stacking order is such that carbon-neutral electricity-delivering sources (hydro, nuclear and renewables excluding biofuels) are at the bottom. When overlaid with the energy demand of the electricity sector (lower hashed area) one observes that carbon-free electricity supply falls far short of electricity demand. Similarly, at the top the two resources that most naturally supply the transportation sector, oil and biofuels, are overlaid by the demand of that sector (upper hashed area), illustrating the continued constraint of the sector, even as biofuels grow. Note: the remaining usage (between the hashed regions) is used in industry, commercial and domestic heating.

about half of the power sector is supplied in a carbon-neutral way, part green (hydro and renewables) and part nuclear. The remainder is met by coal and natural gas. In this light, the often-heard argument appears valid, that new renewables are more productively employed if they replace coal-based power production rather than gasoline and diesel in the transportation sector [6]. The argument for biomass is somewhat more involved; we return to this in Section 15.6.

The argument is further strengthened if we look at the scale at which energy systems are physically integrated. Taking the US and Canada, Western Europe and Japan as examples, we find in all three cases that the regional breakdown is the same as the world-scale breakdown of Fig. 15.1, hence the inference must be the same: there is very limited need or scope for hydrogen as a carrier of "green" energy in the next half a century. The argument only breaks down at national or island levels. A point in case is France, where 85% of electricity demand is met by nuclear power and where, therefore, on a national level – but not at the European level – the argument could be (and is) made that France would benefit from making hydrogen from non-fossil energy resources, both renewable and nuclear. Another example is Iceland, whose commitment to renewable hydrogen is based on an exceptionally rich endowment with renewable energy resources and low population density.

Moving to the transport sector, Fig. 15.1 quantifies the continued oil stress of the transportation sector. Even when a rather aggressive development of biofuels is included, the transportation demand for oil does not go down. This underlines that in all likelihood the energy security issue is as much a long-term issue as the climate change issue. It induces a strong motivation around the world to develop non oil-based transportation fuels, even in the absence of CO_2 emission concerns. Current investments in the area of Gas-to-Liquids (GTL) and plans for Coal-to-Liquids (CTL) attest to this [7].

When the supply to the electricity and transportation sectors is jointly taken into consideration, one is led to conclude that the energy supply diversity is best served by allowing "green electricity" to maximally penetrate the electricity sector and simultaneously swing the deployment of NG and coal to instead serve the transportation sector. (Note that there may well be synergies between hydrogen fuel production and clean power production. These will be briefly touched upon in Section 15.6 for coal below.) The extent to which one thereby accommodates the objective of CO_2 emissions reduction depends on the mode of hydrogen production and distribution, and the extent to which it enables carbon capture and sequestration. To a discussion thereof we now turn.

15.3
Modes of Hydrogen Production and Distribution

As stated at the outset, for hydrogen to contribute to "sustainable transport," it is not sufficient that it reduces transport CO_2 and improves supply diversity, it must also do so in a cost-effective manner vis-à-vis the alternatives. The above discussion of energy availability and CO_2 mitigation forces us to look in first instance to central, industrial-scale production of hydrogen from NG and coal, where one could exploit the point-source nature of the CO_2 production to capture and sequester it. Central production of hydrogen introduces the problem of hydrogen distribution. Pipeline transport is the most obvious means. In places where hydrogen pipelines exits, it has proven both practical and cost-effective to locate hydrogen filling stations near them. However, a pipeline network is unlikely to grant relevant coverage during the early stages of hydrogen rollout. This makes it advisable to consider high-pressure gaseous truck transport and cryogenic liquid truck transport as the two main distribution options.

While today liquid hydrogen truck transport is an accepted mode of distribution in the merchant gas market, it has often been discarded in relation to hydrogen fuel, on the basis of the energy intensity of the process, which seems counterproductive to meeting the aims for which hydrogen fuel will be introduced. As will be underpinned later, we have come to the conclusion that the savings in distribution and retailing cost will offset the cost of hydrogen liquefaction. Also, by practicing carbon sequestration, the higher energy intensity does not translate into higher CO_2 emissions. At the same time it prompted us to consider advanced options for liquefaction, especially focusing on the reduction of the power consumption. One such option that we have identified is the integration of hydrogen liquefaction with an LNG import terminal. At these terminals, LNG is regasified at grid pressure by heat exchange. Oftentimes the exergetic potential of the cold is left unutilized, though in some cases the cold is used for power production and warehousing [8]. Where that is the case, the use for hydrogen liquefaction offers a very attractive opportunity, reducing both the liquefier capital cost as well as the specific power consumption (the latter by 60%). This idea has been discussed in detail by Linde Kryotechnik and Shell [9].

The first alternative for central production that we consider is small-scale, so-called "forecourt" reforming, where hydrogen is produced *in situ* at the hydrogen filling stations, thereby bypassing the distribution problem, but simultaneously sacrificing CCS as a practical option. The second alternative is electrolytic production of hydrogen. Notwithstanding what was said above about availability of (green) electricity, it is relevant to include electrolysis for reference. We consider it natural to examine it in combination with on-site, "forecourt" production of hydrogen. It thereby exploits existing electricity infrastructure and circumvents the hydrogen distribution problem.

15.4
The Cost of Hydrogen Fuel

Excellent assessments of hydrogen production, distribution and retail cost exists. Examples of these include the estimates of SFA Pacific [10], and of the H2A working group [11]. In our experience these data are of a very high quality, especially where it concerns the cost of capital equipment items. Where we differ from the estimates of these consortium estimates is in the total installed cost of equipment. In constructing the values below we have used our best professional judgment and leveraged Shell experience in a wide range of industrial and retail projects. We have also taken care to relate all costs to the same early rollout situation. This case is defined as follows.

15.4.1
Case Definition

The case considers a hydrogen infrastructure for some 100 000–200 000 hydrogen vehicles in a larger metropolitan region, consuming a total 100 t per day (tpd) of hydrogen. Only at such a scale will new production facilities be required, as at smaller scale one can usually use the surplus capacity of the region. Typical examples would be Los Angeles/San Diego and greater New York in the US, The Tokyo/Yokohama metropolis in Japan and the Rhineland region in Europe. In such a setting, the average distribution distance from the central production facility to a retail site is some 75 km. While in some of the cases hydrogen is distributed and stored as a liquid, we have for all cases assumed that the vehicles require high-pressure hydrogen. Here we have assumed that the delivery pressure will be 350 bar, reflecting current practice at most hydrogen refilling stations. It may well be that ultimately a choice will be made for GH_2 retail at 550 or 700 bar. This would increase retail-site cost slightly. But it would do so for all cases, so that it does not change the overall picture.

The cost estimates presented here are based on existing technology. This implies that capital-related cost may come down in the future as technology develops. We note, however, that many of the main capital items, from SMR plants to hydrogen compressors, are essentially mature technology, and the scope for further cost reduction is relatively small. The case may be different for forecourt reformers and forecourt electrolyzers, but, as will be elaborated below, the effect thereof is relatively limited.

On the financial side, we have assumed an internal rate of return on capital of 10%, in line with other generic costing studies [10, 11]. A simple and straightforward way to account for capital cost is by applying a "capital charge". A capital charge of x% implies that fraction of the total capital invested has to be recovered each year from the yearly production. At a tax rate of 35% this leads to a capital charge of 21% for capital with a 10-year economic life, as appropriate for distribution and retail capital, and a capital charge of 15% for industrial capital with an assumed lifetime of 20 years. The costs for operation and maintenance (O&M) have been assumed to be proportional to the capital employed. On the basis of relevant industrial experience, 5% is a good estimate. The energy prices for NG and electricity reflect actual values in the US, mid-last year. A listing of these and other assumptions underlying our case is given in Table 15.1.

Table 15.1 Summary of key assumptions used in the definition of the case, and calculation of costs. Note that different capital charges have been used for industrial capital and distribution & retail capital, respectively, reflecting different asset lifetimes (see text).

Financial premises

Capital charge (industrial): 15%
Capital charge (distr & retail): 21%
O&M charge: 4–5%

Central production

Availability: 90%
Production (average): 100 tpd-H_2
Plant capacity: 110 tpd-H_2

Distribution

Distribution distance (average): 75 km
LH_2 truck delivery: 3500 kg-H_2
GH_2 truck delivery: 350 kg-H_2

Energy prices

NG (industrial): $7 per MMBTU
NG (commercial): $8 per MMBTU
Electricity (industrial): $0.05 kWh^{-1}
Electricity (commercial): $0.11 kWh^{-1}

Retail sites

Total number of retail sites: 100
Retail mode: GH_2: 350 bar
Retail site capacity: 1500 kg-H_2 day^{-1}
Site utilization: 70%
Retail site sales (average): 1000 kg-H_2 day^{-1}

15.4.2
Results

Within this case and with the assumptions of Table 15.1 the retail costs for the five modes of hydrogen production, distribution and retail have been evaluated. The result is shown in Fig. 15.2. What is perhaps most striking is the similarity between the central production base cases of gaseous and liquid (two left-most columns). As seen, savings in distribution in retail essentially offset the cost and energy associated with liquefaction. For distribution this is due to the low delivery capacity of high-pressure gaseous delivery, which, at 350 kg, is an order of magnitude lower than for liquid. Liquid truck delivery makes the logistics of hydrogen effectively similar to that of conventional gasoline and diesel. Further savings accrue at the retail site, where for gaseous delivery on-site recompression is costly both in monetary and energy terms, while in the case of liquid delivery, hydrogen storage (as a cryogenic liquid) is considerably cheaper than gaseous storage, and liquid cryo-pumping considerably reduces the compression costs.

As indicated above, energy efficiency is still a key consideration, especially when considering liquefaction. Based on the work we have done together with Linde Kryotechnik, we believe it is technically feasible to integrate hydrogen liquefaction with an LNG regasification terminal. This reduces the energy expenditure considerably, from 10 kWh per kg LH_2 for stand-alone liquefaction to 4 kWh per kg for the integrated case. In addition, capital savings contribute to lower the

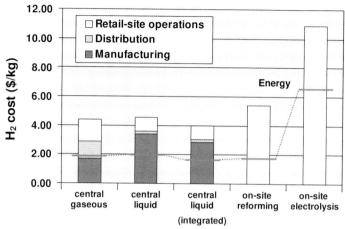

Fig. 15.2 Cost of hydrogen fuel for five different modes of production, distribution and retail. All options other than electrolysis are NG-based; for on-site reforming and electrolysis the cost for hydrogen manufacture is included in the retail-site costs. In all cases – including the liquid cases – actual refueling is gaseous, at 350 bar.

Overlaying the bars that give the breakdown of cost to the operation (manufacturing, maintenance, retail) is the total energy cost. For "central gaseous" it includes an important contribution from delivery (diesel fuel) and retail (electricity for on-site compression). For a more extensive discussion, see the text.

liquefaction cost. Not included in the cost here is a charge for the use of the LNG cold, which may reduce the monetary incentive somewhat, but not the energy savings.

These three cases make clear that when central production is considered, gaseous and liquid distribution are both plausible, and that the actual investment decision for one or the other may be decided by local market conditions, such as the (anticipated) retail density, and practical considerations such as the spatial and safety requirements for gaseous and liquid storage. All three cases come out significantly cheaper than on-site reforming or electrolysis.

In our analysis, hydrogen production via on-site reforming is more costly than central production and distribution, by about $1–1.5 kg^{-1}. This is only partly due to the reformer itself: of the total retail-site capital, some 40% is spent on the reformer, a similar fraction on storage, and 10% on compressors. As the on-site capital cost contributes just over $3 kg^{-1}, the reformer capital charge is about $1.3 kg^{-1}. It follows that cheap reformers by themselves will not make on-site reforming competitive with central production. As a point of further clarification we mention that in the "central gaseous" case the storage cost is effectively part of distribution cost as storage is in the high-pressure tube trailers, which are left on site at delivery. As a consequence, we see on-site reforming having two possible roles: one is to minimize investment risk in the very early phases of roll-out, before central production in regions at the 100 tpd scale can be justified, and a second is for hydrogen filling stations that are beyond the reach of a central production facility. (The latter will be most relevant in case of a gaseous distribution network.)

As to on-site electrolysis, the situation is much the same as for as retail-site capital is concerned, but adding to that is the energy cost associated with hydrogen production. At a realistic $\$0.11$ kWh^{-1}, the 55 kWh kg^{-1} required for electrolysis and 2.7 kWh kg^{-1} for compression makes for an electricity bill of $\$6.35$ kg^{-1}. The often-heard argument that this can be lowered (and the grid stabilized) by employing off-peak electricity works only up to a point: since the electrolyzer capital charge weighs in at about $\$2.3$ kg^{-1}, even at full utilization. As a consequence, the larger electrolyzer that would be required if hydrogen production were restricted to off-peak hours would be significant. In conclusion, while one may argue that the application of $\$0.11$ kWh^{-1} from Table 15.1 leads to a conservative estimate of the cost of electrolytic hydrogen, it is difficult to argue that the cost of electrolysis could be brought down even to the level of $\$6$–$7$ kg^{-1} unless there is a reduction both in the availability of cheap, excess (green) electricity and in the cost of electrolyzers. As argued above, this makes it a long-term rather than a medium-term prospect, albeit that the technology will find applications in the very early phases of roll out for the same reasons as on-site reforming.

15.5
"Clean Hydrogen" and the Scope for CO_2 Reduction

Now that we have established a case for medium-term hydrogen production from fossil fuels with CCS, we should quantify the scope for CO_2 reduction in association with this mode of hydrogen production. We feel that this is generally an under-explored topic. For instance, most well-to-wheel studies do not include it as a separate category. One reason may be the arbitrariness of the level of remaining emissions: does one stop at the CO_2 that is most readily captured? Or does one assume that upstream and downstream activities will also be "clean"? In the presence of these very real questions, we chose here to set our baseline on the premise that 90–95% of the CO_2 produced at the central production (and liquefaction) facility will be captured, but not the upstream and downstream emissions, i.e., those associated with bringing NG to the hydrogen plant (upstream), and not those of hydrogen distribution and retail (downstream). As a note of caution, the analysis below covers foreground CO_2 emissions only, i.e., those of the direct energy usage associated with hydrogen production and delivery; it excludes background emissions that a full life cycle analysis would include, such as those associated with plant constructions, equipment manufacturing, etc.

15.5.1
Scope

For the three NG-based central hydrogen production and distribution pathways that were introduced above we have quantified the CO_2 emissions without and

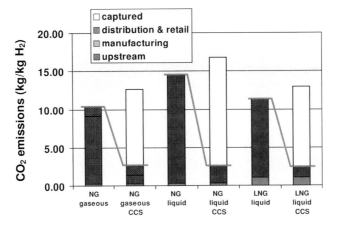

Fig. 15.3 CO_2 emissions associated with the three central hydrogen production cases of Fig. 15.2. In each case the impact of CO_2 capture at the central manufacturing complex is indicated. Note that CO_2 capture is an energy consuming process, causing an increase in the CO_2 *produced* in the CCS cases. The lines between bars serve as guides to the eye, illustrating reduction in total CO_2 *emitted* upon application of CCS.

with carbon capture. The results are shown in Fig. 15.3. Technically, this assessment is based on 90% capture from the plant flue gas, which is very rich in CO_2 (up to 20% in the case of an SMR), using commercial capture processes [12, 13]. In dealing with the energy requirements of the capturing process (which result in additional CO_2 produced), we have assumed both heat and power are produced on-site from NG, with the flue gas thereof being added to the primary SMR flue gas. Thus, the 90% capture is applied to all CO_2 produced at the manufacturing (and liquefaction) site.

The most relevant observation from the figure is that the magnitude of the remaining CO_2 emissions are quite similar for all three production and distribution pathways – about 2.5-kg CO_2 per kg of H_2. But these remaining emissions are quite differently distributed. In the case of gaseous distribution, about half of the CO_2 emissions result from the truck delivery operations – with diesel trucks – and from on-site recompression, powered with grid-average electricity having emissions of 0.5 kg kWh^{-1}. In the case of stand-alone liquefaction, the distribution and retail emissions are negligible for reasons discussed earlier (efficient delivery and a reduction in on-site compression), which offset the increase in the 10% non-captured of the central plant. Note that capturing the CO_2 associated with liquefaction power is required to bring the emissions down to this level, to be achieved for instance by local power generation. In both these cases, the assumed NG source was pipeline NG with upstream CO_2 emissions equivalent to 3% of NG delivered. Since the LNG terminal-integrated liquefaction scheme would naturally use LNG, the upstream emissions are higher due to the energy requirements of the LNG process and shipping. This more or less offsets the reduction in remaining plant emissions for the more efficient integrated liquefaction complex.

As said before, the figure of 2.5-kg CO_2 per kg of H_2 does not necessarily represent a lower limit of what can be achieved, as much of the remaining emissions still come from point sources, from which CO_2 could again be captured. It does indicate a sensible well-to-wheel emission level in the medium term when CCS will not be ubiquitous.

Cost: Technically, the present assessment is based on 90% capture from the plant flue gas, which is very rich in CO_2 (up to 20% in the case of undiluted SMR flue gas), by commercial capture processes [12, 13], and subsequent compression of the CO_2 to 80 bar. The cost of this, based on today's state of the art, at the scale of 350-kton CO_2 per year (corresponding capture of 10 kg per kg-H_2 and 100 tpd hydrogen), and under the financial assumptions of Table 15.1, is about $40 ton^{-1} captured and $50 ton^{-1} avoided CO_2. This leaves the cost for CO_2 transport, injection and monitoring. These costs tend to be very case specific, ranging from "negative costs" when the pure, pressurized CO_2 stream can be used for enhanced oil recovery (EOR) and, according to the best estimated of IPCC, $5–20 net cost per ton of CO_2 when storage and injection is without beneficial use [14]. If we combine these figures with the anticipated scope of CCS in association with the three hydrogen production and distribution pathways (Fig. 15.3), it follows that CCS adds $0.5 kg^{-1} of hydrogen in favorable cases and up to $1 kg^{-1} in less unfavorable locations and for stand-alone liquefaction.

15.5.2
Hydrogen versus Gasoline and Diesel

Let us, at the end of this discussion, set off the 2.5 kg per kg of remaining CO_2 emissions from "clean hydrogen" against the gasoline and diesel alternatives. Such comparisons are never unambiguous, because such a comparison involves fuels derived from different feedstocks (e.g., oil versus NG) and different, future vehicles, for instance a 2015 hybrid ICE vehicle versus a fuel cell vehicle. Nonetheless, there seems a consensus that, when compared with a future diesel-hybrid ICE, the energy use – expressed as MJ km^{-1} – of a future FCV will be some 40% less. If we than choose a reference FCV that uses 1 kg of hydrogen per 100 km, the equivalent reference hybrid-ICE would use 5.7 L diesel per 100 km. The latter has tail pipe emissions of 130 g-CO_2 km^{-1}, to which 25 g-CO_2 km^{-1} must be added for well-to-tank emissions. Obviously, the FCV has zero tailpipe emissions, while the well-to-wheel emissions of "clean hydrogen" are around 25-g CO_2 per km (2.5-kg CO_2 per kg H_2). Thus, when "clean hydrogen" replaces diesel (or gasoline, for which the numbers are not significantly different) it effectuates an 85% reduction in CO_2 emissions. This is a dramatic increase in scope relative to the case where the CO_2 is not captured; in those cases the CO_2 emissions reduction is 33%, even in the best of cases (gaseous distribution and on-site reforming).

15.6
Coal and Biomass

Considering that coal had a prominent place in our discussion of energy availability but not in our discussion of cost and CO_2 reduction potential, it is appropriate that we address the topic still, albeit briefly. As to cost, coal gasification is more capital intensive than SMR, but this is offset by the lower feedstock cost and lower price volatility. Another difference, which sets coal apart from the case defined above, is the scale-sensitivity. Coal gasification, due to its requirement of oxygen and its demands for solids handling is economically viable only at larger scales, e.g., from 300 tpd upwards. At this much larger scale, coal-based hydrogen production costs are on par with that of the NG-SMR route at a NG price of around $7–8 per MMBTU. The remaining CO_2 emissions can be brought down to levels not necessarily much higher than for NG. However, per kg of hydrogen the CCS cost will be about twice as high as for the equivalent NG case, due to the higher CO_2 intensity of coal relative to NG. If this production was coupled with a dense pipeline network, which is conceivable in the longer term – allowing reduced delivery cost relative to our "central gaseous" case – hydrogen from coal could plausibly be delivered at around $5 kg^{-1}, setting a benchmark for hydrogen production that is both "secure" and "clean".

In addition, "clean hydrogen" fuel could be produced in synergy with "clean power." One of the options for "clean power" is so-called pre-combustion CO_2 capture, which essentially consists of a hydrogen manufacturing unit of which the CO_2 is captured and of which the hydrogen used to produce power. Such a power plant would offer the opportunity to use part of the "clean hydrogen" for the fuel market. This would both improve the demand-matching capacity of the power plant and at the same time offer the opportunity to flexibly supply the growing market for transport hydrogen. This could be a significant economic benefit in the early and medium-term phases of hydrogen rollout.

Finally, we dwell briefly on the scope for bio-hydrogen production. It occupies a special place within the realm of renewable energy sources, as it is the only major renewable resource for which the energy is available as chemical energy, rather than as electrical energy (as for solar, wind, hydro, etc.). Because of this, biomass offers non-electrolytic routes to green hydrogen, via gasification or intermediates such as bio-ethanol that are potentially much cheaper than the electrolytic route (*vide supra*). At the same time, and even though estimates for biomass availability vary widely, biomass should in most regions be considered to be a supply-constrained resource, the use of which for hydrogen production competes with alternative uses, not only in the power sector but also in the transportation sector, as carbon-neutral biofuels. Therefore, while the availability and deployment issue is ultimately a political one, the outcome of which may differ between countries and regions, biomass, unlike other "green" sources, offers a path to affordable "green" hydrogen.

15.7
Conclusion

In this chapter we have surveyed the prospects for hydrogen as a transportation fuel, and investigated how it can simultaneously enhance energy supply security and reduce the CO_2 emissions of that sector. From a comparison of estimated primary energy availability and the energy demand of the transportation and electricity sectors, we concluded that energy supply diversity is in the medium term best served by using fossil feedstocks – notably NG and coal – rather than by using renewable resources, with the possible and conditional exception of biomass.

When CCS is practiced in association with central production it also contributes in a very significant way to CO_2 reduction from the transport sector. This "clean hydrogen" could have well-to-wheel emissions of 2.5-kg CO_2 per kg H_2, a level at which the transport-sector emissions would be reduced by 85%, when hydrogen fuel cell vehicles replace hybrid ICE vehicles. We have shown that this emission level can be achieved both for gaseous and liquid distribution of hydrogen, and that the total costs for both distribution modes are similar, offering the possibility to adapt flexibly to local retail preferences and industrial opportunities.

Finally, the medium-term preference for "clean" over "green" is reinforced by cost considerations. We have shown that hydrogen produced from "green" electricity will only become affordable once major cost reductions in both green electricity and electrolyzer cost have been accomplished. Only biomass may – in specific regions – offer earlier opportunities for "green hydrogen".

Acknowledgments

The authors acknowledge the helpful contributions of Leon Rubinstein and Wim Wieldraaijer to this chapter.

References

1 *Mobility 2030: Meeting the Challenges to Sustainability*, World Business Council for Sustainable Development, **2004**, www.wbcsd.org
2 A. Wimmer, T. Wallner, J. Ringler, F. Gerbig, H2-Direct Injection – A Highly Promising Combustion Concept. SAE-World Congress, Detroit, **2005**.
3 D.W. Keith, A.E. Farrell, *Science*, **2003**, 301, 315. This paper quantifies air quality improvement, but concludes that this by itself is insufficient to justify hydrogen vehicles.
4 N. Demirdöven, J. Deutch, *Science*, **2004**, 305, 974.
5 *Energy Needs, Choices and Possibilities*, Shell International, **2001**. www.shell.com/static/media-en/downloads/51852.pdf
6 See, for example, N. Thompson, D. Rickeard, *Transport Fuels for the Future*, Inland Transport Committee Round Table, Geneva (Switzerland), **2002**.

7 *IEA International Energy Outlook*, **2005**, www.eia.doe.gov/oiaf/ieo/special_topics.html
8 An example is Tokyo Gas' Negishi terminal. See http://www.tokyo-gas.co.jp/env/ecorep/e_text/pdf/P10_11.pdf
9 A. Kuendig, K. Loehlein, G.J. Kramer, J. Huijsmans, World Hydrogen Energy Conference, Lyon (France), **2006**, paper 713.
10 D.R. Simbeck, E. Chang, *Hydrogen Supply: Cost Estimate for Hydrogen Pathways – Scoping Analysis*, NREL report, **2002**, NREL/SR-540–32525.
11 DOE H2A Analysis Group, http://www.hydrogen.energy.gov/h2a_analysis.html
12 S. Reddy, J. Scherffius, C. Roberts, Fluor's Econamine FG PLUSSM Technology: an enhanced amine-based CO_2 capture technology, Second National Conference on Carbon Sequestration, Alexandria (USA), **2003**.
13 T. Mimura, T. Nojo, M. Iijima, T. Yoshiyama, H. Tanaka, *Proceedings of the 6th International Conference on Greenhouse Gas Control Technologies (GHGT-6)*, J. Gale, Y. Kaya (eds.), Elsevier Science Ltd, **2003**.
14 See, for example, Carbon dioxide capture and storage, summary for policy makers and technical summary, (IPCC, **2005**) www.ipcc.ch/activity/ccsspm.pdf

16
Solar Photocatalysis for Hydrogen Production and CO_2 Conversion

Claudio Minero and Valter Maurino

16.1
Introduction

The sustainability of the global development of our society is heavily related to the availability of sufficient sources of energy, their environmental impact, and their consumption/restoration ratio. The total annual energy consumption in 1998 was 400 exajoules (EJ) (400×10^{18} J), which corresponds to a continuous power consumption of 12.7 TW [1–3]. An estimated doubling of annual energy needs by the 2050 and a three-fold increase by 2100 is foreseen, namely 850–1050 EJ (27–33 TW) is expected at the present incremental rate. The increment in energy need will be greater than the total energy currently produced.

Some 80% of present global energy consumption is generated by fossil fuel, with an emission of 7 Gt-C per year in the atmosphere as CO_2. If the actual distribution of energy sources is maintained the atmospheric CO_2 concentration will more than double by 2050 (>750 ppm), with only a modest predicted reduction if the composition of energy sources, the efficiency of energy production and consumption, and the control of CO_2 emissions are pursued [4, 5]. The enhancement of the greenhouse effect and the change in the mean global temperature due to a doubling of atmospheric CO_2 (1.5–4.5 K) [6] will be serious, leading to more severe weather patterns, a rise of sea levels, a change in the oceanic thermohaline circulation, and a change of the biological species distribution and habitat patterns.

To meet the demand for 10–20 TW of carbon neutral energy, three options are viable: (a) fossil fuel use in conjunction with carbon sequestration; (b) nuclear power; (c) renewable sources.

The requirement of carbon sequestration is to find protected storage for a potential 25 Gt of CO_2 emitted annually. This emission is 600× the amount currently injected in oil wells to spur production and 20000× the amount of CO_2 stored annually in Norway's Sleipner reservoir, the first implemented example of CO_2 storage [3, 7].

Catalysis for Renewables: From Feedstock to Energy Production
Edited by Gabriele Centi and Rutger A. van Santen
Copyright © 2007 WILEY-VCH Verlag GmbH & Co. KGaA, Weinheim
ISBN 978-3-527-31788-2

The production of 10 TW of nuclear power with the available nuclear fission technology will require the construction of a new 1 GW_e nuclear fission plant every day for the next 50 years. If this level of deployment would be reached, the known terrestrial uranium resources will be depleted in 10 years [3]. Breeder reactor technology should be developed and used. Fusion nuclear power could give an inexhaustible energy source, but currently no exploitable fusion technology is available and the related technological issues are extremely hard to solve.

Renewable sources are [1, 2]:

1. Biomass, from which 7–10 TW would be available by the use of the entire agricultural land of the planet, covering it with the fastest-growing known energy crops, like switchgrass. There are many concerns about the possibility of obtaining a positive energy balance between the energy output of biomass fuels and the energy input needed to plant, fertilize, grow, harvest, process and ferment the biomass [8].

2. Wind on land will supply 2.1 TW from saturating the entire class 3 land area with wind mills, according to IPCC estimates.

3. Hydroelectric will supply additional 1.5 TW by damming all available rivers.

4. The cumulative energy in all tides and ocean currents in the world amounts to less than 2 TW.

5. The total geothermal energy integrated over all land area is 12 TW, of which only a small fraction is practically exploitable.

6. Conversion of solar energy. Solar energy is an unlimited and freely available energy source. The solar constant at the top of atmosphere is 170000 TW (1365 W m^{-2}), of which on average 120000 TW strikes the earth, the remainder being scattered or absorbed by atmosphere and clouds. Assuming from the solar spectrum that half of this energy could be converted into a useful form (thus neglecting infrared), the average on the spherical earth shape in 24 h, and losses due atmosphere absorption and scattering, then 1.2×10^4 TW are available. Less than 0.2% of land would supply 20 TW of power at 100% conversion efficiency of useful energy. This world area is less than 2.6% of the Sahara desert, or the equivalent surface of 2/3 of Italy. Localization on tropical sites (about 4–8× the average irradiation) and lower conversion efficiency (10–25%) lead to about the same land coverage values. From the above discussion, solar energy is the compulsory choice for the future.

However, there are many challenges to exploiting direct conversion of solar energy into electricity and/or chemical fuels. For a comprehensive review of the basic research challenges in the photovoltaic conversion of solar energy see Refs. [2, 3, 9].

For capture and conversion of solar energy, semiconductors possess several key advantages, including relatively high absorption coefficients and the possibility to separate the photogenerated carriers at homo and heterojunctions. The energy harvested by charge carriers generated can be exploited in several ways. In photovoltaic (PV) systems they are separated and collected to produce electricity. Because solar energy is diffuse and intermittent, effective storage methods are critical to match supply and demand. Consequently, the direct generation of solar fuels is also a desirable option. Although not the topic of this chapter, PV cells will be briefly discussed here, as the technological problems and costs related to electricity generation are the same as those encountered in solar fuels cells.

The ultimate thermodynamic limit for the conversion of sunlight to electricity or chemical free energy is 93%, whereas the Shockley–Queisser limit [10] for the conversion efficiency is 31% for single junction devices, as would apply to silicon wafer and most present thin film cells, and using semiconductors with band gap ranging from 1.25 to 1.45 eV. The Shockley–Queisser analysis is based on a single threshold absorber, on the assumption of rapid thermalization of hot electron–hole couples with loss of the energy in excess of the band-gap (thermal equilibrium between electrons and phonons), and assuming that only radiative recombination is operating (perfect crystal with no defects) when the solar cell and the sun are modeled as black bodies at 300 and 6000 K, respectively.

Current PV technology, based on crystalline silicon, offers cell modules with 10–12% conversion efficiency at a cost of $ 3.5 peak watt (W_p, power produced at a 1000 W m^{-2} radiation intensity and AM1.5 spectrum), which corresponds to a cost of $ 350 m^{-2} with 10% efficiency. Considering an infrastructure cost of $ 250 m^{-2}, the present cost of a grid connected PV electricity is around $ 0.30 kWh^{-1}. A reasonable target for solar energy costs could be $ 0.40 W_p^{-1} or $ 0.02 kWh^{-1}, which means $ 125 m^{-2} at 50% conversion efficiency. With the present learning curve for mass production of crystalline silicon PV modules these target costs will be reached too far into the future (20–25 years, depending on production growth rate).

The second generation of PV cells, based on amorphous or polycrystalline thin films, (amorphous hydrogenated silicon, where optical absorption is enhanced by impurity scattering [9], polycrystalline chalcogenides like copper indium/gallium selenide (CIGS) and CdTe [11]), would offer a major reduction in starting material costs. Thin film cells can be fabricated on flexible substrates with cheap production systems like roll to roll production [9]. The main shortcoming is the need for high absorbing direct band gap materials. The predicted lower limit cost is about $ 30 m^{-2}. Combined with the anticipated efficiency limit (15% or 150 W m^{-2}), electricity generation costs could reach the targets ($ 0.02–0.04 kWh^{-1}). However, the present cost (on average $ 100 m^{-2}) and efficiencies (5–10%) are far from these targets. The conversion efficiency must be increased substantially [12]. Stability is also an issue.

Effective storage methods are critical, considering that solar energy is diffuse and intermittent, and that only 15% of the total energy consumption is as electrical energy, the remainder being as fuel feedstock. In this respect, the direct generation of fuels, e.g., in a non-regenerative photoelectrochemical (PEC) cell, is auspicious. The photoactive material should be coupled with suitable catalysts to convert the photogenerated charge into CH_4 (e.g., from CO_2 reduction) or H_2

(e.g., from water splitting). PEC cells add complexity to PV cells, as with the optical energy losses of the photoactive material, they add the electrochemical losses associated with the photoelectrode–electrolyte electron-transfer processes.

16.2
The Photocatalytic Process

Upon excitation of a semiconductor, the electrons in the conduction band and the hole in the valence band are active species that can initiate redox processes at the semiconductor–electrolyte interface, including photocorrosion of the semiconductor, a change in its surface properties (photoinduced superhydrophilicity [13]), and various spontaneous and non-spontaneous reactions [14–19]. These phenomena are basically surface-mediated redox reactions. The processes are depicted in Fig. 16.1. Owing to the slow spontaneous kinetic of the reactions between the

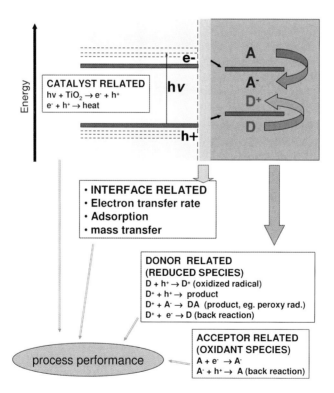

Fig. 16.1 Processes involved in photocatalytic reactions, exemplified for some reactions for TiO_2 in the presence of oxygen. When both redox states of acceptor A and donor D are energetically inside the band gap, the reaction of charge carriers is thermodynamically possible: the kinetically slow reaction of A + D is favored (catalyzed) by intervention of light (photocatalysis, usually exoergonic). When H_2O is both A and D, the net reaction is $2H_2O \rightarrow 2H_2 + O_2$ (water photosplitting, endoergonic).

donor and the acceptor, the reaction between A and D promoted by the intervention of the light and the light absorber is termed *photocatalyzed*. The promoted reaction can be exoergonic or endoergonic. The separated electron–hole couple migrates to the surface, where the carriers are eventually trapped and react irreversibly with acceptor and donor couples present in solution, respectively.

Photocatalysis over oxide semiconductors has been proposed as an effective, mild and environmental friendly technique for the abatement of (refractory) organic contaminants with oxygen as electron acceptor [15]. In this case the reaction is normally exoergonic, as the overall reaction (mineralization) is organics + $yO_2 = xCO_2 + zH_2O$.

The fixation of the absorbed photon energy as chemical energy requires that the electron–hole couple drives an endoergonic process (photoelectrolysis). The result is light conversion through a photoelectrochemical process. The first example of such a process was water photosplitting reported by Fujishima and Honda on a TiO_2 photoanode biased at 0.4 V vs. NHE [20], holding the promise of photocatalysis as a clean and renewable source of fuels. However, photoelectrochemical light fixation involves an interfacial electron transfer associated with large overpotentials when the electrodic process involves many-electron redox couples like water. As a consequence, the semiconductor surface should have opportunely tuned (electro)catalytic properties, or an electrocatalyst should be present on the surface. Early attempts to achieve water photosplitting with colloidal semiconductor oxides were plagued by low photonic efficiency [21], due to the high band gap of the semiconductor used and a poor understanding of the mechanism of interfacial electron transfer.

Our knowledge of the fundamental aspects of photocatalysis has advanced greatly in recent years [22, 23], and the lack of driving force needed to surmount activation barriers can be overcome by properly engineering the catalytic properties of the system at the nanoscale level.

The configuration of a photoelectrochemical cell (PEC) with a semiconductor photoanode for water splitting could greatly improve the process performance. The efficiency limit of a PEC used for splitting water is 41% for a photoanode band gap that just matches the free energy of the water splitting reaction (1.23 eV at 25 °C and 1 atm). In this case, no driving force is left to overcome activation barriers of the electrodic processes. The efficiency decreases to 18% for a bandgap of 2 eV and to 0.05% for 3 eV (the bandgap energies of α-Fe_2O_3 and TiO_2, respectively) [24]. Accounting for the thermodynamic potential needed for water splitting, overvoltage losses, and the energy required for driving the reaction, 1.6–1.8 V have to be provided to produce hydrogen from water. This driving force is provided by a semiconductor material with a bandgap of 1.6 eV < E_g < 2.2 eV immersed in aqueous solution. The use of suitable thin film photoanode would allow efficient charge separation in the space charge layer, as well as spatial separation of the anodic and cathodic processes to avoid back reactions of photogenerated O_2/H_2.

Major barriers are finding efficient light-absorbing materials and catalysts with corrosion-resistance properties in water and energy levels that match the reduction and oxidation half-reactions to reduce overvoltage losses. Major issues concern the finding of a suitable catalyst for the four-electron water oxidation. A two band gaps non-regenerative PEC for water splitting with 18% solar conversion efficiency has been described [25]. New stable photocatalysts with visible sensitization, new concepts for enhanced charge separation and new redox catalysts have also been developed. All these aspects and their integration in new photocatalytic systems for water splitting and CO_2 fixation will be reviewed.

The photocatalytic process can be divided into elementary processes that take place in different *space–time* positions (here exemplified for TiO_2):

1. Catalyst related processes:
    a) light absorption: $h\nu\ (\geq E_g) + TiO_2 \rightarrow e^-_b + h^+_b$
    b) thermalization of hot carriers: $e^-_b \rightarrow e^-_{CB}$; $h^+_b \rightarrow h^+_{VB}$
    c) bulk recombination: $e^-_{CB} + h^+_{VB} \rightarrow$ heat, $h\nu\ (= E_g)$
    d) separation of carriers/migration of carriers to the surface(s):
    $e^-_{CB} \rightarrow e^-_s$; $h^+_{VB} \rightarrow h^+_s$

2. Interface-related processes:
    a) surface trapping: $h^+_s + \equiv Ti-OH \rightarrow \equiv Ti-O^\bullet + H^+$
    $e^-_s + \equiv Ti-OH \rightarrow \equiv Ti(III) + OH^-$
    b) surface recombination: $e^-_s + \equiv Ti-O^\bullet + H^+ \rightarrow \equiv Ti-OH$
    $h^+_s + \equiv Ti(III) + H_2O \rightarrow \equiv Ti-OH + H^+$
    c) transfer of redox active species to the surface (eventual mass transfer limitation)
    d) adsorption of D and A, and foreign species (e.g., ions)

3. Donor/acceptor related processes:
    a) donor reaction with h^+ and trapped h^+: $D + h^+ \rightarrow D^{\bullet+}$
    b) donor back reaction (donor-mediated recombination): $D^{\bullet+} + e^- \rightarrow D$
    c) acceptor reaction with e^- and trapped e^-: $A + e^- \rightarrow A^{\bullet-}$
    d) acceptor back reaction (acceptor-mediated recombination): $A^{\bullet-} + h^+ \rightarrow A$
    e) donor–acceptor radical–ion reaction: $D^{\bullet+} + A^{\bullet-} \rightarrow DA$
    f) D/D or A/A radical–ion reaction (dimerization):
    $D^{\bullet+} + D^{\bullet+} \rightarrow D_2$; $A^{\bullet-} + A^{\bullet-} \rightarrow A_2$

These primary elementary reactions are followed by reaction of intermediates with charge carriers, as well as other solution reactions like hydrolysis, and so the system is, early on, crowded with many species. The following discussion is limited to the analysis of primary events.

Catalyst related fundamental aspects are the light harvesting capability of the semiconductor and the kinetics of the bulk recombination processes. Direct band gap semiconductors, with a band gap in the visible range are desirable. However, in small band-gap materials the stability against photocorrosion is low. The most stable materials for photocatalysis and water splitting are metal oxide semiconductors, with indirect band gap exceeding 2.5 eV. Bulk recombination processes include radiative recombination (negligible in indirect band gap materials), bulk defect mediated (Shockley–Read–Hall) recombination (negligible for perfectly crystalline materials) and Auger recombination (important at high carrier injection rates).

Surface related properties are carrier trapping on intrinsic (due to surface dangling bonds) and extrinsic (related to adsorbates, including donor and acceptor) surface states, carrier recombination mediated by surface states [26], and mass transfer of acceptor and donor and products from/to bulk solution.

Acceptor and donor related processes involve interfacial electron transfer from free/trapped charge carries, including back reactions 3b and 3d [22, 27]. These latter processes are essentially donor/acceptor mediated recombination reactions.

Each of the above processes has its own characteristic kinetic and rate law and, in principle, each responds differently to the process variables (illumination intensity, dopant density, presence of adsorbates, activity of surface potential determining ions, width of and potential drop in the space charge region, position of the band edges).

The overall process performance, as measured by *photon efficiency* (number of incident photon per molecule reacted, like the *incident photon to current conversion efficiency*, or IPCE, for PV cells), depends on the chain from the light absorption to acceptor/donor reduction/oxidation, and results from the relative kinetic of the recombination processes and interfacial electron transfer [23, 28]. Essentially, control over the rate of carrier crossing the interface, relative to the rates at which carriers recombine, is fundamental in obtaining the control over the efficiency of a photocatalyst. To suppress bulk- and surface-mediated recombination processes an efficient separation mechanism of the photogenerated carrier should be active.

16.2.1
Quantum Yield

The relative significance of the various processes depicted in Fig. 16.1 can be assessed through the use of a relevant kinetic model. Recently, a simple kinetic model was shown to be able to describe the dependence of the photocatalytic transformation rates on the operational parameters (light intensity, substrate and catalyst concentration) [22, 23]. The model is also able to quantify the photocatalyst properties related to the dynamics of the photogenerated charge carriers, distinguishing the relative role of recombination and interfacial electron transfer. It is based on the steady state assumption for reactive species involved (charge carriers and primary reaction intermediates), and, contrary to usual assumptions in molecular photochemistry, on second-order kinetics. The model is quite flexible, allows different degrees of complexity (e.g., inclusion of surface adsorption of the donor and acceptor, of a current doubling effect, of the donor/acceptor mediated recombination or back reaction), and is able to predict the Langmuir–Hinshelwood (L-H) like dependence of the reaction rate on the donor concentration observed for the photocatalytic processes.

Figure 16.2 depicts the processes considered, including back reactions, but not current doubling. Even if the photocatalytic process is located at the semiconductor–liquid interface [29], it was demonstrated that the dark adsorption constant of the donor does not match the adsorption constant obtained from the fitting of the photocatalytic degradation data against the L-H model [30]. The L-H

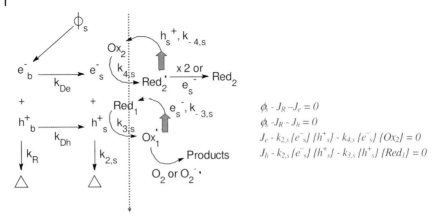

$\phi_s - J_R - J_e = 0$
$\phi_s - J_R - J_h = 0$
$J_e - k_{2,s} \{e^-_s\} \{h^+_s\} - k_{4,s} \{e^-_s\} \{Ox_2\} = 0$
$J_h - k_{2,s} \{e^-_s\} \{h^+_s\} - k_{3,s} \{h^+_s\} \{Red_1\} = 0$

Fig. 16.2 Simplified kinetic model of the photocatalytic process. φ_s represents the light absorbed per unit surface area of the photocatalyst, e^-_b and h^+_b are the photogenerated electrons and holes, respectively, in the semiconductor bulk, k_R is the bulk recombination rate constant and J_R the related flux, whatever recombination mechanism is operating; Δ is the heat resulting from the recombination; k_{De} and k_{Dh} are the net first-order diffusion constants for fluxes J_e and J_h to the surface of e^-_b and h^+_b in the semiconductor lattice, respectively; e^-_s and h^+_s are the species resulting from the surface trapping of electrons and holes, respectively, at the semiconductor surface; $k_{2,s}$ is the recombination rate constant between surface-trapped electrons and holes, $k_{3,s}$ si the rate constant for the oxidation of the adsorbed molecule Red_1 with h^+_s, $k_{4,s}$ is the rate constant for the reduction of the adsorbed molecule Ox_2 with e^-_s. The constants $k_{-3,s}$ and $k_{-4,s}$ refer to the back reactions (thick arrows). When the back and other reactions are not considered (right-hand part of the scheme) the reported simple stationary state approach on charge carrier flux could be assumed.

like profiles are instead due, as the model suggests, to the charge carrier limitation by recombination.

Applying the steady state to the kinetic system showed in Fig. 16.2, and neglecting the back reactions $k_{-3,s}$ and $k_{-4,s}$, one obtains for the quantum yield η_o of the process [23]:

$$\eta_0 = -\gamma + \sqrt{\gamma(\gamma + 2\zeta)} \tag{1}$$

where:

$$\gamma = k_{o,s}\{Red_1\}\{Ox_2\}/\phi_s = \frac{k_{3,s}k_{4,s}}{2k_{2,s}} \frac{\{Red_1\}\{Ox_2\}}{\phi_s} \tag{2}$$

which is the ratio between the actual photocatalytic rate at the surface ({ } are surface concentrations) and the absorbed photon flux, and:

$$\zeta = 1 - J_R/\phi_s = -w + \sqrt{w(w+2)} \tag{3}$$

is the net fraction of light-generated charge carriers that reach the surface. Here J_R is the bulk recombination flux. If, to a first approximation, $J_R = k_R\{e^-{}_b\}\{h^+{}_b\}$, $J_e = k_{De}\{e^-{}_b\}$, and $J_h = k_{Dh}\{h^+{}_b\}$, then $w = k_{De}k_{Dh}(2k_R\varphi_s)^{-1}$ gives a normalized measure of the efficiency of the transfer of charge carriers to the surface with respect to bulk recombination and the absorbed photon flux.

Given a donor and an acceptor, y depends on the *surface properties* of the catalyst, namely the surface recombination rate constant $k_{2,s}$ of charge carriers (mediated by surface traps) and the constants for interfacial electron transfer to acceptors/donors. The value of y can vary from 0 (complete surface recombination) to ∞ (complete interfacial transfer of the charge carrier flux *reaching the surface*). High y values can be attained if the charge carriers reaching the surface are fully scavenged by donor/acceptor species (high surface concentration of acceptor/donor and/or high rates of interfacial electron transfer). w depends only on the *bulk properties* of the catalyst, i.e., the charge carriers' diffusivity and the bulk recombination rate constant; w varies between 0 (high bulk recombination, also induced by high photon absorption, which implies $J_R/\varphi_s = 1$ and $\zeta = 0$) and ∞ (complete diffusion to the surface, with $J_R/\varphi_s \to 0$ and $\zeta = 1$). Clearly, the quantum yield (η_o) increases monotonically as y increases, for $y \to \infty$, $\eta_o \to 1$. The model has important implications for the interpretation of rates and quantum yields in catalyst slurries, given the proper transfer function for surface to volume reactivity and suitable evaluation of the absorbed light [23].

Figure 16.3 shows the dependence of the QY from the non-dimensional variables y and w. The increase of both acceptor and donor concentration, until back reactions become important (outside the simple model hypotheses), or the decrease of the absorbed light φ_s would increase y and the quantum yield. The

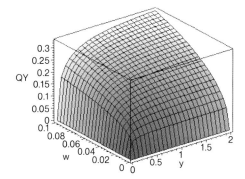

Fig. 16.3 Quantum yield (QY) for electron and hole transfer to solution redox acceptors/donors as a function of the reduced variables y (related to the surface properties of the catalyst, i.e., ratio between interfacial electron transfer rate and surface recombination rate) and w (related to the ratio between surface migration currents of hole and electrons to the rate of bulk recombination), according to the proposed kinetic model [23].

quantum yields $\eta_o \to 1$ in the limit that $\varphi_s \to 0$, which is the case of low charge carrier concentration, and when the recombination rate vanishes.

Given a value of γ, the quantum yield can be increased by (a) minimizing the *catalyst related losses*, i.e., maximizing ζ (see Eq. 3 and Fig. 16.3) – this implies enhancement of the migration of photogenerated charge carriers to the surface and the suppression of bulk recombination; (b) by minimizing the *surface related losses*, either by increasing the donor and acceptor rate constants (by surface classical thermal catalysis) or reducing the surface recombination constant $k_{2,s}$ (by surface modification). Notably, at a fixed w, the QY, or the rate, shows an asymptotic behavior (like that of the Langmuir–Hinshelwood model) independently of any adsorption properties of the substrates. Thus, the apparent adsorption constant, leading to the asymptotic limit of QY at a given w, could depend more on w than on actual surface concentration accounted for in γ.

16.2.2
Catalyst Related Losses

16.2.2.1 Carrier Thermalization

The primary loss of efficiency is related to the light harvesting property of the catalyst. The hot carrier thermalization is the principal loss. For a single band gap system there is a tradeoff between the increase of the band gap energy, which decreases thermalization losses, and solar light harvesting. These losses can be reduced by using multiple band gap systems or by collecting hot carriers, as developed in the design of the third generation high performance solar cells. Many concepts have been demonstrated to overcome the Shockley–Queisser limit. An improvement of efficiency of 2–3× is expected if new fundamental concepts are used.

The major limiting factor is the loss of the excess kinetic energy of hot carriers through phonon emission, which by itself confines maximum efficiency to 44%. The approach to reduce this loss and increase efficiency beyond the 31% limit has been to use a cascade of multiple p-n junctions, i.e., a tandem cell [31], where multiple cells are used with different (and narrowing) band gaps, each cell converting a narrow range of photon energies close to its band gap. Performance increases as the number of cells in the stack increases. In the limit of an infinite stack of band gaps perfectly matching the solar spectrum, ultimate conversion efficiency with direct sunlight increases to 87%. Limiting tandem cell performance is relatively good even with a small number of cells in the stack, increasing from the single cell direct sunlight efficiency of 41% to 56, 64 and 69% as the number of independently operated cells increases to 2, 3 and 4, respectively [32]. Usually, cells are designed so that their current output matches, so that they can be connected in series. This poses additional constraints to tandem cell design, making it very sensitive to the spectral content of the incident light. Besides the needs of materials with appropriate band gap, each cell should generate the same amount of current as the others. Multijunction PV cells are already in commercial production with two different technologies. Double and triple junction cells based on the

GaInP/GaAs/Ge systems have conversion efficiencies approaching 30% [33]. Triple junction amorphous silicon cells, based on the Si:Ge:H alloy, with efficiencies up to 12% have been reported [34].

For a single band gap system, two basic approaches for the conversion of hot carriers into electricity or chemical energy have been proposed to enhance the efficiency of photon conversion: (a) extraction of the hot carries before they cool, with the production of an enhanced photovoltage [35]; (b) production of two or more electron–hole pairs per photon absorbed, with photocurrent enhancement [36, 37].

Carrier cooling rates are very fast in bulk semiconductor, due to phonon emission. To slow carrier cooling, the electron and hole must be isolated from the lattice. An interesting concept to achieve this goal is the use of semiconductor quantum dots (QDs) in which the separation of the discrete exciton quantized energy levels is greater than phonon energies [35, 38]. Experimentally, a deceleration of at least one order of magnitude of the hot electron cooling dynamic in InP QDs has been observed [39]. If phonon-mediated carrier cooling in QDs is effectively inhibited, the formation of multiple excitons through impact ionization, the inverse process of Auger recombination, becomes important. Very efficient multiple exciton creation by one photon has been reported for PbSe QDs [40] and a quantum yield of 300% at a photon energy of $4E_g$ was reported. A possible QD solar cell configuration has been discussed [38]. A limiting efficiency of 85% has been calculated for a cell of band gap approaching zero with carrier multiplication through impact ionization [41].

Most of the systems proposed are based on single band gap materials, with relevant photocatalyst related optical energy losses. Water photosplitting using these systems has low photonic efficiency, requires the use of a wide band gap photoactive material to overcome overpotentials and, in many instances, the presence of sacrificial donor/acceptor species. Efficient water splitting requires the integration of different systems in (nano)structured devices employing more than one photoactive material. This approach overcomes the Shockley–Queisser limit. Multiband cells could be created, putting one or more electronic energy levels in the forbidden band of the bulk semiconductor material through superlattice structures based on a periodic structure of alternating layers of semiconductor materials with wide and narrow band gaps, high concentration impurities such as rare-earths in wide bandgap semiconductors, or by using semiconductors with multiple narrow bands such as I-VII and I_3-VI compounds [42] (see also Section 16.4 on New Materials).

16.2.2.2 Charge Separation

Semiconductor structures that develop space charge layers and contact potentials, like films of proper thickness, films with applied external bias, homo- and hetero-(nano)junctions, permit significant suppression of bulk recombination processes and, potentially, allow high quantum yields. Spatial separation of electron and holes also allows the separation of cathodic and anodic processes in a photoelectrochemical cell (eventually at the micro and nano level), minimizing surface recombination and back reactions, and permitting the use of proper electrocatalysts for both processes.

Optical losses related to the catalyst are effectively suppressed if the charge carriers are generated in the space charge layer, because the recombination current J_R in this region is smaller than in the bulk. The space charge layer at the interface with the solution is strongly controlled by the interfacial properties, depending also on solution composition. Colloidal semiconductors, with dimensions smaller than the space charge layer width, have no fully developed space charge layer [43] and, consequently, scarcely efficient charge separation.

Space charge layers and contact potential for efficient charge carrier separation can be achieved with proper semiconductor structure in several ways. When possible semiconductor structures are considered, the charge separation can be attained in an *active mode*, i.e., by the use of a potential bias in a photoelectrochemical cell, or in a *passive mode*, i.e., with the use of proper contact between different phases.

16.2.2.3 Active Charge Separation

This is the first approach, used in the photoelectrochemical water splitting over an n-TiO_2 anode by Fujishima and Honda [20]. Besides enhancing the electric field in the space charge layer, the variation in the Fermi level of the semiconductor electrode helps to overcome the electrode overpotentials. Several researchers have reported that the effectiveness of the photocatalytic degradation of organics with polycrystalline TiO_2 thin films can be improved by applying a positive bias across the photoanode [44–48]. The improvements result from a decrease in the recombination rates of photogenerated electrons and holes [44, 46]. Increases in photodegradation quantum yield up to 37% were observed [45]. Photoelectrocatalytic reactors with separate anodic and cathodic compartments can operate with the anode under an atmosphere with low oxygen content [44]. The spatial separation of anodic and cathodic processes allows the use of electrons to drive useful reactions; a photoelectrocatalytic reactor was proposed in which at the cathode hydrogen peroxide is produced from dioxygen reduction [49]. Changes in the double layer structure induced by the bias voltage can also overcome the negative effects due to the presence of certain ions in solution [44].

16.2.2.4 Passive Charge Separation

Semiconductor – Electrolyte Interface The electric field in the space charge region that may develop at the semiconductor electrolyte interface can help to separate photogenerated e^-/h^+ couples, effectively suppressing recombination. When a semiconductor is brought into contact with an electrolyte, the electrochemical potential of the semiconductor (corresponding to the Fermi level, E_F of the solid [50]) and of the redox couple (A/A^-) in solution equilibrate. When an n-type semiconductor is considered, before contact the E_F of the solid is in the band gap, near the conduction band edge. After contact and equilibration the E_F will be the same everywhere in the system. The value of E_F in the semiconductor will change more than the electrochemical potential $qE(A/A^-)$ of the redox couple in solution (being $E(A/A^-)$ the redox potential of the couple and q the electron charge) because, even for a dilute concentration of the redox couple, the solution has far more states per unit energy than does the semiconductor in its band gap region. Generally, for an n-type semiconductor, E_F of the solid is located at an energy greater than the electrochemical potential of the solution. So majority carriers (electrons) migrate in solution, leaving an excess positive charge in the semiconductor, arising from ionized donors. The positive charge in the solid is spread out over a *space charge region* or *depletion layer*, which produces an electric field in the solid. The solution has an excess negative charge, which is located in a narrower region that consists of the Helmholtz and Gouy-Chapmann charge layers [51]. A photogenerated e^-/h^+ couple in the space charge region will sense the electric field, the two carriers are separated, and recombination is inhibited. The hole (minority carrier) will be directed toward the surface and injected in solution, the electron migrates in the bulk. The potential drop in the space charge regions results in a spatial dependence of the band edge positions (band bending) and in the development of an interface barrier or Schottky barrier (equal to the band bending) for the majority carriers. The width of the space charge region is proportional to the square root of the dielectric constant ε divided by the donor density in the solid (N_D) [52]. Typical space charge layers in oxide semiconductors used in photocatalysis are of the order of hundreds of nanometer (for TiO_2 anatase, $\varepsilon = 179$, $N_D = 10^{17}$ cm^3). In colloidal semiconductor and very thin films the space charge layer is not fully developed, and, to avoid recombination, the separation of charge carriers must rely on very fast scavenging processes. Thin films with proper thickness and donor density are then mandatory to successfully use this effect.

Semiconductor – Metal Junctions Besides the semiconductor–liquid interface, electron–hole separation can be attained also when the couple is generated in the space charge layer of a homo/heterojunction or semiconductor–metal junction. The metal can also act as electrocatalyst (e.g., for reduction of O_2, H^+ or CO_2). The development of the proper structure, including arrays of multiple junctions in series to enhance photovoltages and efficiently harvest radiation [53] and/or the inclusion of suitable electrocatalysts, is crucial.

Semiconductor–noble metal nanocomposites have been widely investigated in photocatalysis and photochemical water splitting [21, 54–58] and seem to provide one way to increase the semiconductor photoefficiency. Besides charge storage and rectification, the metal can catalyze charge-transfer processes. Acceptor reduction by electrons (milliseconds) is usually much slower than the oxidation of donors by holes (100 ns) in photocatalysis with TiO_2 [17]. Consequently, an increase in the rate of electron transfer to the acceptor (e.g., O_2, H^+) can increase the quantum yield. The rate enhancement depends on the type of metal, the metal islands size, the nature of donor/acceptor and the type of semiconductor. For example, an increase in activity of Pt-TiO_2 nanocomposites for the degradation of phenols depends on the TiO$_2$ colloids used [59] – metallized TiO$_2$ P25 oxidizes methanol and ethanol faster than the pure one, but is less active in the degradation of chloroform [60]. Pt-TiO$_2$ is very active in H$_2$ photoproduction using alcohols as hole scavenger [61]. It seems that there are contradictory trends, and the mechanism of charge separation in metal–semiconductor nanojunctions is not completely understood.

Macroscopic n-type materials in contact with metals normally develop a Schottky barrier (depletion layer) at the junction of the two materials, which reduces the kinetics of electron injection from semiconductor conduction band to the metal. However, when nanoparticles are significantly smaller than the depletion layer, there is no significant barrier layer within the semiconductor nanoparticle to obstruct electron transfer [62]. An accumulation layer may in fact be created, with a consequent increase in the electron transfer from the nanoparticle to the metal island [63]. It is not clear if and what type of electronic barrier exists between semiconductor nanoparticles and metal islands, as well as the role played by the properties of the metal. A direct correlation between the work function of the metal and the photocatalytic activity for the generation of NH$_3$ from azide ions has been made for metallized TiO$_2$ systems [64].

In ZnO QD–metal colloid systems a Fermi level equilibration is reached [65], and for Ag, Au and Cu islands electron capture is inhibited as the redox potential is cathodically shifted, approaching the ZnO conduction band level. Conversely, Pt appears to be an efficient electron sink and no electrons reside on the ZnO of a ZnO-Pt system during extended irradiation. Recent studies have established the unusual catalytic properties of gold nanoparticles, in addition to quantized charging. The apparent Fermi level of TiO$_2$–Au systems depends on the metal particle size [66], shifting towards more negative values as the size of gold nanoparticles decreases. This shift makes the TiO$_2$–Au system more reductive than pristine TiO$_2$ and suggests the possibility of tuning the reducing power of photogenerated electrons.

Heterojunctions Efficient charge separation can be accomplished with appropriate semiconductor heterojunctions. The usefulness of heterojunctions, and in general of heterostructures, is that they offer precise control over the states and motions of charge carriers in semiconductors [67]. An ideal heterojunction consists of a semiconductor crystal in which there exists a plane across which the identity of the atoms participating in the crystal changes abruptly. The ideal structure is approached quite closely in some systems, in which the crystal structure and lattice parameters of the two crystals closely match. In high-quality AlGaAs–GaAs heterojunctions the interface is essentially atomically abrupt [68]. There is an entire spectrum of departures from the ideal structure, in the form of crystalline defects. These defects affect the electrical characteristics of the system by creating localized states that trap charge carriers. If the density of such interfacial traps is sufficiently large, they will dominate the electrical properties of the interface. This is what usually happens at poorly controlled interfaces such as the grain boundaries in polycrystalline materials.

The central feature of an ideal heterojunction is that the bandgaps of the participating semiconductors are usually different. The energy of at least one of the band edges changes when the relevant carriers pass through the heterojunction. Most often, there will be discontinuities in both the conduction and valence band. These discontinuities are the origin of most of the useful properties of heterojunctions. The relative position of the bands and the doping level decide band profiles, bending, surface barriers and discontinuities and, in turn, where the carriers are directed. Also crystallographic orientation and other factors influence the exact arrangement of atoms near the interface and play a role in establishing the relative position of band edges. A space charge layer can also forms at the heterojunction between materials with the same type of majority carriers (e.g., Nn and Pp) [69]. The theoretical prediction of heterojunction band alignments has attracted a good deal of attention and the interested reader is refer to the literature [67, 68, 70, 71].

There are many published examples in which the coupling of two different materials leads to an increase in the photocatalytic activity. Many of them concern coupling and junctions between different nanoparticles, considering also different topologies, like coupled and capped systems [72]. Tentative explanations based on possible heterojunction band profiles are given. However, in-depth analysis of the heterojunction band alignment, the physical structure of the junction, the role of (possible) interfacial traps and of specific catalytic properties of the material is still lacking. Some recently published models and concepts based on (nano)junction between different materials are briefly reviewed here.

n–n Junction TiO_2 Degussa P25 is recognized as one of the most active photocatalyst. Recently, Smirniotis and coworkers have advanced an explanation of this relatively high activity based on the enhancement of charge separation at heterojunctions [59]. TiO_2 P25 powder is composed of 80% anatase (band gap 3.2 eV) and 20% rutile (band gap 3.0 eV), and possibly contains rutile–anatase nanojunctions. The deposition of Pt nano-islands of subnanometer size on colloidal TiO_2 of pure anatase phase (Hombikat UV-100 powder) enhances the photocatalytic degradation rate of phenol. This enhancement can be explained by charge separation at the Pt-TiO_2 contact, as discussed before. However, the photocatalytic activity of pristine TiO_2 P25 is higher than that of pristine and platinized Hombikat UV-100 and platinization of TiO_2 P25 leads to a slight decrease in the degradation rate of phenol with respect to pure P25. The proposed explanation involves band discontinuities at the rutile–anatase nanocontacts possibly present in the P25 material, which create a barrier for the electrons photogenerated in the anatase phase, but promote the migration of the hole from anatase to rutile. These nano-heterojunctions assure a relatively efficient suppression of the recombination process in liquid photocatalytic reactions, and Pt deposition cannot increase further the photoefficiency.

N-n Heterojunction   Another heterojunction that could favor charge separation is the contact between a transparent conductor, in the form of a degenerated semiconductor [like fluorine-doped SnO_2 (FTO) or indium tin oxide (ITO)], and a wide band gap n-type metal oxide semiconductor (like TiO_2). This is an example of an Nn heterojunction. Transparent conductors are widely used as back contacts for photoelectrochemical solar cells, and the contact FTO/TiO_2 is often considered as ohmic [73]. However, much evidence indicates that FTO and ITO contacts are rectifying and their characteristic can be explained if the junction is considered to be an Nn type. A scanning electrochemical microscopy study of the localization of the cathodic and the anodic processes over illuminated TiO_2 thin films deposited onto an ITO substrate showed that most of the reduction process occurs on the exposed ITO portion [74]. Salvador and coworkers have studied the potential distribution at the $FTO-TiO_2$/electrolyte junction by interference reflection spectroscopy [75]. Their results showed that the FTO/TiO_2 cannot be considered as an ohmic but as a rectifying contact, where the FTO behaves as a highly doped n-type semiconductor that absorbs an important part of the equilibrium contact potential in the dark. As a result a contact barrier is built at the $FTO-TiO_2$ interface, which favors electron migration to FTO and forces holes to remain on TiO_2.

p–n Nanodiodes Lee and coworkers have reported the synthesis of a p–n heterostructure in the form of a photocatalytic multiple nanodiode (PCD) by depositing nanoislands of $CaFe_2O_4$ (a metal oxide with p-type semiconductivity) on a microcrystalline substrate formed by an Aurivillius phase layered perovskite [76]. This last material is a niobate of Pb and Bi doped with W to obtain an n-type semiconductor. Their band gaps are 1.9 and 2.75 eV, respectively, allowing visible light photocatalysis. The proposed band alignment of the two materials shows that the heterojunction is rectifying. Photogenerated holes in either material are driven to p-$CaFe_2O_4$, whereas electron go to n-$PbBi_2Nb_{1.9}W_{0.1}O_9$. Owing to this likely enhancement of charge separation, the PCD shows a better photocatalytic activity under visible irradiation for the oxidation of water to O_2 in the presence of a sacrificial oxidant ($AgNO_3$). The reported quantum yields for O_2 evolution in the presence of irradiated PCD, $PbBi_2Nb_2O_9$ and $TiO_{2-x}N_x$ are 38%, 29% and 14% respectively.

16.2.2.5 Mediated Charge Separation

Besides voltage barriers at semiconductor homo or heterojunction, the charge separation process can be accomplished in a *regenerative photoelectrochemical cell* at the semiconductor liquid junction by minority carrier injection to a redox couple in solution. In a modification of this concept, the charge injection originates from an excited dye into the conduction band of a wide band gap oxide semiconductor (commonly TiO_2) with mesoporous or nanocrystalline morphology. The holes in the valence band of the dye are scavenged by a redox couple (I_3^-/I^-) in solution. This is the principle of Dye Sensitized Solar Cell (DSSC), proposed by Grätzel [77]. DSSCs hold a promise of very low costs solar cells for electricity production and water photosplitting using solar light. The main issues related to this device are the stability, linked mainly to dye materials and the electrolyte, the development of new (nanostructured) electron-conducting phases with increased electron transport, the development of non-water based electrolytes with good hole-transport properties to avoid the production of highly reactive species from water oxidation.

Of the photocatalytic systems and structures composed of a single active material, eventually coupled with redox catalysts and/or metals, only a wide band gap oxide semiconductor, like Pt/TiO$_2$, requiring UV irradiation, showed some photoactivity for water photosplitting. Water splitting with visible light requires the irradiation of multiple band gap photoelectrochemical cells (PEC) or Z-scheme systems (like the photosynthesis system of plants etc.).

An interesting Z scheme for water splitting with visible light was proposed by Arakawa and coworkers [78]. They used two photocatalysts (Pt-WO$_3$ and Pt-SrTiO$_3$ Cr- and Ta-doped) suspended in a solution of a redox couple (I$^-$/IO$_3^-$) that mediates charge separation and transport between them. In a Z-scheme-like two-photon process the water is oxidized by valence band (VB) holes from Pt-WO$_3$ and reduced by conduction band (CB) electrons from (Cr, Ta)-doped Pt-SrTiO$_3$. The redox mediator (I$^-$/IO$_3^-$) shuttles electrons from the CB of WO$_3$, which are unable to reduce H$^+$ (CB edge redox potential of WO$_3$ is 0.09 V vs. NHE at pH 7) but can reduce IO$_3^-$ to I$^-$ ($E° = 0.67$ V vs. NHE), to the VB of (Cr, Ta)-doped SrTiO$_3$, where I$^-$ is reoxidized by a photogenerated hole. The process is possible because the oxidation rate of I$^-$ by VB hole from Pt-WO$_3$ is very slow compared with water oxidation, and reduction rate of IO$_3^-$ with CB electron from doped Pt-SrTiO$_3$ is analogously slow compared with water reduction. The production of H$_2$ and O$_2$ is stoichiometric (2:1 ratio) and no sacrificial electron/hole scavengers seem to be needed. The estimated photonic efficiency is 0.1% at 420 nm. The efficiency of the process is fairly low, but demonstrated the possibility to design a Z-scheme multiphoton process for water photosplitting with the proper choice of photocatalysts and redox mediator(s), in which only photons having energies in the visible range are needed.

16.2.3
Surface-related Losses

Semiconductor surfaces and interfaces play a fundamental role in controlling the rate and mechanism of photogenerated charge carrier transfer to solution. The dangling bonds present at the semiconductor–liquid interface, as well as the presence of other defects and/or the presence of adsorbates generate a population of acceptor and donor surface (or interface) states located in the band gap. The interface properties are of paramount importance in defining surface states concentration and their reactivity. Among intrinsic properties is reconstruction of the surface exposed (hydroxylation/protonation). Defects (like steps and kinks), surface charge pH dependent, ion adsorption, and surface complexation influence the nature and density of surface states. These states can act as carrier traps, mediating interface electron transfer to redox active species at the surface and contributing to the total interfacial current density. However, interfacial states can also act as recombination centers. Control of the rate of interfacial carrier transfer and the rate of carrier recombination at the interface, a typical catalysis issue, is fundamental to acquiring chemical control in regenerative and non-regenerative semiconductor/electrolyte based photon conversion devices [28]. From Eq. (2) the quantum yield depends directly on γ, which is proportional to surface efficiency $k_{3,s}k_{4,s}/k_{2,s}$ [23].

When anodic and cathodic sites are not properly spatially separated (e.g., with a space charge layer), the solution near the semiconductor surface is easily crowded by many reactive species, namely oxidized donor (e.g., organic substrate cation radical, OH radical) and reduced acceptor (HO_2 and H_2O_2 in the presence of dioxygen). These species, if present at a relevant concentration near the semiconductor surface, can *back react* with electron and holes, respectively, acting like a surface recombination center and strongly lowering quantum yields (Reactions 3b and 3d in Section 16.2). The effect is larger when these species are confined to the surface, e.g., by complexation.

The metal-oxide–water interface is characterized by a charge build-up whose entity and sign is not only pH dependent but results also from the specific adsorption of ions [79–82]. Charge build-up at the interface changes the potential drop in the Helmholtz double layer, causing band unpinning. The flat band potential of semiconductor oxides, and the band edges positions, depends on the nature and composition of the electrolyte. Specific adsorption of ions can shift the flat band position significantly [83]. Its dependence on the solution pH is Nernstian [26, 43, 84], when H^+ and OH^- are potential determining ions. The adsorption of ions can thus change the driving force of electron transfer, introducing surface states with diverse carrier trapping and recombination abilities, and can inhibit the adsorption of other species.

Recently, Lewis and coworkers demonstrated that, for non-adsorbed, outer sphere redox species and a mono-crystalline ZnO electrode with a surface characterized by a very low density of defects, the ZnO/H_2O junction display a nearly ideal energetic and kinetic behavior [28, 84–86]. The current density vs. potential displayed a first-order dependence on the acceptor and on surface electron concentration. The dependence of the rate constant observed on the driving force and the reorganization energy of the redox couple revealed excellent agreement with the prediction of the Marcus theory of outer sphere interfacial electron transfer. The work of Lewis and coworkers set up the basis for exerting chemical control over forward and reverse rates of light-induced interfacial charge transfer. The verification of the Marcus inverted region, the identification of the maximum outer sphere charge transfer rate constants at optimal exoergocity, the role of the reorganization energy and the effect of pH induced change in the band edges position have been revealed.

Relatively little is understood in the presence of non planar–non ideal interfaces, where electronic levels located in the band gap region act as recombination centers. Colloidal materials, low cost polycrystalline materials and films, interpenetrating networks of absorber and charge collecting phases (e.g., as in the DSSC cells), and the presence of redox active adsorbing species, all give rise to midgap states in which electron transfer is controlled by inner sphere processes. However, some paradigms to chemically eliminate mid-bandgap states were proposed recently. Alkylation of the (111)-oriented surface of Si passivate the surface toward oxidation (hole trapping), as well as electrical recombination processes [87].

The oxide–water interface can be easily modified through surface complexation by ions. Over hydroxylated TiO_2 colloids and films, it is generally recognized that the photogenerated holes are trapped by surface states that can be identified as the surface hydroxyls, and the surface species generated can be regarded as an adsorbed OH radical (Reaction 2a) [88]. Chemical substitution of surface hydroxyls with redox inert groups should suppress charge recombination, at least at surface hole traps. Fluoride ions adsorb efficiently over anatase TiO_2 colloids, substituting surface hydroxyls with almost full surface coverage in the pH range 3–4 [27]. The redox potential of the couple $F^•/F^-$ is 3.6 V [89], so the electronic state of the adsorbed fluoride should overlap with the TiO_2 VB and could not act as a surface hole trap. In this respect, fluoride adsorption passivates the surface of the TiO_2. Besides the suppression of the surface recombination and the decrease in the flat band potential, due to the decrease of the positive charge of the IHP, a marked change in the electron transfer mechanism is also observed. The observed effects are: (a) an increase in the degradation rate of organic substrates that react mainly through a $^•OH$ radical mediated pathway, with a bell-shaped dependence on pH, reflecting the distribution of $\equiv$Ti–F [27]; (b) a kinetic analysis of competition experiments with different OH scavengers allowed the quantification of the relative role of direct electron transfer and mediated oxidation through OH radical in the photocatalytic degradation of phenol, showing that over TiO_2/F the transformation proceeds almost entirely through mediated oxidation by free OH, whereas on naked TiO_2 about 10% is due to a direct hole (inner sphere) oxidation and the other 90% to OH_{ads} [90, 91]. In TiO_2/F systems surface trapped holes, or OH_{ads}, can not form, and the hole is transferred directly to the water, with the formation of free OH radicals, as demonstrated through spin trap experiments [92]. Another feature of surface passivation with fluoride is the inhibition of hole transfer to adsorbable donors, such as carboxylic acids [93], due to suppression of the inner sphere electron transfer route.

All the surface recombination processes, including back reaction, can be incorporated in a heavy kinetic model [22]. The predicted, and experimentally observed, effect of the back reactions is the presence of a maximum in the donor disappearance rate as a function of its concentration [22]. Surface passivation with fluoride also showed a marked effect on back electron transfer processes, suppressing them by the greater distance of reactive species from the surface. The suppression of back reaction has been verified experimentally in the degradation of phenol over an illuminated TiO_2/F catalyst [27].

Surface complexation of TiO_2 with fluoride also shows a relevant effect on dioxygen reduction. Over illuminated TiO_2/F in the presence of dioxygen and an organic donor a sustained production of H_2O_2 is observed, with steady state concentration levels of 1–1.3 mM – nearly 100× the levels reported for naked

TiO$_2$-based photocatalysts [94]. The ability of peroxides to complex Ti(IV) and the surface of titanium dioxide has long been known. Thus, the role of the redox inert ligand is inhibition of the formation of surface superoxo/peroxo species. When these are produced from O$_2$ reduction by e$_{CB}$, in the presence of fluoride ions a release in solution of HO$_2$/H$_2$O$_2$ is achieved, reducing the possibility of back oxidizing them or inhibiting H$_2$O$_2$ degradation.

Similar effects are produced by surface complexation of redox inert cations. The adsorption of metal cations onto TiO$_2$, with an inner sphere complexation mechanism, is pH dependent, and is favored at higher pH [95, 96]. Usually, in a pH interval of two units the metal oxide surface passes from an almost free surface to complete saturation of the inner sphere complexation sites. Despite the numerous studies available on cations adsorption at the metal oxide/electrolyte interface, until recently the influence of this phenomenon on TiO$_2$ mediated photocatalytic processes has been overlooked. The beneficial influence of the presence of Al(III) ions on the photocatalytic degradation of salicylic acid at pH 3 has been reported [97]. The effect was tentatively attributed to the inhibition of the accumulation of degradation intermediates onto the TiO$_2$ surface by Al(III). Recently it was observed that the kinetic of formate photooxidation in the presence of Zn(II) is highly pH dependent and can be faster than onto naked TiO$_2$ or nearly completely inhibited, depending on Zn(II) adsorption. Under these conditions there is also a relevant photoproduction of H$_2$O$_2$ from O$_2$ reduction. The extent of these effects correlates with the extent of Zn(II) adsorption onto TiO$_2$ particles [98].

Surface recombination is also a key point in DSSC operation and the concept of surface passivation is emerging. Slow kinetics for the electron recombination process with the solution hole transporter are essential to allow the charge carriers enough time to be collected into their respective contacts [99]. Recently, the concept of a "kinetic barrier" for interfacial charge recombination processes in DSSCs has been demonstrated by the use of ultrathin Al$_2$O$_3$ or Nb$_2$O$_5$ layers [100], of an ultrathin adsorbent layer [101] deposited onto the surface of the nanocrystalline metal oxide films and of core–shell heterostructures [102], leading to a significant increase in the photocurrent. So, the cell efficiency can be improved by adding surface coatings, adsorbates, or modifying the surface to change, or "passivate", the surface trap states that can mediate recombination with the hole transporter.

The effects on the dynamics of photo-injected electrons where not systematically studied, despite scattered reports on the influence of amines, which induce surface deprotonation, and lower surface charge with a resulting negative shift in band edge position and an increase in the open circuit potential, V_{oc} [103]. The opposite effect is induced by Li$^+$ ions, which intercalate in the oxide structure. Guanidinium ions increase V_{oc} when used as counterions in place of Li$^+$. Other adsorbing molecules that influence both V_{oc} and short circuit current are polycarboxylic acids, phosphonic acids, chenodeoxycholate and 4-guanidinobutyric acid.

Very few studies have been reported on surface modifications until now for water photosplitting.

16.3
Photoelectrochemical Cells

The preceding discussion highlighted that, in colloidal systems redox shuttles, reagents and products, as well as cathodic and anodic sites, are not spatially separated, limiting the yield mainly by back reactions at the same site. Moreover, the different photoactive materials should have CB and VB positioned at correct redox levels, because catalyst nanoparticles are all immersed in the same solution at the same redox potential. Proper spatial separation of cathodic and anodic reactions can be achieved using a photoelectrochemical cell (PEC) design. The use of multiple band gap materials as photoelectrodes in a stacked (series) arrangement, with decreasing band gap from top to bottom, adds further advantages. This configuration allows a better harvesting of the solar photons, decreasing carrier thermalization losses. The series arrangement adds the photovoltages generated in each layer, allowing the attainment of the photovoltage needed for driving water splitting from low band gap materials.

Various configurations of multiple band gap PEC (MPEC) cell are possible. The interested reader is referred to the excellent review of Licht [53]. The simplest MPEC cell contains two photoactive materials with different band gaps coupled in bipolar (series, two photons per electron) or inverted (common node) schemes. In both configurations, the semiconductor–electrolyte interface can be of direct or indirect type. A direct interface can be of ohmic or Schottky type; indirect interface are generally ohmic and built positioning either a metal or an electrocatalyst between the semiconductor and the electrolyte. This last arrangement allows the separation of the semiconductor from the electrolyte, avoiding photocorrosion processes and, eventually, lowering polarization losses through the introduction of an electrocatalyst. The bipolar scheme is conceptually simpler than the inverted one, but generates a large open circuit photopotential, V_{oc}, with larger polarization losses, and needs a current matching among the subsequent layers. Finally, both schemes can be configured as either regenerative (same redox couple at the anode and cathode, electricity production), or non regenerative (different redox couple, storage of chemical energy).

Several storage systems can be implemented, making up water splitting. The electrochemical NiOOH/MH metal hydride storage process has been implemented in a bipolar AlGaAs/Si cell [104].

Operation of a storage PEC cell involves both optical and electrochemical losses. Bipolar band gap cells, having higher photovoltages generated by a series combination of increasing band gaps, can sustain water electrolysis with reduced optical losses. To optimize H_2 and O_2 production, without significant additional energy losses, the water electrolysis must occur at a potential V_{H2O} near the PEC point of maximum power, where V_{H2O} is given by the thermodynamic redox potential of water splitting plus overpotential losses ζ needed to drive the electrolysis current density. An appropriate PEC design is required. Licht and coworkers [25] have developed the bipolar AlGaAs ($E_g = 1.6$ eV)–Si ($E_g = 1.1$ eV) cell, which generates an open circuit and maximum power photopotentials of 1.57 and 1.3 V respectively, which is well suited to the water electrolysis thermodynamic potential of $E°_{H2O} = 1.23$ V. This cell, having photopotential matched with $E°_{H2O}$, is combined, through ohmic contacts, with effective electrocatalysts (Pt black for H_2 evolution and RuO_2 for O_2 evolution), obtaining $\eta_{electrolysis} > 85\%$. The resultant bipolar PEC with indirect ohmic contacts drives sustained water splitting at 18.3% ($\eta_{photo} = 21.2\%$) with AM0 irradiation. Alternative dual band gap systems, based on GaInP/GaAs semiconductors, show $\eta_{photo} > 30\%$, but generate a higher photopotential (2.0–2.1 V at maximum power), so a proper series circuit of solar and electrolysis cell should be used to avoid large overpotential losses.

Problems with these apparatus are related to the costs, due to the use of Pt and single crystalline silicon substrate. Polycrystalline materials and less expensive electrocatalysts can cut costs. Nakato and coworkers have proposed the concept of a polysilicon-TiO_2 photoelectrode as a dual band gap system for water splitting [105], claiming that, in principle, the conversion efficiencies for this less expensive polycrystalline electrode can be greater than 10%. However, several issues have not been solved: (a) the preparation of a stable, doped TiO_2 with absorption extended to 500 nm that can sustain visible light oxygen photoevolution; (b) stabilization of Si in aqueous electrolytes, for which they propose a kind of indirect ohmic contact based on surface methylation and Pt nanodots deposition; (c) comprehension and regulation of the n-Si/TiO_2 junction to obtain correct energy level matching and rectifying properties apt to sustain a biphotonic process. Further complexities arise from a porous particulate TiO_2 layer. In this case band bending at the Si surface is determined mainly by the Si/electrolyte contact.

The integration of a Z-scheme (two photons) for water splitting in a tandem PEC with a nanocrystalline α-Fe_2O_3 photoanode and a dye-sensitized TiO_2 photocathode was recently proposed by Grätzel et al. [106]. The cell has a top nanocrystalline α-Fe_2O_3 anode that will absorb photons having a wavelength less than 600 nm. The photoproduced electron–hole pairs in the iron oxide films are separated. Holes diffuse to the electrode–electrolyte interface and oxidize water. Elcctrons are collected through the conducting glass back-contact and injected into the cathode of a dye-sensitized solar cell (DSSC), reducing the redox shuttle (I^-/I_3^-). The DSSC absorbs light in the red and near-infrared spectral region, which is transmitted by the iron oxide, pumping up electrons. The electrons injected into the TiO_2 CB by the dye have a greater reduction potential than electrons in the α-Fe_2O_3 CB, and are able to reduce water. Problems related to the low photoefficiency of α-Fe_2O_3 were solved by using oriented hematite nanosheets, synthesized by ultrasonic spray pyrolysis. The nanosheets are thinner than the holes diffusion length in hematite. This example also shows how important the type of materials is in achieving acceptable yields.

16.4
New Materials

Basic issues related to the active materials for water photosplitting concern: (a) efficient charge separation, good carrier mobility and lifetime; (b) proper position of band edges with respect to the redox couples involved (e.g., H^+/H_2 and O_2/H_2O); (c) proper light harvesting abilities (optical band gap in the visible); (d) good electrocatalytic properties (low overvoltages for cathodic and anodic processes); (e) spatial separation of anodic and cathodic sites; (f) photocorrosion stability of the semiconductor immersed in aqueous electrolytes.

One phase, cheap and simple active materials, with concurrent optimization of optical and electrolysis yields, are needed. The evolution of O_2 is the key process of a true catalytic system. Few catalysts can decompose water into H_2 and O_2 in a stoichiometric amount under solar light without the presence of a sacrificial scavenger. Probably, a single catalyst having all the required features does not exist. However, fundamental knowledge as to how some materials are able to carry out water photolysis is quite important for future developments.

16.4.1
Crystal Structure and Activity

Other than TiO_2 and titanates, which demonstrated active materials from the early research, promising materials are oxides composed of octahedrally coordinated d^0 transition metal ions (Zr^{4+}, Nb^{5+} and Ta^{5+}, and Ti^{4+}). Several tantalates and niobates demonstrated fairly good activities for water photosplitting, without the presence of a sacrificial donor/acceptor [107–113]. Among tantalates the most active simple material is $LiTaO_3$ [114], which shows a perovskite type structure, in which TaO_6 octahedra are connected by corner sharing. Other tantalates are even more active when loaded with a NiO co-catalyst (0.05–1% w/w). Among them there are $ATaO_3$ (A = Na and K) [108, 110], also with a perovskite structure; $A'Ta_2O_6$ (A' = Ca, Sr, Ba) with a $CaTa_2O_6$-type structure; $Sr_2Ta_2O_7$ with a layered perovskite structure [111]; and $K_2LnTa_5O_{15}$ with a tetragonal tungsten bronze structure [115]. Corner-shared TaO_6 octahedra, including a displacement of Ta^{5+} from the center of the octahedra, are common to all these structures. A quantum yield for water photosplitting of 56% at 270 nm on La-doped $NaTaO_3$ was obtained [108]. These tantalate photocatalysts are also active for the reduction of nitrate ions into N_2 [116].

The activity of these materials was attributed to the ability to promote multiple electron transfer for O_2 evolution and the presence of at least partially separated oxidation and reduction sites [108]. Their major drawback is the high optical band-gap, which in several cases is over 4.0 eV ($\lambda < 310$ nm).

The nature of crystal also influences the position of the band edge, as demonstrated by the differences in activity of NiO loaded tantalate and niobate. NiO co-catalyst acts as H_2 evolution site, collecting electrons from the active material. The energy levels of the top of valence bands of oxide semiconductors composed of transition metals of d^0 configuration are almost the same (2.9 V vs. NHE), because the electronic states at the top of their valence bands consist mainly of O 2p orbitals. The electronic states at the bottom of conduction band consist mainly of metal d orbitals. Tantalates have more negative conduction band energy levels than do Nb-containing materials because the former have larger band gap energies than the latter. Photoexcited electrons can be transferred efficiently to NiO particles on the tantalate pyrochlores. However, such efficient electron transfer seldom occurs on niobate pyrochlores. This difference was attributed to the energy levels of the bottom of conduction bands of tantalates, which are much more negative than those of the niobate [117].

The relevant influence of crystal structure is also evidenced by a series of R_3TaO_7 and R_3NbO_7 rare earths (R = Y, Yb, Gd, or La). In these compounds the crystal structures changed with increasing ionic radius of the R^{3+} ion from fluorite-type cubic structure to pyrochlore-type cubic structure, and finally to a weberite-type orthorhombic structure. Notably, the water splitting into H_2 and O_2 in a stoichiometric ratio proceeded over NiO-loaded La_3TaO_7 and La_3NbO_7 photocatalysts with distorted orthorhombic structures [118, 119]. In the case of NiO loaded titanates both cubic pyrochlore ($Y_2Ti_2O_7$, $Gd_2Ti_2O_7$) and monoclinic perovskite ($La_2Ti_2O_7$) structures are very active for water photosplitting [119].

All of the above-mentioned *active* materials show a crystal structure characterized by a continuous network of corner shared octahedral units of metal cations (TaO_6, NbO_6, TiO_6), which presumably allows for high charge carriers mobilities and efficient charge separation. A relevant role of local electric dipole moments cannot be excluded [111] (see also below).

Recently, it was demonstrated that RuO_2 loaded p block metal ions oxides [120, 121] (MIn_2O_4, with M = Ca, Sr, Sr_2SnO_4, $NaInO_2$, $LaInO_3$, $M_2Sb_2O_7$, with M = Ca, Sr, $CaSb_2O_6$, $NaSbO_3$, $ZnGa_2O_4$, $ZnGeO_4$) and mixed oxides [122] ($LiInGeO_4$) with octahedrally coordinated d^{10} configuration are also good photocatalysts for stoichiometric water photosplitting. The top of the valence band of these compounds is similar to the d^0 compounds VB, in that it is formed by the O 2p orbitals. The d electrons are deep in the VB. The bottom of the conduction band is formed by the p orbitals of the p-block metal (in some cases with contribution of the s orbitals of the co-cations, like in Zn_2GeO_4), as opposed to the d^0 compounds. The activity of these p block compounds depends on the calcination temperature and the type of metal oxide, both affecting the crystal structure, as well as the RuO_2 loading.

The crystal structure of these compounds seems to be strongly correlated with the photocatalytic activity [123]. Active materials are characterized by an orthorhombic structure with distorted MO_6 octahedra, characterized by *high dipole moments*:

- $SrIn_2O_4$, where in the crystal structure there are two kinds of the octahedral InO_6: one has a dipole moment of 2.8 D and the other 1.1 D;
- $Sr_{0.93}Ba_{0.07}In_2O_4$ with dipole moments of 1.70 and 2.58 D;
- $M_2Sb_2O_7$ (M = Ca, Sr), $CaSb_2O_6$ and $NaSbO_3$ with distorted SbO_6 octahedra;
- Ca_2SnO_4 and Sr_2SnO_4 consisting of distorted SnO_6 octahedra;
- Zn_2GeO_4 in which the GeO_4 tetrahedron is heavily distorted to generate a dipole moment inside.

In contrast, distortion-free crystal structures exhibited negligible activity. Among them are: $AInO_2$ (A = Li, Na), which possesses a normal InO_6 octahedron nearly free from distortion, for which the dipole moment is zero; Ba_2SnO_4 with distortion-free SnO_6.

Moreover, in all these materials the conduction band had large dispersion, indicative of a large mobility of photoexcited electrons [121], which when coupled with the local dipole moments allows fast and efficient charge separation. Notably, $BaTi_4O_9$ and $Na_2Ti_6O_{13}$ (with d^0 configuration) possess distorted TiO_6 octahedra (dipole moments of 5.7 and 4.1 D for $BaTi_4O_9$ and 6.7, 5.8, and 5.3 D for $Na_2Ti_6O_{13}$) and show relevant photoactivity [123].

From the framework depicted, it emerges that photocatalytic activity seems strictly related to the dipole moment generated by a distorted crystal structure, namely electron–hole separation upon photoexcitation is promoted by a local electric field due to a dipole moment and, in turn, this promotes *vectorial* movement of electron and holes.

In addition to the catalyst (bulk) reduced losses of these materials, the configuration of the surface, to reduce surface losses, is also important. Some naked materials showed good activity for oxygen evolution. This four-electron process is evidently kinetically favored when simultaneous electron transfer can occur, as in chlorophyll Photosystem II. The clear distinction between cathodic and anodic sites can explain the high activity of NiO-loaded La-doped $NaTaO_3$ [108]. Oxygen evolution sites have been suggested to occur at the bottom of grooves in the layered structures at the catalyst surface, allowing concurrent transfer of multiple electrons.

A new criterion in active material formulation can be set: the crystal lattice should be engineered so that the position of band edges is a controlled spatial function, also in homomaterials, either in the bulk or at the surface, allowing efficient charge separation and spatial distinction of cathodic and anodic sites. This is particularly important because the catalytic sites are by definition good catalysts also for the reverse reaction, which was previously referred to as back reaction. A known example of this is Pt, which is a good catalyst for H^+ reduction, but also for oxygen reduction.

16.4.2
Visible Sensitization

The materials described above, although promising, have wide band gap and no visible activity. Actually, there is no visible active photocatalyst that works efficiently for both oxidative and reductive sides of the water photosplitting. Pt/CdS was demonstrated to be active for H_2 evolution [124] and WO_3 for O_2 evolution in the presence of hole or electron scavengers, respectively. Nevertheless, also in the presence of sacrificial agents, the quantum yield obtained was not larger than 1%. All the reported results suggest that efficient charge separation is urgent for the utilization of the photocatalysis concept for visible light harvesting.

One proposed material is $BiVO_4$ in its monoclinic (scheelite structure) polymorph, which shows a good photocatalytic activity for O_2 evolution in the presence of Ag^+ (9% quantum yield at 450 nm) [125]. The band gap of this vanadate is 2.4 eV, which is smaller than that of WO_3 (2.7 eV). The band gap narrowing is, presumably, due to the contribution of the 6s orbitals of Bi(III) to the top of the VB. A nanocrystalline monoclinic $BiVO_4$ film electrode on conducting glass showed excellent efficiency (IPCE = 29% at 420 nm) for O_2 production under visible light in the presence of Ag^+ [126]. A Z-scheme system for water photosplitting, without sacrificial acceptor, constituted of the Fe^{3+}/Fe^{2+} redox couple as an electron relay and $BiVO_4$ plus $Pt/SrTiO_3$ Rh-doped as powdered heterogeneous photocatalysts showed an apparent quantum yield at 440 nm of 0.3% and is active up to 520 nm [127]. The scheme is very similar to that involving WO_3 [78] discussed earlier.

The valence band top of d^0 or d^{10} stable oxide semiconductor with good photocatalytic activity consists of an O 2p orbital. Commonly, the electrochemical potential of this VB is 2.5–3.5 V vs. NHE, which is considerably more positive than the redox potential of the O_2/H_2O redox couple. Strategies to narrow the band gap of these oxides and better match the redox potentials are:

(1) doping with, or addition of, a cation or an anion with electronic states (either occupied or empty) located in the band gap;
(2) creation of a new band, by a new element as a solid solution or as a mixed oxide, with occupied states that mix with O 2p orbitals, extending the VB toward higher energies, or with empty states that mix with states at the bottom of CB, extending the CB toward lower energies.

A common way to obtain visible sensitization via strategy (1) is doping with electron donors such as transition metal ions with partially filled d-orbitals, forming *localized* electron donor levels in the band gap. This type of doping on TiO_2 results in the development of a color. However, in most cases localized levels of the donor and/or oxygen vacancies associated with charge unbalance act as recombination centers. The result is a marked decrease of the photocatalytic activity even under ultraviolet irradiation. The increase in recombination rate can be counterbalanced with a fast scavenging of the relevant charge carrier. For example, $Pt/SrTiO_3$ co-doped with Cr(III) and Ta(V) evolves H_2 from an aqueous NaI solution [78]. The co-doping with two cations with different charge should maintain the charge balance of the crystal structure and, consequently, eliminate oxygen defects.

A better approach is the introduction in the original crystal of a new element, forming a solid solution, or a new mixed oxide, that adds filled electronic levels over the top of the VB or empty levels just under the bottom of the CB. These extend the delocalized states of VB toward higher energies or of CB toward lower energies. In this case a less defective system, without intraband gap states and continuous bands should be formed. An example of this strategy (2) is silver niobates with a perovskite structure [128]. These compounds evolve O_2 from an aqueous silver nitrate solution under visible light irradiation. $AgNbO_3$ shows a band gap of 2.86 eV, 0.6 eV lower than $NaNbO_3$. Electronic band calculation through a plane-wave based density functional method showed that Ag^+ 4d electronic orbitals contributed to the top of the VB, narrowing the band gap.

Sulfur-based catalysts shows smaller band gaps with respect to the corresponding oxides. A ZnS–CuInS$_2$–AgInS$_2$ solid solution was recently proposed for H$_2$ production under visible irradiation in the presence of aqueous solution containing sulfide and sulfite [129]. The main drawback of sulfide is photocorrosion.

Nitrogen doping is also a very well pursued strategy for the visible sensitization of semiconductor oxides since the publication of the work by Asahi and coworkers in which they demonstrated both theoretically [electronic band structure calculation by the full potential linearized augmented plane wave (FLAPW) method] and experimentally that N-doped TiO$_2$ could be a good visible light active material [130]. A nitrogen concentration dependence on photocatalytic activity of TiO$_{2-x}$N$_x$ powders has been reported [131, 132]. In the perovskite material Sr$_2$Nb$_2$O$_7$, nitrogen doping redshifted the light absorption edge into the visible light range, and induced visible light photocatalytic activity. An optimum amount of nitrogen doping is reported. The intermediate phase still maintains the original layered perovskite structure, but with a part of its oxygen replaced by nitrogen and oxygen vacancy to adjust the charge difference between oxygen and nitrogen [133].

The above discussion suggests that a good strategy for the development of new materials capable of visible activity could be the concurrent combination of anion and cation substitution. This approach was pursued in the tuning of In(OH)$_y$S$_z$ catalyst. In(OH)$_3$ is a wide band gap material ($E_g = 5.17$ eV). Sulfur doping lowers the band gap to 2.19 eV, shifting up the VB top, but the activity for H$_2$ production in the presence of a Na$_2$S–Na$_2$SO$_3$ solution was reported to be quite low, due to the low thermodynamic driving force. The CB bottom of In(OH)$_y$S$_z$ is very similar to the redox potential of the H$^+$/H$_2$ couple. Zinc doping increases slightly the band gap (2.64 eV), moving up the CB bottom, allowing more driving force for H$^+$ reduction and enhancing the photocatalyst activity [134]. Among multiple-cation/anion oxides, PbBi$_2$Nb$_2$O$_9$ seems quite promising. This material has a band gap of 2.88 eV. The quantum yield reported for oxygen production from water containing AgNO$_3$ is 29%, as well as 1% of H$_2$ evolution from aqueous methanol solution [135]. Interestingly, the material is active on both redox sides, but only in the presence of sacrificial donor/acceptor. This implies a strong negative role of solution species, causing recombination.

16.5
Conclusions

Despite recent achievements in active materials, PEC configuration and thermal catalysis, hydrogen production through direct water photosplitting with good solar photon efficiency and low cost apparatus is still far from practical exploitation.

Powders give statistically mixed phases and, possibly, spatially unseparated reduction and oxidation sites, as well as poor space charge layers for carrier separation. This leads to high rates of bulk and surface recombination, as well as solution species back reactions. Light scattering losses add a further decrease in photon efficiency. Much research is still needed on (electro)catalysis, material design and synthesis, proper light harvesting, and proper structure for charge separation.

A key issue is the improvement of solar light harvesting. New active materials with high optical absorption in the visible and good photostability are needed. Implementation of carrier multiplication through impact ionization in quantum dots arrays could mitigate the losses related to carrier thermalization. The alternative approach is the development of vertically stacked tandem systems of increasing band gap active materials, which effect H^+ reduction and water oxidation on opposite sides.

The surface roughness increases normally the scattering, decreasing light harvesting, and increasing the mean residence time of surface generated chemical species, enhancing product mediated recombination. Besides the elimination of surface defects, surface passivation procedures through chemical modification, e.g., with complexation by anions and cations, can generate new ways to suppress surface recombination and avoid back reaction with solution species.

Structuring of material as thin films could help to suppress charge carrier recombination, and to build multi-bandgap materials. This can be achieved through the control of space periodicity and/or the control of space charge layer thickness. The material could be built both with a series of layers (vertical arrangement as in PV cells), in which the charge carrier flow is directed perpendicular to the surface, and, more importantly, as alternating or regularly spatially mixed islands (horizontal arrangement). This last surface configuration is more suited when both charge carriers have to interact with the solution. Two materials with appropriate band gaps and Fermi level could horizontally set up a heterojunction where charge carriers are horizontally separated. In this way they could react with solution species at different locations, avoiding surface recombination, the back reactions, and partially also the back reactions of chemical products. The horizontal arrangement also permits, possibly, a "one pot" synthesis, leading to a reduction of photocatalytic material costs.

Substantial research efforts should be dedicated also to the development of low cost multiple-electron transfer catalysts for oxygen production. Electrochemical losses related to O_2 evolution are a considerable part of the overall inefficiencies.

Finally, an interesting concept, recently advanced, is the implementation of active materials as nanotube arrays. These systems have high surface area to optimize contact between semiconductor and electrolyte, and good light trapping properties. Their inner space could also be filled with catalysts or sensitizers and/or pn junctions to obtain charge separation and facilitate electron transport [136].

Acknowledgments

Financial support from University of Torino (Ricerca Locale), Hysyvision Project – Regione Piemonte (DOCUP ob.2 – periodo di programmazione 2000/2006, Deliberazione della Giunta Regionale 7 novembre 2005, n. 2–1322) and project

D34-Regione Piemonte (Nanostructured polymeric materials for the fabrication of functional coatings) are acknowledged.

References

1. World Energy Assessment: Energy and the Challenge of Sustainability; United Nations, New York, **2000** (http://www.undp.org/energy/activities/wea/draft-start.html).
2. *Basic Research Needs for Solar Energy Utilization*, Report on the Basic Energy Sciences Workshop on Solar Energy Utilization, April 18–21, **2005**, US Department of Energy, http://www.sc.doe.gov/bes/reports/files/SEU_rpt.pdf.
3. R. Eisenberg, D.G. Nocera, Overview of the forum on solar and renewable energy, *Inorg. Chem.* 44 (**2005**) 6799–6801.
4. T.M.L. Wigley, R. Richels, J.A. Edmonds, Economic and environmental choices in the stabilization of atmospheric CO_2 concentrations, *Nature* 379 (**1996**) 240–243.
5. M.I. Hoffert, K. Caldeira, A.K. Jain, E.F. Haites, L.D. Harvey, S.D. Potter, M.E. Schlesinger, T.M.L. Wigley, D.J. Wuebbles, Energy implications of future stabilization of atmospheric CO_2 content, *Nature* 395 (**1998**) 881–884.
6. K. Caldeira, A.K. Jain, M.I. Hoffert, Climate sensitivity uncertainty and the need for energy without CO_2 emission, *Science* 299 (**2003**) 2052–2054.
7. IPCC Special Report on Carbon dioxide Capture and Storage, http://arch.rivm.nl/env/int/ipcc/pages_media/SRCCS-final/IPCCSpecialReportonCarbondioxideCaptureandStorage.htm
8. D. Pimentel, D.W. Patzek, Ethanol production using corn, switchgrass and wood; biodiesel production using soybean and sunflower, *Nat. Resources Res.*, 14 (**2005**) 65–74.
9. A. Shah, P. Torres, R. Tscharner, N. Wyrsch, H. Keppner, Photovoltaic technology: The case for thin-film solar cells, *Science* 285 (**1999**) 692–698.
10. W. Shockley, H.J. Queisser, Detailed balance limit of efficiency of p-n junction solar cells, *J. Appl. Phys.* 32 (**1961**) 510–519.
11. A.N. Tiwari, Thin film chalcogenide photovoltaic materials, *Thin Solid Films*, 480–481 (**2005**) 1 and following papers.
12. M.A. Green, Third generation photovoltaics: Ultrahigh conversion efficiency at low cost, *Prog. Photovolt: Res. Appl.* 9 (**2001**) 123–135.
13. R. Wang, K. Hashimoto, A. Fujishima, M. Chikuni, E. Kojima, A. Kitamura, M. Shimohigoshi, T. Watanabe, Light induced amphiphilic surfaces, *Nature* 388 (**1997**) 431–432.
14. M.A. Fox, M.T. Dulay, Heterogeneous photocatalysis, *Chem. Rev.* 93 (**1993**) 341–357.
15. E. Pelizzetti, C. Minero, Metal oxides as photocatalysts for environmental detoxification, *Comments Inorg. Chem.* 15 (**1994**) 297–337.
16. E. Pelizzetti, N. Serpone (eds.), *Photocatalysis, Fundamental and Applications*, J. Wiley and Sons, New York, **1989**.
17. M.R. Hoffmann, S.T. Martin, W. Choi, D.W. Bahnemann, Environmental applications of semiconductor photocatalysis, *Chem. Rev.* 95 (**1995**) 69–96.
18. O. Carp, C.L. Huismann, A. Reller, Photoinduced reactivity of titanium dioxide, *Progr. Solid State Chem.* 32 (**2004**) 33–177.
19. P.J.K. Robertson, D.W. Bahnemann, J.M.C. Robertson, F. Wood, in: P. Boule, D.W. Bahnemann, P.K.J. Robertson (Eds.), *Environmental Photochemistry Part II*, The Handbook of Environmental Chemistry, Vol. 2M, Springer, Berlin, **2005**, pp. 367–423.
20. A. Fujishima, K. Honda, Electrochemical photolysis of water at a semiconductor electrode, *Nature* 238 (**1972**) 37–38.

21 E. Borgarello, J. Kiwi, E. Pelizzetti, M. Visca, M. Grätzel, Sustained water cleavage by visible light, *J. Am. Chem. Soc.* 103 (**1981**) 6324–6329.

22 C. Minero, Kinetic analysis of photo-induced reactions at the water semiconductor interface, *Catal. Today* 54 (**1999**) 205–216.

23 C. Minero, D. Vione, A quantitative evaluation of the photocatalytic performance of TiO_2 slurries, *Appl. Catal. D: Environ.* 67 (**2006**) 257–269.

24 C.E. Byvik, A.M. Buoncristiani, B.T. Smith, Limits to solar power conversion efficiency with applications to quantum and thermal systems, *J. Energy* 7 (**1983**) 581–588.

25 S. Licht, B. Wang, S. Mukerji, T. Soga, M. Umeno, H. Tributsch, Efficient solar water splitting, exemplified by RuO_2-catalyzed AlGaAs/Si photoelectrolysis, *J. Phys. Chem. B* 104 (**2000**) 8920–8924.

26 S.R. Morrison, *Electrochemistry at Semiconductor and Oxidized Metal Electrodes*, Plenum Press, New York, **1980**, pp. 189–229.

27 C. Minero, G. Mariella, V. Maurino, E. Pelizzetti, Photocatalytic transformation of organic compounds in the presence of inorganic anions. 1. Hydroxyl-mediated and direct electron-transfer reactions of phenol on a titanium dioxide-fluoride system, *Langmuir* 16 (**2000**) 2632–2641.

28 N.S. Lewis, Chemical control of charge transfer and recombination at semiconductor photoelectrode surfaces, *Inorg. Chem.* 44 (**2005**) 6900–6911.

29 C. Minero, F. Catozzo, E. Pelizzetti, Role of adsorption in photocatalyzed reactions of organic molecules in aqueous titania suspensions, *Langmuir* 8 (**1992**) 481.

30 J. Cunningham, G. Al-Sayyed, Factors influencing efficiencies of TiO_2-sensitised photodegradation. Part 1. Substituted benzoic acids: Discrepancies with dark-adsorption parameters, *J. Chem. Soc., Faraday Trans.* 86 (**1990**) 3935–3941.

31 A. De Vos, Detailed balance limit of the efficiency of tandem solar cells, *J. Phys. D: Appl. Phys.* 13 (**1980**) 839–846.

32 A. Marti, G.L. Araujo, Limiting efficiencies for photovoltaic energy conversion in multigap systems, *Sol. Energy Mater. Solar Cells* 43 (**1996**) 203–222.

33 N.H. Karam, J.H. Ermer, R.R. King, M. Hadda, L. Cai, D.E. Joslin, D.D. Krut, M. Takahashi, J.W. Eldredge, W. Nishikawa, B.T. Cavicchi, D.R. Lillington. High efficiency $GaInP_2$/GaAs/Ge dual and triple junction solar cells for space applications. Office for Official Publications of the European Communities, Luxembourg, *Proceedings of the 2nd World Conference on Photovoltaic Solar Energy Conversion*, Vienna, July 6–10, **1998**, pp. 3534–3539.

34 J. Yang, A. Banerjee, K. Lord, S. Guha, Correlation of component cells with high efficiency amorphous silicon alloy triple junction solar cells and modules. Office for Official Publications of the European Communities, Luxembourg, *Proceedings of the 2nd World Conference on Photovoltaic Solar Energy Conversion*, Vienna, July 6–10, **1998**, pp. 387–390.

35 A.J. Nozik, Spectroscopy and hot electron relaxation dynamics in semiconductor quantum wells and quantum dots, *Ann. Rev. Phys. Chem.* 52 (**2001**) 193–231.

36 P.T. Landsberg, H. Nussbaumer, G. Willeke, Band-band impact ionization and solar cell efficiency, *J. Appl. Phys.* 74 (**1993**) 1451–1452.

37 S. Kolodinski, J.H. Werner, T. Wittchen, H.J. Queisser, Quantum efficiencies exceeding unity due to impact ionization in silicon solar cells, *Appl. Phys. Lett.* 63 (**1993**) 2405–2407.

38 A.J. Nozik, Exciton multiplication and relaxation dynamics in quantum dots: Application to ultrahigh-efficiency solar photon conversion, *Inorg. Chem.* 44 (**2005**) 6893–6899.

39 J.L. Blackburn, R.J. Ellingson, O.I. Micic, A.J. Nozik, Electron relaxation in colloidal InP quantum dots with photogenerated excitons or chemically injected electrons, *J. Phys. Chem. B* 107 (**2003**) 102–109.

40 R.J. Ellingson, M.C. Beard, J.C. Johnson, P. Yu, O.I. Micic, A.J. Nozik, A. Shabaev, A.L. Efros, Highly efficient multiple

exciton generation in colloidal PbSe and PbS quantum dots, *Nano Lett.* 5 (**2005**) 865.

41 P. Würfel, Solar energy conversion with hot electrons from impact ionization, *Solar En. Mat. Solar Cell* 46 (**1997**) 43–52.

42 A. Luque, A. Marti, Increasing the efficiency of ideal solar cells by photon induced transitions at intermediate levels, *Phys. Rev. Lett.* 78 (**1997**) 5014–5017.

43 M. Grätzel, Colloidal semiconductors, in *Photocatalysis, Fundamental and Applications*, E. Pelizzetti, N. Serpone (eds.), J. Wiley and Sons, New York, **1989**, pp. 123–157.

44 D.H. Kim, M.A. Anderson, Photoelectrocatalytic degradation of formic acid using a porous titanium dioxide thin-film electrode, *Environ. Sci. Technol.* 28 (**1994**) 479–483.

45 R.J. Candal, W.A. Zeltner, M.A. Anderson, Effects of pH and applied potential on photocurrent and oxidation rate of saline solutions of formic acid in a photoelectrocatalytic reactor, *Environ. Sci. Technol.* 34 (**2000**) 3443.

46 K. Vinodgopal, S. Hotchandani, P.V. Kamat, Electrochemically assisted photocatalysis: titania particulate film electrodes for photocatalytic degradation of 4-chlorophenol, *J. Phys. Chem.* 97 (**1993**) 9040–9044.

47 K. Vinodgopal, I. Bedja, P.V. Kamat, Nanostructured semiconductor films for photocatalysis. Photoelectrochemical behavior of SnO_2/TiO_2 composite systems and its role in photocatalytic degradation of a textile azo dye, *Chem. Mater.* 8 (**1996**) 2180–2187.

48 H. Hidaka, Y. Asai, J. Zhao, K. Nohara, E. Pelizzetti, N. Serpone, Photoelectrochemical decomposition of surfactants on a TiO_2/TCO particulate film electrode assembly, *J. Phys. Chem.* 99 (**1995**) 8244–8248.

49 W.H. Leng, W.C. Zhu, J. Ni, Z. Zhang, J.Q. Zhang, C.N. Cao, Photoelectrocatalytic destruction of organics using TiO_2 as photoanode with simultaneous production of H_2O_2 at the cathode, *Appl. Catal. A: General* 300 (**2006**) 24–35.

50 H. Reiss, The Fermi level and the redox potential, *J. Phys. Chem.* 89 (**1985**) 3783–3791.

51 S.R. Morrison, *Electrochemistry at Semiconductor and Oxidized Metal Electrodes*, Plenum Press, New York, **1980**, pp. 153–186.

52 Y. Pleskov, *Semiconductor Photoelectrochemistry*, Springer, Berlin, **1986**, Ch. 2.

53 S. Licht, Multiple band gap semiconductor/electrolyte solar energy conversion, *J. Phys. Chem. B* 105 (**2001**) 6281–6294.

54 B. Kraeutler, A.J. Bard, Heterogeneous photocatalytic preparation of supported catalysts. Photodeposition of platinum on titanium dioxide powder and other substrates, *J. Am. Chem. Soc.* 100 (**1978**) 4317–4318.

55 K. Kalyanadundaram, M. Grätzel, E. Pelizzetti, Interfacial electron transfer in colloidal metal and semiconductor dispersions and photodecomposition of water, *Coord. Chem. Rev.* 69 (**1986**) 57–125.

56 P.V. Kamat, Photophysical, photochemical and photocatalytic aspects of metal nanoparticles, *J. Phys. Chem. B* 106 (**2002**) 7729–7744.

57 A. Sclafani, M.N. Mozzanega, J.M. Herrmann, Influence of silver deposits on the photocatalytic activity of titania, *J. Catal.* 168 (**1997**) 117–120.

58 H. Gerischer, A. Heller, Photocatalytic oxidation of organic molecules at TiO_2 particles by sunlight in aerated water, *J. Electrochem. Soc.* 139 (**1992**) 113–118.

59 B. Sun, A.V. Vorontsov, P.G. Smirniotis, Role of platinum deposited on TiO_2 in phenol photocatalytic oxidation, *J. Phys. Chem. B* 19 (**2003**) 3151–3156.

60 J. Chen, D.F. Ollis, W.H. Rulkens, H. Bruning, Photocatalyzed oxidation of alcohols and organochlorides in the presence of native TiO2 and metallized TiO2 suspensions. Part (I): Photocatalytic activity and pH influence, *Water Res.* 33 (**1999**) 661–668.

61 P. Pichat, M.N. Mozzanega, J. Disdier, J.M. Herrmann, Room temperature hydrogen production from aliphatic alcohols uv-illuminated Pt/TiO_2 catalysts, *Int. J. Hydrogen Energy* 9 (**1984**) 397–403.

62 J.S. Curran, D.J. Lamouche, Transport and kinetics in photoelectrolysis by semiconductor particles in suspension, *J. Phys. Chem.* 87 (**1983**) 5405–5411.

63 J. Gerischer, A mechanism of electron hole pair separation in illuminated semiconductor particles, *J. Phys. Chem.* 88 (**1984**) 6096–6097.

64 Y. Nosaka, K. Norimatsu, H. Miyama, The function of metals in metal-compounded semiconductor photocatalysts, *Chem. Phys. Lett.* 106 (**1984**) 128–131.

65 A. Wood, M. Giersig, P. Mulvaney, Fermi level equilibration in quantum dot–metal nanojunctions, *J. Phys. Chem. B* 105 (**2001**) 8810–8815.

66 V. Subraniam, E. Wolf, P.V. Kamat, Catalysis with TiO_2/gold nanocomposites. Effect of metal particle size on the Fermi level equilibration, *J. Am. Chem. Soc.* 126 (**2004**) 4943–4950.

67 D.V. Morgan, R.H. Williams (eds.), Physics and technology of heterojunction devices, *IEE Mater. Devices Ser. 8*, IET, **1991**.

68 R.J. Matyi, in *Heterostructures and Quantum Devices* (VLSI Electronics: Microstructure Science), W.R. Frensley, N.G. Einspruch (eds.), Academic Press, San Diego, **1994**, Ch. 2.

69 D. Neamen, *Semiconductor Physics and Devices*, 3rd edn, McGraw-Hill, New York, **2002**, Ch. 9.

70 H. Kroemer, Heterostructure devices: A device physicist looks at interfaces, *Surf. Sci.*, 132 (**1983**) 543–576.

71 J. Tersoff, in *Heterojunction Band Discontinuities, Physics and Device Applications*, F. Capasso and G. Margaritondo (eds.), North-Holland, Amsterdam, **1987**, p. 3.

72 I. Bedjat, P.V. Kamat, Capped semiconductor colloids. Synthesis and photoelectrochemical behavior of TiO_2-capped SnO_2 nanocrystallites, *J. Phys. Chem.* 99 (**1995**) 9182–9188.

73 D. Cahen, G. Hodes, M. Grätzel, J.F. Guillemoles, I.J. Riess, Nature of photovoltaic action in dye-sensitized solar cells, *J. Phys. Chem. B* 104 (**2000**) 2053–2059.

74 H. Maeda, K. Ikeda, K. Hashimoto, K. Ajito, M. Morita, A. Fujishima, Microscopic observation of TiO_2 photocatalysis using scanning electrochemical microscopy, *J. Phys. Chem. B* 103 (**1999**) 3213–3217.

75 M. Turrion, B. Macht, H. Tributsch, P. Salvador, Potential distribution and photovoltage origin in nanostructured TiO_2 sensitization solar cells: an interference reflection study, *J. Phys. Chem. B* 105 (**2001**) 9732–9738.

76 H.G. Kim, P.H. Borse, W. Choi, J.S. Lee, Photocatalytic nanodiodes for visible light photocatalysis, *Angew. Chem. Int. Ed.* 44 (**2005**) 4585–4589.

77 B. O'Reagan, M. Grätzel, A low cost high efficiency solar cell based on dye sensitized colloidal TiO_2 films, *Nature* 353 (**1991**) 737–740.

78 K. Sayama, K. Mukasa, R. Abe, Y. Abe, H. Arakawa, Stoichiometric water splitting into H_2 and O_2 using a mixture of two different photocatalysts and an IO_3^-/I^- shuttle redox mediator under visible light irradiation, *Chem. Commun.* (**2001**) 2416–2417.

79 W. Stumm, *Chemistry of the Solid-Water Interface*, Wiley Interscience, New York, **1995**.

80 D.A. Dzombak, F.M.M. Morel, *Surface Complexation Modeling*, Wiley Interscience, New York, **1990**.

81 R.O. James, T.W. Healy, Adsorption of hydrolyzable metal ions at the oxide – water interface. II. Charge reversal of SiO_2 and TiO_2 colloids by adsorbed Co(II), La(III), and Th(IV) as model systems, *J. Colloid Interface Sci.* 40 (**1972**) 53–64.

82 B.P. Nelson, R. Candal, R.M. Corn, M.A. Anderson, Control of surface and ζ potentials on nanoporous TiO_2 films by potential-determining and specifically adsorbed ions, *Langmuir* 16 (**2000**) 6094–6101.

83 X.G. Zhang, *Electrochemistry of Silicon and its Oxides*, Kluwer/Plenum, New York, **2001**, Ch. 3.

84 T.W. Hamann, F. Gstrein, B.S. Brunschwig, N.S. Lewis, Measurement of the driving force dependence of interfacial charge transfer rate constants in response to pH changes at n-ZnO/H_2O interfaces, *Chem. Phys.* 326 (**2006**) 15–23.

85. T.W. Hamann, F. Gstrein, B.S. Brunschwig, N.S. Lewis, Measurement of the free-energy dependence of interfacial charge-transfer rate constants using ZnO/H_2O semiconductor/liquid contacts, *J. Am. Chem. Soc.* 127 (**2005**) 7815–7824.
86. T.W. Hamann, F. Gstrein, B.S. Brunschwig, N.S. Lewis, Measurement of the dependence of interfacial charge-transfer rate constants on the reorganization energy of redox species at ZnO/H_2O interfaces, *J. Am. Chem. Soc.* 127 (**2005**) 7815–7824.
87. T.W. Hamann, N.S. Lewis, Control of the stability, electron transfer kinetics, and pH-dependent energetics of Si/H_2O interface through methyl termination of Si(III) surfaces, *J. Phys. Chem. B* 110 (**2006**) 22291–22294.
88. D. Lawless, N. Serpone, D. Meisel, Role of OH. radicals and trapped holes in photocatalysis. A pulse radiolysis study, *J. Phys. Chem.* 95 (**1991**) 5166–5170.
89. D.M. Stanbury, Reduction potentials involving inorganic free radicals in aqueous solution, *Adv. Inorg. Chem.* 33 (**1989**) 69–138.
90. C. Minero, G. Mariella, V. Maurino, E. Pelizzetti, Photocatalytic transformation of organic compounds in the presence of inorganic ions. 2. Competitive reactions of phenol and alcohols on a titanium dioxide-fluoride system, *Langmuir* 16 (**2000**) 8964–8972.
91. C. Minero, V. Maurino, E. Pelizzetti, Mechanism of the photocatalytic transformation of organic compounds, in V. Ramamurthy, K.S. Schanze (eds.), *Semiconductor Photochemistry and Photophysics*, Vol 10 of Molecular and Supramolecular Photochemistry, Marcel Dekker, New York, **2003**, pp. 211–229.
92. M. Mrowetz, E. Selli, Enhanced photocatalytic formation of hydroxyl radicals on fluorinated TiO_2, *Phys. Chem. Chem. Phys.* 7 (**2005**) 1100–1102.
93. C. Minero et al. work in progress.
94. V. Maurino, C. Minero, G. Mariella, E. Pelizzetti, Sustained production of H_2O_2 on irradiated TiO_2–Fluoride Systems, *Chem. Commun.* (**2005**) 2627–2629.
95. Z. Zhang, P. Fenter, L. Cheng, N.C. Sturchio, M.J. Bedzyk, M. Predota, A. Bandura, O.J.D. Kubicki, S.N. Lvov, P.T. Cummings, A.A. Chialvo, M.K. Ridley, P. Benezeth, L. Anovitz, D.A. Palmer, M.L. Machesky, D.J. Wesolowski, Ion adsorption at the rutile-water interface: Linking molecular and macroscopic properties, *Langmuir* 20 (**2004**) 4954–4969.
96. J.K. Yang, A.P. Davis, Competitive adsorption of Cu(II)-EDTA and Cd(II)-EDTA onto TiO_2, *J. Colloid Interf. Sci.* 216 (**1999**) 77–85.
97. M.I. French, J. Peral, X. Domenech, J.A. Ayllon, Aluminum(III) adsorption: a soft and simple method to prevent TiO_2 deactivation during salicylic acid photodegradation, *Chem. Commun.* (**2005**) 1851–1853.
98. V. Maurino, C. Minero, E. Pelizzetti, G. Mariella, A. Arbezzano, Influence of Zn(II) adsorption on the photocatalytic activity and the production of H_2O_2 over irradiated TiO_2, *Res. Chem. Interm.* 33 (**2007**) 319–332.
99. L.M. Peter, E.A. Ponomarev, G. Franco, N.J. Shaw, Aspects of the photoelectrochemistry of nanocrystalline systems, *Electrochim. Acta* 45 (**1999**) 549–560.
100. E. Palomares, J.N. Clifford, S.A. Haque, T. Lutz, J.R. Durrant, Control of charge recombination dynamics in dye sensitized solar cells by the use of conformally deposited metal oxide blocking layers, *J Am. Chem. Soc.* 125 (**2003**) 475–482.
101. N. Kopidakis, R.N. Neale, A.J. Frank, Effect of an adsorbent on recombination and band-edge movement in dye-sensitized TiO_2 solar cells: Evidence for surface passivation, *J. Phys. Chem. B* 110 (**2006**) 12485–12489.
102. M. Law, L.E. Greene, A. Radenovic, T. Kuykendall, J. Liphardt, P. Yang, ZnO–Al_2O_3 and ZnO–TiO_2 core-shell nanowire dye-sensitized solar cells, *J. Phys. Chem. B* 110 (**2006**) 22652–22663.
103. N. Kopidakis, R.N. Neale, M. Gratzel, A.J. Frank, Effect of a coadsorbent on the performance of dye-sensitized TiO_2 solar cells: Shielding versus band-edge movement, *J. Phys. Chem. B* 109 (**2005**) 23183–23189.
104. B. Wang, S. Licht, T. Soga, M. Umeno, Stable cycling behavior of the light

104 invariant AlGaAs/Si/metal hydride solar cell, *Sol. Energy Mater. Sol. Cells* 64 (**2000**) 311–320.

105 S. Takabayashi, R. Nakamura, Y. Nakato, A nano-modified Si/TiO$_2$ composite electrode for efficient solar water splitting, *J. Photochem. Photobiol. A: Chem.* 166 (**2004**) 107–113.

106 A. Duret, M. Grätzel, Visible light-induced water oxidation on mesoscopic α-Fe$_2$O$_3$ films made by ultrasonic spray pyrolysis, *J. Phys. Chem. B* 109 (**2005**) 17184–17191.

107 H. Kato, A. Kudo, New tantalate photocatalysts for water decomposition into H$_2$ and O$_2$, *Chem. Phys. Lett.* 295 (**1998**) 487–492.

108 H. Kato, K. Asakura, A. Kudo, Highly efficient water splitting into H$_2$ and O$_2$ lanthanum-doped NaTaO$_3$ photocatalysts with high crystallinity and surface nanostructure, *J. Am. Chem. Soc.* 125 (**2003**) 3082–3089.

109 H. Kato, I. Tsuji, A. Kudo, Strategies for the development of visible-light-driven photocatalysts for water splitting, *Chem. Lett.* 33 (**2004**) 1534–1539.

110 H. Kato, A. Kudo, Water splitting into H$_2$ and O$_2$ on alkali tantalate photocatalysts ATaO$_3$ (A = Li, Na, and K), *J. Phys. Chem. B* 105 (**2001**) 4285–4292.

111 A. Kudo, H. Kato, S. Nakagawa, Water splitting into H$_2$ and O$_2$ on new Sr$_2$M$_2$O$_7$ (M = Nb and Ta) photocatalysts with layered perovskite structures: Factors affecting the photocatalytic activity, *J. Phys. Chem. B* 104 (**2000**) 571–575.

112 R. Abe, M. Higashi, K. Sayama, A. Yoshimoto, H. Sugihara, Photocatalytic activity of R$_3$MO$_7$ and R$_2$Ti$_2$O$_7$ (R = Y, Gd, La; M = Nb, Ta) for water splitting into H$_2$ and O$_2$, *J. Phys. Chem. B* 110 (**2006**) 2219–2226.

113 R. Abe, M. Higashi, Z. Zou, K. Sayama, Y. Abe, H. Arakawa, Photocatalytic water splitting into H$_2$ and O$_2$ over R$_3$TaO$_7$ and R$_3$NbO$_7$ (R = Y, Yb, Gd, La): Effect of crystal structure on photocatalytic activity, *J. Phys. Chem. B* 108 (**2004**) 811–814.

114 A. Kudo, Photocatalyst material for water photosplitting, *Catal. Survey Asia* 7 (**2003**) 31–38.

115 A. Kudo, H. Okutomi, H. Kato, Photocatalytic water splitting into H$_2$ and O$_2$ over K$_2$LnTa$_5$O$_{15}$ powder, *Chem. Lett.* 29 (**2000**) 1212–1213.

116 H. Kato, A. Kudo, Photocatalytic reduction of nitrate ions over tantalate photocatalysts, *Phys. Chem. Chem. Phys.* 4 (**2002**) 2833–2838.

117 S. Ikeda, M. Fubuki, Y.K. Takahara, M. Matsumura, Photocatalytic activity of hydrothermally synthesized tantalite pyrochlores for overall water splitting, *Appl. Cat. A: Gen.* 300 (**2006**) 186–190.

118 R. Abe, M. Higashi, Z. Zou, K. Sayama, Y. Abe, H. Arakawa, Photocatalytic water splitting into H$_2$ and O$_2$ over R$_3$TaO$_7$ and R$_3$NbO$_7$ (R = Y, Yb, Gd, La): Effect of crystal structure on photocatalytic activity, *J. Phys. Chem. B* 108 (**2004**) 811–814.

119 R. Abe, M. Higashi, K. Sayama, Y. Abe, H. Sugihara, Photocatalytic activity of R$_3$MO$_7$ and R$_2$Ti$_2$O$_7$ (R = Y, Gd, La; M = Nb, Ta) for water splitting into H$_2$ and O$_2$, *J. Phys. Chem. B* 110 (**2006**) 2219–2226.

120 J. Sato, N. Saito, H. Nishiyama, Y. Inoue, New photocatalyst group for water decomposition of RuO$_2$-loaded p-block metal (In, Sn, and Sb) oxides with d^{10} configuration, *J. Phys. Chem. B* 105 (**2001**) 6061–6063.

121 J. Sato, H. Kobayashi, K. Ikarashi, N. Saito, H. Nishiyama, Y. Inoue, Photocatalytic activity for water decomposition of RuO$_2$-dispersed Zn$_2$GeO$_4$ with d^{10} configuration, *J. Phys. Chem. B* 108 (**2004**) 4369–4375.

122 H. Kadowaki, J. Sato, H. Kobayashi, N. Saito, H. Nishiyama, Y. Simodaira, Y. Inoue, Photocatalytic activity of the RuO$_2$-dispersed composite p-block metal oxide LiInGeO$_4$ with d^{10}-d^{10} configuration for water decomposition, *J. Phys. Chem. B* 109 (**2005**) 22995–23000.

123 J. Sato, H. Kobayashi, Y. Inoue, Photocatalytic activity for water decomposition of indates with octahedrally coordinated d^{10} configuration. II. Roles of geometric and electronic structures, *J. Phys. Chem. B* 107 (**2003**) 7970–7975.

124 N. Serpone, E. Borgarello, M. Graetzel, Visible light induced generation of

hydrogen from H_2S in mixed semiconductor dispersions; improved efficiency through inter-particle electron transfer, *J. Chem. Soc., Chem. Commun.* **1984**, 342–344.

125 A. Kudo, K. Omori, H. Kato, A novel aqueous process for preparation of crystal form-controlled and highly crystalline $BiVO_4$ powder from layered vanadates at room temperature and its photocatalytic and photophysical properties, *J. Am. Chem. Soc.* 121 (**1999**) 11459–11467.

126 K. Sayama, A. Nomura, Z. Zou, R. Abe, Y. Abe, H. Arakawa, Photoelectrochemical decomposition of water on nanocrystalline $BiVO_4$ film electrodes under visible light, *Chem. Commun.* (**2003**) 2908–2909.

127 H. Kato, M. Hori, R. Konta, Y. Shimodaira, A. Kudo, Construction of Z-scheme type heterogeneous photocatalysis systems for water splitting into H_2 and O_2 under visible light irradiation, *Chem. Lett.* 33 (**2004**) 1348–1349.

128 H. Kato, H. Kobayashi, A. Kudo, Role of Ag^+ in the band structures and photocatalytic properties of $AgMO_3$ (M: Ta and Nb) with the perovskite structure, *J. Phys. Chem. B* 106 (**2002**) 12441–12447.

129 I. Tsuji, H. Kato, A. Kudo, Visible-light-induced H_2 evolution from an aqueous solution containing sulfide and sulfite over a ZnS–$CuInS_2$–$AgInS_2$ solid-solution photocatalyst, *Angew. Chem. Int. Ed.* 44 (**2005**) 3565–3568.

130 R. Asahi, T. Ohwaki, K. Aoki, Y. Taga, Visible light photocatalysis in nitrogen doped titanium oxides, *Science* 293 (**2001**) 269–271.

131 H. Irie, S. Watanabe, N. Yohino, K. Hashimoto, Visible-light induced hydrophilicity on nitrogen-substituted titanium dioxide films, *Chem. Commun.* (**2003**) 1298–1299.

132 J.S. Jang, H.G. Kim, S.M. Jia, S.W. Bae, J.H. Jung, B.H. Shond, J.S. Lee Formation of crystalline $TiO_{2-x}N_x$ and its photocatalytic activity, *J. Solid State Chem.* 179 (**2006**) 1067–1075.

133 S.M. Ji, P.H. Borse, H.G. Kim, D.W. Hwang, J.S. Jang, S.W. Bae, J.S. Lee, Photocatalytic hydrogen production from water–methanol mixtures using N-doped $Sr_2Nb_2O_7$ under visible light irradiation: Effects of catalyst structure, *Phys. Chem. Chem. Phys.* 7 (**2005**) 1315–1321.

134 Z. Lei, G. Ma, M. Liu, W. You, H. Yan, G. Wu, T. Takata, M. Hara, K. Domen, C. Li, Sulfur-substituted and zinc-doped $In(OH)_3$: A new class of catalyst for photocatalytic H_2 production from water under visible light illumination, *J. Catal.* 237 (**2006**) 322–329.

135 H.G. Kim, D.W. Hwang, J.S. Lee, An undoped, single-phase oxide photocatalyst working under visible light, *J. Am. Chem. Soc.* 126 (**2004**) 8912–8913.

136 Y.S. Chen, J.C. Crittenden, S. Hackney, L. Sutter, D. Whand, Preparation of a novel TiO_2-based p-n junction nanotube photocatalyst, *Environ. Sci. Technol.* 39 (**2005**) 1201–1208.

Conclusions, Perspectives and Roadmap

Gabriele Centi and Rutger A. van Santen

1
Introduction

The increase in energy prices and the need to secure future energy needs by increasing diversity of the energy supply mix have fostered the necessity of introducing renewable raw materials (RRM) and sources of energy (RSE).

Increasing the use of RRM is an opportunity for a more sustainable use of resources, with consequent benefits for the environment and our quality of life. For example, it has been estimated that direct and indirect contribution from industrial use of RRM could be about 30 million tonnes of CO_2 equivalent per annum by 2010 [1]. Further environmental benefits arise from (a) improved soil and water quality through a reduction in toxic effluents from manufacturing processes, (b) reduced pressure on landfill sites through special waste management systems such as the composting of bio-degradable materials, and (c) an increase in bio-diversity on farms, deriving from the introduction of alternative, non-food crops into agriculture. There are also political reasons for fostering the use of RRM, because they provide (a) direct social benefits by rejuvenating rural communities through the establishment of local industries and (b) provide farmers with additional sources of income, thereby securing their jobs. In addition, a driver for RRM as industrial feedstock is the possibility of obtaining tailor made and sometimes improved performance compared, or in combination, with conventional materials derived from fossil fuels.

The same motivation holds for the area of energy with respect to the use of RRM for chemical production. The amount of energy involved is a factor of about $20\times$ more in terms of use of fossil fuel resources. Realizing an eco-efficient and environmental friendly usage of energy requires important progress in various areas such as (a) the introduction of alternative energy sources, (b) improved energy conservation, and (c) more efficient energy storage and transportation. This book is limited to the use and perspective of catalysis in using RRM and RSE. The various chapters deal with (a) options for catalytic processes to biofuels, including the catalytic upgrading of bioproducts in these processes and (b) the role

Catalysis for Renewables: From Feedstock to Energy Production
Edited by Gabriele Centi and Rutger A. van Santen
Copyright © 2007 WILEY-VCH Verlag GmbH & Co. KGaA, Weinheim
ISBN 978-3-527-31788-2

of catalysis in the production of fine and oleo-chemicals from renewable resources such as oils from oilseed crops, starch from cereals and potatoes, and cellulose from straw and wood. Selected topics related to RRM are also covered. Sub-topics and examples provide a background to the new directions of catalysis research in the field we discuss. For example, very relevant in this area of RRM is the integration of biocatalysis with chemocatalysis. For this reason a specific chapter is dedicated to this topic.

In addition to the chapters discussing the various aspects of bio-energy, two chapters are dedicated to hydrogen production and fuel cells. A second book in this series, based on a second workshop, "Catalysis for Sustainable Energy Production" (organized by IDECAT – the European Network of Excellence on catalysis, see Preface), will discuss these aspects in more detail.

The final chapter of this book is dedicated to solar energy as a source of hydrogen and for CO_2 conversion. This chapter introduces the main concepts of the field and highlights the role of catalysis for using solar energy.

After a short discussion of some general aspects that complement the technical issues discussed throughout the book, this concluding chapter provides a possible roadmap for catalysis research in the RRM and RSE. The roadmap is based on several sources – the contributions, discussion and round table session of the workshop "Catalysis for Renewables", organized in May 2006 by IDECAT (see Preface), the more general discussion within the network of excellence IDECAT and the catalysis community as well, and also the indications given in documents (Strategic Research Agenda and Implementation Action Plan) prepared by the European Technology Platform for Sustainable Chemistry (SusChem, see www.suschem.org).

2
Driver for a Biomass Economy

The three main driver towards a biomass economy are (a) security of supply, which derives from pressure caused by depleting reserves, political instability and rising prices; (b) global warming, with the pressure of implementing the Kyoto protocol along with increasing social perception of climate catastrophes; and (c) the local economy, which is affected by a combination of oil imports and the agricultural economy.

The potential of biomass resources is questionable. It depends on "ecological guardrail" considerations on suitable land and human "appropriation" of terrestrial biomass to be dedicated to the provision of food, fodder, and biomaterials. Notably, however, large amounts of biomass waste are already available worldwide. We may distinguish between (a) primary by-products, e.g., at the source (sugar beet tops, straw, verge grass, prunings, greenhouse residues, etc.), (b) secondary by-products, e.g., later in the production chain (potato peels, sugar beet pulp, sawdust, etc.) and (c) tertiary by-products, e.g., one that have already been used (used frying oil, slaughterhouse waste, manure, household organic wastes,

Table 1 Bioenergy production potentials for selected biomass types in 2050. (Adapted from Hunt [2]).

Biomass type	Potential (EJ)
Agricultural residues	15–70
Organic wastes	5–50
Forest residues	30–150
Energy crops (agricultural land)	0–700
Energy crops (marginal land)	60–150

used paper, demolition wood). In addition, specific crops, such as rape, energy grain, Miscanthus, switchgrass, SRC, sugar beet, etc., could be cultivated in unused lands specifically for a biomass-economy. Often, this is an opportunity to support the economy in rural areas and there are, thus, social incentives to subsidize this production.

Table 1 summarizes the long-term potential contribution of biomass to the global energy supply in the year 2050 [2]. In the most optimistic scenarios, bioenergy could provide over twice the current global energy demand, without competing with food production, forest protection efforts, and biodiversity. In the least favorable scenarios, bioenergy could supply only a fraction of current energy use. However, in both cases bioenergy will be a significant component in reducing the cost of fossil energy imports, and in diversifying the energy supply mix. A mediated balance between the different scenarios indicates that bioenergy has a potential to provide up to some 30% of the global energy consumption in the near future. The implementation of this potential depends on technological improvements, including in the field of catalysis for renewable, innovation capabilities and the market conditions in both developing and industrialized countries.

3
Main Issues and Perspectives on Bioenergy and Biofuels in Relation to Catalysis

3.1
Biofuels

We may differentiate between direct conversion of biomass into bioenergy (electricity and heat, solid fuels from biogenic wastes and residues, biogas, etc.) and biofuels. Catalysis has a minor role in the first case but is a critical element in the production of biofuels. However, notably, there are also potentially interesting developments related to bioenergy.

Biogas can be produced from dedicated crops (e.g., corn without kernels), lignocellulosics wastes of biorefineries, as well as from plant and animal wastes from the food industry. Gas production is a quite mature technology, but the change in the use of the raw gas for valuable products such as BGtL (bio-gas to liquid) needs the development of catalysts for gasification reactors and for gas upgrading such as tar cracking [3].

1. The transformation of straw and agrofood residues with high sulfur and ash content requires the development of materials for sulfur abatement at high temperature, tar cracking and as monolith for syngas production by exothermic or autothermal processes thanks to catalysts supported on materials with a high thermal conductivity.

2. Bio-hydrogen is a possible, interesting output of biogas production, increasing the overall efficiency of the process through efficient and stable catalysts towards the main contaminants.

Regarding biofuels, it is common to differentiate between 1st and 2nd (or next) generations.

3.1.1 First Generation Biofuels

Ethanol from biomass, often called bioethanol, is currently the main biofuel on a global scale. It may be produced from various sources, if they contain high amounts of sugar, or starch, which are then transformed catalytically into simple sugars and then fermented into ethanol. Sugar cane as a the feedstock is already used in large amounts to produce ethanol in Brazil. Other crops that can be converted into ethanol are cassava, maize, sorghum, and wheat. More plant species could be suitable feedstocks for ethanol production, including perennial crops, but their yields, costs, and farming features are still not well know. Other possible feedstocks are potatoes and sugar beets. The conversion of their starch content into sugar has a high energy demand, so that the cost of the product is quite high, but the selling cost depend also on the revenue from by-products.

Biodiesel can be produced from various oilseed-yielding plants like castor, cotton, jatropha, palm, rape, soy, etc. The straight vegetable oils (SVO), which can be derived by physical and chemical treatment (milling/refining), are then converted into fatty acid methyl esters (FAME), also known as biodiesel. Similar to ethanol, these routes are established and proven, and their costs depend heavily on two factors:

- costs of the feedstock (>90% for SVO, some 85% for FAME);
- revenues from by-product utilization (cake, glycerin).

Another route for biodiesel is to "hydrotreat" unprocessed bio-oils (from castor, cotton, palm, soy, etc.) so that no transesterification is needed to stabilize the biodiesel.

First generation biofuels are a direct substitute for fossil fuels in transport and can readily be integrated into fuel supply systems. They may also help prepare the way for further advanced developments, such as hydrogen. Although most biofuels are still more costly than fossil fuels their use is increasing in around

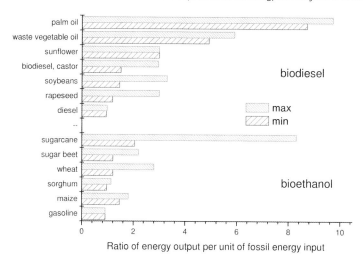

Fig. 1 Ratio of energy output per unit of fossil energy input for various crops. (Adapted from Hunt [2]).

the world. Encouraged by policy measures, global production of biofuels is now estimated to be over 35 billion liters.

The biofuel yield of different crops is differs, but a more correct analysis also requires consideration of the (fossil) energy inputs needed to seed, harvest, and process the crops, etc. Clearly, the results depend significantly on regional and technology details, but an overall comparison of net energy yield factors is given in Fig. 1.

Although technologies for both bioethanol and biodiesel are well established, improved catalysts are still needed.

In bioethanol production a critical step is the hydrolysis of the biomass, e.g., the breakdown of carbohydrates into its component sugars. Biomass is first pretreated to solubilize the hemicelluloses and expose the cellulose for subsequent enzymatic degradation. The cellulose then undergoes enzymatic hydrolysis to produce glucose, which can be converted into biofuels and chemicals by fermentation. It is necessary to create a new generation of cheap enzymes for hydrolysis of cellulose and lignocellulose to fermentable sugars (ones that can complete the biomass hydrolysis during fermentation).

The biocatalysts used for bioethanol production must be able to fully convert the carbohydrates into ethanol and other fuels, be robust, and tolerant of the toxic compounds formed during the pretreatment (hydrolysis) process. They must be able to withstand the stress of high ethanol and substrate concentrations, low pH, etc. At present no such strains are available and significant challenges lie ahead to develop such robust biocatalysts. It is also necessary to expand the substrate usage spectrum of microorganisms (e.g., C_5-sugars), and to increase tolerance to industrial conditions (e.g., high product tolerance, fast growth, high yield and productivity).

For biodiesel, catalytic or bio-catalytic technologies need to be either developed or improved in the following areas:

1. In biomass transesterification it is necessary to develop (basic) catalysts with high conversion efficiency operating under heterogeneous conditions, which avoid the presence of catalyst residues in the final product and allow a cleaner product with concurrent savings of catalyst, as well as simpler separation of the reacted materials from the reactants, e.g., integration of catalysts and membrane in a new advanced reactor design.

2. The conversion of rapeseed oil and recycled oil from food production and cooking requires the development of heterogeneous basic catalysts with high stability towards water and impurities.

It is also necessary to develop enzymatic trans-esterification processes that may find applications for waste material such as rape seed oil cake and for glycerol (propanediol, GTBE).

The production of these biofuels requires in general an efficient integration of biocatalysts into industrial processes, biocatalytic reaction engineering and the integration of chemo- and bio-catalysis, either in subsequent steps or in heterogeneous catalysis, e.g., through catalyst or engineering design. Common to all areas of a bio-based economy [3] is thus both the need of integrating bio-, homo- and hetero-geneous catalysis, and of integrated catalyst and reactor design. Process simplification through advanced catalyst design is also a must to decrease the cost of production of biofuels – the key economic element. Advances may derive from the ability to realize in one process step ("one-pot") complex multi-step reactions. This includes, for example, the development of cascade reactions featuring bio- and chemio-catalysts working in tandem, or the design of a new solid catalyst integrating bio-, homo- and hetero-catalytic functionalities into a nanoreactor design of the catalyst surface.

3.1.2 Second (or Next) Generation Biofuels

The "next" generation of biofuel processes should differ from the first in (a) utilizing the whole plant as a feedstock and (b) the use of "non-food" perennial crops (woody biomass and tall grasses) and lignocellulosic residues and wastes (woodchips from forest thinning and harvest residues, surplus straw from agriculture).

The use of fast-growing perennial crops such as "short-rotation" wood and tall grass crops allows use of a wider range of soils, including marginal or degraded land and requires less agrochemical inputs. However, the real impact on soil and water of this fast growing cellulosic biomass still requires more detailed evaluation. In addition, the conversion of this cellulosic biomass into biofuels is more difficult than conventional sugar, starch and oilseed crops.

Possible options for the conversion of these lignocellulosic plant materials include:

1. Biocatalytic conversion of lignocellulose into bioethanol, which requires upgrading of existing processes of fermenting sugars by using enzymatic-enhanced pretreatment of (hemi)cellulose. New, improved biocatalysts are needed for this route.
2. Thermochemical "biomass-to-liquid" (BtL) conversion, involving thermal gasification of the biomass and subsequent synthesis of biofuels by the Fischer–Tropsch process. Various aspects of the use of catalysis in this process are discussed in the several chapters.

Alternative pathways, also discussed in part in the various chapters, include (catalytic) pyrolysis, "flash" or "fast" processes for wet biomass without pre-drying, hydro-thermal upgrading (HTU), conversion of solid biomass more or less directly into a natural-gas equivalent called substitute natural gas (SNG), or even to hydrogen.

Combinations of 1st and 2nd generation conversion routes and technological coupling of biofuel and electricity conversion ("hybrids") are potential options in the near future. For example, the process efficiency for a combined cycle (CC) is typically around 50% and could be improved to about 58% using a combination of BtL and CC or to around 70% using bioethanol produced from lignocellulose combined with BtL and CC [2].

In conclusion, the economically competitive, non-subsidized production of liquid biofuels requires (a) the use cheaper and more reliable sources of renewable raw material; (b) efficient conversion, with minimum waste, of cellulosic, fiber or wood-based, waste biomass into fermentable sugars; (c) significantly improved efficiency of the production processes; and (d) use by-products (e.g., glycerol in biodiesel production). Several of these aspects are discussed in details in various chapters.

Notably, several types of liquid biofuels exist or are under development and have the potential to replace fossil fuels, especially in the transportation sector. The focus is on organic fuels such as ethanol, butanol, methanol and their derivatives ETBE, MTBE, which can be produced by fermentation, but also biodiesel and liquid biogas, which can provide interesting biomass-based alternatives to diesel and LPG.

First-generation biofuels can be already used in low-percentage blends with conventional fuels in most vehicles and can be distributed through the existing infrastructure. Developing a substitute for diesel is of particular importance in the European context given that the EU is currently a net importer of diesel, while it exports petrol. However, even using the most modern technologies, the cost of EU-produced biofuels, for example, is still not competitive with that deriving from fossil fuels, even if the price differential is progressively reducing. With the technologies currently available, EU-produced biodiesel breaks even at oil prices around € 60 per barrel, while bioethanol becomes competitive with oil prices of about € 90 per barrel [4].

Biofuels can be used as an alternative fuel for transport, as can other alternatives such as liquid natural gas (LNG), compressed natural gas (CNG), liquefied petroleum gas (LPG) and hydrogen. Subsiding the use of currently available biofuels is considered an intermediate step to reduce greenhouse gas emissions, to diversify transport energy sources, and to prepare the economy for other alternatives in the transport sector that are not yet mature. However, on a longer term perspective, lower production costs are a must for continuing the use of biofuels. The supply of feedstocks is also crucial to the success of the biofuel strategy.

The most promising second-generation biofuel technology – ligno-cellulosic processing – is already well advanced. In Europe, for example, three pilot plants have been established, in Sweden, Spain and Denmark. Other technologies to convert biomass into liquid biofuels (BtL) include Fischer–Tropsch biodiesel and bio-DME (dimethyl ether). Demonstration plants are in operation in Germany and Sweden.

The actual higher costs of biofuels could be balanced by the environmental benefits, but any biofuel strategy has to focus on:

1. optimizing greenhouse gas benefits for the expenditure made;
2. avoiding environmental damage linked to the production of biofuels and their feedstocks;
3. ensuring that the use of biofuels does not give rise to environmental or technical problems.

Catalysis could play a relevant role in limiting the environmental impact associated with biofuel production. In fact, it has been estimated, for example, that significantly more waste water arises in the production of biofuels than in the production of fuels using fossil resources. New (catalytic) processes for handling waste water deriving from biofuel production need to be developed.

3.2
Biorefineries

Biomass, both from residues/wastes and dedicated crops, can be converted not only into bioenergy (electricity, heat) and biofuels for transport but also into bulk chemicals or materials that are nearly equivalent to, or sometimes even better than, those derived from fossil hydrocarbons.

A biorefinery maximizes the value derived from the complex biomass feedstock by (a) optimal use and valorization of feedstock, (b) optimization and integration of processes for better efficiency, and (c) optimization of inputs (water, energy, etc.) and waste recycling/treatment. Integrated production of bioproducts, especially for bulk chemicals, biofuels, biolubricants and polymers, can improve their competitiveness and eco-efficiency. However, although a few examples of biorefineries already exist (Chapters 3 and 6), many improvements are still needed to enhance the process [5]:

- total use of the plant (better fragmentation and fractionation);
- development of processes to add value to all fractions of the plant and also to valorize by-products of other industrial systems (e.g., black liquor in wood/paper industry, glycerol from biodiesel, whey from cheese production, etc);

- downstream processing strategies (low cost recovery and purification);
- development of closed-cycle sustainable systems, etc.

It is also necessary to study the whole value chain as well as the "biorefinery value chain" for optimization of costs, CO_2 reduction, and energy usage.

Catalysis plays a relevant role in several of these areas of development but, generally, an innovative effort to find new catalytic materials and their integration into advanced reactor technologies is required, e.g., to combine reaction and separation to reduce the overall costs of the process. More specific needs include:

- Development of new catalytic routes to disrupt plant materials and subsequent fractionation with little energy input.
- Valorization, retreatment or disposal of co-products and wastes from biorefinery by catalytic treatments. This includes the utilization of plant and biomass fractions that are residual after the production of, for example, bioethanol and from other production chains (e.g., production of methane).
- Development of new and/or improved routes for chemical building blocks for polymers, lubricants and fine chemicals, including through the integration of the biorefinery concept and products into the existing chemical production chain.

Efficient technology could also be developed based on catalytic biomass pyrolysis for the conversion of biomass into clean and renewable liquid bio-oil. This would facilitate its introduction into the energy market as a renewable fuel or as source of high value chemicals. It is possible to produce stable liquid biofuels from biomass flash pyrolysis, in a single stage catalytic process, although further developments are necessary.

Before 1900, a large share of the chemical "industry" was based on biomass; it served as a feedstock for chemicals made from wood, sugars, starches, and fats. To convert these feedstocks into useful chemicals, mainly fermentation, chemical modification or thermochemical methods were applied. However, these processes were later abandoned in favor of the more economic and efficient processes based on fossil resources, in particular oil. Easier transport and more stable chemical composition (biomass feedstocks are highly diverse, depending on the source) are two relevant additional factors in favor of fossil fuels. Therefore, although the concept of biorefinery is attractive, there are several barriers to economically feasible.

As discussed in this book (Chapter 2, for example) a main difference between fossil fuels and biomass as feedstocks is that in the former case the functionalization of base chemicals obtained from the oil (ethylene, propylene, aromatics, etc.) occurs essentially by introduction of heteroatoms, while in the case of biomass-derived based chemicals (glycerol, for example) it is necessary to eliminate heteroatoms (oxygen, in particular). Consequently, the catalysts required to develop a petrochemistry based on bio-derived raw materials need to be discovered and cannot simply be translated from existing ones, even if the knowledge accumulated over many years will make this discovery process much faster than that involved in developing the petrochemical catalytic routes.

Biomass is complex in composition, consisting of starch, cellulose, hemicellulose and lignin and small amounts of fats. In the past, typically only one of these constituents of the biomass was converted, and the rest discarded. The operations were thus highly inefficient when compared with fossil hydrocarbon refinery.

Different routes for converting biomass into chemicals are possible. Fermentation of starches or sugars yields ethanol, which can be converted into ethylene. Other chemicals that can be produced from ethanol are acetaldehyde and butadiene. Other fermentation routes yield acetone/butanol (e.g., in South Africa). Submerged aerobic fermentation leads to citric acid, gluconic acid and special polysaccharides, giving access to new biopolymers such as polyester from polylactic acid, or polyester with a bio-based polyol and fossil acid, e.g., "biopolymers".

Dextrose, obtained from starch, is the raw material for sorbitol and other sugar alcohols and polyols. Isolated starch (usually from corn) can be chemical modified and is used in large amounts as an inexpensive binder in the textile and paper industry. Chemically modified starch can be used as super-absorbers or in polymer manufacturing.

Animal fats and oilseeds (soybean, cottonseed, coconut, palm) are important feedstocks for plasticizers and cosmetics. Destructive distillation of biomass yields methanol ("wood alcohol"), acetic acid, tar and charcoal, and was the backbone of the "chemical industry" in former centuries.

However, most of these routes are still economically unattractive and the possibility of creating an equivalent petrochemistry based on biomass, which depends on raising the conversion efficiency and establishing "cascades" in which the residues of one product serve as inputs for another, still suffers from the relatively unattractive products derived from hemicellulose and lignin. Therefore, to bring back biomass into the "chemical business", the utilization of biomass must be enhanced by integrating it into biorefinery (Fig. 2).

The biomass is first processed by thermochemical methods (gasification, pyrolysis), for example, to form synthesis gas, which can be processed further to methanol or Fischer–Tropsch (FT-) hydrocarbons. To gasify solid biomass, both circu-

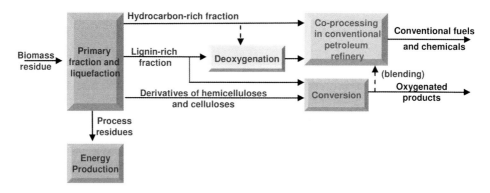

Fig. 2 Schematic flowsheet of a biorefinery. (Adapted from Schaverien [5]).

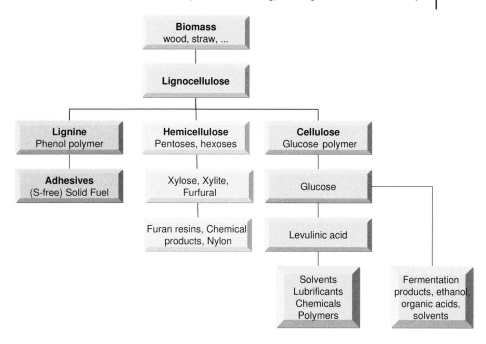

Fig. 3 Schematic flowsheet of ligno-cellulose transformations in a biorefinery.

lating fluidized-bed and entrained-flow systems could be used. The alternative approach is to break the hemicellulose into sugars to make hemicellulose available for fermentation, i.e., to yield ethanol. Figure 3 shows a scheme of the possible transformations of ligno-cellulose in a biorefinery.

Figure 4 gives an alternative scheme of possible biomass conversion pathways. Depending on the type of available biomasses and the objective products, each biorefinery will implement a different production and conversion scheme.

The thermochemical route is well known, as the basic technology was developed for coal and lignite and brought on-stream in Germany during World War II, even though further developments could be necessary, as discussed in Chapters 6 to 8. FT synthesis is also well known and has been implemented on a large scale in South Africa. The FT process has been further elaborated by Shell in its plant in Indonesia, using natural gas as a feedstock, and almost all oil and chemical companies have on-going developments.

The main drawback with the thermochemical route for biomass utilization is the strong dependence on scale-up. To be competitive, the capacity has to be of the order of a small oil refinery (approx. 1 million tonnes per year), but there then exists the problem of the cost of transporting the biomass relatively long distances to this production capacity. Pyrolysis or related technologies ("flash" or "fast") could transform biomass into liquid products that are more easily transported, and these liquid products could then be the input for a large, centralized

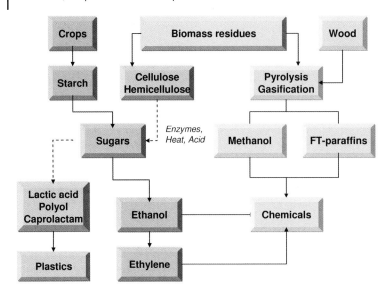

Fig. 4 Schematic flowsheet of possible biomass conversion pathways in a biorefinery.

gasification plant (Chapter 7). In addition, the economics rely on low feedstock costs. FT systems based on biomass would also require sites near refineries, and sophisticated logistics to handle the massive annual feedstock inflow.

The alternative pathway is the biochemical route. It processes starches/sugars into ethanol, a standard technology with installations world-wide, but in a biorefinery the start is the whole-plant material or biomass residues containing hemicellulose, which is broken into sugars that then can be fermented to ethanol and/or other alcohols such as butanol. As mentioned before, there is the need to develop novel and/or improved biocatalysts for alternative organic fuels, such as biobutanol, by fermentation processes.

To break up cellulose/hemicellulose, it is treated physically (milling), with heat, and hydrolyzed (sulfuric acid + enzymes). Also in this case, improved (bio)catalytic hydrolysis processes for cellulose/hemicellulose are needed. The sugar can then serve as feedstock for standard fermentation plants.

The technology of using hemicellulose is old and has been used with wood chips by the Swiss during the Word War II. Residues of the Kraft pulp process (Kraft liquor) have been used as a feedstock for ethanol production in Sweden since 1908. The major draw-back is the energy and capital intensive milling, the large amount of sulfuric acid (and, hence, neutralization agent), and high steam requirements to run the system. Furthermore, sulfuric acid at elevated temperature is quite corrosive. Currently, this technology is not competitive with fossil-fuel-based products.

New biocatalysts (genetically modified bacteria) could break up cellulose/hemicellulose, but it is necessary, on one hand, to decrease the cost of enzyme production and, on the other hand, to improve reactor and process technology, also to reduce costs. The enzymatic biochemical route needs sophisticated process control, which in turn leads to larger plants to meet economies of scale. NREL [6] calculates the necessary capacity of this technology for corn stover as feed to be in the 1 million tonnes per year range, or an equivalent output of some 200 000 tonnes ethanol per year. This capacity can only be used at sites with a high supply of biomass residues and an elaborate logistical infrastructure.

Therefore, in the absence of significant improvements also on the (bio)catalysis side, the competitiveness of an enzymatic biochemical route must is still uncertain.

In conclusion, as with the second-generation biofuel technologies (and respective hybrids), with which biorefineries are closely related, it is not possible to estimate in detail today how in future the biorefinery concept will be implemented, what the cost of production will be, and which products they are going to deliver to the market. The concept is promising insofar as oxygenated chemicals are becoming more prominent, and biochemical conversion knowledge is benefiting from developments of "white" biotech.

3.3
Use of By-products Deriving from Biomass Transformation

The valorization of by-products in biomass conversion is a key factor for introducing a biomass based energy and chemistry. There is the need to develop new (catalytic) solutions for the utilization of plant and biomass fractions that are residual after the production of bioethanol and other biofuels or production chains. Valorization, retreatment or disposal of co-products and wastes from a biorefinery is also an important consideration in the overall biorefinery system, because, for example, the production of waste water will be much larger than in oil-based refineries. A typical oil-based refinery treats about 25 000 t d^{-1} and produces about 15 000 t d^{-1} of waste water. The relative amount of waste water may increase by a factor 10 or more, depending on the type of feed and production, in a biorefinery. Evidently, new solutions are needed, including improved catalytic methods to eliminate some of the toxic chemicals present in the waste water (e.g., phenols).

An example of the valorization of by-products discussed in this book is the case of glycerol produced in biodiesel by trans-esterification. Glycerol may be converted into traffic fuels (Chapter 10) or chemicals (Chapter 11). There are different possible routes in upgrading glycerol to chemicals, such as selective oxidation of glycerol (Fig. 5), glycerol carbonate as a new solvent and product, glycerol as a component of new polymers, selective reduction processes and biochemical transformations. The decrease in the cost of glycerol as raw material will thus open up various possibilities. As pointed out in Chapter 11, old and forgotten reactions will be reinvestigated to meet current needs, while many unique and inventive catalytic reactions are to be expected in the next few years, parallel to the price decrease of glycerol. Although several possibilities already exist, further research into the catalytic conversion of glycerol is needed.

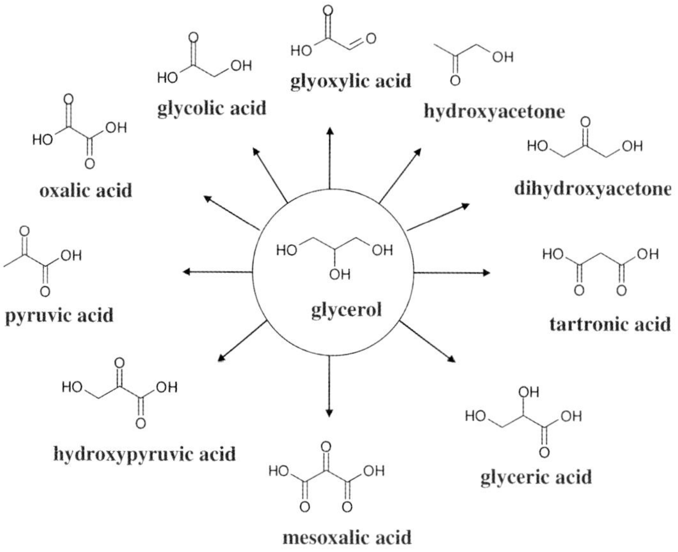

- Selective oxidation of glycerol
- Glycerol carbonate as a new solvent and product
- Glycerol as a component of new polymers
- Selective reduction processes
- Biochemical transformations

Fig. 5 Scheme of the possible products obtained from glycerol by selective oxidation, with an indication also of other possible routes for upgrading glycerol.

Similarly, there is interest in developing new routes for conversion and upgrading of bioethanol (Chapter 9), and in general for converting the possible products obtained by current and future fermentation processes.

The production of H_2 from some by-products of biomass conversion is also a possible option, which also requires the development of, new, more stable, more efficient catalysts that operate directly in the liquid phase. The catalytic production of hydrogen from more valuable products, such as bioethanol, should be reconsidered with appropriate economic assessments that take into account the alternative possible uses of these products.

3.4
Biomass as Feedstock for Chemical Production

Biomass feedstocks are already used in some processes, such as in oleochemistry (Chapter 4). In addition, biopolymers are somewhat well established, although their use and production still needs to increase and new improved (catalytic) methods for their production are required. Fine chemicals from renewables (Chapter 5) is another area in which various examples already exist, demonstrating that natural products play an important role in chemical production. Catalytic methods, chemo- as well as bio-catalysis, are of vital in the conversion of natural products into derivatives (semi-synthesis). In chemo-catalysis conventional catalysts, such as mineral acids, are being replaced by recyclable solid catalysts. Further progress is also expected in cascade processes in which synthesis steps are combined to give "one pot" methods.

A thorough analysis of value chains and the development of alternative value chains starting from biomass derived feedstocks, including assessment of the economic viability of the transformation of the chains, is required. This should be followed by the identification of easy entry points for the implementation of novel value chains. Technical key issues are generic methods to cope with the variability of raw materials derived from biomass and higher susceptibility to contamination by microorganisms and suitable catalysts for biorefineries.

Several developments in the field of catalysis are required for the introduction of biomass-based raw materials in the chemical industry:

- Development of novel synthetic routes for efficient conversion of biomass derived raw materials with high performance, stability and selectivity, by integrating bio-, chemical and catalytic processes. Synthetic pathways in which the complexity needed in a target molecule is already preformed in the biomolecule are especially favorable.

- Development of robust processes based on new catalytic reactor engineering solutions to improve energy and process efficiency and economics, and which combine reaction and separation. Special attention needs to be paid to cope with the natural variability in the quality of raw materials.

Major market opportunities and potential applications for renewable raw materials (RRM) fall into four main areas: chemicals, speciality chemicals, industrial fibers and industrial oils [7].

- *Chemicals*: e.g., chemical intermediates and polymers.
- *Speciality chemicals*: adhesives, solvents, surfactants (e.g., in personal care products and soaps & detergents), specialized organics and pharmaceutical and healthcare products (e.g., vitamins).
- *Industrial fibers*: paper & board, composites, textile fibers and bulk fibers.
- *Industrial oils*: lubricants, e.g., hydraulic oils, motor & gear oils, transmission fluids and metalworking fluids.

Table 2 reports the expected market development of major RRM-based products [7]; notably, the potential market may significantly enlarge if the progressive introduction of biorefineries decreases the production costs, on one side, and increases the number of biomass-based products on the other side.

The main current products in the field of biopolymers are (a) composite materials for special and high quality performance applications, (b) fashion and comfort/hygiene where their silky feel, breath ability and anti-static characteristics create a value over and above the environmental aspects (e.g., sports wear

Table 2 Estimated EU potential for major RRM-based products. (Adapted from [7]).

Market sector	Total market (1998) (ktonne)	Share in 1998 (%)	Potential share in 2010 (%)
Polymers	33000	0.1	1.5
Lubricants	4240	2.4	5.0
Solvents	4000	1.5	12.5
Surfactants	2260	50.1	52.0

and carpets made from poly-lactic acid (PLA), and the diaper liners and working overalls made from starch-based materials), (c) materials for agriculture, such as biodegradable mulch films, (d) plastic bags for the collection and composting of biodegradable waste, (e) food and industrial packaging materials, and (f) catering products, i.e., plates, cutlery etc.

The expected contribution of catalysis in this area will derive both from the availability, at low processing costs, of new monomers obtained from biomasses and from the development of an optimized combination of biotechnology processes with classical and new biocatalytic processes. Research priorities for catalysis in the area of polymers from renewable materials for packaging, furniture, domestic water purification and recycling include the need to develop novel catalysts, e.g., for functionalization of polymeric and dendrimeric materials, with side-chain photoactive molecular switches (to be used as smart materials), or the development of multifunctional materials, combining, for example, nanofiltration with catalytic reactivity.

Polymer matrix composites (PMCs) are part of a versatile and widely used class of engineering materials with many diverse applications, including transportation, electrical and electronics as well as construction. Their use is rapidly expanding. Traditionally, synthetic fiber (of which glass fiber is by far the most common) is used as reinforcement in these materials, but there is now great interest in new nanocomposites, e.g., with carbon nanofibers, or with layered materials, etc. These reinforcement materials may act also as catalyst for the curing of the thermosetting polymer used as matrix (e.g., during the forming process by extrusion). In general the properties of the nanocomposite greatly depend on the (catalytic) processes at the interface between the polymer matrix and the reinforcement material. There is thus the need for a better understanding of these aspects to develop new, high-performance composites for applications ranging from cars and aerospace to electronic.

The optimization of biocatalysts for the production of bio-based performance materials and nanocomposites is another necessary development to produce new composites. It also includes the combination of nano-biotechnological methodologies with catalysis for the construction of sensor surfaces and microelectronic technologies for the read out of biomolecular interactions. Examples are new bio-electronic devices (biochips) for the recognition of nucleic acids and proteins, biocompatible coatings for microsized batteries (lithium-ion batteries and for primary lithium) for implanted biomedical devices, detectors based on synthetic biological sensors for early cancer diagnosis, and/or for anti-bio-terrorism protection. The understanding of bio- and chemo-catalytic functionalities, their integration in recognizing materials (doped materials, membranes, tubes, conductive materials, biomarker detection, etc.) and the development of smart composite materials (e.g., bio-polymer–metal) are all necessary elements to reach above objectives. It is thus necessary to create the conditions to realize a cross-fertilization between scientific areas such as catalysis, membrane technology, biotech materials, porous solids, nanocomposites, etc., which so far have had limited interaction. Synergic interactions are the key factor to realizing the advanced nanoengineered devices cited above.

Similarly to catalysis, the properties of these composite materials are also determined by a hierarchy of structures on very different length/time scales. Therefore, linking mesoscale molecular models and continuum descriptions is relevant for their understanding and optimization. Together with advanced synthesis methods and functional testing, it is thus necessary also to develop new improved computational methods to provide an understanding of materials' properties and to assist in the development of new functional materials.

In the lubricant sector, oleochemically-based fatty acid esters have proved to be powerful alternatives to conventional mineral oil products. For home and personal care applications a wide range of products, such as surfactants, emulsifiers, emollients and waxes, based on vegetable oil derivatives have proved to provide extraordinary performance benefits to the end-customer. Selected products, such as the anionic surfactant fatty alcohol sulfate, have been investigated thoroughly with regard to their environmental impact compared with petrochemical based products by life-cycle-analysis. Other product examples include carbohydrate-based surfactants as well as oleochemical based emulsifiers, waxes and emollients. The catalysts used in the synthesis of these molecules need further development.

3.5
Use of Solar Energy

The actual world energy demand is about 20 TW (2006), while the energy provided by solar light to the world may be estimated to be about 600–1000 TW at practical sites. In other words, if about 2–3% of the solar energy can be captured and stored efficiently, the problem of energy could be solved. Clearly, there is the problem of the efficiency and cost of the solar photon conversion, but these data show the long-term potential of converting solar energy into a form that can be stored and transformed.

The are two possibilities, the direct solar into electricity conversion (photovoltaic) and the solar into chemical energy conversion (solar fuels), e.g., the production of H_2 by water splitting or the photocatalytic conversion of CO_2 into chemicals that may be then used as fuels or converted into electricity in fuel cells. These aspects are discussed in detail in Chapter 16.

The use of solar energy is complementary, not alternative, to the use of biomass for energy production. In fact, reliance on a few limited sources of energy is detrimental not only to the environment but also to future economic stability. The future lies in diversification of energy sources, tailored to the requirements and resources of each country.

Significant research is needed to develop new routes for photovoltaic solar cells, to raise their efficiency and decrease the energy cost of production to competitive levels. Part of these activities are quite relevant also for catalysis, to satisfy the need for improved (photo)catalysts, advanced hybrid catalytic materials and reactive thin films, and to control the interfacial processes. Research into new nanostructured material systems is expected to lead to low cost solar cells, and also to new catalytic materials. Progress in organic or hybrid systems promise new production technologies, e.g., printing, that could also be used to prepare tailored catalytic surfaces. However, in general terms, the development of next-generation photovoltaic cells has several aspects in common with catalysis, but is a distinct field of activity, apart from the transferability of results to the development of advanced photocatalytic materials.

There is a more direct link between catalysis and the development of solar light-based devices for the production of H_2 from water or fuels/chemicals from CO_2. Not withstanding that these reactions are still a challenge, we should not further delay support of a significant research effort in this area, which represents a potential solution for the future of renewable energy. To improve the exploitability of such reactions, which are still at a laboratory proof-of-concept scale, a better understanding and control of the following aspects is necessary: (a) surface recombination processes and effect of doping and heterojunctions, (b) back reactivity of the species in solution, (c) poor charge space layers, (d) solar light harvesting and efficiency with visible light, (e) photostability, etc. Nanostructured thin films, which need the development of suitable synthesis procedures that are also useful for preparing advanced catalysts, can possibly overcome the current production of no more than one exciton per particle (powders), control of space periodicity, and of space charge layer thickness.

Research priorities are also in the development of (a) multiple electron transfer catalysts for oxygen production, (b) advanced methods for surface complexation by anions and cations to avoid back reactions and improve charge separation, and (c) new materials' nanoarchitecture to improve light harvesting. The latter priority involves the development of (a) materials for visible absorption, (b) methods for efficient and stable sensitization, (c) methodologies to improve charge transfer and separation, e.g., by creating nanoheterojunctions, (d) reduction of light scattering, e.g., by controlling surface roughness, and (e) controlling nanoarchitecture, e.g., by creating spatially separated oxidation/reduction sites, etc.

Therefore, although research on the use of solar energy, especially for solar to chemical energy conversion, is a long-term objective, it is of paramount relevance to already push research activities in this direction.

4
Conclusions

The chapters of this book have been selected to provide an introduction to the catalytic issues of biomass conversion processes. The introductory chapters make clear the political decisions, especially in the EU, that drive biomass conversion technology, its prospects compared with other options for renewable energy, and the main technological options for conversion of biomass into secondary energy carriers.

A general conclusion is that incentives to convert conventional fossil fuel based technology into biomass based technologies are large, but implementation will be slowed down for several reasons. Firstly, there is the availability of biomass itself. The preferred feedstock should not compete with food production. Processes to convert such biomass practically into secondary energy carriers are not yet commercially available. In the mean time, technologies converting food related biomass will be implemented. But this will only occur for a limited period in the near future.

Conversion of such biomass into chemicals may be expected to have a much longer future perspective. Most chapters in this book are committed to the catalysis of biomass feedstock to bulk or fine chemicals. Here one notes the need to define platform molecules and their conversion technologies as well as the need for more insights in the fundamental catalysis of these processes.

A second reason for the slow substitution of biomass for fossil fuels is an economic one. The economic issues of hydrogen production are dealt with in Chapter 15. Hydrogen produced from fossil fuel under conditions in which carbon dioxide is sequestered is compared with the economics of hydrogen produced from biomass.

This remains a competitive option with the consequence that now hydrogen has to be stored, transported and converted. A chapter on fuel cell technology has also been added (Chapter 14).

A third reason why, in the very long term, the use of biomass technology for energy conversion will remain limited is the much more favorable land use efficiency of solar energy. This is explained in Chapter 2.

High-temperature processes, based on pyrolysis, gasification or combustion of biomass are the preferred conversion routes for non-food competing biomass conversion processes to secondary energy.

Upgrading of the bio-oils that result from pyrolysis or related processes is in need of significant improvements that require substantial catalytic innovation – providing a great challenge to the catalytic community. Also, our understanding of the chemical processes that occur in pyrolysis or gasification has to be improved. Such chemical understanding that relates to the chemistry of biomass feed deoxygenation to increase the hydrogen–carbon ratio will benefit strongly from the design of new catalytic procedures and systems.

In addition, biorefining based on bio-oils needs new robust catalysts that are not yet available. In the long term, solar energy should be used directly to produce fuels and chemicals instead of though biomass. This is a long term perspective; a more intense research effort in this direction is needed. It was for this reason that the chapter on solar energy was integrated into this book, which is dedicated mostly to biomass.

It is our view that the biomass conversion challenge is a great new opportunity for catalysis science and technology to develop new materials and catalytic process designs of benefit to future generations. This concluding chapter has also tried to give a unified view of the various aspects discussed in the book and to define selected research priorities that constitute a starting roadmap for catalysis for renewables. We are, however, well aware that this area is in rapid evolution and thus any possible roadmap has to be continuously revised and upgraded. Table 3 summarizes the research priorities highlighted here.

Table 3 A roadmap for catalysis for renewables (see text for more details).

Main area	Research priorities	Timeline (0 – 5 – 10 years)
BGtL process (biogas to liquid)	More robust and stable catalysts for gasification and gas upgrading (tar cracking)	0–2
Bioalcohols	New catalytic approaches for the pretreatment and hydrolysis of (ligno)cellulose biomass to fermentable sugars	0–3
	Robust and poison-tolerant (bio)catalysts	4–6
	Biocatalysts able to use a large spectrum of substrates and/or to produce alcohols other than ethanol (butanol, in particular)	4–7
Biodiesel	Solid base catalysts with a high conversion efficiency	0–1
	Efficient coupling of catalysts and membrane	2–4
	Trans-esterification catalysts stable in the presence of water and impurities	3–5
	(Bio)catalysts for trans-esterification of waste products	4–6

Table 3 *(continued)*

Main area	Research priorities	Timeline 0—5—10 years
Thermochemical BtL (biogas to liquids)	Catalytic pyrolysis for wet biomass	0–5 years
Biorefineries	New catalytic pretreatment of plant materials	0–3 years
	Valorization, pretreatment or disposal of co-products and wastes from biorefinery by catalytic treatments	0–3 years
	New and/or improved catalytic processes for chemicals production through the integration of the biorefinery concept and products into the existing chemical production chain	0–5 years
	New advanced catalytic solutions to reduce waste emissions (solid, air and, especially, water)	0–2 years
	New catalysts to selectively de-oxygenate products from biomass transformation	3–7 years
	Catalysts to selectively convert chemicals in complex multicomponent feedstocks	3–7 years
	New biomimetic catalysts able to operate under mild conditions	3–7 years
	Small catalytic pyrolysis process to produce stabilized oil for further processing in larger plants	2–4 years
Valorization of by-products	New routes for glycerol and bioethanol upgrading	0–2 years
	Improved and more efficient catalysts to produce H_2 from by-products and waste in liquid phase	0–2 years
Chemicals from biomass	Optimized combination of biotechnology processes with classical and new (bio)catalytic processes to produce, at low cost, new monomers	0–2 years

Table 3 *(continued)*

Main area	Research priorities	Timeline 0 — 5 — 10 years
	Catalytic methods to produce smart and/or advanced materials by functionalization of polymeric and dendrimeric materials	▬▬ (0–5)
	New catalysts for polymer matrix nanocomposites	▬ (0–2)
	Nano-biotech catalytic methods to produce high-performance materials	▬▬ (0–5)
Solar energy	Advanced design and architecture in nanostructuring photocatalytic thin films	▬▬ (0–5)
	New photoelectrocatalytic solutions to separate the production of H_2 and O_2 by water splitting without sacrificial agents	▬▬ (0–5)
	Production of fuels or chemicals by photocatalytic conversion of CO_2	▬▬▬ (5–10)
	Improved understanding of the surface recombination processes, back reactivity of the species in solution, and role of charge space layers	▬ (2–5)
	Low cost multiple electron transfer catalysts for oxygen production	▬ (2–5)

General issues

Integration of biocatalysts into industrial processes and biocatalytic reaction engineering	New research and conceptual tools to allow chemo- and biocatalysis to work efficiently in tandem and to perform complex multi-step reactions	▬▬ (0–5)
	Enhancing existing or new microorganisms to reach optimum production capacities under industrial conditions	▬▬ (0–5)
	New and easily-scalable bio-catalytic reactors	▬ (2–5)

Table 3 *(continued)*

Main area	Research priorities	Timeline 0 — 5 — 10 years
Process eco-efficiency	Novel catalytic routes for efficient conversion of biomass derived raw materials with high performance, stability and selectivity, and which preserve the complexity already preformed in biomolecules	0–5 years
	New catalytic reactor engineering solutions to improve energy and process efficiency	0–5 years
	Process simplification by effective integration of catalysis and separation in new advanced processes	0–5 years
	New catalysts able to cope with the natural variability in the quality of raw materials.	0–5 years
Cross-fertilization	Advanced nanoengineered devices by integration of catalysis, membranes, biotech materials, porous solids, and nanocomposites concepts	0–5 years
Research tools and fundamental understanding	New catalyst design for effective integration of bio-, homo- and heterogeneous catalysis	0–5 years
	New approaches to realize "one-pot" complex multistep reactions	0–2 years
	Understanding catalytic processes at the interface in nanocomposites	0–5 years
	New routes for nano-design of complex catalysis, hybrid catalytic materials and reactive thin films	5–10 years
	New preparation methods to synthesize tailored catalytic surfaces	0–5 years
	New theoretical and computational predictive tools for catalysis and catalytic reaction engineering	0–5 years

Finally, we conclude this book with remarks made on the options and opportunities in catalysis for renewables by Dr. Tom Kieboom (General Science Manager at DSM Food Specialties R&D) in the final round-table discussion of the workshop "Catalysis for Renewables" (Rolduc, The Netherlands, May 16–18, 2006; see Preface):

- *Biomass conversion for energy:*
  - focus should be on agricultural and wood waste as feedstock;
  - focus should be on catalytic conversion (i.e., low T and P processes to give a really low-energy-consuming processes);
  - multi-catalytic approaches are required to convert such complex and tough materials into liquid low-molecular-weight materials;
  - integration of bio- and chemo-catalytic processing as a novel cascade way of processing these difficult feedstocks (e.g., cooperation of enzymes, base, peroxides and surfactants for cleaning laundry).

- *Carbohydrate conversion for chemistry:*
  - Sugar, starch, cellulose, lactose and inulin are still attractive renewable feedstocks for conversion into other high-oxygen-containing chemicals and materials. Note that most present-day chemical products and materials from fossil oil have a high oxygen content (brought in by partial oxidation), while carbohydrates already have a high oxygen content (i.e., have a better energy balance).
  - Studies of new combinations of bio- and chemo-catalysis, easily performed in aqueous medium, can lead to completely new conversion processes (under ambient conditions).

- *Plant oil conversion for chemistry:*
  - Triglycerides have great potential for both low- and high-oxygen chemicals and materials.
  - The fatty acid part may be considered as complementary to carbohydrates as feedstock (low versus high oxygen content); they are still considered as "not-fancy" instead of being considered by (catalytic) chemists and biochemists as challenging starting materials for new process development.
  - The same as above holds for the (by-product) glycerol, a really challenging molecule for cascade conversions in water.

- *Matching chemical procedures with biological synthesis:*
  - The general basis for success in the three above-mentioned areas is the development of chemical and chemo-catalytic procedures that are mutually compatible (temperature, pressure and solvent). In addition, to exploit fully the power of enzymatic conversions, this compatibility is best sought within the range of reaction conditions microorganisms and enzymes are able to work in.
  - When there are more conversions available that are mutually compatible, process technological solutions (e.g., membranes) can further fully exploit low-energy cascade-type reaction sequences (instead of the other way around – non-compatible conversion steps will always demand high separation costs).

The final comment of Tom Kieboom was that with respect to all energy-related discussions made during the workshop, "converting biomass first into a useful product or material, followed by burning after its use" is a much better way than "directly converting or burning biomass" to gain energy. So, with respect to renewables the sequence [Biomass] → [Product] → [Energy] should be promoted more.

We are thus facing a significant and rapid change in our oil-based economy to progressively introduce renewable raw materials. The development of new catalysts concepts and solutions will be a key component in this transition, as it was in the past in the change to the current oil-based economy. Investment in a sustained research effort on catalysis implies an investment in a better and more sustainable future for our society.

Acknowledgments

We acknowledge Dr. Ad Kolen (NRSC-Catalysis, Eindhoven The Netherlands) for his invaluable effort in managing the cited workshop and this book. We also thank all scientists that attended the workshop, "Catalysis for Renewables" (Rolduc, The Netherlands, May 16–18th, 2006; see Preface), and actively contributed to its success and to this discussion. A special thanks also to all lecturers at the workshop and, especially, to Professor Johannes Lercher (TUM, München, Germany), who successfully organized the final round-table discussion.

References

1 J. Ehrenberg (Ed.), Status Report on *Current Situation and Future Prospects of EU Industry using Renewable Raw Materials*, European Commission – DG Enterprise, Brussels, February **2002**.

2 S. Hunt, Report *Biofuels for Transportation – Global Potential and Implications for Sustainable Agriculture and Energy in the 21st Century*; WWI (Worldwatch Institute)/gtz (Deutschen Gesellschaft für technische Zusammenarbeit GmbH), Washington DC (**2006**).

3 *SusChem (European Technology Platform for Sustainable Chemistry) Implementation Action Plan*, published by CEFIC (European Federation of Chemical Industries), Dec. **2006**.

4 *An EU Strategy for Biofuels*, European Commission Report Document, COM(2006) 34 final, Brussels (**2006**).

5 C. Schaverien, Presentation of Shell Global Solutions at the 1st International Bio-refinery Workshop organized by the EU and US-DOE, Washington, DC July 20–21, 2005 (**2005**).

6 NREL (National Renewable Energy Laboratory). Lignocellulosic biomass to ethanol process design and economics utilizing co-current dilute acid pre-hydrolysis and enzymatic hydrolysis for corn stover. Golden, CO US (2002). Report available on web: http://www.nrel.gov/docs/fy02osti/32438.pdf

7 Status Report, *Current Situation and Future Prospects of EU Industry using Renewable Raw Materials*, prepared by European Renewable Resources & Materials Association (ERRMA) for the European Commission – DG Enterprise, Brussels, **2002**.

Index

a
acrolein 224
7-ADCA 288 f.
AFC, see fuel cell
air
– bleed 323
– secondary 177
– separation 306
alcohol
– oxidase 286
– polyhydric 231 f.
aldol reaction 279
aldose 27
alkaline fuel cell (AFC), see fuel cell
alkaloid 108
– cinchona 114
alkene, glycerol 228
alkoxylation, limonene 105
alkyl polyglycoside 89 ff.
– carboxylate 91
– emulsifier 93
alkylglucoside 62
aluminium 262
– AlGaAs-GaAs heterojunction 364
– Al_2O_3 layer 370
– bipolar AlGaAs/Si cell 371
Amberlyst-15 247
amino acid 88
analytical method, in situ 290
anhydro-sugar, see sugar
anode 322 ff.
antimony, $M_2Sb_2O_7$ 374
aqueous-phase reforming (APR) 35, 219 f.
arabinonic acid 59
arabinose 24
arabitol 59
Arrhenius kinetic 167 f.
artimisinine 115 f.
asthaxanthin 103
auto-oxidation 264

autothermal reforming (ATR) 158
avian flu 116
azelaic acid 80 ff.

b
band bending 363
Bayer-Villiger oxidation 108
betaine 103
bilayer 323
bio enatioselectivity 285
bio-bio cascade 277 f.
bio-chemo cascade 281 ff.
– glucose into mannitol 282
bio-chemo integration 276 ff.
bio-diesel 13, 121, 217 ff., 223, 251
– glycerol 209 ff., 223
– NExBTC 221
– process 218 ff.
bio-ethanol 121, 183 ff., 348
– diesel blend 195
– fuel cell 199
– gasoline blend 193
– production 188 ff.
– upgrading 200
– valorization 200
bio-gas 40 ff.
bio-oil 136, 148
– liquid phase 56
biobased oleochemical 75 ff.
– industrial development 75 ff.
biocatalysis 273 ff.
bioconversion 193
bioenergy 389
Biofine 37
biofuel 22 f., 45, 193, 389
– bioconversion 193
– first-generation 23
– second-generation 399
– thermochemical platform 190
– yield 184

Catalysis for Renewables: From Feedstock to Energy Production
Edited by Gabriele Centi and Rutger A. van Santen
Copyright © 2007 WILEY-VCH Verlag GmbH & Co. KGaA, Weinheim
ISBN: 978-3-527-31788-2

Index

biomass 8 ff., 23, 56, 120 ff., 190 ff., 348, 352
- aqueous-phase 56
- bio-ethanol 190
- by-product 399
- combustion 148
- composition 122
- conversion 12, 119 ff., 163 ff.
- economy 388
- feedstock 400
- fuel 119 ff., 179
- gasification 129 f., 149
- grate furnace 175
- hydrothermal liquefaction 135
- lignocellulose 43, 123
- pre-treatment 126, 148
- pyrolysis 56, 133
- thermal conversion 147 ff., 163 ff.
Biomass Technology Group BV (BTG) 31
bioprocessing 41
- consolidated 41
bioproduct 53 ff.
biorefinery 55, 125, 394
biosynthesis 275 ff.
bismuth, $BiVO_4$ 375
butadiene 231
iso-butane 264
n-butanol 41, 203
tert-butanol 264
tert-butyl hydroperoxide (TBHP) 260

c

cadmium telluride 353
calcium oxide 311 f.
campholenic aldehyde 108
camposterol 113
Candida antarctica 261
capsaicin 104
carbohydrate 27, 274
- conversion 285
carbon
- C=C bond 267
- deposition 159, 329 f.
- formation 308 f.
carbon dioxide 2 ff., 120, 147
- capture 22, 301 ff., 313
- capture and storage (CCS) 301 ff., 337 ff.
- conversion 351 ff.
- emission 22, 300, 346 ff.
- pre-combustion CO_2 capture 348
- reduction 345
- release 164 ff.
- sequestration 22
- sorbent 310 ff.

carbon oxide tolerance 323 f.
carboxylic acid 369
β-carotene 102
cascade conversion 273 ff.
- glycerol 279
cascade reaction, one-pot 277
castor oil 78, 266
catalysis 119, 389 ff.
- biocatalysis 273 ff.
- cascades on industrial scale 294
- enantioselective 113
- fatty acid 257 ff.
- heterogenous 261
- homogenous 260
- in concert 273 f., 290
- in water 294
- integration 291
- renewable 407
- selective epoxydation 257 ff.
- thermochemical conversion of biomass into fuel 119 ff.
- transformation of glycerol 223 ff.
- transition-metal 294
catalyst 130, 213 ff.
- aluminium 262
- bimetallic 239
- bulk property 359
- cobalt 241 ff.
- combined action 281 ff.
- chromium 241 ff.
- design 140
- electrocatalyst 363
- gold-based 237 ff.
- high-temperature shift (HTS) 303
- homogenous 285
- loss 360
- low-temperature shift (LTS) 303
- manganese 241 ff.
- metal 281 ff.
- metal-activated chromium oxide type 244
- monometallic 239
- nickel–based 152
- palladium phosphine 249
- photocatalytic process 356
- Pt/Bi 234, 251
- Pt/C 234
- ruthenium 241 ff.
- titanium-based 264, 369
- yttriumoxide 250
catalytic technology, renewable 1 ff.
catechol 104
cathode 324 ff.

cell
– efficiency 370
– factory design 294
– voltage 316
cellulose 15, 24, 35, 123
– conversion 139
– enzyme 37
– native 109
cephalexin 288
CFB, *see* gasifier
CFD, *see* computational fluid dynamics
chalcogenide 353
char 31 ff.
charge
– carrier flux 359
– layer 363
– space charge region 363
charge separation 361
– active 362
– mediated 366
– passive 362
chemical product 125
– low volume, high value 125
chemical source term 176
chemicals 23
– conversion 203
– enantiomerically pure 285
– fine 102, 285
– production 46, 400
chemistry
– human 274
– natural 275
chemo-chemo conversion 278 ff.
chemocatalyst 273
chiral compound 113
chitin 103
m-chloroperbenzoic acid (MCPBA) 259
cholesterol 103 ff.
CHOREN 35
CHP, *see* combined heat and power *or see* cumylhydroperoxide
chromium 325
– poisoning 331
CIGS, *see* copper indium/gallium selenide
(−)-cinchonidine 114
circulating fluidized bed (CFB) 152
citronella oil 108
(+)-citronellal 107
Clostridium acetobutylicum 41
CNG, *see* natural gas
coal 306, 348
– particle 170
– pyrolysis 165
– to liquid (CTL) 340

coconut oil 76, 90 ff.
codeine 109
column chemistry 293
combined heat and power (CHP) 147
combustion 177 ff.
– forward 170
– reverse 170
– turbulent 180
compartmentalization 278, 290
compatibility 278
composite material 403
computational fluid dynamics (CFD) 163 ff.
contaminant 317
conversion 53 ff., 127
– bio 193
– biomass 12, 119 ff., 163 ff., 193, 268
– carbohydrate 285
– carbon dioxide 351 ff.
– cascade 273 ff.
– catalytic 53 ff.
– cellulose 139
– chemicals 203
– chemo-chemo 278 ff.
– enzymatic 276
– FAME 268
– front 173
– glycerol 209 ff.
– lignocellulosic biomass 127
– microbial 294
– multi-step 273 f.
– one-pot, three step 285
– simultaneous catalytic 282
– technology 10
copper indium/gallium selenide (CIGS) 353
coriander oil 266
corn 43
cracking 135 f.
– zeolite 137
p-cresol 106
crosslinking 290
crude sulphate turpentine (CST) 105
crystal structure 373
CTL, *see* coal
cumylhydroperoxide (CHP) 260
current doubling effect 357
γ-cyclodextrine 113
p-cymene 58 ff., 106

d
10-deacetylbaccatin III 102
decarbonylation 27
decarboxylation (DCO) 136

degradation 328
dehydration 29
– catalytic 226, 251
– glycerol 224 ff.
dehydroepiandrosterone 112
dehydroxylation 247
– selective 242
deoxygenation 26 ff., 47, 232
– polyol 232
– selective 135 ff.
4-deoxyglucose 285
depletion layer 363 f.
depolymerization 29
devolatilization 168
dextrose 90
dialkyl carbonate 95
dicarboxylic acid ester 84
dichlorohydrin 251
dicodide 111
Diels-Alder reaction, micro-wave heating 112
diesel 25, 195, 347
diether 229
diglycerol 227
diglyceryl alkyl ether 231
dihydrocarvone 108
dihydromorphine 111
dihydroxyacetone 231 ff.
– selectivity 235
dilaudid 111
dimer acid 80 ff.
dimer diol 82
– polyether 82
dimerization 81
dimethyl ether (DME) 159
diode laser, tunable 164, 180
diosgenin 112
dipole moment, MO_6 374
dolomite 312
donor/acceptor
– concentration 357
– recombination 357
– related process 356
DSSC, see solar cell

e

efficiency, electrical 321
elaidic acid 261
electricity 5, 45
– renewable 300, 317
electrocatalyst 363
electrochemistry 314
– biocatalysis 292

electrode 317 ff.
electrolysis 338
– alkaline 316
– solid oxide (SOE) 318
electrolyte 325 ff.
– interface 362
electrolyzer 315 f.
– efficiency 316
– proton exchange membrane (PEME) 317
– solid polymer (SPE) 317
electron transfer 364
emission 163
emollient 95
emulsifier 86 ff.
energy
– availability 339
– bio 389
– carrier 338
– consumption 22
– cumulative 352
– efficiency 26, 43
– fossil 5
– geothermal 352
– hydro-electric 352
– long-term 339
– nuclear 10
– primary 5
– renewable 21 f., 352, 387
– secondary 6
– solar 9 ff., 352, 403
– wind 352
enol 27
enzyme 276 ff.
– aggregate 293
– organic solvent 294
EOR, see oil
ephedrine 114
epichlorohydrin 251
epoxidation 62, 81
– chemoenzymatic 260
– fatty acid 259 ff.
– non-catalytic 259
– selectivity 262 ff.
– sharpless 114
– solvent-free condition 268
epoxide 83
epoxy/polyurethane multilayer system 84
Esmerone® 113
esterification 266
ethanol 39 ff., 139
– bio-ethanol 121, 183 ff.
etherification
– catalyst 213 ff., 226

– glycerol 212 ff., 226 ff.
– isobutene 212 ff.
– kinetics 216
(*R*)-ethyl lactate 114
ethyl pyruvate 114
ethyl-(*R*)-4-cyano-3-hydroxybutyrate 280
etorphine 111
evolution, directed 114

f

fatty acid 13, 86, 257 ff.
– condensate 88
– emulsifier 86
– ester 61, 84
– long-chain 76
– lubricant 84
– methyl ester (FAME) 259 ff.
– selective epoxydation 257 ff.
– short-chain 76
– surfactant 86
– tall oil (TOFA) 82
– unsaturated 81
fatty alcohol 86
– sulfate (FAS) 88
fatty ester 269
Fcc (face centered cubic) 137
FCV, *see* fuel cell
feed cost 43
feedstock
– biomass 400
– green chemical 251
– pre-treatment 45
– renewable 223
fermentation 37 ff., 277, 288 ff.
ferulic acid 103 f.
FGM, *see* flamelet generated manifold
Fischer-Tropsch (F-T) 159
– catalyst 34
– process 17, 33 ff.
fixed bed reactor 170 ff.
flame, turbulent piloted 177
flamelet generated manifold (FGM) 175
FLAPW (full potential linearized augmented plane wave) 377
flow, turbulent reactive 177
formic acid 36
fossil energy carrier 5
fossil fuel 4, 45
fossil oil 1
fossil resource 3
fractionation 126
front propagation model 180
front velocity 172 f.

fructose 282
– L-fructose from glycerol 287
FTO, *see* tin
Fuajasite 135
fuel 123 ff.
– additive 193
– bio-fuel 22 f., 45, 193
– component 200, 217
– glycerol 218 ff.
– hydrogen 341 ff.
– liquid transportation 125
– low value, high volume 125
– nitrogen 163
– oxygenated 123
– starvation 324
– traffic 218 ff.
– upgrading 136
fuel cell 159, 301, 314 ff., 326
– alkaline (AFC) 160
– bio-ethanol 199
– methanol 231
– molten carbonate (MCFC) 160
– phosphoric acid (PAFC) 160
– proton-exchange membrane (PEMFC) 160, 319 ff.
– solid oxide (SOFC) 160, 326 ff.
– vehicle (FCV) 338
furan 29
– derivative 60
furfural 36 f., 61

g

galactose 24, 290
– oxidase 285 f.
gallic acid 110
gas
– clean-up 153
– liquefied natural (LNG) 338
– natural (NG) 302 ff., 337 ff.
– recirculating 177
– to liquid (GTL) 340
gasification 33 ff., 48, 56, 304, 348
– biomass 129, 179
– catalytic 129 f.
– chemistry 130
– hot compressed water 131
– integrated gasification combined cycle (IGCC) 147
– product 177
– pyrolysis oil 129
gasifier 151
– circulating fluidized bed (CFB) 152
– downdraft 151

gasifier (cont.)
– entrained bed   152
– updraft   151
gasoline   25, 193, 347
glucose   24, 89, 280, 293
– hydrogenation   59
glyceraldehyde   237
glyceric acid   234 ff.
glycerine carbonate   250
glycerol   61, 210 ff., 399
– acrolein   224
– alkene   228
– bio-diesel   210
– cascade conversion   279
– dehydration   224
– dehydroxylation   247
– electrochemical oxidation   232
– ester   84
– ether   210 ff.
– etherification   212 ff.
– fatty acid ester   61
– fuel   218 ff.
– gas-phase   233
– hydrogen   249 f.
– hydrogenolysis   241 ff.
– isomerase   282
– mannitol   282
– oligomerization   226, 251
– oxidation   64, 231 ff., 241
– polymerization   226
– production   210
– property   210
– reforming   219, 249
– selectivity   227 ff.
– sub/supercritical condition   225
– terbutylation   229
– use   210
glyceryl unsaturated dioctyl ether   230
glyceryl unsaturated octyl ether   230
glycidol   250
glycol   231
glycolic acid   238
glycolysis pathway   276
gold-based catalyst   237
Gouy-Chapman charge layer   363
grain   43
grate
– CFD   175
– moving   169
grate furnace   163 ff., 175 ff.
– biomass   175
greenhouse gas   147

grid reactor   165, 180
GTL, see gas
Guerbet alcohol   96

h
HDO, see hydrodeoxygenation
heating value   26 ff.
Helmholtz charge layer   363
hemicellulose   15, 24, 35, 123
heroine   109
heterojunction   364 f.
heteropolyacid   106
2-hexyldecanol   96
high-temperature shift (HTS)   303
Hombikat UV-100 powder   365
hydrocarbon   5, 25
hydrodeoxygenation (HDO)   136
hydrogen   5, 147, 196, 337 ff.
– clean   337 ff.
– cryogenic liquid   343
– distribution   340
– economy   338
– energy chain   300
– evolution   317
– fuel   341
– green   337 ff.
– internal combustion engine (H2-ICE)   338
– oxidation   315, 329
– production   249, 299 ff., 316 ff., 330, 340, 351 ff.
– sustainable transport   340
hydrogen sulfide   158
hydrogenation   110
– enantioselective   114
– glucose   282
– Pd/C   110
hydrogenolysis   242 ff.
– glycerol   245 ff.
– poly-ol   243
– selectivity   246
hydrolysis   35 ff., 139
hydroperoxide, organic   260
hydropyruvic acid   231 ff.
hydrotalcite   310
HydroThermal Upgrading (HTU)   32
(S)-4-hydroxy-oct-7-en-2-one   287
– aldol condensation   287
hydroxylation   367
hydroxymethylfurfural   36 f.
4-hydroxyphenylpropane   141
hydroxypropanone   225
3-hydroxypropionaldehyde   244
3-hydroxypropionic acid   41

i

impurity 324 f.
indium
– copper indium/gallium selenide (CIGS) 353
– In(OH)$_3$ 377
– SrIn$_2$O$_4$ 374
– tin oxide (ITO) 366
integrated gasification combined cycle (IGCC) 147
interface related process 356
Iogen 40
ionic liquid 38
IPCE, see photon efficiency
iron, α-Fe$_2$O$_3$ 372
isobutene 212 ff.
isomerase 282
(−)-isopulegol 107
isosorbide 67
ITO, see indium

k

ketomalonic acid 235
ketose 27
kinetic
– dynamic 284
– first-order 167
– in concert 284
kinetic barrier 370
Kraft process 38
Kyoto protocol 22

l

lactic acid 39, 60
lactobionic acid 287
β-lactoglobulin 287
lactose 287 f.
– oxidation 59
lactylamine 287 f.
Langmuir-Hinshelwood (L-H) 284, 357
lanolin 103
lanthanum manganite, strontium-doped (LSM) 318, 331
lauric oil 76
lauryl glucoside 91
lead, PbSe 361
lecithin 103
levoglucosan 135
levulinic acid 16, 36 f., 60
lignin 15, 24 f., 35, 123
– vanillin 104
lignocellulose 21 ff., 48, 126
– bioconversion platform 193
– biomass 43
– enzyme 37
– ethanol 193
– polymer 127
– thermochemical conversion 127
light intensity 357
limestone 312
limonene 105
– alkoxylation 105
linalool 106
linoleic acid 267
lipase 284
lipase B 261
liquefaction 126 f., 346
– biomass 132 ff.
– hydrothermal 135
liquid interface 367
liquid-phase reforming 249
LNG, see gas
LSM, see lanthanum manganite
low-temperature shift (LTS) 303
lubricant 62, 84, 403

m

malaria 115
mannitol 282
– bio-chemo cascade 282
mannose 24
manufacturing cost 44
MCFC, see molten carbonate fuel cell
MCM-41 264
MCPBA, see m-chloroperbenzoic acid
MDI, see methylene di(phenylisocyanate)
medium density fiber board (MDF) 164
medium engineering 293
– catalytic cascade 293
membrane 292
– reactor 307 ff.
menthol 107
– (−)-menthol 107
– one-pot synthesis 108
metal hydride storage process 371
metal junction 363
metal-oxide water interface 368
metathesis reaction 63
methanol 33 ff., 159
– fuel cell 231
methyl tert-butyl ether (MTBE) 183
methyl elaidate 265
methyl β-D-galactopyranoside 285
methyl oleate 265
methylene di(phenylisocyanate) (MDI) 84
N-methylglucamide 90
methylglyoxal 233
Michaelis-Menten 284

mineralization 355
mixture fraction 175
molten carbonate fuel cell (MCFC) 160
monocarboxylic acid ester 84
monoterpene 105
morphine 109 ff.
MPEC, see photoelectrochemical cell
MTBE, see methyl *tert*-butyl ether
muscle relaxant 113
mutarotation 282 ff.
myrcene 106 f.

n

naloxone 110
naltrexone 110
nanocomposite 363
nanodiode
– photocatalytic multiple (PCD) 366
– p-n 366
narcotic drug 109
natural gas (NG) 302 ff., 337 ff.
– compressed (CNG) 337
– liquefied (LNG) 338
NExBTL 221
nickel 317 ff.
– metal hydride storage process 371
– Ni-YSZ 329
– zirconia 318
niobium
– Nb_2O_5 layer 370
– niobate 373
– niobate pyrochlores 374
nitrogen oxide (NO_x) 163, 180
– emission 175
Novozym 435 261
nuclear energy 10

o

octanol 94
octyl α-D-glucopyranoside 293
oil
– enhanced oil recovery (EOR) 347
– epoxidized 80
– fossil 1
– refinery 46
oleic acid 84, 267
oleochemical 269
– dicarboxylic acid 80
– industrial development 75 ff.
– polymer application 79
– polyunsaturated 261
oligomerization
– glycerol 226
– selectivity 226

opium 109
organosolv pulping 126
oseltamivir phosphate 116
overpotential 314
oxazole 80
oxidation 81, 279
– alkaline glycerol 232
– auto 264
– Bayer-Villiger 108
– electrochemical 231 f.
– gas-phase 233
– partial (POX) 159
– polyhydric alcohol 231 f.
– preferential (PrOx) 159
– Pt/Bi catalyst 234
oxide, multiple cation/anion 377
oxycodone 110 f.
oxygen 234
– evolution 317
– reduction 315, 331
oxygenate 25
– additive 200
ozonolysis 81

p

paclitaxel 102
PAFC, see phosphoric acid fuel cell
palladium
– Pd-Ag 308 ff.
– Pd-Au/C 239
– Pd/C hydrogenation 110
palm kernel oil 76, 90 ff.
paracodin 111
PCD, see nanodiode
PEC, see photoelectrochemical cell
pelargonic acid 84
PEMFC, see proton-exchange membrane fuel
   cell
pentaerythritol ester 93
peppermint oil 107
Perovskite type structure 373
peroxopolyoxometalate (POM) 260
petroselinic acid 267
phosphoric acid fuel cell (PAFC)
   160
phosphorylation 279
photocatalysis 351 ff.
– one pot synthesis 378
photocatalyst 365
photocorrosion 356
photocurrent 370
– enhancement 361
photoefficiency 363
photoelectrocatalytic reactor 362

photoelectrochemical (PEC) cell   353 ff., 371 ff.
– multiple band gap (MPEC)   371
photoelectrolysis   355
photon absorption   359
photon conversion   361
photon efficiency   357
– incident photon to current conversion efficiency (IPCE)   357
photosplitting   374
photovoltage   361
photovoltaic (PV) system   353
phytase   279
α-pinene   105
β-pinene   105
plant cost   42
platform molecule   57
platinum   314 ff.
– Pt-Au/C   239
– Pt/Bi catalyst   234
– PtBiCeO$_2$/C   235
– Pt/C catalyst   234
– PtCo   325
– PtMo   322 f.
– PtRu   322 f.
– Pt-TiO$_2$ nanocomposite   363
poly(3-hydroxybutyric acid)   41
polyglycerol   64
polyglycerol ester   92 f.
– emulsifier   93
polylactide   60
polymer application   79
– oleochemicals   79
polymerization, glycerol   226
polyol   83
– dehydrogenation   232
– ester   84 ff.
POM, see peroxopolyoxometalate
power density   321
POX, see oxidation   159
pristine TiO$_2$ P25   365
probability density function (PDF), flamelet   175
product purification   45
L-proline   114
– homogenous catalyst   285
propanal   225
1,2-propanediol (1,2-PDO)   64, 241 ff.
1,3-propanediol (1,3-PDO)   41, 65, 101, 241 ff.
protein
– acylated   88
– crosslinking   287
– fatty acid condensate   88

proton-exchange membrane fuel cell (PEMFC)   160, 319 ff.
protonation   367
PrOx, see oxidation
pulping process   38
– organosolv   126
PV, see photovoltaic system
pyrolysis   17, 28 ff., 42 ff., 127 ff., 154 f., 164
– ablative   157
– biomass   132, 154 f.
– catalytic   134 f.
– fast   156
– flash   156
– fluidized bed   157
– non-catalytic   132
– slow   156
– vacuum   157
pyrolysis oil   30, 48, 136 ff., 155
– zeolite   138

q

quantum dot (QD) semiconductor   361
quantum yield   357
quinic acid   103, 116
quinine   109 ff.

r

Raney copper   244
Raney nickel   132
raw material   75
reaction progress variable   175
reaction source term   172
reactor   307
– sorption-enhanced   307
reactor design
– cascade conversion   292
recirculating gas   177
recombination   356
– Auger   356
– bulk   356 ff.
– bulk-defect mediated   356
– donor/acceptor mediated   357
– radiative   356
recovery
– enhanced oil recovery (EOR)   347
– step   277
redox cycling   329
reformate   319
reforming   338
– aqueous phase   35, 219 f.
– autothermal (ATR)   158
– forecourt   341
– liquid-phase   249
– membrane   313

reforming (cont.)
- sorption-enhanced (SER) 310 ff.
- steam 219 f., 301 ff.
renewable catalytic technology 1 ff.
renewable electricity 300
renewable fuel 184
renewable resource 352, 387
- ecological compatibility 77
- toxicity 77
renewable raw material (RRM) 387
renewables 101 ff.
- catalysis 407
resource, fossil 3
retro Aldol reaction 27
retro-Aldol decomposition 29
retro-Michael reaction 27 ff.
riboflavine 280
ricinoleic acid 78, 267
rocuronium bromide 113
RRM, see renewable raw material
RSE, see source of energy

s
saccharification 41
Saccharomyces cerevisiae 39
saccharose 89
Schotten-Baumann condition 88
Schottky barrier 363 f.
sebacic acid 80 ff.
semiconductor 356 ff.
- electrolyte interface 362
- liquid interface 367
- metal junction 363
- quantum dot (QD) 361
- surface 368
sensitization 375
SER, see reforming
sertraline 281
Sherwood number 171
shikimic acid 116
Shockley-Queisser limit 360
Shockley-Read-Hall 356
silica, titanium containing 264 ff.
sintering process 159
sitagliptin 281
sitosterol 113
SMR, see steam methane reforming
SOE, see electrolysis
solar cell, dye-sensitized (DSSC) 372
solar energy, see energy
solid oxide fuel cell (SOFC) 160, 326 ff.
solvent, supercritical 32
sorbitol 67, 89
source of energy, renewable (RSE) 387

soya-bean oil 260
- FAME mixture 269
SRM, see steam reforming
stability 326
starch 13, 66
- modification 69
- oxidation 67
steam explosion 40
steam methane reforming (SMR) 302 ff.
steam reforming (SRM) 158, 219 f., 301 ff., 327 ff.
steroid 112
sterol 112
succinic acid 41
sucrose 89
sugammadex 113
sugar 13, 27, 41 ff.
- anhydro 29
- cane 43
- derivative 37
sulfur 304 f.
- catalyst 377
- doping 377
sulfur trioxide gas 88
sunflower oil, high-oleic (HO) 266
superhydrophilicity, photoinduced 354
surface 368
- adsorption 357
- loss 360 ff.
- passivation 369
- property 359
surfactant 86
- carbohydrate-based 89
syn(thesis) gas 30 ff., 48, 148, 302 ff., 318
- cleaning 153
synthesis
- multi-step 274 ff.
- organic 274 ff., 288
- precipitation-driven 294

t
Takasago process 107
Tamiflu® 116
tantalate 373
- pyrochlores 374
tar 33, 148
(R,R)-(+)-tarfaric acid 114
tartronic acid 240
TBHP, see tert-butyl hydroperoxide
telomerization 69, 231
- butadiene 69, 231
- glycol 231

TEMPO (2,2,6,6-tetramethylpyperidine-
   1-oxyl)   240
– silica-entrapped   240
terbutylation   229
terpene   58
α-terpineol   106
α-terpinyl acetate   106
(−)-thebaine   110
thermal conversion front   169
thermalization, carrier   360
tin
– Ca_2SnO_4 374
– fluorine-doped SnO_2 (FTO)   366
titanium
– $BaTi_4O_9$ 375
– $Na_2Ti_6O_{13}$ 375
– Ti-HMS   263
– Ti-MCM-41   263 ff.
TiO_2 Degussa P25   365
– TiO_2-Au system   364
– TiO_2/F   369
– Ti-ZSM-5   263
TOFA, see fatty acid
torrefaction   126
toxicity   77
transesterification   61
Trichoderma   41
triglyceride   259
triglycerol   227
tubocurare   112
tubocurarine chloride   112

v
vanilla bean   103
vanillin   103 f.
– lignin-vanillin   104
– Rhodia vanillin process   104
– wood-sourced   104
valorization   200
– by-product   399

vaporization   133
vegetable oil   43, 86, 266 f.
vitamin C   113

w
water
– electrolysis   315
– gasification   131
– photosplitting   372 ff.
water-gas shift (WGS) reaction   33 f., 159,
   302 ff., 322
– catalyst   313
Weizmann process   41
wheat   43
Wittig reaction   103
wood pyrolysis   165

x
xylose   24

y
YSZ, see zirconium oxide

z
zeolite   137
– aluminium   262
– cracking   137
– large-pore   135 ff., 214
– pyrolysis oil   138
– Sn-Bea (Sn-beta)   108
– Ti-Bea (Ti-beta)   108, 262 ff.
zinc
– doping   377
– mono-crystalline ZnO   368
– Zn_2GeO_4 374
– ZnO quantum dot metal colloid system
   364
zirconium oxide
– yttrium stabilized (YSZ)   326

Related Titles

V. Parmon, A. Vorontsov, D. Kozlov, P. Smirniotis

Photocatalysis

Catalysts, Kinetics and Reactors

2008
ISBN 978-3-527-31784-4

A. Züttel, A. Borgschulte, L. Schlapbach (Eds.)

Hydrogen as a Future Energy Carrier

2008
ISBN 978-3-527-30817-0

F. Endres, D. MacFarlane, A. Abbott (Eds.)

Electrodeposition in Ionic Liquids

2007
ISBN 978-3-527-31565-9

R. A. Sheldon, I. Arends, U. Hanefeld

Green Chemistry and Catalysis

2007
ISBN 978-3-527-30715-9

G. A. Olah, A. Goeppert, G. K. S. Prakash

Beyond Oil and Gas: The Methanol Economy

2006
ISBN 978-3-527-31275-7

B. Kamm, P.R. Gruber, M. Kamm (Eds.)

Biorefineries – Industrial Processes and Products

Status Quo and Future Directions

2006
ISBN 978-3-527-31027-2

B. Cornils, W. A. Herrmann, I. T. Horvath, W. Leitner, S. Mecking, H. Olivier-Bourbigou, D. Vogt, D. (Eds.)

Multiphase Homogeneous Catalysis

2005
ISBN 978-3-527-30721-0